T0282325

Climate Governance in International and Comparative Perspective

A volume in
Public Administration & Social Equity
Richard Gregory Johnson III, *Series Editors*

Public Administration & Social Equity

Richard Gregory Johnson, III, *Series Editors*

Multi-Sector Partnerships for the Public Good (2023)
Samuel L. Brown and Richard Greggory Johnson, III

Climate Governance in International and Comparative Perspective

Issues and Experiences in Africa, Latin America, and the Caribbean

edited by

Peter F. Haruna
Texas A&M International University

Laila El Baradei
The American University in Cairo

Liza van Jaarsveldt
University of South Africa

Abraham D. Benavides
The University of Texas at Dallas

Cristina M. Stanica
Northeastern University

INFORMATION AGE PUBLISHING, INC.
Charlotte, NC • www.infoagepub.com

Library of Congress Cataloging-in-Publication Data

A CIP record for this book is available from the Library of Congress
http://www.loc.gov

ISBN: 979-8-88730-642-1 (Paperback)
       979-8-88730-643-8 (Hardcover)
       979-8-88730-644-5 (E-Book)

CONTENTS

PART I

INTRODUCTION

PART II

REGIONAL PERSPECTIVES

PART III

NATIONAL PERSPECTIVES

PREFACE

There is growing global interest in Africa, Latin America, and the Caribbean (ALAC) and the matter of how to improve the quality of life of their people. This is for a good reason: The world cannot afford to ignore the climate change impacts that have been occurring across these regions, impacts with tremendous implication for the institutionalization of climate governance for the routine tasks of governing. Although the book covers a limited number of countries in these Global South regions, the research findings presented in it will serve the needs of those who wish to learn about the future direction of policy, research, and scholarship there. The focus of the book is to present research findings related to governance levels, actors, sectors, and concepts within the framework of public service institutional building. The comprehensive empirical descriptions and analyses of governance and administration contexts and cultures written by some of the best minds in the sub-field will inform interested scholars about the latest research findings in the main field.

Admittedly, the book was born from our interest in public administration and governance as disciplines and professional fields of practice. Focused research and scholarship on public administration and climate governance in the Global South is lagging. Much of the discourse centres almost exclusively on the substance of climate governance in the Global North, which gives good insight into global climate governance through the prism of the wealthiest nations. Of course, the Intergovernmental Panel on Climate Change periodically publishes reports with specific sections on the status of climate change in the Global South which obviously enrich understanding and show patterns of opportunities and challenges there. What is lacking, however, is a single, "one-stop" volume that provides conceptual and empirical analysis on matters devoted to public administration and climate governance in the context of the poorest and least emitting regions. This book addresses a dire need and provides essential reading for that purpose.

Specifically, readers will find answers to several questions, such as:

- How do ALAC nations approach public administration and climate governance?
- How have climate governance and administration evolved and changed in ALAC nations?
- What does climate governance look like in ALAC nations?
- What role does the administrative state play in climate governance in ALAC nations?
- How do multiple climate actors interact in the context and culture of development?
- What challenges does climate governance face in ALAC nations?
- What lessons have ALAC nations learned about climate governance?

In this research lead book, the focus is to provide new knowledge about the processes of developing professionally relevant ideas about climate governance. It is about knowledge and scholarship only. Political issues are not the target of discussion and chapter contributors do not make statements about matters of national sovereignty. The term ALAC nation is used here to refer specifically to existing geographical areas and national boundaries in which formal and informal climate governance, public administration, and professional socialization are conducted. Given the above caveat, chapter authors were selected from within the ALAC regions and instructed to address local, national, regional, and global experiences with public administration and climate governance while tying the discussions back to the common body of knowledge and scholarship defining the fields at national, regional, and global level, where appropriate. In addition, they were instructed to apply comparative analysis to distil cross-cultural knowledge.

With these instructions and the stimulating discussions and analyses provided, the book is logically organized in five parts as follows: Part I lays the foundation and rationale for the book by discussing the global context of climate governance, and by making the case for a focus on climate governance in the Global South regions of Africa, Latin America, and the Caribbean. Part II explores regional perspectives, focusing on the East African Community, the Cayman Islands, and Central America. Part III, backbone of the book, analyses diverse national perspectives from seven countries— Egypt, Ghana, Mexico, Nicaragua, South Africa, Suriname, and Zimbabwe. Local government is the focus of Part IV, analysing experiences from four countries. Part V closes the book by synthesizing ideas and teasing out lessons learned in the ALAC regions. We hope that:

- The book blends experiential and theoretical knowledge of climate governance.
- Based on the case analyses, the book distils relevant knowledge, tools, skills, and abilities for conducting climate governance-related activity in diverse cultural contexts.
- The book offers guidance to scholars, researcher, and practitioners striving for updated and detailed understanding of public administration and climate governance in ALAC nations.
- A comprehensive book dealing with emerging trends in public administration and climate governance in ALAC nations has been developed.

Finally, we wish to express our gratitude to our publisher, Information Age Publishing, for providing us with the opportunity to share our insights with readers. Their review processes and the advice provided to the editors and individual authors greatly enhanced the final product. We also thank individual experts who volunteered to review the chapters to enhance their quality.

—**Peter F. Haruna**
Laila El Baradei

ACKNOWLEDGMENTS

A book of this nature is only possible through the dedication of the editors, authors, and those who share in its goal and purpose. For authors, they will find satisfaction in the opportunity to tell their part of the story about climate governance in Africa, Latin America, and the Caribbean nations. As editors and production staff we have had the honour of assisting them to accomplish their goal. On our part as editors, we especially want to acknowledge the work and professionalism of the book series coordinator and project supervisor at Information Age Publishers. Without their guidance and review, this book could not have become exposed. Beyond these, an untold number of people made it possible for us to work on this book through their support. This includes the leadership of our universities, reviewers, colleagues, and students who have helped us beyond words. We deeply appreciate their contributions. We are indeed a global scholarly community.

—Peter F. Haruna
Laila El Baradei
Abraham D. Benavides
Liza Van Jaarsveldt
Cristina M. Stanica

PART I

INTRODUCTION

CHAPTER 1

CLIMATE GOVERNANCE IN AFRICA, LATIN AMERICA, AND THE CARIBBEAN

Peter F. Haruna
Texas A&M International University

Cristina Stanica
Northeastern University

Climate change is an existential threat, and scholars agree that it is "one of this century's most difficult challenges demanding new strategies to steer societies towards common transformational goals" (Sapiains et al., 2020, p. 46; Filho et al., 2021). At the core of such goals is the notion that human ideas and institutions are not fixed (Foucault, 1966). Thus, the United Nations Framework Convention on Climate Change imposes emission reduction on the world's richest and most polluting nations. However, developing countries of Africa, Latin America, and the Caribbean face a moment of reckoning. Do these countries have the ability to coordinate the actions of state and non-state actors to fend off the impact of climate change as they struggle to build liberal democracies? And the issue is whether they can

Climate Governance in International and Comparative Perspective, pages 3–17
Copyright © 2024 by Information Age Publishing
www.infoagepub.com

contain it while protecting their most vulnerable populations and groups made worse off by frequently occurring sudden onset of weather-related events. The answer to these questions is at best unclear mainly because literature and knowledge production tend to focus on Global North countries and Westernized epistemic approaches without integrating "the Global South's particularities" (Sapiains et al., 2020, p. 46).

THEMES AND QUESTIONS

Three ideas undergird this book. The first is demonstration through description and analysis of what climate governance and its various facets—cultures, institutions, mechanisms, systems, processes, state/non-state involvement, and activism—look like and imply for fighting climate change in the Global South's countries of Africa, Latin America, and the Caribbean.

Such description and explanation are necessary to tease out information about the quality of climate governance in the world's poorest and ecologically hot zones by examining questions of decision-making goals, societal roles, processes, and policy outcomes. In so doing the book addresses a research need and questions the tendency in academia to reproduce the epistemic schema of area studies in the Global North's countries. The contention is that the fight against "anthropogenic climate change" must also confront Eurocentrism, a trap for public governance as both a research framework and a reform program. Rooted in failures to rescue public governance from the "epistemic dominance of the western academic community" (Haque & Turner, 2013, p. 244), the resistance is gaining momentum (Francis et al., 2020; Agrofuel, 2007). And yet the issue is not just tensions within the Global North-South dialogue but also the inability or reluctance to embrace an "expansive view" of public administration's purpose (Roberts, 2019, p. 1).

The second and related idea driving the book is an inquiry into how the Global South's regions can learn from each other through South-South knowledge sharing and learning about climate governance approaches, using the United Nations Framework Convention on Climate Change (UNFCCC) as a reference point. Adapted in 1992 for the first time, the UNFCCC gives clear direction and guidance to create a strategically resilient world. Its core mandate is to curtail human activity—humanity's global ecological footprint—that harms the global atmosphere and human welfare (Encyclopedia of the Anthropocene, 2018). Nowhere is this noble mission more prescient than in the Global South countries caught between saving the planet on the one hand and their need for innovative agricultural practice, land-use norms, energy and food production systems that can transition them fairly to low carbon economies on the other (World Bank, 2022;

Zeufack & Cust, 2023). Because of the public interest reflected in the contributions of the Intergovernmental Panel on Climate Change (IPCC), the UNFCCC is critical to the book's purpose of analyzing climate governance assumptions and their related challenges.

Of interest here is the 2022 IPCC Report that offers insights into climate change by examining ecosystems, biodiversity, and human communities both at global and regional levels (ipcc.ch/report/ar6/wg2/). In addition, it reviews and highlights various human societies' vulnerabilities, damage, and adaptive capacities. However, and as Mahony and Hulme (2016) argue, IPCC authors come mostly from the Organization of Economic Cooperation and Development (OECD) countries with sparse representations from the Global South. If indeed knowledge production around climate governance, as Sapiains et al. (2020) and Kane and Boulle (2018) argue, comes mostly from the Global North with sparse Global South perspectives and epistemologies, climate action is unlikely to reflect context-related situations in regard to climate change measures. In sum, Global South theorizing based on decolonial studies (Ndlovu-Gatsheni, 2020) is necessary otherwise climate action in Global South countries may suffer due to the underrepresentation of Indigenous knowledge production in mainstream literature. Left far behind most parts of the world on several measures, the goal is to discover how these regions that are so frail and, in many ways, so endowed can chart a path to fair climate governance.

The third idea undergirding the book is anchored to a comparative approach to promote South-South knowledge sharing and learning. While interest in comparative public administration and policy has a long history (Peters, 2021; Pollitt & Bouckaert, 2017; Heady, 2001; Riggs, 1964), research on comparative climate governance with a focus on Global South countries is sparse. A function of the nature of climate governance itself—encompassing complex institutional building, organizational alignment, and behavioral change for formulating policies (Norris-Tirrell & Clay, 2010)—few researchers have an appetite for it and even fewer have resources to invest in its research and scholarship. And yet comparativists must contend with structural features of public sectors across countries and state political traditions, making the task of comparison difficult but necessary. It is correct that public sectors consist of large numbers of different organizations interlinked through various instruments such as legal, judicial, personnel, and financial systems (Vehoest et al., 2016). But this is not to talk about influential international organizations, public policy networks, civil society, and corporate actors, as well as transnational civil society actors involving think tanks, NGOs, social movements, humanitarian relief, and networked public-private partnerships (Vukasovic, 2018; Ladi, 2019; Moloney & Stoycheva, 2018). Thus, the questions of interest to comparative researchers include not just how countries differ in terms of their institutional structures/

processes and societal and community-wide influences, but also how transnational and international actors impact climate governance.

INTERREGIONAL OPPORTUNITY AND CHALLENGE

When it comes to the Global South's regions of Africa, Latin America, and the Caribbean with over a quarter of the world's population, opportunities and challenges exist that make comparative approaches to climate governance compelling. At its core, governance is a human enterprise (Ford & Ihrke, 2019). The deep-rooted historical, cultural, and social connections between these regions date back to the fifteenth century (French, 2021)— and must inform research and scholarship. In much of the literature, the term governance implies a departure from the traditional focus on internal government to increase attention on the blurred boundaries between government institutions and other societal actors (Pierre, 2000). From that perspective, the government becomes a facilitator, able to use new methods to steer and govern in a world where collective action is also defined by self-governing networks (Stoker, 1998). The interactions between civil society and government, whether formal (institutional processes and frameworks) or informal (political culture, parties, and the media), are what define the main character of 'governance' in a particular context (Wilson, 2000). Therefore, knowing who governs, with what purpose, and through what mechanisms is the starting point of comparing and analyzing climate governance in the various contexts included in this book.

Studies in the humanities and social science disciplines argue that cultures in these regions influence each other based on their shared experience to overcome colonialism and achieve social, cultural, economic, and political independence. For example, Abidde et al. (2018, p. 1) argue the need to increase bilateral and multilateral cooperation between these regions because of their "shared history of slavery, colonialism, and underdevelopment." Thus, partnership building is self-evident within the framework of the 2006 first African-South American summit for "establishing interregional partnerships" (Al Jazeera, 2006). But such initiatives must dig deep to discover underlying connections based on situated knowledge.

Nonetheless, and in spite of the historical ties and current initiatives that bring Africa and its cultures to Latin America and the Caribbean, the regions share a mutual lack of knowledge of each other in terms of climate governance. This is due, first, to perceptions based on linguistic and cultural differences that hamper communication and intra- and inter-regional exchanges. Granted, with its geographical, political, legal, economic, and cultural diversity, Africa lacks a single policy frame for engaging with other regions. But countries such as Angola, Cape Verde, Equatorial Guinea,

Guinea Bissau, Mozambique, São Tomé and Príncipe, and Western Sahara with Portuguese-Spanish influences similar to those in Latin America and the Caribbean (Peters, 2021; Moreno, 1969) offer openings and opportunities for engagement.

Second, the mutual lack of knowledge between the regions is due partly to sparse comparative research analyzing approaches to climate governance that confront the planetary crises of climate change, greenhouse gas emissions, and losses due to deforestation and degradation across the regions. This is not to downplay the significance of sources of climate data from national and multinational institutions, including United Nations, European Union, World Bank, International Monetary Fund, Interamerican Development Bank, African Union, to name but a few. Nor is it to overlook case studies aimed at guiding efforts to design effective and sustainable adaptation initiatives (Maino & Verner, 2022; Filho et al., 2021; Kronik & Verner, 2010). To be sure, one of the most recent publications on climate change management takes a global perspective, presenting substantial research and field projects from these regions and worldwide (Filho et al., 2021). In addition to highlighting successful cases, this six-volume series emphasizes the roles of research, leadership, and transformation to resilience, covering training and education while exemplifying capacity building in specific sectors. The current book builds on these ideas by exploring ways in which to connect sectors and improve climate governance.

Third, studies on transatlantic migration and Black diasporan experience expose the issues of enslavement, racism, and anti-colonial resistance (Davies, 1994; Klein, 1986)—a staple of history and humanities scholars—based on historical, sociological, geographical, and in some cases political and economic perspectives. In some ways, this kind of work overlaps with the goals of this book especially as it underpins the *International Decade for People of African Descent (2015–2024)*, as well as individual state obligations to guarantee the fundamental rights of people of African descent (Economic Commission for Latin America and the Caribbean, 2020). In addition, it overlaps with the theme of this book as it relates to human rights, equality, and inclusion in regions where environmental human rights defenders face increasing forms of repression (Montoya, 2022). Nonetheless, this book differs by focusing explicitly on climate governance as a topic of research interest. Thus, in dealing with the specific subject of climate governance, the book emphasizes not just how these regions organize themselves to fight adverse climatic conditions but also how they assess action and progress on climate policy.

Finally, the literature strongly suggests the worrisome mismatch between imported European-style state structures inherited from colonial rule and the decentralized social, cultural, political, and economic structures of society in these regions (Francis et al., 2020; Magbabelo, 2020; Ijeoma & Nzewi,

2018; Badie, 1992). There is hardly any evidence that the forces of digitalization, urbanization, regionalization, and globalization (O'Neil, 2022) can neutralize the effect of the mismatch on the quality of governance. Neither is there evidence that ongoing neoliberal market or quasi-market reform initiatives based on models of competition addresses the mismatch except attempts at decentralization and devolution. And yet such a mismatch has implications for creating external value through climate governance. Oriented toward various European administrative traditions classified as "Anglo-American, Germanic, Napoleonic, and Scandinavian" models (Peters, 2021, p. 26) with distinct values, norms, and behaviors, state structures and governance systems in these regions have no organic relationship with society. While a contractual relationship exists in varying degrees through constitutional-legal and institutional arrangements, it is tenuous at best due to disruptive changes and/or misinterpretation of constitutional requirements. Thus, lacking an organic link to and only dependent on a weak contractual relationship with society, the states and governing systems in these regions have a fuzzy framework within which to frame climate governance approaches.

As noted above, the concept of administrative tradition, its defining features, and how it impacts the conduct of public administration and governance has substantial relevance for framing approaches to climate governance in these regions. If as Peters (2021, p. 25) argues, administrative "tradition involves what people, especially political and administrative elites, think about administration," what tradition(s) or model(s) do political and administrative elites in these regions follow as they design climate governance approaches especially after enduring multiple colonial rule? Historians such as Adelman (1999) and Daly (1968) analyze the legacies of colonialism shaping states and public administration in Latin America, demonstrating a strong colonial heritage in continued state development there in both positive and negative ways. Likewise, Tignor (1999) argues that officials under Spanish rule maintain positions through personal ties and connections. However, the region has a varied history ranging from indigenous rule, military dictatorship, to democratization (Benavides, 2017). Other scholars note political-administrative reforms based on the Anglo-American tradition and agree that questions about the quality of governance remain (Velarde et al., 2014). Such questions must factor into how Latin America and the Caribbean consider climate governance policy options.

The case of Africa presents an interesting challenge of its own, having undergone multiple colonial rules, experienced military dictatorships and one-party rule, and implemented varied reforms with implications for framing climate governance approaches. This leaves the question of administrative tradition ill-addressed and makes it more complicated than meets the eye. For example, Olivier de Sardan (2009) argues that administrative traditions in Africa arise from two "ruptures," involving colonization and bureaucratization on the one hand and the creation of self-governing states and nationalism on the other. Likewise, Young (2012) identifies three trajectories in state development in Africa, including nationalism, radicalization, and democratization in that order. While these characterizations are helpful, they tell only half of the story, suggesting, incorrectly, a linear but simplistic transition from one legacy to another.

However, as Peters (2021) and Staniland (1971) argue, colonial footprints are visible in varying degrees, but traditional administration and informal structures remain in what Brinkerhoff and Goldsmith (2005, p. 200) describe as "institutional dualism." In this scenario, westernized rulebound systems aimed at improving responsiveness and accountability run side-by-side with embedded local understandings and practices influencing and shaping governance.

But much of the public administration literature often takes a narrow, technocratic, and government "know-how" approach to governance, one that does not address underlying and structural root causes of policy failures and maladministration in these regions. Such narrow approaches result in narrow prescriptions. This is a challenge to write about because the way the book intends to portray it diverges from traditional notions of what public administration entails. For example, McCourt (2013) takes a problem-solving perspective in discussing public service reform in Africa, emphasizing the questions of how to make government orderly and how to get it closer to the grassroots, among others. Focusing on organizational and managerial solutions, McCourt suggests the development of structures to enhance institutional capacity building.

Similarly, in the case of Brazil and other "late democracies," Berman and Puppim de Oliveira (2021, p. 1183) argue that "shortcomings in state building inflict heavy costs on society," which is correct. But their call for "another round of reforms," one that rebalances "the roles of accountabilities of politicians and the bureaucratic apparatus" (pp. 1188–1189) reflects the prevailing organizational mindset. While there is nothing inherently wrong with organizational and managerial approaches, they fail to get at the type of change that climate governance requires to be effective. Following Roberts (2019, p. 1), we argue that for climate governance approaches to be sustainable, they must be "bold" enough to transcend the concerns for efficiency and to embrace the broader public interest of "strategic governance."

CLIMATE GOVERNANCE CHALLENGE

Climate change is an existential threat, and the problems it poses to Africa, Latin America, and the Caribbean are self-evident—it is causing extensive economic damage and worsening extreme poverty (World Bank, 2021;

World Food Program, 2020). While a "loss and damage" fund negotiated at COP27 to help highly vulnerable countries is reassuring, there is no mechanism yet for its implementation. It remains a planning nightmare: Who contributes to the fund? How much does each nation contribute? Where is it housed? Which nations benefit from it? Without legal teeth, vulnerable nations such as those in these regions, view the fund as an inadequate climate measure from the most polluting nations. If experience with other sources of climate funding—emission reduction and adaptation money—is anything to go by, the loss and damage fund will generate only a paltry sum. Thus, climate action is facing a severe challenge.

But while climate action initiatives are underway in these countries, the conceptual frames and tools they deploy—perpetrators of structural inequities and institutional dysfunctions—remain in place. The same frames and tools, including coloniality, top-down decision making, inclusive exclusion, organizational and managerial approach, and market-based interventions that fail to address under-development play a central role in climate action measures. For example, in *A Roadmap for Climate Action in Latin America and the Caribbean 2021–2025*, the World Bank (2021) proposes a well-conceived model for integrating climate and development through transformative climate action, but the proposal is based in and informed by the orthodox top-down and market-based language. A couple of examples suffice: one of the priorities for the World Bank's engagement in the two regions is to build on "consultations with sector and country experts" (p. 7), reinforcing the orthodoxy and the construct of top-down and privileged consultation that excludes diverse views and voices from the bottom-up.

At the same time, the World Bank's proposal advocates "protecting the communities most vulnerable to climate change," by "strengthening health systems and physical access in rural areas, combined with social protection and contingent sources of finance" (p. 10). This is laudable, but if experience with previous World Bank and IMF-sponsored interventionist programs teaches anything, it is that social programs are the first to go in a period of austerity measures. And there is little to no confidence in protecting social programs in regions where the existing structures disadvantage underrepresented communities and groups. Neither is there a grounded understanding of how rural communities in these regions survive and how climate investments can create value for them and improve lives. Like the orthodox notion of public administration and policy, which makes it difficult to embed equity and inclusion in practice, teaching, and research (Blessett et al., 2019), the Bank's proposal is not meant to be inclusive.

Nonetheless, the World Bank's proposal is not an isolated example. The *Draft Africa Climate Change Strategy 2020–2030* is equally problematic if not more so. Its guiding philosophy is rightly based in *Africa Agenda 2063*, which advocates "united efforts, self-reliance, Africa financing its own climate smart, all-inclusive, people-driven development and

Africa speaking with one voice in global fora" (p. 7), but its focus on institutions and the minimal attention paid to local communities, women, and young people is astounding. It is correct that Africa "faces the greatest impacts from climate change" and that African countries have done their part by submitting "their Nationally Determined Contributions" and by "implementing national climate actions" (pp. 6–7). Likewise, it is appropriate that they seek and incorporate feedback from regional economic communities, academia, and the African Ministerial Conference on the Environment. But broad-based participation and collective action, which are lacking must be the north star to guide Africa's climate action initiatives.

By emphasizing the "African Union Commission, Regional Economic Communities and their agencies together with many other partners" (p. 7), the strategy reflects a familiar approach, one aimed at achieving not transformative, but incremental change within the prevailing socioeconomic and institutional structures and processes. This approach reinforces the continued hierarchical perspective of public administration and climate governance and downplays if not ignores the increasing diversity of actors and approaches to climate mitigation. And it barely considers that climate mitigation action is durable and effective when it is conducted in the "context of sustainable development, equity, and poverty eradication, and rooted in the development aspirations of the societies within which they take place" (IPCC WGIII, 2022).

Despite the growing public awareness of the need for disaster risk reduction, mainstream public administration literature continues to exclude at risk populations—those living in rural communities and the informal labor work force for whom social protection programs are mere temporary measures (Onyishi et al., 2020; Banerjee & Jackson 2017). While the reform of institutional frameworks is creating spaces for participation and accountability through decentralized governing systems, such reform hardly keeps pace with the demand for change.

Africa's youth are yearning for opportunities to shape their future in a substantive fashion (Umutoniwase, 2021). And if rural communities, vulnerable groups, and the youth become substantive partners in climate action measures, the values underpinning climate governance will be transformed. Such values will be based on the needs of underrepresented communities and not just on market-based programs. Alternatives to market and profit-driven climate investment ventures such as community-based self-help groups, neighborhood help initiatives, and farmers groups exist across the regions. Because they develop organically based on lived rather than westernized experience, they have a chance to strengthen climate resilience (Bateman, 2011).

SUMMARY

The theme tying the book's chapters together is the shared understanding that climate governance and administrative studies, as well as policy creation, and its implementation in these regions have particularities intertwined with lived experience and global trends. Chapter authors have instructions to provide deep descriptions and analyses of climate governance and its consequences in these developing contexts, drawing freely from the existing literature and interweaving them with local, national, and global experiences. The goal is to advance comparative administrative study and promote cross-cultural and regional learning in the context of the existential threat that climate change poses. By examining, comparing, and connecting climate governance patterns, systems, institutions, processes, and outcomes to development policy in these regions, chapter authors expose policy gaps and distill lessons for improving climate mitigation, strengthening adaptive capacity, and protecting vulnerable populations.

The book offers chapter authors the opportunity to ask big questions about inclusive climate governance that existing approaches overlook. If climate governance models are to be inclusive and sustainable, they must provide opportunities for communities and people from all social-cultural backgrounds to tell their own stories and discuss their understandings about conservation and environmental protection. The authors have the opportunities to engage with the central issue of how people in these regions define their own realities, their methodologies for doing so, and the ways in which they diverge and/or converge with the existing westernized approaches to public administration and climate governance. By doing so, the authors contribute to enhance a better understanding and framing of climate governance approaches beyond the Eurocentric perspective. As Francis et al. (2020) argue in *Inequality Studies in the Global South*, where one studies a problem matters for how one understands and addresses it.

For the purpose of coherence and, as summarized below, the book is structured in five parts. Part I, Introduction, consists of two chapters that lay out the rationale for the book. In Chapter 1, Peter Haruna and Cristina Stanica argue that more research attention needs to be paid to the Global South countries faced with more frequent and more severe weather extremes. They ask the critical question of whether African, Latin American, and the Caribbean countries have the ability to coordinate governance at multi levels and with multi-actors in the face of climate change as they struggle to protect their most vulnerable populations. In Chapter 2, Abraham Benavides and Liza Van Jaarsveldt examine governance frameworks from multiple perspectives that both illuminate opportunities for and pose challenges to Global South countries.

Part II, Regional Perspectives, provides region-wide climate perspectives beyond individual states. In Chapter 3, Henry Mburu and Joseph Macheriu explore climate financing, climate monitoring, reporting and regulatory programs within the Eastern African Community of seven states faced with high climate vulnerability. After examining each state separately, they provide comparative analyses that highlight the degree of vulnerability within the community, especially in regard to the intersectionality of food insecurity, conflicts, and climate change impacts. In Chapter 4, Geneve Philip-Durham confronts climate governance in the Cayman Islands, examining the agenda-setting process and reviewing key policy measures while evaluating institutional awareness and public opinion sensitization efforts. In Chapter 5, Michael Yoder compares Costa Rica and Panama's climate governance approaches based on reforestation policy and management. While both countries have made progress, he argues the need for including small-scale farmers and indigenous communities to improve participation and inclusivity in climate governance. In Chapter 6, Hugo Renderos addresses public transportation issues in the major cities of Guatemala, Honduras and El Salvador with a focus on the pollution effect of school bus systems there. He concludes with a recommendation that the region stands to benefit by adopting the Salvadoran approach with emphasis on eliminating black carbon.

Part III consists of eight chapters, focusing attention on climate governance at the national level with mutually reinforcing themes. In Chapter 7, Peter Haruna and Hugo Renderos analyze and compare national policies and programs in Ghana and Nicaragua, arguing that climate governance is ineffective because their policies barely connect with their respective societies and beliefs about nature and preservation. Likewise, Brighton Shoniwa argues in Chapter 8 for better aligning climate action with new public governance principles to enhance commitment to civic participation in climate change policy initiatives. While advocating more public involvement in disaster management in Chapter 9, Manlio Salas and Heidi Smith emphasize the role local governments play in disaster prevention and management in Mexico. In Chapter 10, Syanda Mthuli amplifies the theme of public involvement, arguing for the need to foster climate change understanding at the lowest level of governance in South Africa.

In Chapter 11 and Chapter 12, Thean Potgieter calls attention to ways in which Ocean Health and the blue economy can be improved. He connects Ocean Health to the welfare of communities in South Africa, arguing for the need to protect aquatic life as a way of helping to save the planet. Highlighting the South African experience, he argues that to sustain ocean life requires reasonable regulation of harvesting ocean resources with minimal polluting effect. In Chapter 13, Laila El Baradei and Shaimaa Shabbah analyze the Egyptian experience with climate governance, especially in the lead up to COP27. They highlight the governance architecture for climate change mitigation and adaptation while suggesting ways for improvement. In Chapter 14, Kalim Shah shines light on climate governance in Suriname by describing the national adaptation planning and institution building process. While Suriname's model of climate adaptation planning is untested, Shah argues that it offers lessons for other developing nations. In Chapter 15, Laila El Baradei discusses challenges the sea-level rise poses to the historic city of Alexandria, Egypt. She argues that effective climate governance entails addressing policy implementation gaps and loopholes that the Egyptian experience exposes. In Chapter 16, Abraham Benavides and coauthors apply a governance perspective to analyze and highlight the environmental effects of water quality and the risks associated with unsustainable solid waste practices. In Chapter 17, Victor Ferreros and coauthors use Ostrom's polycentric environmental governance theory to explore Kenya's climate governance, emphasizing solar energy in cities.

The final three chapters of the book focus on public and community engagement in climate governance along the lines suggested by the current and previous IPCC reports. In Chapter 18, N. S. Matsiliza advocates the involvement of varied organizations in South Africa's climate change mitigation initiatives. In Chapter 19, Redemption Chatanga and Mareve Biljohn explore ways in which communities in South Africa understand and perceive climate change with the purpose of nudging community participation in climate governance. To close the book in Part V, Peter Haruna and Abraham Benavides provide a critical summary and an assessment of the quality of governance in Chapter 20. They highlight intellectual and policy lessons learned in climate governance, emphasizing that more research is needed to understand what adaptation means and how to achieve it through community-based solutions.

REFERENCES

Al Jazeera. (2006, November 6). *Africa–South American summit begins.* Retrieved March 14, 2023, from https://www.aljazeera.com/news/2006/11/30/africa-south-american-summit-begins

Abidde, S. O. (2018). *Africa, Latin America, and the Caribbean: The case for bilateral and multilateral cooperation.* Lexington Books.

Adelman, J. (1999). *Colonial legacies: The problem of persistence in Latin American history.* Routledge.

Banerjee, S. B., & Jackson, L. (2017). Microfinance and the business of poverty reduction: Critical perspectives from rural Bangladesh. *Human Relations, 70*(1), 63–91.

Bateman, M. (2011). *Confronting microfinance: Undermining sustainable development.* Kumarian Press.

Benavides, A. (2017). Transforming Latin America—One innovation at a time. *Public Administration Review, 77*(3), 469–472.

Berman, E., & Puppim de Oliveira, J. A. (2021). Exposing the unfinished business of building public administration in late democracies: Lessons from the COVID-19 response in Brazil. *Public Administration Review, 81*(6), 1183–1191.

Brinkerhoff, D. W., & Goldsmith, R. A. (2005). Institutional dualism and international development: A revisionist interpretation of good governance. *Administration & Society, 37*(2), 199–224.

Davies, D. J. (1994). *Slavery and beyond: The African impact on Latin America and the Caribbean.* Jaguar Books.

Economic Commission for Latin America and the Caribbean. (2020). *People of African descent in Latin America and the Caribbean: Developing indicators to measure and counter inequalities.*

Daly, G. (1968). Prolegomena on the Spanish American political tradition. *Hispanic American Historical Review, 48,* 37–58.

Dellasala, D. A., & Goldstein, M. I. (2017). *Encyclopedia of the anthropocene* (5-volume set). Surfbird Consulting.

Filho, W. L., Luetz, J. M., & Ayal, D. Y. (2021). *Handbook of climate change management: Research, leadership, transformation.* Springer.

Foucault, M. (1966). *The order of things: An archeology of the human sciences.* Vintage Books.

Ford, M. R., & Ihrke, D. M. (2018). Perceptions are reality: A framework for understanding governance. *Administrative Theory & Praxis, 41*(2), 129–147. https://doi.org/10.1080/10841806.2018.1512337

Francis, D., Valodia, I., & Webster, E. (2020). *Inequality studies from the global south.* Routledge.

French, H.W. (2021). *Born in Blackness: Africa, Africans, and the making of the modern world, 1471 to the Second World War.* W. W. Norton and Company.

Grosfoguel, R. (2007). The epistemic decolonial turn: Beyond political economy paradigms. *Cultural Studies, 21*(2–3), 211–223.

Gulrajani, N., & Moloney, K. (2011). Globalizing public administration: Today's research, tomorrow's agenda. *Public Administration Review, 72*(1), 78–86.

Haque, M., & Turner, M. (2013). Knowledge-building in Asian public administration research, education, and practice: Current trends & future challenges. *Public Administration and Development, 33*(4), 243–248.

Heady, F. (1998). *Public administration: A comparative perspective.* Marcel Dekker, Inc.

Ijeoma, E. O., & Nzewi, O. I. (2018). *Culture, philosophies, and reforms in public administration for the globalizing world: A reflection on local, regional and international perspectives.* AOSIS.

Intergovernmental Panel on Climate Change. (2022). *Climate Change 2022: 6th assessment report: Impacts, adaptation and vulnerability.* United Nations.

Klein, H. S. (1986). *African slavery in Latin America and the Caribbean.* Electronic Books.

Kronik, J., & Verner, D. (2010). *Indigenous peoples and climate change in Latin America and the Caribbean.* World Bank.

Ladi, S. (2019). European studies as a tributary of global policy and transnational administration. In D. Stone, K. Moloney (Eds.), *Oxford handbook of global policy and transnational administration* (pp. 293–309). Oxford University Press.

Magbabelo, J. O. (2020). A critique of Willy McCourt's "Models of Public Service Reform." *World Affairs: The Journal of International Issues, 24*(3), 142–147.

Maino, R., & Emrullahu, D. (2022). *Climate change in Sub-Saharan Africa's fragile states: Evidence from panel estimations.* International Monetary Fund.

Mamdani, M. (1999). *Citizen and subject: Contemporary Africa and the legacy of late colonialism.* Princeton University Press.

Mahony, M., & Hulme, M. (2016). Epistemic geographies of climate change knowledge. *Environmental Humanities, 10*(1), 330–337.

McCourt, W. (2013). *Models of public service reform: A problem-solving approach.* Policy Research Working Paper 6428. World Bank.

Montoya, M. (2022). *Defending the environment shouldn't be deadly.* Retrieved March 17, 2023 from https://www.worldpoliticsreview.com/land-defenders-environmental-activists-indigenous-rights/

Moloney, K., & Stoycheva, R. (2018). Following the money: How transparent are the world's international organizations? *Issues Brief.* Retrieved March 14, 2023 from https://www.internationalaffairs.org.au/australianoutlook/io-transparency

Moreno, E. J. (1969). *Legitimacy and stability in Latin America: A study of Chilean political culture.* Beacon Press.

Norris-Tirrell, D., & Clay, J.A. (2020). *Strategic collaboration in public and nonprofit administration: A practice-based approach to solving shared problems.* Taylor & Francis.

Olivier de Sardan, J. (2009). State bureaucracy and governance in francophone West Africa: An empirical diagnosis and historical perspective. In G. Blundo & P. Y. Le Meur (Eds.), *The governance of daily life in Africa* (pp. 39–72). Brill.

Onyishi, C. J., Ejike-Alieji, A., Ajaero, C., Mbaegbu, C., Eziebe, C., Onyebueke, V., Mbah, P., & Thaddeus Nzeadibe. T. (2020). COVID-19 pandemic and informal urban governance in Africa: A political economy perspective. *Journal of Asian and African Studies, 56*(6), 1–25. https://doi.org/10.1177/0021909620960163

Peters, B. G. (2021). *Administrative traditions. Understanding the roots of contemporary administrative behavior.* Oxford University Press.

Pollitt, C., & Bouckaert, G. (2017). *Public management reform: A comparative analysis—Into the age of reform.* Oxford University Press.

Riggs, F. (1964). *Administration in developing countries: The theory of prismatic society.* Houghton Mifflin.

Sapiains, R., Ibarra, C., Jimenez, G., O'Ryan, R., Blanco, G., Moraga, P., & Rojas, M. (2021). Exploring the contours of climate governance: An interdisciplinary systematic literature review from a southern perspective. *Environmental Policy and Governance, 31*(1), 46–59.

Staniland, M. (1971). Colonial government and populist reform: The case of the Ivory Coast, part I. *Journal of Administration Overseas, 10*, 33–42.

Tignor, R. L. (1999). Colonial Africa through the lens of Colonial Latin America. In J. Adelman (Ed.), *Colonial legacies: The problem of persistence in Latin American history.* Routledge.

Umutoniwase, C. (2021). *Conversation with the UN Deputy Secretary-General.* Nairobi, Kenya. https//ourfutureagenda.org/2021/06/dialogue-with-the-deputy-secretary-general

United Nations. (n.d.). *UN framework convention on climate change.* Retrieved, March 14, 2023 from https://enb.iisd.org/negotiations/un-framework-convention-climate-change-unfccc

Velarde, J. C., Lafuente, M., & Sangines, M. (2014). *Serving citizens: A decade of civil service reforms in Latin America (2004–13).* Washington, DC, Inter-American Development Bank.

Verhoest, K., Van Thiel, S., Bouckaert, G., & Laegrad, P. (2016). *Government agencies, and practices from 30 countries.* MacMillan

Vukasovic, M., & Stensaker, B. (2018). University alliances in the Europe of Knowledge: Positions, agendas and practices in policy processes. *European Educational Research Journal, 17*(3), 349–364.

World Bank Group. (2022). *A roadmap for climate action in Latin America and the Caribbean, 2021–2025.* World Bank.

Zeufack, A., & Cust, J. (2023). *Africa's resource future: Harnessing oil, gas, and minerals for economic transformation during the low carbon transition.* World Bank.

CHAPTER 2

CLIMATE GOVERNANCE IN GLOBAL CONTEXT

Abraham David Benavides
The University of Texas at Dallas

Liza Ceciel van Jaarsveldt
University of South Africa

ABSTRACT

This chapter begins with an explanation and definition of governance in the context of climate change in the Global South. The emphasis is to set the foundation for what governance is and how it is manifested in various situations in the Southern Hemisphere. Various definitions are considered and governance in general is explained. How countries have governed and managed to address the various public policy issues surrounding climate change is a major challenge explored in this book. Therefore, the reader is presented with various studies that have looked at governance and how governance has been used and defined in a number of disciplines. A better understanding of pushing forward public policy issues of concern to societies as a whole is essential. Finally, a summary is provided of how governance can affect positive change for many developing countries.

Climate Governance in International and Comparative Perspective, pages 19–37
Copyright © 2024 by Information Age Publishing
www.infoagepub.com

The United Nations Sustainable Development Goals (UNSDG) provide a broad road map to guide our society towards a better way of life for all nations and people. For instance, some of the seventeen goals include no poverty, zero hunger, quality education, clean water and sanitation, reduced inequalities, affordable and clean energy, sustainable cities and communities and a number of other qualities of life initiatives. The title of UNSDG Goal 13 is "Take urgent action to combat against climate change and its impacts." Its targets and indicators include several practical and action-oriented steps that nations and individuals can take to prepare for projected catastrophes See Table 2.1. For instance, Target 13.1 encourages all countries to "Strengthen resilience and adaptive capacity to climate-related hazards and natural disasters." Target 13.2 promotes the adoption of measures to "integrate climate change into national policies, strategies and planning." And finally, Target 13.3 suggests an improvement in "education, awareness-raising and human and institutional capacity on climate change mitigation, adaptation, impact reduction and early warning."

The United Nations Sustainable Development Goal 13 with its targets and indicators is a practical way to address a public policy issue that has real consequences for the people it is trying to serve. Climate change in its many forms and configurations has been debated, with respect to its reality or authenticity, by scientist, politicians, administrators, and academics for a number of years. The point of this chapter is not to argue the merits of the debates, but to look at how nation states and public and nonprofit organizations have come to grips with governing a phenomenon that is having real impacts on society. The perceived reality is, that something is happening, and although our experts may not agree as to the exact nature of the various changes in our climate, the recognized considerations are that something must be done. Following prudent guidelines like those offered by the United Nations is a step in the right direction.

Another significant element that can be learned from the United Nations Sustainable Development Goal 13 is the care and concern demonstrated for individuals living in areas of the world that would be considered still developing. For instance, Target 13.a notes the need to secure funds to help meet the "needs of developing countries." Similarly, Target 13.b suggests promoting "mechanisms for raising capacity for effective climate change-related planning and management in least developed countries and small islands."

The language here: "least developed countries" and "developing countries" connotes a reality that in our world, there are nations that are not as developed or industrialized as others and merit assistance. Some scholars group these countries into categories and use terms such as Third World (an older term not used as frequently anymore), underdeveloped, less developed, or

TABLE 2.1   United Nations Sustainable Development Goal 13: *Take urgent action to combat against climate change and its impacts—*Targets and Indicators

Target 13.1   Strengthen resilience and adaptive capacity to climate-related hazards and natural disasters in all countries

Indicators

13.1.1   Number of deaths, missing persons and directly affected persons attributed to disasters per 100,000 population

13.1.2   Number of countries that adopt and implement disaster risk reduction strategies in line with the Sendai Framework for Disaster risk Reduction 2015–2030

13.1.3   Proportion of local governments that adopt and implement local disaster risk reduction strategies in line with national disaster risk reduction strategies

Target 13.2   Integrate climate change measures into national policies, strategies and planning

Indicators

13.2.1   Number of countries with nationally determined contributions, long-term strategies, national adaptation plans and adaptation communications, as reported to the secretariat of the United Nations Framework Convention on Climate Change

13.2.2   Total greenhouse gas emissions per year

Target 13.3   Improve education, awareness-raising and human and institutional capacity on climate change mitigation, adaptation, impact reduction and early warning

Indicators

13.3.1   Extent to which (i) global citizenship education and (ii) education for sustainable development are mainstreamed in (a) national education policies; (b) curricula; (c) teacher education; and (d) student assessment

Target 13.a   Implement the commitment undertaken by developed-country parties to the United Nations Framework Convention on Climate Change to a goal of mobilizing jointly $100 billion annually by 2020 from all sources to address the needs of developing countries in the context of meaningful mitigation actions and transparency on implementation and fully operationalize the Green Climate Fund through its capitalization as soon as possible

Indicators

13.a.1   Amounts provided and mobilized in United States dollars per year in relation to the continued existing collective mobilization goal of the $100 billion commitment through to 2025

Target 13.b   Promote mechanisms for raising capacity for effective climate change-related planning and management in least developed countries and small island developing States, including focusing on women, youth and local and marginalized communities

Indicators

13.b.1   Number of least developed countries and small island developing States with nationally determined contributions, long-term strategies, national adaptation plans and adaptation communications, as reported to the secretariat of the United Nations Framework Convention on Climate Change

Source: Adapted from United Nations Sustainable Development Goal 13 Targets and Indicators

developing countries—as used in the literature of the United Nations—or more recently, the favored term is the Global South.

Chris Calimlim, (2023) refers to the Global South as "various countries around the world that are sometimes described as developing, less developed, or underdeveloped. Many of these countries—although by no means all—are in the Southern Hemisphere, largely in Africa, Asia and Latin America" He goes on the say that they are generally poorer "have higher levels of income inequality and suffer lower life expectancy and harsher living conditions than countries in the Global North—that is, richer nations that are located mostly in North America and Europe, with some additions in Oceania and elsewhere." In general, there is a difference between countries in these two hemispheres and the focus of this book is how climate change has affected the countries in the Global South. Specifically, a theme that runs through many of these chapters is how these countries in the Global South have addressed climate change through what is called governance or a systematic process or structure or systems for making decisions.

This chapter begins with an explanation and definition of governance in the context of climate change in the Global South. How countries have governed and managed to address the various public policy issues surrounding climate change is a major challenge explored in this book. Next the reader is presented with various studies that have looked at governance and how governance has been used and defined in a number of disciplines. The chapter provides additional insights into governance in the international context with specific attention to the Global South. A better understanding of pushing forward public policy issues of concern to societies as a whole is essential. Finally, a summary is provided of how governance can affect positive change for many developing countries. This chapter made use of a conceptual study that analyzed the role, nature and importance of governance. The reading and analyzing of content specific to articles, reports on climate change, books and policy papers formed the bases of this chapter. This process of content analysis assisted to determine the important role that governance plays in climate change specifically for countries from the global south.

DEFINING GOVERNANCE

Although governance is not a new word or term, for many it appears to be synonymous with government. This is partly due to the fact that both words are derived from the same root word of govern or in Latin "gubernare" which means to "direct, rule, steer, or to be at the helm" (Britanica, n.d.). The Oxford Advanced Learner's Dictionary (2005: 646) further helps us understand the word govern by including the adjectives of "reign, control, influence, conduct, or supervise; conduct the policy, actions, and affairs of a state, organization, or people"). Over the years, the word government has taken on the notion of the structure of government or in other words, those components such as the legislature, the executive, the judiciary, the laws, codes, statues, ordinances, and its component constitutions which defines how—mechanically and fundamentally—a government is structured and works. This structure does to a degree affect how decisions are made and in many countries the government structure and governance are synonymous. The structure allows for decisions that involve multiple organizational partners and the work of governing is carried out to the benefit of the sovereign. However, in many cases this is not the norm and the structure of government prevents cooperation with multinational organizations, nonprofits, and agency to agency cooperation. The bureaucratic structure hinders the ability of the government to fully govern. Governance on the other hand, is not defined so much by how the government is structured but by how a government carries out its functions or processes to support its sovereign. For instance, how are decisions made? Is there accountability and transparency? Are fairness and responsibility manifest in public policies?

In the publication *Governance Today* (2023), it defines governance as

> The system by which entities are directed and controlled. It is concerned with structure and processes for decision making, accountability, control and behavior at the top of an entity. Governance influences how an organization's objectives are set and achieved, how risk is monitored and addressed and how performance is optimized. Governance is a system and process, not a single activity and therefore successful implementation of a good governance strategy requires a systematic approach that incorporates strategic planning, risk management and performance management. Like culture, it is a core component of the unique characteristics of a successful organization. (p. 1)

Similarly, the United Nations Educational Scientific and Cultural Organization (UNESCO 2023) defines governance as

> The processes that are designed to ensure accountability, transparency, responsiveness, rule of law, stability, equity and inclusiveness, empowerment, and broad-based participation. Governance also represents the norms, values and rules of the game through which public affairs are managed. Governance therefore can be subtle and may not be easily observable. In a broad sense, governance is about culture and institutional environment in which citizens and stakeholders interact among themselves and participate in public affairs. (p. 1)

These definitions provide a solid foundation for understanding governance and its relationship to and differences from government. Table 2.2 further illustrates the differences by noting the response of government

TABLE 2.2 Differences Between Governance Leadership and Government Management

	Governance Leadership	Government Management
Vision/Ownership	Set norms, strategic vision and direction, and formulate high level goals and policies	Run the organization in line with the broad goals and direction set by the government body
Execution	Oversee management and organizational performance to ensure the organization is working in the best interests of the public	Make operational decisions and policies, keep the government bodies informed and educated
Implementation	Direct and oversee management to ensure the organization is achieving the desired outcomes and to assure the organization is acting prudently, ethically, and legally	Implement the decisions within the context of the mission and strategic vision
Decision-Making	Decisions are based on the needs of stakeholders who are served by the organization's mission. Transparency, equity and accountability are key criteria to follow	Makes decisions based on line of authority and existing laws and ordinances, responsive to requests for additional information

Source: Adapted from UNESCO Concept of Governance Notes (2023).

management as opposed to governance leadership. Governance Leadership is to be understood as a process whereby leaders make commitments to take action that is beneficial for the individuals being served or for public policy issues being supported as well as for the organization. In terms of climate change, governance leadership would manifest itself as leadership that recognizes that something needs to be done; it would mobilize resources to assure the education and implementation of policies that would promote a safe environment; transparency and responsiveness would be high on the leadership's agenda; decisions made would be participatory and empower individuals to act responsibly, and public policies enacted would have gone through a fair processes in which all parties would have had an opportunity to opine on any legislation being put forward.

THE LITERATURE ON GOVERNANCE

Governance has been viewed, studied, and examined through a number of lenses and by a number of disciplines (Braganza, & Lambert, 2000; Vigoda-Gadot, 2002; Meier, 2010; Sørensen, & Torfing, 2020; Di Giulio, & Vecchi, 2021). This mosaic of governance literature and study (Robichau, 2011) has caused some contentions and generated lively debate with respect to a perfectly acceptable definition of governance. Peters and Pierre (2007) note that the "concept of governance has come to be used more commonly in the discussion of public administration, but the meaning of the term is not always clear" (1). From the simple definitions of governance as noted above to the more complex notions of governance as network structures of organizations coming together to solve, manage, and direct issue of public policy (Klijn, & Skelcher, 2007; Klakegg, Williams, Magnussen, & Glasspool, 2008; Sørensen, & Torfing, 2009; Lewis, 2011; Barwick, & Gross, 2019;) we are left to parcel out how we look at and understand governance.

Katsamunska (2016) argued that by the end of the last century and the beginning of the twenty-first century the theory of governance took on a significant degree of importance because of its ability to "cover the whole range of institutions and relationships involved in the process of governing." In other words, the concept lent itself to capture the broad latitude that included not only the structure of government but also the complicated arrangements that had developed as governments devolved and transferred some of their responsibilities to the private and nonprofit sectors through contracting, mobilizing, advocacy and developmental activities (Choudhury and Ahmed (2007). This practice of lean government or Hollow State (Milward and Provan, 2000) has found itself in a number of countries around the world as they have attempted to incorporate third-party providers in the delivery of services. The popularity of governance as a broad term and its flexibility as a theory to bend and shape to various scholars research interest, has had some argue that a paradigm shift has occurred in the field and that public management (a concept closely related to governance) has replaced public administration as a field of study (Olowu, 2002; Bryson, Crosby, and Bloomberg, 2014; McDonald III, Hall, O'Flynn, and van Thiel, 2019). Providing government services directly by government agencies or the bureaucracy is classical public administration. Governance provides the opportunity and freedom to move beyond the restrictions of the rigid administrative apparatus and explore new methods of delivery that are suitable for a 21st century society.

Although several scholars have written about governance and its origins (Stillman 1990; Frederickson, 2010; Farazmand, 2017) no identifiable event is present as a point of departure. Nevertheless Stillman (1990) notes that the "historical roots of public administration in the United States [coupled with] a unique decentralized approach to public administration, which prioritized individualism, voluntarism, and limited government intervention, [contributed to] contemporary governance" (1) as we know it today. In other words, its historical roots and heritage had its founding in public administration. Fenger and Bekkers (2007) note that George Fredrickson writes about Harlan Cleveland saying in 1972 that "what the people want is less government and more governance."

Another concept associated with governance is that of collaborative governance (Ansell, and Gash, 2008; Skelcher, Mathur, and Smith, 2015; Emerson, 2018; Robertson, and Choi, 2012; Emerson, Nabatchi, and Balogh, 2012). Collaborative governance subscribes to the notion that multiple actors share in the responsibility to assure that various services are delivered or that diverse public policies are carried out in spaces in which they all operate. In other words, governments most often represented by public agencies, meet with their private sector counterparts, and nonprofit organizations to work together as one unit to engage in "consensus-oriented decision-making" (Ansell, and Gash, 2008 1) which eventually provides the desired service for the public. Skelcher, Mathur and Smith (2015) note that "multiple public, private, and non-profit actors join together to shape, make and implement public policy. Partnerships are organizational manifestations of institutional design for collaboration. They offer flexibility and stakeholder engagement but are loosely coupled to representative democratic systems."

Finally, there are a number of ways that one could define governance and a simple look at the internet reveals hundreds of words with governance attached to end such as participatory governance, multi-level governance, regulatory governance, environmental governance, corporate governance, public governance, private governance, organizational governance, global governance, nonprofit governance, and many, many more. Finally, the United Nations Office of the High Commissioner Human Rights (2023) defines good governance as "processes of governing the institutions and practices through which issues of common concern are decided upon and regulated. Good governance adds a normative or evaluative attribute to the process of governing. From a human rights perspective, it refers primarily to the process whereby public institutions conduct public affairs, manage public resources and guarantee the realization of human rights." They further note that the key attributes of good governance are: transparency, responsibility, accountability, participation, and responsiveness to the people in need.

GOVERNANCE IN INTERNATIONAL PERSPECTIVE

Governance as a theory and as a way of understanding our world and complex social systems has gained popularity worldwide. The uses and concepts of governance to describe the ins-and-outs of the new public management movement all over the world have been discussed by several authors (Kettl 2000; Lynn, Heinrich, and Hill, 2000). Part of its appeal is in its ability to make sense of how coordination or even harmonization can exist between various governmental and non-governmental organizations that come together to address related social issues. The complexities and intricacies of our world today mandate that governments use various means necessary to deliver services and respond to public policy issues that confront societies. If that means that countries use the private sector or a group of nonprofits to deliver those services through new structural frameworks, then those partnerships should be implemented. Some scholars have suggested that governance can be vertical in terms of its focus either international or from one level of government to another. Or horizontal a move away from government executives to semi-public autonomous and private organizations (Van Kersbergen and Van Waarden, 2009). For instance, in Norway and the United Kingdom researchers looked at how public investment project governance frameworks conducted business. Particularly they looked at the impact of how embedded governance principles functioned in practice and if their effects were consistent with their aims (Williams, Klakegg, Magnussen, and Glasspool (2010). Similarly, other scholars have looked at coordination and collaboration issues with respect to social systems in Britain; complex public sector partnership in the South East of England; networked community governance in the United Kingdom; and neighborhood governance in England (Flinders 2002; Freeman and Peck, 2007; Durose, 2009; Lowndes and Sullivan, 2008). Finally, in South Korea, Yoo and Kim (2012) looked at governance mixture and varying patterns to find any underlying variation between hierarchy, markets, and networks. In other words, the attempt was made to capture quantitatively if all three governance modes exited. Their results confirmed their hypothesis that all three governance modes were situationally being used in conformance to the situation.

Governance in the international realm and in the global south in general is very much tied to and influenced by international organizations. Lyne, Nielson, and Tierney (2006) argue that national governments delegate to international organizations much in the same way as they delegate to domestic governmental and non-governmental agencies. Khan (2017) further suggests that national governments delegating to international organizations refer to this type of governance as an example of a "holistic theory of public administration and public management." According to Salamon (2002), understanding public administration as governance, rather than government, is a shift in the unit of analysis from public agencies and individual programs to generic program tools. In essence, what this means is that public problem solving is no longer confined to a specific public agency and/or program but approached by and divvied up among a variety of technologies belonging to various governmental and non-governmental programs and actors. Tools are not only technical but also political and cultural. While the United States may prefer more market-based tools, countries in the Global South may be more inclined to rely on the state for the traditional delivery of service.

GOVERNANCE AND THE GLOBAL SOUTH

As noted previously, the Global South is defined as countries which are in the southern hemisphere of our world and have various developmental and governmental structural issues to overcome. They are mostly concentrated in Central and South America, Africa, some countries of the Middle East, India and Southeast Asia. In general, the term is used to describe countries that lack fully viable economies, have low per capita income, lack industrialization and have excessive unemployment (World Population Review, 2023). Although this definition has its exceptions as some countries in the southern hemisphere do not fit this definition (Australia and New Zealand), likewise there are some nations in the northern hemisphere that are struggling and would fit the description perfectly (Ukraine, Pakistan and Afghanistan). Some scholars agree with the following

> The Global South has multiple definitions. The Global South has traditionally been used to refer to underdeveloped or economically disadvantaged nations. These countries are those who tend to have unstable democracy, are in the process of industrializing, and have historically frequently faced colonization by Global North countries (especially by European countries). The second definition uses the Global South to address populations that are negatively affected by capitalist globalization. Based on either of these definitions, the Global South is not the same as the geographical south. Still, to avoid confusion, inaccuracy, and possible offense, many scholars prefer to use the terms "developing countries" or "low-income economies." (World Population Review, 2023)

As an example, some scholars have focused on particular issues in the Global South and the effects of governance on those challenges. For instance, Khan (2015) starts from a premise that there is a lack of good governance in the Global South and then looks into the issues of failed states, globalization, bureaucratic capacity, and the Human Development Index. Adams, Zulu, and Ouellette-Kray (2020) examine more particularly the literature on urban water insecurity in the Global South, specifically focusing on community-based water governance. It is based on 75 case studies from East Africa, South Asia, Southeast Asia, Central America, South America, Southern Africa and West Africa. The studies considered were published between 1990 and 2019. Similarly, Harris, Goldin, and Sneddon (2013) focused on key water governance challenges in the Global South. They examine questions of hegemony, specifically with respect to scarcity and crisis, marketization and privatization, and participation. Finally, Arriagda, et. al. (2018) discuss the polycentric climate governance arrangements in the country of Chile in South America. Specifically, they look at the creation of the Chilean Ministry of Environment through Act Number 20417 of 2010 that created the ministry and provides its legal jurisdiction. In essence, the Act gives the ministry legal authority to act on behalf of the federal government while encouraging collaboration with legal and local authorities. The authors conclude that a polycentric governance can be improved with stronger inclusion of local level actors.

Another important source of understanding how climate change and climate governance affects the Global South are international organizations. For instance, the UN Framework Convention on Climate Change (UNFCCC) process, The Kyoto Protocol, The Organization for Economic Cooperation and Development (OECD), and the Paris Agreement, with near universal participation, have all led to some improvements and policy developments. These efforts although unenforceable, establish some standards to which most countries abide by. However, in some cases, as shown most recently by the United States, some nations can pull out of agreements or simply fail to self-regulate their behavior in terms of climate change execution or administration. For instance, recently, the International Panel on Climate Change (IPCC) 2023 reported on the "state of knowledge of climate change, its widespread impacts and risks, and climate change mitigation and adaptation, based on the peer-reviewed scientific, technical and socio-economic literature" in its sixth assessment report (AR6). Specifically, the report integrates various reports and recognizes:

> The interdependence of climate, ecosystems and biodiversity, and human societies; the value of diverse forms of knowledge; and the close linkages between climate change adaptation, mitigation, ecosystem health, human well-being and sustainable development. Building on multiple analytical frameworks, including those from the physical and social sciences, this report identifies opportunities for transformative action which are effective, feasible, just and equitable using concepts of systems transitions and resilient development pathways. Different regional classification schemes are used for physical, social and economic aspects, reflecting the underlying literature. (p. 38)

One of the arguments of the report is that climate change "has affected human and natural systems across the world with those who have generally least contributed to climate change being most vulnerable." In other words, countries with "lower emissions per capita" generally tend to be the "most vulnerable." An example of this include Pakistan where one third of the country was flooded in 2022 due to a glacier melting heat wave and rainfall that killed 1,700 citizens while the country only produces 1% of the worlds greenhouse gasses. The same applies to Africa that has countries that produce the lowest greenhouse gas emissions internationally, yet the continent has nine of the world's ten most climate venerable countries that include Chad, Eritrea, Sudan, Niger, Liberia and Somalia. The report goes on to state that investing in these countries would yield a positive benefit in terms of mitigating the negative effects of climate change. The information collected and reported by these international organizations can be informative to better understand the affects and consequences of climate change on different countries.

Another concept generally used by those who work in the field of the Global South and governance is Multilevel Climate Governance (MCG). In essence, this term refers to multiple state actors and subnational governments that participate both nationally and internationally in activities surrounding climate change and its effects on people and how governments are governing the situation. These interactions are sometimes called paradiplomacy (de Macedo et al., 2023). For instance, de Macedo, Jacobi, and de Oliveira, (2023) use Brazil as a case study and examine four cities and their response to climate change and how their MCG or paradiplomacy effects their climate policies and their influence nationally and internationally. They note that their selection of cities in the Global South was purposeful in that in recent years more research has examined how climate change has affected nations in the southern hemisphere. A simple cursory search on the internet revels a number of studies on climate change and the Global South for instance, "Upscaling climate change adaptation in small and medium-sized municipalities: current barriers and future potentials," "An emerging governmentality of climate change loss and damage," "Exploring the influence of multidimensional variables on access to electricity in rural areas of the Global South," and "Green growth and innovation in the Global South: a systematic literature review" just to mention a few.

To further accentuate the effect of governance, climate change, and the Global South, Boasson, Burns and Pulver (2023) look at three climate change models—market failure, socio-technological transition, and public support. Within each model they look at themes of influence, decision-making and implementation. In addition, they looked at six actor groups that consistently participate in governance in one form or another. These actor groups included politicians, business organizations, climate advocacy organizations, anti-climate action groups, and two emerging groups—Indigenous peoples' organizations and labor unions. The authors concluded that all three models provide information and insights in helping to understand the impact of governance and climate change over the years. However, they also conclude that "for capturing the range of actors engaged in domestic climate mitigation the public support model is most helpful as it captures and depicts the increasingly contested nature of climate governance" Although all three models provide unique perspectives and some models highlight what others do not, there is still work to be done as to who "does climate governance" with respect to influence, decision-making, and implementation.

Finally, in the 2022 South America Report of the *Lancet* Countdown on Health and Climate Change: Trust the Science. Now That We Know We Must Act (Hartinger et. al. 2023), we are informed concerning various health repercussions for the impact of climate change in the Global South, specifically for South America. The report consists of four main sections highlighted by the bullet points below.

- Climate change is harming the health of South Americans, it's time to take prompt action
  The adverse health effects of climate change are accelerating and dispropor-tionately affecting the most vulnerable populations in South America. For the past ten years, populations in every country in the region have seen their health increasingly affected by climate change-related hazards. This trend will only continue if prompt action is not taken.

- South American countries must increase their preparedness to pro-tect populations from the health impacts of the climate crisis
  Understanding, assessing, and tracking the health impacts of climate change and health co-benefits of climate actions are critical for the develop-ment of adaptation plans and policies that can protect the health of South America populations from the increasing climate-related health hazards and maximize their positive impact.

- South America must continue and accelerate efforts toward the race to zero-carbon transition
  South America must continue and accelerate efforts to mitigate its Green-house Gas emissions, reduce land use change linked to deforestation, de-carbonize its energy and transport system, and increase its use and produc-tion of renewable energies. Doing so will not only help the region meet its commitments under the Paris Agreement but will also deliver major health gains from improved air quality, reduced energy poverty, reduced inequities in access to transportation, and more active lifestyles.

- South American countries require serious financial commitments to respond to the challenges imposed by climate change
  Implementing climate change adaptation policies and actions for the health and wellbeing of populations is a no-regrets investment that requires the support of governments, with transparent financial commitments and con-crete budget allocation.

The report provides evidence to support its claims and to show how these goals can be met. For instance, the report notes that there was an increase of 12.3 million "person-days of heatwave exposure in older adults above 65 in South America, an increase in heat-related mortality of 160%, economic losses associated with the mortality are equivalent to the average income of 485,000 local workers. South America also experienced a sharp increase in climate suitability for dengue, a disease that entails a major public health concern in the region, with a 35.3% increase for all countries except Chile. Countries like Peru and Ecuador showed the highest variability regarding this indicator in close association with "El Niño" events that must be understood as a potential added risk to that posed by climate change alone." The report focuses health risks and threats, the deficiency or lack of "health adaptation plans," and scarce funding for various programs to address climate change in South America. These items are crucial for the continent to be prepared for the expected change to come. It goes on to say that it must double "its effort to create resilient health systems and prepare to change its future. Its current trajectory of climate inaction will only lead to more inequality, poverty, and vulnerability.

CONCLUSION

Some issues in our world community are far too difficult to maneuver and manage by any one nation, government agency or nonprofit organization alone, and require a unified front from several institutions to address the challenge appropriately. This intersection is governance and these multiple issues requires a type of governance that has the capacity to involve multiple organizations thus coordinating a response that would deliver a viable sustainable answer or resolution. Norris-Tirrell and Clay (2010) suggest that the environment and the climate is just such an issue that requires complete attention.

For example, in terms of climate change in general and governance in particular, Meiklejohn, Moloney, and Bekessy (2021) apply a practical lens to local government climate change governance and propose how to rethink engagement practices. They propose that in addition to community engagement, local governments should promote appropriate levels of household heating and cooling they say: "climate governance practices performed by local governments include regulation, infrastructure provision, service delivery and advocacy. Regulatory practices include land use planning provisions and guidance, local traffic laws and ecologically sustainable design requirements for new building approvals." Infrastructure provision may include "local sustainability centers that provide direct demonstrations of low or zero carbon forms of household practices, as well as supportive built forms, such as walking and cycling paths." Service delivery includes "domestic waste collection and recycling that directly addresses the production of household greenhouse gas emissions arising from consumption practices." Finally, they note that "advocacy practices, such as engaging in formal consultations with state and federal governments and seeking supportive policy frameworks and resourcing" should be encouraged to take full advantage of governance.

This chapter looked at various definitions of governance and how a better understanding of governance and its relationship to climate change in the global south would provide an increased understanding of the various issues. A number of examples were shared to highlight the importance of governance—the collaboration of a number of agencies to accomplish a goal—and how that networking facilitated the accomplishment of a goal. The United Nations Sustainable Development Goal 13 "Take urgent action to combat against climate change and its impacts" was highlighted as a viable way to move forward to address issues of concern regarding the management or governance of climate change.

As our world moves forward in the twenty first century combating various challenges that affect all individuals in this world, it is essential to remember that solutions to our problems can be achieved when governments, nonprofits, and the private sector come together to address issues. In the case of climate change in the global south, governance appears to be a major contributor to helping organizations address the difficulties that are being confronted and will surely be encountered in the future.

REFERENCES

Adams, E. A., Zulu, L., & Ouellette-Kray, Q. (2020). Community water governance for urban water security in the Global South: Status, lessons, and prospects. *Wiley Interdisciplinary Reviews: Water, 7*(5), e1466.

Alegre-Bravo, A., & Anderson, C. L. (2023). Exploring the influence of multidimensional variables on access to electricity in rural areas of the Global South. *Applied Energy, 333*, 120509. https://doi.org/10.1016/j.apenergy.2022 .120509

Ansell, C., & Gash, A. (2008). Collaborative governance in theory and practice. *Journal of Public Administration Research and Theory, 18*(4), 543–571. https://doi .org/10.1093/jopart/mum032

Ansell, C., Sørensen, E., & Torfing, J. (2020). The COVID-19 pandemic as a game changer for public administration and leadership? The need for robust governance responses to turbulent problems. *International Review of Administrative Sciences, 86*(4), 632–651. https://doi.org/10.1177/0020852320960953

Arriagada, R., Aldunce, P., Blanco, G., Ibarra, C., Moraga, P., Nahuelhual, L., & Gallardo, L. (2018). Climate change governance in the anthropocene: Emergence of polycentrism in Chile. *Elementa: Science of the Anthropocene, 6*.

Barwick, C., & Gross, V. (2019). The circulation of public officials in a fragmented system: Urban governance networks in Paris. *Public Administration, 97*(4), 892–909.

Boasson, E. L., Burns, C., & Pulver, S. (2023). The politics of domestic climate governance: making sense of complex participation patterns, *Journal of European Public Policy, 30*(3), 513–536. https://doi.org/10.1080/13501763.2022.2096102

Braganza, A., & Lambert, R. (2000). Strategic integration: Developing a process–governance framework. *Knowledge and Process Management, 7*(3), 177–186.

Britanica. (n.d.). Govern. In *The Britannica Dictionary.* Retrieved March 1, 2024, from https://www.britannica.com/dictionary/govern#:~:text=1,by%20a%20 10%2Dmember%20council

Bryson, J. M., Crosby, B. C., & Bloomberg, L. (2014). Public value governance: Moving beyond traditional public administration and the new public management. *Public Administration Review, 74*(4), 445–456. https://doi.org/10.1111/puar.12200

Calimlim, C. (2023). The Global South is on the rise—But What exactly is the Global South?, *The Conversation.* https://theconversation.com/the-global-south-is-on-the-rise-but-what-exactly-is-the-global-south-207959

Choudhury, E., & Ahmed, S. (2007). The shifting meaning of governance: Public accountability of third sector organizations in an emergent global regime. *Public Administration Review, 67*(3), 561–588. https://doi.org/10.1111/j.1540-6210.2007.00718.x

de Macedo, L. S. V., Jacobi, P. R., & de Oliveira, J. A. P. (2023). Paradiplomacy of cities in the Global South and multilevel climate governance: Evidence from Brazil. *GPPG 3*, 86–115. https://doi.org/10.1007/s43508-023-00060-7

Dictionary.com. (n.d.). Govern. In *dictionary.com.* Retrieved March 1, 2024 from https://www.dictionary.com/browse/govern#

Di Giulio, M., & Vecchi, G. (2021). Implementing digitalization in the public sector. Technologies, agency, and governance. *Public Administration, 38*(2), 206–223. https://doi.org/10.1177/09520767211023283

Durose, C. (2009). Front-line workers and 'local knowledge': Neighbourhood stories in contemporary UK local governance. *Public Administration, 87*(1), 35–49.

Emerson, K., Nabatchi, T., & Balogh, S. (2012). An integrative framework for collaborative governance. *Journal of Public Administration Research and Theory, 22*(1), 1–29. https://doi.org/10.1093/jopart/mur011

Emerson, K. (2018). Collaborative governance of public health in low-and middle-income countries: lessons from research in public administration. *BMJ Global Health, 3*(Suppl 4), e000381.

Farazmand, A. (2017). Governance reforms: The good, the bad, and the ugly; and the sound: Examining the past and exploring the future of public organizations. *Public Organization Review, 17*(3), 267–284. https://doi.org/10.1007/s11115-017-0376-3

Fenger, M., & Bekkers, V. (2007) The governance concept in public administration *Governance and the democratic deficit: Assessing the democratic legitimacy of governance practices,* 13–33.

Flinders, M. (2002). Governance in Whitehall. *Public Administration, 80*(1), 51–75.

Frederickson, H. G. (2010). Whatever happened to public administration? Governance, governance everywhere. *Public Administration Review, 70*(s1), s6–s19. https://doi.org/10.1111/j.1540-6210.2010.02226.x

Freeman, T., & Peck, E. (2007). Performing governance: a partnership board dramaturgy. *Public Administration, 85*(4), 907–929.

Fünfgeld, H., Fila, D., & Dahlmann, H. (2023). Upscaling climate change adaptation in small-and medium-sized municipalities: Current barriers and future potentials. *Current Opinion in Environmental Sustainability, 61*, 101263. https://doi.org/10.1016/j.cosust.2023.101263

Governance Today. (2023). *What does governance mean?* https://governancetoday.com/GT/GT/Material/Governance__what_is_it_and_why_is_it_important _.aspx#:~:text=What%20does%20Governance%20mean%3F,the%20top%20of%20an%20entity.

Harris, L. M., Goldin, J. A., & Sneddon, C. (2013). *Contemporary water governance in the global South.* Routledge.

Hartinger, S. M., Yglesias-Gonzalez, M., Blanco-Villafuerte, L., Palmeriro-Silva, Y. K., Lescano, A. G., & Stewart-Ibarra, A. (2023). The 2022 South America report of the *Lancet* countdown on health and climate change: Trust the science. Now that we know we must act. *The Lancet Regional Health Americas, 20*, 100470 https://doi.org/10.1016/j.lana.2023.100470

International Panel on Climate Change. (2023). *Climate change 2023 synthesis report.* https://www.ipcc.ch/report/ar6/syr/downloads/report/IPCC_AR6_SYR_LongerReport.pdf

Jackson, G., N'Guetta, A., De Rosa, S. P., Scown, M., Dorkenoo, K., Chaffin, B., & Boyd, E. (2023). An emerging governmentality of climate change loss and damage. *Progress in Environmental Geography, 2*(1–2), 33–57. https://doi.org/10.1177/27539687221148748

Katsamunska, P. (2016). The concept of governance and public governance theories. *Economic Alternatives, 2*(2), 133–141.

Kettl, D. F. (2000). *The global public management revolution: A report on the transformation of governance.* Brookings Institution Press.

Khan, H. A. (2015). *The idea of good governance and the politics of the global south: An analysis of its effects.* Routledge.

Khan, H. A. (2017). *Globalization and the challenges of public administration: Governance, human resources management, leadership, ethics, e-governance and sustainability in the 21st century.* Springer.

Klakegg, O. J., Williams, T., Magnussen, O. M., & Glasspool, H. (2008). Governance frameworks for public project development and estimation. *Project Management Journal, 39*(1_suppl), S27–S42.

Klijn, E. H., & Skelcher, C. (2007). Democracy and governance networks: compatible or not?. *Public Administration, 85*(3), 587–608.

Kyle S. Herman (2023) Green growth and innovation in the Global South: A systematic literature review. *Innovation and Development, 13*(1), 4369. https://doi.org/10.1080/2157930X.2021.1909821

Lewis, J. M. (2011). The future of network governance research: Strength in diversity and synthesis. *Public Administration, 89*(4), 1221–1234.

Lowndes, V., & Sullivan, H. (2008). How low can you go? Rationales and challenges for neighborhood governance. *Public Administration, 86*(1), 53–74.

Lyne, M. M., Nielson, D. L., & Tierney, M. J. (2006). Who delegates? Alternative models of principals in development aid. *Delegation and Agency in International Organizations, 44.*

Lynn Jr, L. E., Heinrich, C. J., & Hill, C. J. (2000). Studying governance and public management: Challenges and prospects. *Journal of Public Administration Research and Theory, 10*(2), 233–262.

Meiklejohn, D., Moloney, S., & Bekessy, S. (2021). Applying a practice lens to local government climate change governance: Rethinking community engagement practices. *Sustainability, 13*(2), 995.

McDonald III, B. D., Hall, J. L., O'Flynn, J., & van Thiel, S. (2019). The future of public administration research: An editor's perspective. *Public Administration Review, 79*(1), 3–7. https://doi.org/10.1111/puar.12926

Milward, H. B., & Provan, K. G. (2000). Governing the hollow state. *Journal of Public Administration Research and Theory, 10*(2), 359–380. https://doi.org/10.1093/oxfordjournals.jpart.a024273

Meier, K. J. (2010). Governance, structure, and democracy: Luther Gulick and the future of public administration. *Public Administration Review, 70*(2), 297–306. https://doi.org/10.1111/j.1540-6210.2010.02205.x

Norris-Tirrell, D., & Clay, J.A. (2010). The Promise of Strategic Collaboration. In D. Norris-Tirrell, & J. A. Clay (Eds.), *Strategic collaboration in public and nonprofit administration* (pp. 1–16). CRC Press.

Olowu, D. (2002). Introduction—Governance and public administration in the 21st century: A research and training prospectus. *International Review of Administrative Sciences, 68*(3), 345–353.

Oxford Advanced Learners Dictionary. (2005). Oxford University Press.

Peters, B. G., & Pierre, J. (2007). Governance without government? Rethinking public administration. *Journal of Public Administration Research and Theory, 18*(suppl_2), 275–293. https://doi.org/10.1093/jopart/mum017

Robertson, P. J., & Choi, T. (2012). Deliberation, consensus, and stakeholder satisfaction: A simulation of collaborative governance. *Journal of Public Administration Research and Theory, 22*(3), 489–516. https://doi.org/10.1093/jopart/mus020

Robichau, R. W. (2011). The mosaic of governance: Creating a picture with definitions, theories, and debates. *Policy Studies Journal, 39*(1), 119–141. https://doi.org/10.1111/j.1541-0072.2010.00394.x

Salamon, L. M. (2002). *The tools of government: A guide to new governance.* Oxford University Press.

Skelcher, C., Mathur, N., & Smith, M. (2015). The public governance of collaborative spaces: Discourse, design, and democracy. *Public Administration, 93*(4), 873–892. https://doi.org/10.1111/padm.12173

Sørensen, E., & Torfing, J. (2009). Making governance networks effective and democratic through metagovernance. *Public Administration, 87*(2), 234–258.

Stillman, R. J. (1990). The peculiar "stateless" origins of American public administration and the consequences for government today. *Public Administration Review, 50*(2), 156–167. https://doi.org/10.2307/976863

United Nations Educational Scientific and Cultural Organization. (2023). https://www.ibe.unesco.org/en/geqaf/technical-notes/concept-governance

United Nations Office of the High Commissioner Human Rights. (n.d.). https://www.ohchr.org/en/good-governance/about-good-governance

United Nations Sustainable Development Goals. (2017). https://sdgs.un.org/goals

Van Kersbergen, K., & Van Waarden, F. (2009). 'Governance' as a bridge between disciplines: Cross-disciplinary inspiration regarding shifts in governance and problems of governability, accountability and legitimacy. In *European Corporate Governance* (pp. 64–80). Routledge.

Vigoda-Gadot, E. (2002). From responsiveness to collaboration: Governance, citizens, and the next generation of public administration. *Public Administration Review, 62*(5), 527–540. https://doi.org/10.1111/0033-3352.00184

Williams, T., Klakegg, O. J., Magnussen, O. M., & Glasspool, H. (2010). An investigation of governance frameworks for public projects in Norway and the UK. *International Journal of Project Management, 28*(1), 40–50.

World Population Review. (2023). *Global South countries 2024.* https://worldpopulationreview.com/country-rankings/global-south-countries

Yoo, J. W., & Kim, S. E. (2012). Understanding the mixture of governance modes in Korean local governments: an empirical analysis. *Public Administration, 90*(3), 816–828.

PART II

REGIONAL PERSPECTIVES

CHAPTER 3

CLIMATE GOVERNANCE IN THE EAST AFRICAN COMMUNITY

Henry Kimani Mburu
The Catholic University of Eastern Africa

Joseph Ngunjiri Macheru
The Catholic University of Eastern Africa

The World Bank defines climate change governance as the use of institutions to address governance failures, to strengthen incentives, and to build capacity for climate action. To enhance climate change governance, several actions have been taken at the global level. Notwithstanding these actions, in 2018, at the Bali Annual Meeting, the UN Secretary General, Antonio Guterres, called for a novel framework that could integrate financing, reporting, and budgeting for climate change action. As a follow up on this call, the Finance Ministers from around the world signed the Helsinki Principles in April 2019. These principles would allow governments to work within a national framework and guide action in mitigating climate change.

Climate Governance in International and Comparative Perspective, pages 41–64
Copyright © 2024 by Information Age Publishing
www.infoagepub.com

In this chapter we focus on a comparative analysis of climate change governance within the East African Community (EAC). The EAC comprises Kenya, Uganda, Tanzania, Rwanda, Burundi, South Sudan, and The Democratic Republic of Congo as the Partner States. We organize our chapter as follows. We first present country-specific case studies that give a general overview of climate governance in the country. We then delve into a detailed comparative analysis of climate change governance among the Partner States. For this purpose, we apply four lenses: the legitimacy lens, the Notre Dame Global Adaptation Index (ND-GAIN) lens, the reporting and monitoring lens, and the financing lens.

The EAC provides a unique setting for the themes in this book due to the multiplicity of factors that put communities in the region at considerable risk of climate change effects. These factors include violent conflict, high dependence on rain-fed agriculture, and pastoralism for livelihoods (World Wide Fund For Nature (WWF), 2006), widespread poverty, and low adaptive capacity (van Baalen & Mobjork, 2018). All the Partner States have been ranked as having either alarming or serious risk of hunger according to the Global Hunger Index (Concern Worldwide, 2022).[1] The EAC Partner States have a four-way approach to climate change issues: the global approach (as per UNFCCC), the African Union (AU) approach, the EAC approach, and the individual Partner State approach. We use the UNFCCC approach as our point of take-off at the global level.

A comparative analysis like this one contributes to climate change policy in several ways. Some of these as outlined by Radin and Weimer (2018) include: identifying helpful interdependencies among the Partner States, assessing the effectiveness of different policies in achieving climate change goals, identifying alternatives to climate change action, availing an opportunity to direct political efforts at national and regional level towards identified policy gap areas. For climate change policy researchers, this comparison also helps to widen the scope of research on the North-South debate and understand what global and local actions have successfully or not successfully been applied in the EAC region. Moreover, as Tosun, Jordan, and Maor (2017) indicate, researchers are also interested in identifying instances of over- or under-reactions to climate policies. Other contributions to policy makers are the understanding of similarities and differences in climate change action in the region, avoiding replication of failures, maximizing on resource utilization, and speedy climate action (Geva-May, Hoffman, & Muhleisen, 2018), success and performance of different climate governance networks (Koliba, Meek, & Zia, 2011) and the extent to which the Partner States have embraced the UNFCCC prescriptions as examples of North-generated interventions (Rios & Urbano-Canal, 2023) for application in the Global-South.

We organize our chapter as follows. In Section 2, we give an overview of the EAC region. We also provide a historical perspective of climate change governance in the region and the approaches to climate change governance. We make our comparative analysis in Section 3 and conclude in Section 4.

THE EAST AFRICAN COMMUNITY

The East African Community (EAC) is a regional intergovernmental organization established in 1999. The EAC has since grown to have seven Partner States (See Table 3.1) and has a total land area of approximately 49% of the size of the United States (See Table 3.2). In addition, there has been a drastic decline in the natural forest cover in the region since year 2000. This is a major climate change concern going by the importance of forests in mitigating the climate change issue (Esteban et al., 2014). In terms of land use, the EAC has not had a notable change in the proportion of land under cultivation since the year 2000 but has experienced a continuous decline in the available agricultural land over the same period. The region has, however, posted a gradual growth in GDP in the last decade.

TABLE 3.1 Partner States Joining the EAC	
Partner State	**Date became a full member**
Kenya, Uganda, Tanzania	November 30, 1999
Rwanda	July 1st, 2007
Burundi	July 1st, 2007
South Sudan	September 5th, 2016
Democratic Republic of Congo (DRC)	July 11th, 2022

TABLE 3.2 East African Community Land Area (in '000 Km²)				
Partner State	**Including water bodies**	**% of EAC area**	**Excluding water bodies**	**% of EAC area**
Burundi	27.8	0.6	25.0	0.5
DRC	2,344.9	48.3	2,267.1	49.0
Kenya	582.7	12.0	580.7	12.5
Rwanda	26.3	0.5	24.2	0.5
South Sudan	692.8	14.3	644.3	13.9
Tanzania	939.3	19.3	886.3	19.2
Uganda	241.6	5.0	200.51	4.3
East African Community	4,855.4	100.0	4,628.11	100.0

Source: This table is adapted from the "EAC facts and figures" report of 2016.

Climate Change Governance in the EAC: The Journey

The EAC region has experienced effects of climate change manifested like in other regions around the world. These effects have led to floods, droughts, wildfires, heat, and cold waves (Hunor, et al., 2022) and brutal human health issues (Leal-Filho, Nagy, & Ayal, 2020). Effects of climate change in the region had been predicted to cause increased precipitation (Hulme et al., 2001), increased sea surface temperatures (Funk et al., 2005), disappearance of Mount Kilimanjaro glaciers by 2020 (WWF, 2006), extreme weather events (IPCC, 2001; WWF, 2006), and significantly alter biodiversity as species struggle to adapt to changing conditions (Lovett, et al., 2005). These predictions have since come true and have been accompanied by consequences like food insecurity, flooding, rising lake levels, among others. These effects are expected to increase (Kamau & Mwaura, 2013; WWF, 2006). To mitigate these adverse climate change phenomena, the EAC has been, as part of the global community guided by UNFCCC, involved in various national and international adaptation, mitigation, and resilience strategies. We now focus on these strategies.

All the EAC Partner States, albeit at diverse times and extent, have been involved in national, regional, and global strategies to adapt and mitigate climate change. They have all ratified the United Nations conventions that operationalize climate change action. These include the Kyoto Protocol,[2] the Paris Agreement (PA),[3] and the Helsinki Principles. Table 3.3 shows the dates when each Partner State signed and ratified the PA and the Kyoto Protocol. Although the EAC Partner States are very low emitters of greenhouse gases, ratification of the Kyoto Protocol shows their commitment to climate change mitigation. Furthermore, experts place EAC at high or extreme risk of the impacts of climate change attributable to greenhouse gas emissions (Myers, 2022, p. 80).

TABLE 3.3 Ratification of the Paris Agreement and Kyoto Protocol by EAC Partner States

Partner State	Date signed the PA	Date ratified the PA	Date ratified the Kyoto Protocol
Burundi	April 22, 2016	January 17, 2018	October 18, 2001
DRC	April 22, 2016	December 13, 2017	March 23, 2005
Kenya	April 22, 2016	December 28. 2016	February 25, 2005
Rwanda	April 22, 2016	October 6, 2016	July 22, 2004
South Sudan	April 22, 2016	February 23, 2021	n/a
Tanzania	April 22, 2016	May 18, 2018	August 26, 2002
Uganda	April 22, 2016	September 21, 2016	March 25, 2002

TABLE 3.4 EAC Partner States National Determined Contribution (NDC) Status

Partner State	Current Version	Date submitted	Remarks
Burundi	2	October 5, 2021	First updated version
DRC	2	December 28, 2021	First updated version
Kenya	2	December 28, 2020	First updated version
Rwanda	2	May 20, 2020	First updated version
South Sudan	2	September 21, 2021	Second version
Tanzania	2	July 30, 2021	First updated version
Uganda	3	September 12 2022	Updated

All the EAC Partner States have submitted the Nationally Determined Contributions (NDCs) as shown in Table 3.4 according to the NDC registry.[4] Furthermore, they have also prepared their national adaptation plans (NAPs) or national adaptation plans of action (NAPAs) with the assistance of GCF funding (Table 3.5). The NAPs have been lauded as a strategy to shift from a top-down approach to climate change action, to a more multi-level approach (Sovacool & Van de Graaf, 2018, p. 317). As a community, the EAC Partner States came up with the East African Community Climate Change Master Plan (EACCCMP) for period 2011–2031. The plan provides a long-term vision and a comprehensive framework for adapting and mitigating climate change. The EAC Council of Ministers approved the master plan operational modalities in August 2013. By 2015 other policy tools to make the implementation of this plan possible had been established.

The EAC being part of the African Union, also benefits from the climate change initiatives laid out by the African Union (AU) as a continental umbrella body. To this end, the AU came up with the African Union Climate

TABLE 3.5 EAC Partner States Establishment and GCF Funding of National Adaptation Plan (NAP)

Partner State	Date NAP established	GCF funding in U.S. million dollars
Burundi	January 2007	16.1
DRC	November 2021	68.4
Kenya	July 2016	190.0
Rwanda	December 2006	54.5
South Sudan	November 1, 2016	0.3
Tanzania	July 2015	171.1
Uganda	2007	77.0

Change and Resilient Development Strategy and Action Plan (2022–2032). The plan envisions a sustainable, prosperous, equitable, and climate-resilient Africa by providing a continental framework for collective action in addressing climate change issues.

Hence, as mentioned earlier in the introduction, the EAC Partner States have a four-way approach to climate change issues: the global approach outlined by the UNFCCC, the AU approach, the EAC approach, and the individual Partner State approach. We interpret this as what has been referred to as "polycentric systems" approach to climate change governance (Sovacool & Van de Graaf, 2018, p. 320; Widerberg & Pattberg, 2017). However, there are still particularities of climate change governance pertinent to individual Partner States to consider (Myers, 2022). We highlight some of these in the case studies below.

CLIMATE CHANGE GOVERNANCE IN EAC: A COMPARATIVE ANALYSIS

Consistent with the comparative policy analysis research (Geva-May, Hoffman, & Muhleisen, 2018; Peters, 1994; Radin & Weimer, 2018; Rios & Urbano-Canal, 2023; Sovacool & Van de Graaf, 2018), we apply a comparative case-study method. We use all the EAC Partner States as our target units of comparison. This is reasonable given that the Partner States have a unified regional framework for mitigation and adaptation to climate change besides having their individual policy frameworks and local initiatives. We acknowledge the limitation that all the Partner States did not join the EAC at the same time. However, the effect of this limitation on the analysis is greatly attenuated by the fact that the ratification to the global climate change governance tools are based on individual countries and not regions.

We combine case-based and historical analysis, textual and content analysis, and qualitative analysis. We also employ quantitative analysis in which we limit ourselves to use of descriptive statistics.[5] Our dependent variable is climate change governance. We use multiple independent variables including: time, policy establishment and implementation, climate vulnerability index, climate reporting, and climate financing in our comparisons. We collect our data from the AU, EAC, and Partner States' policy documents, budget statements, reports from international organizations (including World Bank, World Meteorological Organization, UNEP, UNDP, UN Habitat, UNECA, AfDB, and UNFCCC affliated organizations), research documents, Notre Dame Global Adaptation Index database, Climate Change Laws of the World Database, press reports, conference reports, speeches from climate change meetings, and non-governmental organization reports (e.g., WWF, Concern Wordwide, Climate Action Tracker, etc.).

Our analysis is structured by a conceptual framework that focuses on certain "theoretically specified aspects of reality while neglecting others"—interpretive case studies (Sovacool & Van de Graaf, 2018). We apply four different lenses for our analysis: the legitimacy lens, Notre Dame Adaptation Index lens, reporting, regulatory, and monitoring framework lens, and the financing lens.

EAC Partner States Case Studies

Contemporary researchers in climate change impact and adaptation have documented that universal climate governance may not succeed because of the diversity in impacts, outcomes, and resilience building capacity of the countries in Africa (Myers, 2022, p. 82). This has been demonstrated in other parts of the world. For example, in Malaysia, fishermen were found to have high level of adaptation based on their environmental awareness, attitudes, and beliefs, and the knowledge of local environment (Shaffril, Samah, D'Silva, & Yassin, 2013). It has also been documented that violent conflicts in East Africa increase when the weather conditions become unfavorable for agriculture and pastoralism (van Baalen & Mobjork, 2018, p. 558). Political processes and institutions have also been shown to influence climate change governance in East Africa. The low trust in insitutions like the police, judiciary, and the political systems make people seek alternatives to settle issues related to scarce resources through violence (van Baalen & Mobjork, 2018, p. 568). In addition to these, climate change vulnerabilities have been exarcebated by the COVID-19 pandemic (African Climate Policy Centre (ACPC), 2020). It is for these reasons that we now focus on brief case studies of country-specific climate change governance that bring out the unique circumstances of each Partner State. In each case we identify the key climate change vulnerability, the national climate change framework, local climate action projects, and climate govenance challenges.

Kenya

The Kenyan economy relies heavily on agriculture, energy, and tourism; it is, therefore, very sensitive to climate change. Kenya has experienced extended drought, floods, irregular rainfall patterns, and locust attack, among other effects, in recent years. All these have left crops and pastures adversely affected leading to famine (Climate Action Tracker, 2020). According to a report by the Climate Action Tracker (2020), the institutional framework is well developed at the national level but lacks effective ministerial and government level coordination. Climate change action is still a reserve of state actors. However, there is potential for a more democratic and participatory climate change adaptation in the country. As Myers (2022,

p. 88) points out, Nairobi being the headquarter of both UN-Habitat and the UNEP there has been a great focus on research and analysis.

Community-based projects in Nairobi like the Kibera Public Space Project (KPSP) and Mathare (Thorn, Thornton, & Helfgott, 2015) which work on innovative and participative green solutions in conjunction with the local government have great potential of being part of a comprehensive climate change adaptation programme. However, scaling up such local initiatives to national level remains a challenge. In the Thorn, et al. (2015) study, the authors document how a new generation of residents in a slum dwelling have transitioned from the generic strategies of dealing with floods (like evacuation) to creating gated communities and saving schemes that maintain and improve the settlement and illustrate how community-level experience can be applied to enhance development of adaptation planning. In another case, Osano, et al., (2013) document how use of conservancies and payment for ecosystem services have been used to reduce the effect of drought on livestock income by the pastoral communities living in the Masai land for climate change adaptation.

There has been notable increase in livestock-related conflicts in the Turkana district of Northern Kenya during the dry months when water and pasture are depleted. Similarly, there has been more pastoral community violence in the vicinity of water wells in Northern Kenya (van Baalen & Mobjork, 2018, p. 559). To deal with these issues, van Baalen and Mobjork (2018) found that the pastoral communities deploy such methods as traditional conflict mitigation forums, elder councils, and local peace communities that reduce the risk of violent conflict. In recent years, the government has been making some budget allocations to climate action projects (The National Treasury and Economic Planning, 2023). These efforts are complimented by World Bank funded projects.

Uganda

It is estimated that Uganda has one of the highest forest cover losses at 2.6% annually (The World Bank, 2019). According to Kakumba (2022), climate change is by state actors and majority of Ugandans have not heard about it. The government with the support of the World Banks is adopting progressive climate change initiatives (The World Bank, 2019). These initiatives include enancting environmentally friendly policies, climate risk screening, and budget tagging (The World Bank, 2019). In addition, climate change projects that involve both state and non-state actors have been initiated. In Northern Uganda, for example, government officials have facilitated tree and grass planting, use of energy saving technologies, solar systems, sustainable soil, and land management solutions (The World Bank, 2019). Some of these projects have been successful while others have not been successful (van Baalen & Mobjork, 2018, p. 568) ending in heightened conflict.

Democratic Republic of Congo (DRC)

The DRC, one of the most populated countries in Africa, has abundance of natural resources (UNDP Democratic Republic of Congo, 2020), a vast diversity of ecosystems including rainforests, open woodland forests, savannah, cloud, and gallery forests (UNDP Democratic Republic of Congo, 2023). It has the world's second largest tropical forest and its economy is informal. The country has experienced worsening social-economic conditions, political instability, and civil war in recent times. These have caused huge environmental damage and degradation.

Many indigenous people in the DRC are highly vulnerable to climate change effects (UNDP Democratic Republic of Congo, 2020) with women being most affected. In particular, drought and heat waves lead to poverty and food insecurity among these people. Communities like the Pygmies who rely on hunting and harvesting fruits in the forest are highly vulnerable. Women among the indigenous communities in the DRC have no right to speak in meetings, making the climate change vulnerability and adaptation to have a gender perspective. While Word Bank funded projects have been initiated, not much has been reported in terms of climate change action at the national level.

Rwanda

In Rwanda, the main climate change challenges have been landslides, floods, severe rainstorms, and droughts (Rwanda Environment Management Authority, 2019).[6] Despite this, Rwanda has been viewed as a pace setter in response to climate change issues (Ighobor, 2022) in the region. For instance since 2012, Rwanda created the Green Climate Fund, which by 2022 had raised $247 million. According to Mr. Ojielo, the UN Resident Coordinator in Rwanda, Rwanda is a "potential investors prototype ground" and the climate change programs involve both state and non-state actors especially the youth. The government of Rwanda has in recent years allocated funds to climate initiatives. These have been involving both men and women (Rwanda Environment Management Authority, 2019). Rwanda has also benefited from World Bank funded projects.

Tanzania

Tanzania is the second largest of the EAC Partner States after the Democratic Republic of Congo in land area although it is the largest in population. Tanzania has recently experienced serious climate change challenges. These include increased temperatures, rising sea levels, intensified rainfall patterns, and droughts (National Adaptation Plan Global Support Programme (NAP-GSP), 2020). These factors adversely affect the electricity and natural gas infrastructure and complicate the climate problems (Sugar, Kennedy, & Hoomweg, 2013). Only less than half of Tanzanians (Afrobarometer, 2023) have heard about climate change which is mainly led by state actors. Furthermore, even at local government level, there has been little awareness or readiness for climate change action (Myers, 2022, p. 87). In 2012, the government of the Republic of Tanzania established the Climate Change Strategy. Later, enabling institutions were set to provide the policy guidance and coordination of cross-sectoral participation (NAP-GSP), 2020).

Despite these efforts, Myers (2022, p. 85) points out that the current governance trends in Tanzania are not favorable for implementing climate change adaptation policies that involve stakeholders at the grassroots level. Nevertheless, ordinary citizens are able, by themselves, to use community-based organizations and activism to meet their basic needs (Myers, 2022, p. 86). They are not, however, able to scale up those solutions to meaningful regional or national levels. Tanzania has also benefited from World Bank funded projects.

Burundi

Burundi is one of Africa's smallest countries with 98% of its population vulnerable to climate change impacts (Tall & Dampha, 2023). Climate change related flooding and landslides are the main causes of internal population displacement in Burundi (Tall, et al., 2022). There is a peculiar and high rate of land degradation in Burundi due to the many hills (*collines*). The state of climate in Africa report for 2021 (WMO, 2022) documents that the rising water level of Lake Tanganyika exposed the Burundian communities adjacent to the lake to major flooding. Furthermore, Tall and Dampha (2022) report that Burundi loses 1.6% of its GDP to land degradation and 5.2% of its land area annually due to soil degradation. For this reason, there is need for urgent investment to restore landscape and productivity of agricultural land, secure land rights, and build resilience. To mitigate the high risk of landslides, nature-based solutions (NBS) are being employed in Burundi coupled with increasing the vegetation cover. In addition, climate-smart agriculture could help to restore lands, protect ecosystems, and boost crop productivity. Different geographical areas need tailor-made solutions because of the varied profiles.

Overall, Burundi has institutional, policy, financial, and knowledge barriers to address climate change effects and build resilience (Tall, et al., 2022). So far, the country lacks capacity and any organized strategies to deal with these barriers. Moreover, there is no evidence of any direct budget allocations to climate change initiatives by the government of Burundi from the National Budget Brief for FY 2022/23 (UNICEF–Burundi, 2022).

South Sudan

South Sudan is a unique case of the challenges of climate change within the EAC. South Sudan has been classified as a country affected by fragility, conflict, and violence (FCV), by the Global Facility for Disaster Reduction and Recovery (2022). For instance, in 2022, South Sudan experienced the worst floods since the 1960s (The Government of Netherlands, 2022) leaving 835,000 people displaced complicating further the efforts to build resilience in the country. This has increased the frequency of conflicts as floods and drought wreck havoc on the communities especially on women, girls, and the internally displaced persons (IDPs).

Against this background, there have been internationally funded but contextualised responses to climate change challenges in South Sudan. These include planting easily stored crops like groundnuts to ensure access to food supplies even when other crops are destroyed, and rehabilitating locally built dikes to protect crops and shelters from flooding (The Global Facility for Disaster Reduction and Recovery, 2022). International partners like the World Bank, the Netherlands Development Organization (SNV), The United Nations Environmental Programme (UNEP), The Food and Agriculture Organization (FAO) have financed different projects to assist the government of South Sudan deal with these issues.

The Improving Resilience in South Sudan (IRISS) project funded by the DFID's Building Resilience and Adaptation to Climate Extremes and Disasters (BRACED) had major successes both at national and local levels. This project, which was led by the Sudd Institute[7] focused on enhancing resilience to drought and floods especially among women and girls (Kingston, 2018; Villanueva, Itty, & Sword-Daniels, 2018).

The IRISS project applied several models to improve resilience at the community level. These included the agro-pastoral field school (APFS) model, the community animal health worker (CAHW) model, the value chain development (VCD) model, the village savings and loan associations (VSLAs) model, the community resilience planning committees (CRPCs) model, and the school environmental clubs (SECs) model (Kingston, 2018; Kingston & Matturi, 2018a). Another model focused on climate forecasting where farmers and agro-pastoralists were given capacity to forecast and interpret weather information (Kingston & Matturi, 2018b). The last model applied in the IRISS project was the climate-smart technologies in which communities were introduced to energy-saving stores, treadle pump irrigation kits, use of drought or waterlogged resistant crops, quick maturity crop varieties, crop rotation, intercropping, mulching, and cover crops.

Political policies have also played out in issues of climate change governance in South Sudan. The expansion of mechanized farming displaced many farmers and pastrolists leading to increased conflicts over water and land. Drought conditions with abnormally high temperatures lead to sale of low-quality animals at through-away prices. This loss of income among the herders leave them more prone to conflict and joining armed groups (van Baalen & Mobjork, 2018, p. 558).

Using the Legitimacy Lens

Legitimacy is a political science concept that is used to assess the functioning and success of political mechanisms. The legitimacy concept has been applied to global governance systems with two dimensions (Scharpf, 1997; Stupak, Mansoor, & Smith, 2021). The output dimension of legitimacy focuses on the effectiveness in solving problems as they arise. The input dimension focuses on how relevant stakeholders are incorporated in the system, and whether it is transparent and accountable to them.

This definition to climate change governance where emphasis is placed on effectiveness for output legitimacy, and on participation for input legitimacy has been applied by Esteban, et al., (2014). On output, effectiveness is viewed as the "actual activities such as issuing regulation, producing reports, conducting research or organizing meetings" as in Szulecki, et al., (2011, pp. 716–717). We adopt Esteban et al.'s (2014) definition to better understand and compare climate change governance in the EAC Partner States. We use data from Climate Change Laws of the World Database (available from https://climate-laws.org/) for regulations and laws, and other online sources to assess both input and output legitimacy for each Partner State.[8]

Table 3.6 shows that from a legitimacy viewpoint, the EAC has made satisfactory progress in terms of laying down the necessary institutional, legal, and policy frameworks to deal with climate change issues except for South

TABLE 3.6   Climate governance Legitimacy Assessment of the EAC Partner States Since Year 2000

Partner State	Input legitimacy (participation)		Output legitimacy (effectiveness)			
	International partners, government, and academia	Women, civil society, youth, and others	No. of Laws and regulations	Climate fund	Climate Change Research center	Reports[a]
Burundi	✓	Low	11	X	X	✓
DRC	✓	Low	10	X	X	✓
Kenya	✓	Low	19	✓	✓	✓
Rwanda	✓	Low	12	✓	✓	✓
South Sudan	✓	Low	3	X	X	✓
Tanzania	✓	Low	14	X	X	✓
Uganda	✓	Low	10	✓	X	✓

[a] These reports may be by private sector, international partners, or government departments.

Sudan which is lagging. Participation in these processes is, however, by state actors leaving out non-state stakeholders. Only three (two) of the Partner States have established national funding mechanisms (climate-related research facilities). The Partner States have many climate-related reports especially from UNFCCC, UN, and World Bank affiliated organizations but few or none from ministries, devolved governments, local organizations, and communities. Notwithstanding the notable successes in the input legitimacy, the effectiveness aspect of climate governance legitimacy is quite low. This comparison also shows that except for the work done to meet the requirements of UNFCCC, little has been achieved at the national level.

Using the Notre Dame Global Adaptation Index Lens

We compare climate change adaptation among EAC Partner states using the Notre Dame Global Adaptation Index (ND-GAIN) (available from https://gain.nd.edu/our-work/country-index/). The index ranges from 0–100; the higher the index, the better the country adaptation capability to deal with climate change vulnerabilities. The ND-GAIN indices for the seven EAC Partner States[9] range from 31.1 (rank 178) for DRC Congo to 42.0 (rank 125) for Rwanda. These scores and rankings are quite low compared to the top five countries (all in Europe) with scores between 71.0 (Denmark) and 75.4 (Norway). The two most vulnerable EAC countries are South Sudan and Democratic Republic of Congo while the two least vulnerable countries are Tanzania and Kenya.

We use the data provided by ND-ECI to forecast[10] the indices for the year 2030. Our forecasts show that all Partner States except for Rwanda and Kenya would decrease by at least one percentage point by year 2030 (See Table 3.7). The ND-GAIN forecast index for Rwanda would increase by more than five percentage points. These findings corroborate the current challenges in implementing climate change initiatives in the Partner States.

Using Climate Change Reporting, Regulatory, and Monitoring Framework Lens

Climate change reporting may be viewed from the global perspective of submitting mandatory reports according to the requirements of UNFCCC. This is notwithstanding that these requirements have been blamed for not providing an effective accountability framework (Azia & Koliba, 2011). An alternative perspective is that of voluntary country disclosures which include corporate disclosures as a third viewpoint. In terms of UNFCCC

TABLE 3.7   Notre Dame Global Adaptation Index and Vulnerability for EAC Partner States

Partner State	2020 rank	2020 ND-GAIN index	2020 Vulnerability rank in EAC (out of 182 countries)	2030 ND-GAIN forecast index[a]
Burundi	165	35.5	5 (23)	34.5
DRC	178	31.1	2 (10)	28.0
Kenya	149	38.7	6 (39)	38.6
Rwanda	125	42.0	3 (11)	47.2
South Sudan (data is for Sudan)	177	32.3	1 (5)	31.7
Tanzania	145	39.1	7 (45)	38.4
Uganda	166	35.4	4 (13)	34.6

[a] Forecasts are based on ND-GAIN index data from 1995–2020 available from https://gain .nd.edu/our-work/country-index/rankings/ (June 21, 2023)

requirements, all the EAC Partner States have so far submitted their NCs as shown in Table 3.8. The general picture is that the EAC Partner States are not doing badly although South Sudan is lagging.

Considering voluntary country-level reporting, Rwanda is the only EAC Partner State with a reporting framework for climate change. The government first prepared a national vulnerability report in 2015 (Rwanda Environment Management Authority, 2019). The report provided a detailed assessment of climate change vulnerabilities[11] in all the 30 districts to the household level classifying them in terms of severity for every geographical area. Following this report, in the year 2021, a ministerial order was issued giving the procedure of preparing the national report on climate change and the responsibilities of the various organs involved. Overall, Rwanda

TABLE 3.8   EAC Partner States National Communication Status

Partner State	First NC	Second NC	Third NC
Burundi	2001	2010	2019
DRC	2001	2009	2015
Kenya	2002	2015	n/a
Rwanda	2005	2009	2018
South Sudan	2003	n/a	n/a
Tanzania	2003	2014	n/a
Uganda	2002	2014	2022

has a strong Environmental, Social, and Governance (ESG) framework that provides guidelines for climate change reporting (Ighobor, 2022).

Corporate climate change governance reporting is the third perspective to consider. Climate change issues have become corporate governance issues besides being ethical or environmental (Simic, 2016). Arising from this change, some countries have established climate change reporting protocols as national standards for climate change reporting. Countries like Australia (Cotter, Najah, & Wang, 2011; Institute of Chartered Accountants in Australia (ICAA), 2009) and the UK (Tang & Demeritt, 2018) have established corporate reporting guidelines. Furthermore, there have been efforts to harmonize reporting guidelines and practices (Yang & Farley, 2016). Climate change reporting instruments like Energy Reporting and Green Reporting frameworks have been established in different countries.

In EAC Partner States, corporate climate reporting tools have not been developed. However, in Kenya, the Central Bank has recently issued guidelines for reporting climate related risks (Njoroge, 2021. Besides, the National Treasury has been tracking and reporting climate related expenditures in Kenya for both state and non-state actors in line with Article 13 of the Paris Agreement (The National Treasury and Planning, 2020). Research has also shown that there is private climate change reporting (Solomon, Solomon, Norton, & Joseph, 2011). This kind of reporting that relies on risk and risk management dialogue among the institutional investors is not currently practiced in the EAC region.

As per the UNFCCC requirements, all the EAC Partner States have monitoring, regulatory, and verification (MRV) frameworks. Beyond that, however, not much has been done in this area as has been done in other countries like Romania (Hunor, et al., 2022). Hunor, et al (2022) document how local governments use early forecasts so that they are able to assess interventions that can mitigate climate change. Data collection and analysis for climate change is nascent and in most cases absent among the EAC Partner States. The EAC needs to move swiftly to collect data, build the required domestic capacity, map out storage areas of carbon, and establish common legal and regulatory framework in the region.

Uganda has developed an Integrated Monitoring, Reporting and Verification (MRV) tool which is an innovative approach to climate change reporting (UNDP, 2022). Kenya has also developed a similar tool. This tool enables monitoring, reporting, and tracking of all sectoral contributions to climate change action (UNDP, 2022).

Using Climate Change Financing Lens

The EAC Partner States have been struggling with a financing gap in their budgets despite some decline over the last two decades. This is important to note because finance is one of three main enablers for accelerated climate action. The other two are technology and international cooperation. For the EAC to achieve her climate action objectives jointly and individually in the Partner States, there is need to accelerate the provision of finances. While doing so, the EAC is faced with several barriers such as lack of joint regulatory, institutional, and access to market framework, lack of a common currency notwithstanding. These barriers contribute significantly to the economic vulnerability and high debt levels in most of the EAC Partner States. For example, most of the EAC Partner States have no accredited financial institution to receive funds from global climate funds like the GCF.

EAC Partner States face the same financing challenges as faced by other countries in emerging markets and developing economies. These include attracting and scaling private sector climate change financing and the high upfront costs related to climate investments (Prasad, Loukoianova, & Feng, 2022). Some researchers like Songwe and Adam (2023) believe that huge investments in climate resilience can bring major changes not only in Africa's economic model but also in sustainable value chains. Others have indicated that climate financing remains a critical aspect in achieving low carbon emissions, building climate resilience, and economic development (Khatibu, Msami, Mchallo, & Gontako, 2022). Despite these challenges, some of the EAC Partner States have started to make budgetary allocations towards climate change initiatives (United Nations Economic Commission for Africa, 2015).

Challenges external to the financial sector include a growing regional mismatch between available climate finances and the climate related investment needs coupled with unbalanced and limited local capital markets. Furthermore, these markets have weak regulatory frameworks and incapable of ensuring market safeguards and standardization required for commercial green investments (Holtz & Heitzig, 2021). Within the EAC, only the Nairobi Securities Exchange (NSE) has issued guidelines for green financing (African Business, 2022). This is so despite the EAC having attractive and untapped capacity of projects that can use green bond finance (African Business, 2022; Banga, 2019). Moreover, EAC Partner States stand a good chance to attract financing via green bonds given that their fundamental areas of mitigation and resilience building would relate to flooding, solar and wind installations, food security, and sustainable forest management (Reichelt, 2010) in line with the main climate change challenges in the region.

Climate finance access for the EAC Partner States is estimated at 5% of climate financing to developing countries globally (Khatibu, et al., 2022). Of this financing, Kenya receives 41%, Tanzania 20%, and Uganda 19%

(Khatibu, et al., 2022). Financing sources have been bilateral partners (36%), multilateral development banks (57%) and climate funds (7%). Khatibu, et al., (2022) further report that in 2021, 74% of these funds were loans, 24% were grants, and less than 1% was in form of equity. This funding model is unlikely to support long-term climate change actions and at the same time achieve economic growth as alluded to by William Ruto, the President of the Republic of Kenya (Republic of Kenya, 2023) who said, "... no country should be forced to choose between eradicating poverty and preserving the planet."

The extent to which the EAC Partner States have aligned themselves with the Helsinki Principles set out in 2019 by the global finance ministers can also help assess the situation of climate change financing. Helsinki Principle No. 6 is aimed at mobilizing private sources of climate finance and facilitating investments that support climate mitigation and adaptation. As of March 2023,[12] only three of the seven Partner States–Kenya, Rwanda, and Uganda–were members of the Helsinki Principles.

This is an exogeneous signal of either, the ability or commitment, to mainstream climate change into the national fiscal management policy and the entire planning framework of the individual Partner States. A draft framework and guide[13] have been developed by the World Bank Group and the IMF for this purpose to be discussed in June 2023. We showcase country examples from the national budgeting frameworks that indicate that the finance ministers are making efforts to mainstream climate financing.

Kenya

At the industry level, Njoroge (2021) documents three major steps that Kenya has achieved towards financing climate change initiatives. The first is the launching of the Sustainability Finance Initiative (SFI) in 2015. The second was the issue of the first corporate green bond in the greater East African region in January 2020. The third achievement was the accreditation of Kenya Commercial Bank (KCB), by the United Nations Green Climate Fund (GCF) as the first financial intermediary for green financing in East Africa. Additionally, Kenya has established a green bond market and issued listing and trading guidelines at the securities market (NSE, 2019).

The Kenya budget for FY 2022/23 totaling 3,342.8 billion KShs, allocated Ksh 7.0 billion to the Kenya Climate Smart Agriculture Programme and 6.05 billion to the Kenya financing locally led climate action. These two allocations comprised about 0.4 % of the total budget (The National Treasury and Planning, 2022).

Uganda

Out of a total budget of UShs 48.13 trillion, USh 0.659 trillion (about 1.4%) was allocated to the natural resources, environment, climate change, land, and water during the 2022/23 FY.

Tanzania

The resources allocated (or spent) in climate change mitigation and ad-aptation programs remain quite low compared to the efforts put in the governance frameworks. For example, the budget allocations to climate change initiatives under the Vice-Presidents' office in Tanzania have been increasing in the last couple of years. From 300 TShs million in the 2021/22 FY to 1 trillion TShs (233% increase) in the 2023/24 FY (United Republic of Tanzania Ministry of Finance and Planning, 2023). The expenditures in the same vote head increased from 47.8 TShs million in the 2020/21 FY to 103.3 TShs million (116% increase) in 2021/22 FY. Tanzania does not have an accredited institution to receive GCF financing but has identified the need to create a national climate fund (Khatibu, et al., 2022).

Rwanda

During the 2022/23 FY, a program for strengthening climate resilience of rural communities (SCRNRP) was allocated 7.5 billion Frw which was approximately 1.5% of the total budget allocations. Similarly, under the transformation governance pillar smart-crop production and productivity programs were allocated 831,687,490 Frw out of a total of 4,658.4 trillion Frw budget (about 0.2%).

The EAC Partner States, having ratified the PA, are recipients of the GEF financial support. The projects are funded at both national and region-al level. Each of the Partner States is represented at the GEF Council. As of April 2023, the status of GEF funding to the EAC Partner States was as shown in Table 3.9.

Overall, although some work has been done, as documented by Holtz and Heitzig, (2021) climate change financing in Africa (and in East Africa) lags behind other regions. We identify examples of tangible steps towards

TABLE 3.9 GEF Funding to EAC Projects			
Partner State	Projects Funded	GEF Funding (USD million)	GEF Fund
Burundi	20	66,437,433	GET, LDCF
DRC	38	167,903,407	GET, LDCF
Kenya	43	190,389,369	GET, SCCF
Rwanda	23	93,176,632	GET, LDCF
South Sudan	9	33,409,683	GET, LDCF
Tanzania	48	277,767,288	GET, LDCF
Uganda	40	210,733,944	GET, LDCF
Total	221	1,039,817,756	

Source: https://www.thegef.org/projects-operations/recipient-countries
(April 26, 2023)

alleviating this challenge. First, under the Africa Regulatory Support Programme, a £2.975 million project financed by FSD Africa, Kenya, Uganda, and Rwanda are covered. The project, which is expected to end in 2023, is aimed at enhancing the capacity of key financial institutions like the ministries of finance, central banks, and the capital market regulators to develop sustainable long-term financing solutions (FSD Africa, 2023). Second, IPCC (2023) recognizes that financial flows required for climate change initiatives have fallen short of expectation in Sub-Saharan Africa and suggests increasing public finance and public mobilization of private finance from the Global North.

CONCLUSIONS

We conclude that although the EAC Partner states have done a great deal in line with the UNFCCC, there is still a lot that needs to be done. Focus on the existing challenges (NAP-GSP), 2020) would go a long way to get concrete benefits from the already laid down infrastructure. A clear shift in climate change governance from state to non-state actors as has been observed in other regions where cities and corporations commit to climate change action and targets (Widerberg & Pattberg, 2017, p. 70) has not yet been observed in the EAC. This is an important missing link for the success of climate change initiatives because as Nasiritousi, Hjerpe, and Linner, (2016, p. 123) conclude, different non-state actors bring varied comparative advantages to climate change governance. Unity and commitment of political leadership is also a critical aspect of these efforts as alluded to by William Ruto, the President of the Republic of Kenya in his speech during the 3rd Regional Symposium on Greening Judiciaries in Africa in Nairobi (Republic of Kenya, 2023).

There is need for better coordination of actors beyond the "traditional binaries" of state and non-state actors. This is not to mention, as observed by Fozzard (2019), the need to to mainstream climate strategies and commitments into government operations, which remains a hinderance climate change governance within the EAC. We interpret this as a disproportionality where there are under-reactions to climate change policy as explained by Tosun, Jordan, and Maor, (2017, p. 596). Moreover, each EAC Partner State has her own peculiar circumstances despite the cross-cutting challenges. These peculiarities mean that one-fit-for-all strategy can not work in climate change adaptation and mitigation when dealing with vulnerabilities. This is due to, among others, the localized effects of climate change (Nwedu, 2020). Additionally, the potential of community-level knowledge base to deal with climate change and innovatively scaling them up to regional or national levels is yet to be explored and exploited. This is partly due to the lack of non-Western epistemologies (Okoliko & David, 2021). The challenging balance between economy, climate change, and governance (Fagerberg, Laestadius, & Martin, 2016) in Global North also lingers in the EAC besides the reality of violent conflicts and food insecurity.

NOTES

1. The 2022 Global Hunger Index report is available from https://www.concern.net/knowledge-hub/2022-global-hunger-index.
2. The Kyoto Protocol is focused on limiting and reducing greenhouse gases by developed economies.
3. The Paris Agreement is a legally binding international treaty on climate change. 196 Parties at COP 21 in Paris adopted it, on 12 December 2015 and entered into force on 4 November 2016.
4. According to the PA Article 4, paragraph 12 NDCs submitted by Parties are maintained in a public registry by the UNFCCC secretariat.
5. These methods are explained in detail in Geva-May, et al. (2018, p. 28).
6. Ozonnia Ojielo, UN Resident Coordinator in Rwanda
7. The Sudd Institute is a local research organization in South Sudan. Other South Sudan partners in the project were the University of Juba, and local experts and organizations.
8. We do not include the preparation of NAPs in this assessment as this is guided by the UNFCCC and other partners for technical and financial support. We also do not consider the mandated reports to the UNFCCC.
9. The database only reports indices for South and not South Sudan.
10. We forecast using the MS Excel Forecast Sheet functionality.
11. The government, business, civil society, citizens both men and women, rural and urban were all involved in this assessment.
12. Data from https://www.financeministersforclimate.org/member-countries.
13. Draft is available from https://www.financeministersforclimate.org/.

REFERENCES

African Business. (2022, May 16). *Africa poised for green bond growth.* https://african
.business/2022/05/finance-services/

African Climate Policy Centre (ACPC). (2020). Climate change and development in Africa post Covid-19: Some critical reflections. *Discussion Paper.* Addis Ababa, Ethiopia: African Climate Policy Centre.

African Development Bank. (2023, April 26). *The African Development Fund's new climate action window, a pivotal instrument for strengthening Africa's climate resilience.* https://www.afdb.org/en/news-and-events/

African Union Commission. (2022). *African Union climate change and resilient development strategy and action plan (2022–2032).*

Afrobarometer. (2023, April 5). *Tanzanians say climate change is making life worse, demand collective action to fight it.*

Azia, A., & Koliba, C. (2011). Accountable climate governance: dilemmas of performance management across complex governance networks. *Journal of Comparative Policy Analysis: Research and Practice, 13*(5), 479–497.

Banga, J. (2019). The green bond market: a potential source of climate finance for developing countries. *Journal of Sustainable Finance & Investment, 9*(1), 17–32.

Capital Markets Authority. (2019, January). *Policy guidance notes on green bonds.*

Climate Action Tracker. (2020). *CAT climate governance series–Kenya.*

Concern Worldwide. (2022). *2022 global hunger index.*

Cotter, J., Najah, M., & Wang, S. S. (2011). Standardized reporting of climate change information in Australia. *Sustainability Accounting, Management and Policy Journal, 2*(2), 294–321.

Esteban, D., Visseren-Hamakers, I. J., & de Jong, W. (2014). The legitimacy of certification standards in climate change governance. *Sustainable Development, 22*(6), 420–432.

Fagerberg, J., Laestadius, S., & Martin, B. R. (2016). The triple challenge for Europe: The economy, climate change, and governance. *Challenge, 59*(3), 178–204.

Fozzard, A. (2019). *Climate change and governance: opportunities and responsibilities, Governance Notes No. 14.* The World Bank Group.

Geva-May, I., Hoffman, D. C., & Muhleisen, J. (2018). Twenty years of comparative policy analysis: A survey of the field and a discussion of topics and methods. *Journal of Comparative Policy Analysis: Research and Practice, 20*(1), 18–35.

Holtz, L., & Heitzig, C. (2021, March 26). *Africa's green bond market trails behind other regions.* Brookings Institute. https://www.brookings.edu/articles/africas-green-bond-market-trails-behind-other-regions/.

Hulme, M., Doherty, R., Ngara, T., New, M., & Lister, D. (2001). Africa climate change: 1900–2100. *Climate Research, 17*, 145–168.

Hunor, V., Bogdan, S., Camelia, M., Oana, S.-M., Gabriel, S., Teodor, M., & Nicoleta, M.-S. (2022). Climate change effects and risk monitoring in Romania. *Lucrari Stiintifice Management Agricol, 24*(2), 137–143.

Ighobor, K. (2022, November 17). Climate action: Rwanda is a laboratory of innovative ideas. *Africa Renewal.*

Institute of Chartered Accountants in Australia. (2009, October). Climate change reporting. *Charter*, p. 70.

Intergovernmental Panel on Climate Change. (2001). *The third assessment report (TAR) of Working Group 1 of the Intergovernmental Panel on Climate Change (IPCC).*

Intergovernmental Panel on Climate Change. (2023). *AR6 synthesis report: Climate change 2023.*

Kakumba, M. R. (2022, September 6). Climate change worsens life in Uganda: Citizens want collective action to mitigate it. *Afrobarometer, Dispatch No. 547.*

Kamau, J. W., & Mwaura, F. (2013). Climate change adaptation and EIA studies in Kenya. *International Journal of Climate Change Strategies and Management, 5*(2), 152–165.

Khatibu, F. H., Msami, J., Mchallo, I. A., & Gontako, J. (2022, December 19). *Climate finance availability and access in Tanzania.* Retrieved from Research on Poverty Alleviation Programme. https://www.repoa.or.tz/

Kingston, N. (2018, October). A synthesis of lessons from the Improving Resilience in South Sudan project. *Knowledge Matters, 21,* 4–10.

Kingston, N., & Matturi, K. (2018a, October). Experiences with using the community aninal health worker model. *Knowledge Matters, 21,* 23–27.

Kingston, N., & Matturi, K. (2018b, October). Experiences with using the SHARP model. *Knowledge Matters, 21,* 30–33.

Koliba, C., Meek, J. W., & Zia, A. (2011). *Governance networks in public administration and public policy.* CRC Press.

Leal-Filho, W., Nagy, G. J., & Ayal, D. Y. (2020). Viewpoint: Climate change, health, and pandemics—A wake up call from Covid-19. *International Journal of Climate Change Strategies and Management, 12*(4), 533–535.

Lovett, J. C., Midgely, G. F., & Barnard, P. B. (2005). Climate change and ecology in Africa. *African Journal of Ecology, 43,* 279–281.

Matturi, K. (2018, October). Building climate resilience in South Sudan: Summary of research findings. *Knowledge Matters, 21,* 13–14.

Myers, G. (2022). Urban govenance dynamics and climate change in East Africa: A comparison of Dar Es Salaam and Nairobi. *Journal of International Affairs, 74*(1), 80–101.

Nasiritousi, N., Hjerpe, M., & Linner, B.-O. (2016). The role of non-state actors in climate change governance: Understanding agency through governance profiles. *International Environmental Agreements: Politics, Law & Economics, 16,* 109–126.

National Adaptation Plan Global Support Programme. (2020). *National adaptation plans in focus: Lessons from Tanzania.*

Njoroge, P. (2021, November). *Greening Kenya's banking sector.* https://www.centralbank.go.ke/uploads/speeches

Nwedu, C. N. (2020). Towards a prioritized climate change management strategy: a revisit to mitigation and adaptation policies. In L. Filho, W. Nagy, G. Borga, C. M. Munoz, & A. Magnuszewski (Eds.), *Climate change, hazards and adaptation options* (pp. 351–367). Springer.

Okoliko, D. A., & David, J. O. (2021). Ubuntu and climate change governance: Moving beyond a conceptual conundrum. *Journal of Public Affairs, 21*(3) , 1–9.

Osano, P. M., Said, M. Y., de Leeuw, J., Moiko, S. S., Ole Kaelo, D., Schomers, S.,...Ogutu, J. O. (2013). Pastoralism and ecosystem-based adaptation Kenyan Masailand. *International Journal of Climate Change Strategies and Management, 5*(2), 198–214.

Peters, B. G. (1994). Theory and methodology in the study of comparative public administration. *Comparative Public Management: Putting US public policy and implementation in context,* 67–92.

Prasad, A., Loukoianova, E., & Feng, A. X. (2022). *Mobilizing private climate change financing in emerging markets and developing economies, IMF Staff Climate Note 2022/007.* International Monetary Fund.

Radin, B. A., & Weimer, D. L. (2018). Compared to what? The multiple meanings of comparative policy analysis. *Journal of Comparative Policy Analysis: Research and Practice, 20*(1), 56–71.

Reichelt, H. (2010). *Green bonds: A model to mobilise private capital to fund climate change mitigation and adaptation projects.* The World Bank.

Republic of Kenya. (2023, April 3). *Speeches.* https://www.president.go.ke/president-ruto-africa-should-fight-climate-change-together.

Rios, C. D., & Urbano-Canal, N. (2023). The World Bank and education policy in Colombia: A comparative analysis of the effects of international organizations' learning on domestic policy. *Journal of Comparative Policy Analysis: Research and Practice, 25*(1), 101–117.

Rwanda Environment Management Authority. (2019). *Assessment of climate change vulnerabilty–2018.* Government of Rwanda.

Scharpf, F. W. (1997). Economic integration, democracy and the welfare state. *Max Planck Institute for the Study of Societies, Working Paper No. 96/2.*

Shaffril, H. A., Samah, B. A., D'Silva, J. L., & Yassin, S. M. (2013). The process of social adaptation towards climate change among Malaysian fishermen. *International Journal of Climate Change Strategies and Management, 5*(1), 38–53.

Simic, M. (2016, December). Climate change on the corporate governance landscape. *Governance Directions,* pp. 650–653.

Solomon, J. F., Solomon, A., Norton, S. D., & Joseph, N. L. (2011). Private climate change reporting: And emerging discourse of risk and opportunity? *Accounting, Auditing & Accountability Journal, 24*(8), 1119–1148.

Songwe, V., & Adam, J.-P. (2023, February). Delivering Africa's great green transformation. *Working Paper #180.9.* Center for Sustainable Development at Brookings.

Sovacool, B. K., & Van de Graaf, T. (2018). Building or stumbling blocks? Assessing the performance of polycentric energy and climate change networks. *Energy Policy, 118,* 317–324.

Stupak, I., Mansoor, M., & Smith, C. T. (2021). Conceptual framework for increasing legitimacy and trust of sustainability governance. *Energy, Sustainability and Society, 11*(5), 1–57.

Sugar, L., Kennedy, C., & Hoomweg, D. (2013). Synergies between climate change adaptation and mitigation in development. *International Journal of Climate Change Strategies and Management, 5*(1), 95–111.

Szulecki, K., Pattberg, P. H., & Biermann, F. (2011). Explaining variations in the effectiveness of transnational energy partnerships. *Governance, 24*(4), 713–736.

Tall, A., & Dampha, N. K. (2023, March 28). Burundi: Scaling up climate resilience in the land of 3,000 hills. *Africa Can End Poverty.*

Tall, A., Dampha, N. K., Ndayiragije, N., Von Berg, M., Raina, L., & Manirambona, A. (2022). *Tackling climate change, land degradation and fragility—Diagnosing dreivers of climate and environmental fragility in Burundl's colline landscapes.* The World Bank.

Tang, S., & Demeritt, D. (2018). Climate change and mandatory carbon reporting: impacts on business process and performance. *Business Strategy and the Environment, 27,* 437–455.

The Global Facility for Disaster Reduction and Recovery. (2022). *Results in resilience: informating and driving reslience-buidling at the DRM-FCV nexus in South Sudan.* United Nations Office for Disaster Risk Reduction.

The Government of Netherlands. (2022). *Flooding in South Sudan: 'Simply building dikes is not the solution'.* United Nations Office for Disaster Risk Reduction.

The National Treasury and Economic Planning. (2023, February). *2023 budget policy statement*. Government of Kenya.

The National Treasury and Planning. (2020, July 1). *National Treasury Circular Non. 13/2020*. The Republic of Kenya.

The National Treasury and Planning. (2022). *The Mwananchi guide financial year 2022/23 budget*. Government of Kenya.

The World Bank. (2019, May 31). *Ugandan government steps up efforts to mitigate and adapt to climate change*. https://www.worldbank.org/en/news/

Thorn, J., Thornton, T. F., & Helfgott, A. (2015). Autonomous adaptation to global environmental change in peri-urban settlements: evidence of a growing culture of innovation and revitalisation in Mathare Valley slums, Nairobi. *Global Environmental Change, 31*, 121–131.

Tosun, J., Jordan, A., & Maor, M. (2017). Governing climate change: the (dis)proportionality of policy responses. *Journal of Environmental Policy & Planning, 19*(6), 596–598.

UNDP. (2022, October 6). *Uganda's approach to reporting climate actions from all sectors*. https://www.adaptation-undp.org/scala

UNDP Democratic Republic of Congo. (2020, August 5). *Climate change impacts on indigenous peoples of the DRC*. https://www.adaptation-undp.org/

UNDP Democratic Republic of Congo. (2023, June). *DR Congo*. https://www.adaptation-undp.org/explore/middle-africa/

UNICEF–Burundi. (2022). *National budget brief: Burundi*. Bujumbura.

United Nations Economic Commission for Africa. (2015). *Climate governance and climate policy in Africa*. https://www.uneca.org/acpc

United Republic of Tanzania Ministry of Finance and Planning. (2023). *Estimates of public expenditure consolidated fund services and supply votes (ministerial)*. United Republic of Tanzania.

van Baalen, S., & Mobjork, M. (2018). Climate change and violent conflict in East Africa: Integrating qualitative and quantitative research to probe the mechanisms. *International Studies Review, 20*, 547–575.

Villanueva, P. S., Itty, R. P., & Sword-Daniels, V. (2018, October). Routes to resilience: Insights from BRACED. *Knowledge Matters, Issue 21*, 11–12.

Widerberg, O., & Pattberg, P. (2017). Accountability challenges in the transnational regime complex for climate change. *Review of Policy Research, 34*(1), 68–87.

World Meteorological Organization. (2022). *State of the climate in Africa 2021 WMO-No. 1300*. World Meteorological Organization.

World Wide Fund For Nature. (2006). *Climate change impacts on East Africa: A review of the scientific literature*.

Yang, H. H., & Farely, A. (2016). Convergence or divergence? Corporate climate-change reporting in China. *International Journal of Accounting and Information Management, 24*(4), 391–414.

CHAPTER 4

TROUBLE IN PARADISE

Climate Governance
in the Cayman Islands

Genève Phillip-Durham
University College of the Cayman Islands

ABSTRACT

Combined, Caribbean small island developing states (*SIDS*) and sub-national
island jurisdictions (*SNIJs*) have historically contributed the least to global
greenhouse gas emissions. Notwithstanding, the global climate crisis has necessitated policy maneuverability which urgently confronts the imminent associated threats that exacerbate the vulnerabilities of SIDS and SNIJs. The
Caribbean's collective climate advocacy has its origins in the 1994 Barbados
Programme of Action (BPoA). Since then, piecemeal global efforts toward
burden-sharing, continues to have far-reaching implications for the development of the political economy in Caribbean small island territories—The
Cayman Islands, being no exception. Thus, the policies, institutions and legislative frameworks which manage adaptation and mitigation must be sufficiently robust to build resilience *and* catalyze a whole-of-system approach to
combatting climate change at the national level. This Chapter will examine
the agenda-setting trajectory of the Cayman Islands Government in relation

Climate Governance in International and Comparative Perspective, pages 65–79
Copyright © 2024 by Information Age Publishing
www.infoagepub.com
65

to climate governance in a localized context. In so doing, some of the key policy measures and actions, institutional responses and public awareness and sensitization efforts will be reviewed in an effort to highlight some of the nuances and island-nominated challenges and opportunities, that inform the policy-making process and governance of non-traditional security threats such as climate change.

Small island territories offer important insights about policy space, policy practice and policy maneuverability. They present unique case studies which capture the essence of nuanced policy making, governance, and institution building. Nonetheless, previous research has found that preconceived 'best practice' solutions to the multi-faceted challenges with which countries in the Global South are confronted have resulted in very little benefit (Girvan 2006; Andrews et al., 2012). For small island territories this has resulted in unoriginal policy formulation, as well as a progressively shrinking space for policy creation, policy practice, and policy implementation (Girvan 2006; Andrews et al., 2012; Bishop et al., 2012; Turner et al., 2015). Additionally, the lack of significant empirically driven comparative research on the governance practices and policy pursuits of small island developing states and sub-national island jurisdictions—specifically the ones in the Caribbean region, is well noted (Oostindie and Klinkers 2003; de Jong 2009; Baldacchino 2013a; Veenendaal 2015; Veenendaal 2016). An appreciation of the policy and governance practices of small island territories with different constitutional arrangements, institutional arrangements, and development trajectories, is necessary in order to understand how they adapt to survive in a highly globalized setting and the lessons they learn, if any, from each other in so doing.

Countries in the Global South have diverse political, social, and economic capacities and these capacities determine the extent to which creative enterprise and endogenous approaches tare developed. Remedying locally nominated and defined governance policy, and institutional challenges has been an issue. Similarly, there has been and continues to be a balancing act between indigenous knowledge production and policy creation. The preservation of status quo based on metropolitan frames of references, which are entrenched in the legacy of imperialism, are characteristic of many countries in the Global South. What then are the implications of these factors for context-specific policy making, governance, and knowledge production that is representative of the lived-realities of constituents? It is also important to consider if and how these state actors are able to contribute to agenda-setting in the multilateral and global governance arena—in a way that brings meaningful attention to their uniqueness and distinctiveness. Some would argue that privileging the voices of persons who live these realities and are required to proactively plan for and respond to non-traditional security threats such as climate change should be considered. These environmental challenges, disrupt their lives and livelihoods and are at the center or are byproducts of the ill effects of climate change.

Combined, Caribbean small island developing states (SIDS) and subnational island jurisdictions (SNIJs) have historically contributed the least to global greenhouse gas emissions. Notwithstanding, the global climate crisis has necessitated policy maneuverability which urgently confronts the imminent associated threats. These have exacerbate the intrinsic conditions of vulnerability to factors that often tend to be beyond the control of SIDS and SNIJs (Sutton 2011). Vulnerability is defined as the exposure and openness to the impact of external events, circumstances, and shocks which are difficult to control or manage (Briguglio 1995). It is the overarching premise of the 1994 Barbados Programme of Action (BPoA) which sought to highlight the unique challenges faced by small island territories in achieving sustainable development. The Caribbean's collective climate advocacy, has its origins in the BPoA and today, the advocacy has evolved into a human-rights based and justice seeking approach, with both Caribbean state actors and non-state actors at the helm of these advocacy movements. It is important to note that since then, piecemeal global efforts toward burden-sharing continues to have far-reaching implications for the political economy and development of Caribbean small island territories—the Cayman Islands, being no exception.

Within this context of development in the Caribbean Islands, this chapter examines the agenda-setting trajectory of the Cayman Islands Government. Next, it looks at its relationship to climate governance in a localized context. In so doing, some of the key policy measures and actions, institutional responses, and public awareness and sensitization efforts are reviewed. At this point, an effort to highlight some of the nuances and island-nominated challenges and opportunities are explored. With this information, the policy-making process and the governance of non-traditional security threats such as climate change are examined.

LIVED EXPERIENCES

The Cayman Islands, a Caribbean British Overseas Territory, comprises three islands- Grand Cayman, Cayman Brac and Little Cayman. The islands are low-lying and insular, with challenges linked to among other things; rising sea levels, coastal erosion, lack of natural resources and endowments. From time to time there is also need to manage its reputation because of grey listing and blacklisting by global financial regulatory bodies. Yet, the Cayman Islands have proven that "smallness is . . . not . . . an insurmountable problem" and that its size, a total area of around 100 square miles across the three islands, is not a major limitation to integration into the global political economy (Lee and Smith 2010, 1094). The Cayman Islands are one of those small island territories which fit neatly into Persaud's (2011) analysis:

> Small states can make changes that will have impact in a relatively short space of time. The world's successful small states, such as Singapore, Hong Kong, Bermuda, Mauritius and Luxembourg, do not share geography or history, but they have all moved from near the bottom of the economic pile to the top in just one generation, something that is much harder for large states to contemplate. SIDS with limited natural resources and limited prospects for developing lucrative industries, are creating more knowledge and service-based economies. (p. 3)

As the sun set on a once thriving industry—seafaring, the Cayman Islands needed to consider how it would continue to build and sustain its economy and what would replace seamen's earning's which had significantly contributed to the islands' revenues until the 1960s (Johnson, 2001). In the monograph "As I See It," a detailed chronicle of how the Cayman Islands became one of the leading financial services centers in the world is presented. Two historical epochs are referenced, which serve to contextualize Cayman's ascent.

First, a pioneer of Cayman's financial services industry, Sir Vassel Johnson (2001), details Cayman's economic structure prior to World War II, which was primarily based on the seamen's industry. Following World War II and in his capacity as Financial Secretary, Sir Johnson was tasked with the responsibility of diversifying the economy of the Cayman Islands. Since there were no possibilities for a manufacturing industry, nor were there natural resources with export potential, the twin service industries of offshore financial services and tourism were considered "the only way out" (p. 209). He details his thought process as follows:

> Early on the morning of Boxing Day 1965 while I was still lying in my bed, my thoughts went into action. First was the most important and crucial decision to be settled: 'Should the government move on with the building of a financial industry?' The answer was positively 'yes'... The estimated distance through the tunnel in terms of time representing the growth period to maturity was 20 years. (Johnson 2001, p. 209)

With this goal being accomplished, the Cayman Islands, unlike its Caribbean regional counterparts, has managed over time to not only carve out but develop and sustain a more than comfortable niche for itself, and has experienced long term economic growth and development which in many respects outpaces some of its Global North counterparts. Cayman thus represents an anomaly, in the sense that while being geographically located in the Caribbean and by extension the Global South, it does not share some of the same challenges linked to balance of payment deficits and economic vulnerabilities which are consistent with SIDS, SNIJs and developing countries in general. In fact, Cayman remains one of the few countries in the Caribbean region which consistently records budget surpluses and is considered a high-income country.

It should be noted that regardless of their level of development, however, *vulnerability* continues to be at the forefront of discussions on size and survival among small island territories. Their ability to thrive, rebound and recover is intricately linked to this state of being. For instance, in spite of having a robust economy, Hurricane Ivan in 2004 resulted in CI$2.8 billion/US$3.36 billion in damages. Similarly, COVID-19 resulted in major contractions in the tourism and construction sectors which are the two other major revenue sources apart from financial services. Exogenous and endogenous events or shocks to the system have the ability to reverse years of economic growth and development and it is therefore important for high income countries like the Cayman Islands to recognize this. While small island territories like the Cayman Islands have consistently been seeking ways to counter the notion of vulnerability, which often characterizes them, it is important to consider along with particularistic development trajectories, concerns around various issues. For instance, land usage, energy efficiency, coastal erosion, marine pollution and conservation. How the Cayman Islands, with its aggressive growth and development path seeks to balance competing business and public and private sector interests, environmental protections, and preservation and behavior and norm change will now be examined.

TROUBLE IN PARADISE?

Today, climate change remains one of the biggest North–South cooperation challenges and climate governance issues on a global level remains problematic at best. It has consistently compromised and undermined several national efforts in the developing world to mitigate and manage its impacts. For example, limiting temperature increase to 1.5 degrees Celsius as set by The United Nations Framework Convention on Climate Change (UNFCCC), seems to have gone largely ignored by industrialized developed countries in the Global North, while countries in the Global South continue to negotiate their futures and contemplate adjustments to ensure their survival. The power brokerage around governance of the global climate has seen the voices of countries in the Global South repeatedly silenced or ignored, with hegemonic powers doing the bare minimum to advance climate justice and equitable burden-sharing. In this vein, it was suggested that an alternative annex system be developed by the United Nations Framework Convention on Climate Change (UNFCCC) which would serve to capture a more nuanced and justice-oriented distinction between countries based on current emissions and historical responsibility (Geck et al., 2013).

Constituents in the Global South have long maintained that industrialized countries owe their present prosperity to years of historical emissions, which have accumulated in the atmosphere since the start of the industrial revolution (Schneider, 2002). Hence, activists and advocates which represent the interests of developing countries remained adamant about their belief that the UNFCCC should not impose binding obligations on them since they were not responsible for inducing climate change (Mintzer et al., 1994). It should be noted that the eventual division of the UNFCCC's annexes was "based upon an acknowledgement of historic responsibilities for the release of GHGs into the atmosphere" (Geck et al., 2013, p. 2). Notwithstanding, in the early 2000s, Guyana, a developing country in the Caribbean, gained notoriety and raised its international profile for its climate leadership and advocacy, promoting one of the only low carbon development strategies (LCDS) in the Caribbean region at the time. To date, Guyana has received over US$100 million for carbon credits (https://lcds.gov.gy/). However, as a parallel oil and gas sector is being developed in Guyana, it is left to be seen how the LCDS would evolve and whether it will continue to align with global climate goals. It is essential to consider, what type of lessons can be gleaned from the Guyana experience and how Caribbean countries in the Global South can build resilience, while having to make trade-offs between industries from which substantial economic benefits may derive and environmental protection. In terms of the principle of fairness, some have argued that developing countries, being late participants to western style development, should be given a fair chance to pursue their development goals. Many of these goals have had to be sidelined as their countries sought to invest finances in adaptation initiatives instead of development programs.

However, what is the cost benefit of such trade-offs? And what happens when aggressive development pursuits far outpace mitigation and adaptation strategies? It has been argued that "the climate crisis is a central trait of contemporary capitalism" and with the Cayman Islands thrust toward urbanization and commercialization, it is important to recognize that climate change exposure remains a risk. Although the Islands continue to experience positive growth rates, which may allow for protracted recovery periods in the aftermath of a disaster (Mendes, 2022: 1) caution is advised. While climate adaptation is a costly undertaking, the fact remains that regardless of their levels of development; mechanisms, policies, and institutions must be developed in the Global South contexts in order to address climate change at the local level.

On the one hand, climate change in many ways highlights the North-South divisions which characterize global issues ranging from trade to other environmental threats (Bulkeley et al., 2010). On the other, it compels national governments to play a crucial role in climate governance by developing and implementing adaptation and mitigation policies to reduce greenhouse gas emissions and promote sustainable development and sustainable futures. The threat of climate change is evident for island nations like the Cayman Islands. However, the exponential infrastructural growth, development and demand for depleting land resources to satisfy the needs of a steadily increasing population can prove to be detrimental to the sustainability agenda that is needed in the national context. Thus, the policies, institutions and legislative frameworks which manage adaptation and mitigation must be sufficiently robust to build resilience. In this way, it can catalyze a whole-of-system approach to combatting climate change at the national level and should include the government, private enterprise, and civil society.

The Cayman Islands' approach to climate governance involves all of these actors as well as regional and international non-state actors. But like Guyana, could there be contradictions in Cayman's development model and its climate governance thrust? In a recent public hearing or consultation, some attendees voiced the concern that "unless rampant development and questionable planning decisions are addressed, a policy that aims to protect Cayman from rising sea levels and flooding is unlikely to have an impact" (Cayman Compass, 2023, p. 10). Since 2011, the first Cayman Islands Climate Change Policy was drafted. At the time, the National Climate Change Committee was the driving force behind climate governance in the Cayman Islands. The three broad goals focused on adaptation, mitigation, and governance:

- Goal 1—Reduce Cayman's vulnerability and enhance its resiliency to climate change (*adaptation*)
- Goal 2—Promote sustainable low and zero carbon economic activity (*mitigation*)
- Goal 3—Establish a governance framework for climate action that is future-focused, fair to all, accountable and transparent (*governance*).

More than a decade after its first drafting, the Ministry of Sustainability and Climate Resiliency, is now charged with the task of implementing the mechanisms and strategies entailed in a revised Climate Change Policy. This updated plan spans from 2023–2040. As one of the only such Ministries in the Caribbean region, this can be considered to be emblematic of the Cayman Islands Government's commitment to building a climate-resilient country. The Ministry will appropriately serve as the policy development arm, implementation partner and advocate for related initiatives. According to Premier Wayne Panton who also manages that Ministerial portfolio:

> This updated Policy accounts for new insights and shifting circumstances, particularly the climate risks identified in the Cayman Islands Climate Change Risk Assessment 2022. A diverse group of stakeholders and local and international subject experts were consulted in the drafting of this Policy's core concepts and strategies to ensure that current and future needs of this country are addressed. The strategic measures aim to lower the risks that climate change poses to key sectors and groups by reducing vulnerabilities, adapting and embracing sustainable, low carbon activities. The Policy also establishes a governance framework for climate action that is fair and accountable. An all-hands-on deck approach is necessary for the success of the Policy, which includes cooperation and coordination across ministries and departments, the private and public sectors, and civil society. (Public Consultation Draft 2023)

Nevertheless, in spite of the Minister's reassurance, at the public hearing or consultation on the revised policy, concerns were raised regarding "how the policy would address issues such as planning and construction permits given to developers too close to the coastline, general over-development on Grand Cayman, the destruction of the mangroves and seagrass, and pollutants from cruise ships" (Cayman Compass, 2023, p. 10).

In spite of the renewed calls for climate awareness and the movement toward a more aggressive agenda-setting by the current administration, it is important to consider the extent to which the time lapse since 2011, may have contributed to the negative public perception around the advancement of the new policy. For instance, one public hearing or consultation attendee alluded to the "poor history" of enforcing existing laws and policies on key matters like littering and dumping and questioned what resources were in place to enforce elements of the Climate Change Policy (Cayman Compass, 2023). The participant goes on to indicate that the lack of enforcement may have negatively impacted momentum in building and sustaining awareness, with the possibility that apathy and fatigue with delayed decision-making may have set in. Another audience member expressed that she felt jaded with governments that undertake policies that are then dropped by the next administration (Cayman Compass, 2023). She felt that this lack of continuity in the transition of power resulted in the need to develop new public awareness and sensitization campaigns due to the frequency of migratory flows in and out of Cayman. These and other comments may be further compounded by the fact that there exists a prevailing public perception that Caymanian institutions are being eroded by private sector interests. These are the same institutions responsible for legislative frameworks, policy making, and policy implementation around the issue of climate change (Bodden, 2020; Dunning, 2021).

The Cayman Islands with a large expatriate community has the responsibility of protecting the fiscal assets and interests of the affluent and all its residents. It is therefore understandable when concerns arise about policy interests and policy space. Reflecting on this concern, Bodden (2020) argues that "We are not in control of the forces that run the economy. And we know that he who pays the piper calls the tune. The agenda is in reality not set by the elected legislature, but by the movers and shakers who shape the economy. The perennial question for me is: for whom are we developing?" A legitimate point of contention then is whether against the current development backdrop and trajectory, which is predicated on an unrestrained capitalist model, sustainability can truly be achieved. This has been a historical challenge for the Cayman Islands, not only in terms of climate governance but in terms of balancing the needs and interests of indigenous populations and the expatriate community.

It is plausible to surmise that there is opportunity for an environmental thrust in the political landscape of the Cayman Islands. Climate change in and of itself remains not only a scientific issue but is also a political one and a sufficiently sensitized populace can advocate to their political representatives for policies and practices which are responsive to the Cayman community. In spite of the economic challenges associated with climatic events, from an economic standpoint, there are also feasible opportunities which can be explored amidst the climate crisis. In the Cayman Islands, it has been suggested for example, that the financial services sector and the real estate sector be treated:

> As if we are an oil exporting nation, as if this sector is a finite and depleting resource and the earnings from this sector, these sectors will not always be so buoyant. Then what we should do is perhaps set some of the revenue aside in an *independent sovereign wealth fund* with the highest levels of transparency and governance standards and use that fund to finance Cayman's vision. (Dukharan 2023)

While the Cayman Islands is still well-positioned in the global political economy, a sovereign wealth fund though it will not be a panacea in the face of a major climatic disaster, can aid in ensuring that future generations can benefit from the current wealth being accrued in lucrative sectors. Additionally, for persons who inhabit the smaller islands of Cayman Brac and Little Cayman as well as the more remote parts of Grand Cayman; in the northern and eastern jurisdictions, it will be important to make specific investments to guarantee social protection. For instance, in these communities the government could build resilience and implement comprehensive adaptation and mitigation plans. "The causes of climate change are implicated in everyday acts of production and consumption and relate to the ways in which societies organize their transportation, housing, energy, water, and food systems" (Bulkeley et al., 2014, p. 1). An analysis of the aforementioned in relation to the Cayman Islands, linked to the prolonged dependence on the financial services, tourism and construction sectors, illustrates how vulnerability can manifest itself in the wake of political, social, economic and ecological disturbances. While it remains a daunting policy challenge, it is important that the Cayman Islands to position itself to fully support and invest in sustainability and resilience building policies to manage the looming climate crisis at the national level.

POLICY CYCLES

Some components of the Draft 2011 Climate Policy have made it onto the legislative agenda namely, the National Conservation Act- 2013 and the National Energy Policy which were approved by Cabinet in 2017. The focus now will be on the National Energy Policy. The National Energy Policy 2017–2037 outlines a 20-year plan for strategies to be considered by the Cayman Islands Government to ensure that energy and environmental governance is part of the islands framework for sustainable development (www.energy.gov.ky). The policy which represents the end product of multistakeholder hearings or consultations, outlines the Cayman Islands commitment to a reduction in greenhouse gas emissions and the advancement of environmentally friendly sources of energy. The policy focuses on renewable energy, conservation methods and the promotion of efficient energy. The creation of an Energy Policy Council and a 4–5-year review to track and monitor policy implementation, is a notable feature of the 20-year plan. The plan serves to ensure that the momentum to achieve a 70% target of renewable energy by 2037 is sustained.

It should be noted that while the conversations and hearing or consultations around this policy commenced in 2013, it was not until 2020–2021 that the Public Education Campaign Marketing Strategy was developed in an effort to advance the strategic goals and vision of the policy. Since the 2015 Paris Accord, the Cayman Islands has refreshed its commitment to and prioritization of the National Energy Policy which had been tabled in the Legislative Assembly since 2013. It is important again here, to acknowledge the implementation bottle necking and delays that are customary and characteristic of a small island context as is evidenced by the time lapse between 2013 and 2017. Undeniably, sustaining momentum and ensuring continued buy-in for such policy priorities becomes more challenging when focus shifts, time lapses, and political, socioeconomic and environmental circumstances change. Similarly, as administrations change with election cycles (during a 4 to 5-year period), it can be expected that priorities shift and with the reallocation of resources to other areas of interest, important policy measures such as these can be relegated to the backburner.

Nevertheless, a Public Education Campaign in 2017 around the National Energy Plan was launched to bring attention to the goals and objectives of the policy through myriad of sensitization efforts. These efforts involved partnership with private sector stakeholders and business interest groups to support a sustainable trajectory. With the recognition of this important focus for the Cayman Islands consideration was given to its status as a small island territory, prone to similar ecological and environmental vulnerabilities. Other small island developing states and sub-national island jurisdictions also deal with similar areas of concern. The Goals of the National Energy Policy are outlined as: Knowledge and Education, Destination of Excellence, Energy Security and socioeconomic and environmental sustainability. Each goal is supported by a comprehensive set of strategies to support their successful implementation. A major component of the public awareness campaign was a comprehensive strategy to keep the Cayman community educated and informed. Primarily, the effort was to inform on of the short, medium, and long-term implications of climate change and the direct and indirect impact it could have on the Cayman community. Annual reporting and publication of greenhouse gas emissions was identified as one of the main ways in which the community would be kept abreast of the progress being made in line with achieving the 2037 target. Challenges linked to lack of meaningful progress with moving the policy ahead since its launch in 2017 were identified. This recognition could contribute to a degree of apathy in relation to the sensitization efforts.

Another challenge at the outset was the lack of data on the communities levels of awareness and understanding on issues related to climate change and environmental sustainability. The starting point, therefore, was to ensure that information on widespread best practice on mitigation of climate change, and efficient energy use was incorporated into the messaging. Due to the fact that energy consumption bills on the islands were comparatively high, using this fact as a launchpad for creating messaging could have been helpful. A realistic view and overview of timelines and visible progress was also important in order to sustain interest and to get buy in and confidence in the efforts toward an energy secure and climate resilient island. As a result of initial delays in launching the campaign, the public awareness campaign was in direct competition with public health campaigns which may have worked to a disadvantage, as the primary preoccupation of the Cayman community at the time was COVID-19. Removing much of the technical and scientific jargon that was consistent with high level climate change global discussions was also a necessary step toward getting buy in for public support. Key messages around the campaign were to have been targeted toward youth, businesses and residential customers, while social media marketing was identified as one of the key strategies to reach a wide and diverse demographic.

To further augment the marketing efforts around the campaign, the Ministry launched an energy audit competition—the end result of which would be energy retrofitting for the selected home/s. A number of sponsors in the business community supported initial efforts toward community education and sensitization including the Caribbean Utilities Company (CUC). The initial energy audit was completed on a total of 8 homes to promote publicity for climate change. In 2017 a research study was commissioned to develop a roadmap for future resource decisions for the CUC—the sole public electricity provider in Grand Cayman. The scope of the roadmap was to cover issues around transitioning the generation portfolio from a largely fossil based to a renewable dominated portfolio. Additionally, the need for natural gas, value of storage, and baseload renewable generation technologies were also explored. Apart from the public education and sensitization campaign, which was enshrined in the policy document, there was accompanying legislation to support the implementation of the policy and to ensure that the right institutions were at the forefront. The Legislation on the duty waiver for electric vehicles was revised in 2019 and custom duties on electric and hybrid vehicles for personal and commercial use was revisited as well as customs duties on electric motorcycles and bicycles (www.energy.gov.ky). Whereas it can be determined that considerable progress has been made with the National Energy Policy and there is now renewed interest in and an unprecedented level of urgency to prioritize the Climate Change Policy, the perennial problem with policy practice in SIDS and SNIJs is the need to adequately address policy implementation.

KEEPING A.P.A.C.E

The Cayman Islands has developed its agency based on a buoyant economy that continues to exceed all expectations. Notwithstanding, it would do well for lessons to be drawn from the historical experiences of other island nations such as Trinidad and Tobago, that were once resource-rich and delayed diversification efforts, resulting in catastrophic fiscal outcomes when faced with exogenous shocks. Their once thriving hydrocarbon sector began to experience a decline and they were ill prepared for the economic consequences. Positive and creative options are being explored as the Cayman Islands seeks to make progress with the implementation of various climate and energy related policies. One such effort aligns with the acronym APACE which stand for assessing need, policy implementation and timely execution, agency transparency, capacity, and education. Each of these are further elaborated below. While prescriptive policy measures have historically had limited success in the context of small island territories, understanding how public and private interest may shape policy discussions in contexts such as Cayman is essential. So too is a comprehensive understanding of how historical, political and socio-economic factors may shape knowledge creation, production and reproduction, which may then inform endogenous policy and governance priorities.

During a recent survey, one commentator pointed out that "with each change of government comes a change of ministry names and responsibilities" and questioned whether the Ministry of Sustainability and Climate Resiliency would disappear under a future administration (Cayman Compass, 2023, p. 10). Such concerns are not uncommon when we reflect on governance systems and structures throughout the Caribbean region. How do such frames of reference then, inform how decisions are made which foster trust, community and support for context-specific policies that affect lives and livelihoods regardless of changes in political administrations? With competing public and private interests being a major concern, it is important for the Cayman Islands Government to build and sustain trust in the mechanisms being considered and implemented. It would be unfortunate, for example, if the local scenario, were to mirror what plays out in the international context—with marginalized actors feeling disadvantaged as a result of disequilibrium in the burden-sharing regime.

Keeping A.P.A.C.E. for the Cayman Islands and ensuring consistent and continued buy-in for policies around climate change and energy efficiency must involve:

- Assessing needs—of the local community and proactively planning
- Policy implementation and timely execution—of initiatives which flow from policies. It is worth noting that in many small island contexts, there is no shortage of policy, but the implementation deficit is obvious
- Agency transparency—exercising fair manner which prioritizes public interest and reduces concerns about private interest driving policy agendas
- Capacity—institutional capacity and responsiveness need to be addressed in order to manage the delays in decision making which further contributes to apathy and erodes trust
- Education—momentum must be sustained with public education and sensitization and the inclusion of climate change and sustainable development in curriculum at all levels must be considered if norm and behavior change is to be realized on a holistic national level.

Small island nations like the Cayman Islands are not responsible for causing climate change. They feel, however, an obligation to do their part as a community of nations to alleviate the ill effects to our environment. Initiatives such as A.P.A.C.E and increased effort for a continuity of programs, through multiple election cycles, that have proven to be effective are encouraged. Economic development will continue and sustained efforts to communicate with its residents through various marketing campaigns will highlight efforts to be good stewards of the environment. Governance as a tool to work with multiple levels of governments and private and nonprofit organizations—including international organizations will surely make a difference for the Cayman Island going forward.

REFERENCES

Andrews, M., Pritchett, L., & Woolcock, M. (2012). *Escaping capability traps through problem driven iterative adaptation* (PDIA). Working Paper No. 299. Center for Global Development.

Baldacchino, G. (2013). History and identity across small islands: A Caribbean and a personal journey. *Miscellanea Geographica—Regional Studies on Development, 17*(2), 5–11.

Bishop, M. L., Heron, T., & Payne, A. (2012). Caribbean development alternatives and the CARIFORUM–European Union economic partnership agreement. *Journal of International Relations and Development, 16*(1), 82–110.

Bodden, R. (2020). Cayman development "spinning out of control." *The Royal Gazette.* https://www.royalgazette.com/other/business/article/20200213/cayman -development-spinning-outofcontrol/#:~:text=%E2%80%9CWe%20are%20 not%20in%20control,shakers%20who%20shape%20the%20economy.

Briguglio, L. (1995). Small island developing states and their economic vulnerabilities. *World Development, 23*(9), 1615–1632.

Bulkeley, H., & Newell, P. (2010). *Governing climate change.* Routledge.

Bulkeley, H., Andonova, L., Betsill, M., Compagnon, D., Hale, T., Hoffmann, M., ... VanDeveer, S. (2014). *Transnational climate change governance.* Cambridge University Press.

Cayman Compass. (2023). *Climate change policy expected to be passed in July.*

de Jong, L. (2005). Extended statehood in the Caribbean: Definition and focus. In L. de Jong & D. Kruijt (Eds.), *Extended statehood in the Caribbean: Paradoxes of colonialism, local autonomy and extended statehood in the USA, French, Dutch and British Caribbean.* Rozenberg Publishers.

Dunning, K. (2021). Adaptive governance: The proposed port expansion in the Cayman Islands and its impacts to coral reefs. *Marine Policy, 124.*

Geck, M., Weng, X., Bent, C., Okereke, C., Murray, T., & Wilson, K. (2013). Breaking the impasse: Towards a new regime for international climate governance. *Climate Policy, 13*(6), 777–784. https://doi.org/10.1080/14693062.2013.823308

Girvan, N. (2006). The search for policy autonomy in the Global South. In P. Utting (Ed.), *Reclaiming development agendas: Knowledge, power and international policy making.* Palgrave Macmillan.

Johnson, V. (2001). *As I see it.*

Lee, D., & Smith, N. (2010). Small state discourses in the international political economy. *Third World Quarterly, 31*(7), 1091–1105.

Oostindie, G., & Klinkers, I. (2003). *Decolonising the Caribbean: Dutch policies in a comparative perspective.* Amsterdam University Press. http://www.oapen.org/download?type=document&docid=408881

Mintzer, I., Leonard, A., & Chadwick, M. (1994). *Negotiating climate change: The inside story of the Rio Convention.* Cambridge University Press.

Persaud, A. (2011). Fostering growth and development in small states through disruptive change a case study of the Caribbean. Centre for International Governance Innovation (CIGI) Caribbean Papers, No. 11. https://www.cigionline.org/publications/fostering-growth-and-development-small-states-through-disruptive-change-case-study

Schneider, S., Rosencrantz, A., & Niles, J. (2002). *Climate change policy: A survey.* Island Press.

Sutton, P. (2011). The concept of small states in the international political economy. *Round Table, 100*(413), 141–153.

Turner, M., Hulme, D., & McCourt, W. (2015). *Governance, management and development: Making the state work.* Palgrave Macmillan.

Veenendaal, W. (2015). The Dutch Caribbean municipalities in comparative perspective. *Island Studies Journal, 10*(1), 15–30.

Veenendaal, W. (2016). Smallness and status debates in overseas territories: Evidence from the Dutch Caribbean. *Geopolitics, 21*(1), 148–170.

Vinícius M. (2022). Climate smart cities? Technologies of climate governance in Brazil. *Urban Governance, 2*(2), 270–281. www.energy.gov.ky

www.radiocayman.ky. *Economist recommends independent sovereign wealth fund for Cayman* (gov.ky)

CHAPTER 5

FORESTRY AND EVOLVING CLIMATE GOVERNANCE OF COSTA RICA AND PANAMA, 1969–2022

Michael S. Yoder
University of Texas at Austin

ABSTRACT

For many tropical countries of the Global South, including Costa Rica and Panama, deforestation and degradation of forests represent their greatest contributions to greenhouse gas emissions. Given this reality, multinational institutional efforts to halt global warming are urging, through diplomacy and incentives, countries such as Costa Rica and Panama to focus on combating deforestation and forest degradation, and to engage in reforestation and agroforestry, to meet their commitments to reducing anthropomorphic greenhouse gas emissions. Foremost among the programs to accomplish the stated goals that emanated from the United Nations Framework Convention on Climate Change is the REDD+ Program, which assists with funding options through the World Bank. This chapter traces the evolution since the 1960s of reforestation and forest management programs in each country, and the

Climate Governance in International and Comparative Perspective, pages 81–104
Copyright © 2024 by Information Age Publishing
www.infoagepub.com

administration of such programs. It provides a review of academic and policy-related literature, draws on data from government web sites of each country, and includes content analysis of web sites of NGOs involved in promoting and carrying out carbon reduction in both countries. The chapter notes the differences between each country's approaches to climate governance with respect to forests and provides a brief analysis of approaches that are more effective and potentially transferable to climate governance in other developing countries with large forest reserves as well. In both countries, forest management began in a more piecemeal and home-grown fashion in the 1990s, but is subsequently guided by a hybrid of local initiatives and the overarching neoliberal (multilateral, global) paradigm. The chapter concludes that both countries have made steady progress toward reaching climate goals, though there have been some shortcomings in terms of inclusion of small-scale farmers and indigenous communities.

OVERVIEW

For many countries of the Global South, forest management and reforestation represent the primary means to strive to achieve carbon reduction, to comply with global climate change mitigation agreements. The latter include the Paris Climate Agreement and the annual Conferences of the Parties (COPs). Collectively, in developing countries of the tropics, deforestation and forest degradation account for more than half of greenhouse gas emissions; therefore, reduction of deforestation and the halting of forest degradation are the most effective means by which such countries can reach climate-related goals, followed by reforestation and afforestation, which require more time to be effective (Forest Carbon Partnership, 2022).

In the cases of Costa Rica and Panama, such forest-related initiatives particularly stand out, given that the two countries comprise the isthmus that connects the North and South American land masses, and serves as a natural land bridge for plant and animal species to flourish. Both countries are small but contain important swaths of tropical rainforests important in absorption of greenhouse gasses. The land cover associated with such forests is vast enough in each of the two countries that their capacity for carbon absorption and storage outweigh any potential reductions in emissions from buildings, transportation and agriculture. In part because of their natural advantages and in part because of their respective climate governance, both have been successful in carbon reduction compared to other countries, and arguably serve as models for other developing countries to follow as they mitigate climate change.

Despite the bio-geographic similarities between Costa Rica and Panama, which together span only four degrees of latitude such that elevation plays a bigger role in climate variation than do distances from the Equator, the two countries' carbon-reduction policies and governing initiatives differ sufficiently enough to warrant a comparison. Likewise, there are multiple similarities. Both countries are active and committed members of the UN Climate Summits, referred to as COPs, and both are participants in regional reforestation initiatives. Each offers incentives to investors in forests to produce forest products such as timber produced under sustainable conditions as part of their respective climate change mitigation goals. Both countries pursue robust conservation efforts through highly regulated land use in and around national parks, which are valued for their abilities to absorb and retain carbon. Despite the similarities between the two countries' climate governance objectives, there are some notable differences that warrant investigation, as elaborated in the "Similarities and Differences in the two Countries' Approaches" section of this chapter.

Through a review of relevant literature related to climate and environmental governance in the two countries, this chapter provides such a comparative case study, with a particular focus on reforestation, conservation, and other elements of forest management. Additionally, the chapter provides a brief overview of carbon-reduction policies in the realms of agriculture, transportation, and energy production and usage. The literature consulted for the chapter largely derives from academic and news articles, and documents appearing on web sites of federal government entities, NGOS and international financial and environmental organizations, including the United Nations, World Bank and the World Economic Forum. Because of the inclusion of perspectives of environmental NGOs and international financial institutions, the literature included in the research covers sufficiently broad viewpoints. Moreover, this comparative case study is appropriate on the grounds that the GDPs (gross domestic products) of both countries are quite similar. Data by statista.com indicate a GDP in 2021 of $64.42 billion for Costa Rica and $63.61 billion for Panama. And while forests represent 54% of Costa Rica's land area as of 2015, the corresponding figure for Panama is 62.1% (Forest Carbon Partnership, 2022).

The chapter is primarily intended to provide a basic description of policies and related legal mechanisms that the two countries pursue to confront global climate change. In part it seeks to provide ideas for other countries that similarly enjoy natural rainforests, that they might follow in their own climate governance. The chapter additionally mentions the impacts of forest management on marginal groups within the respective national populations, such as small-scale farmers and indigenous communities. Following a discussion of global and regional forestry promotion, the chapter explores each country's corresponding environmental policies. It briefly outlines similarities and differences between them and identifies weaknesses. It then provides a brief explanation of how each country's approaches to climate governance are appropriately regarded as variations on neoliberal policy and practice.

The chapter concludes with a summary of the countries' policies that may be helpful to other Latin American countries well-endowed with forests, and cultural (and historical) propensities to pursue conservation.

The study generally follows a comparative policy approach that simultaneously examines similarities and differences between selected countries. That is, the analysis is based loosely on both a most similar systems design (MSSD) and a most different systems design (MDSD), research tools commonly used in comparative politics. While MSSD facilitates comparisons of a small number of countries (in this case two adjacent countries) that in nonetheless exhibit different policy outcomes, MDSD highlights differences between countries that otherwise exhibit similar policy outcomes (Steinmetz, 2021). As noted, Costa Rica and Panama exhibit similarities in ecosystems while their political histories differ, thus justifying such a comparative approach.

Table 5.1 provides a summary of selected forest cover data for the two countries.

GLOBAL AND REGIONAL FORESTRY PROMOTION AGREEMENTS

Both Panama and Costa Rica have been active in global-scale environmental meetings and agreements. Both participated in the UN Conference on Environment and Development (UNCED) in 1992 in Rio de Janeiro, also known

TABLE 5.1   Costa Rica and Panama: Selected Forest Cover Data, 2000–2020

	Costa Rica	Panama
Total Land Area, Square Km.	51,060	74,340
Urban Population, % of Total (2022)	80%	68%
Total Tree Cover (>30% Threshold), 2000, ha.	3,913,052	5,700,503
Tree Cover as % of Total Land Area	76.6%	76.7%
Loss of Tree Cover 2001–2020, ha.	253,188	441,438
Tree Cover Loss, 2001–2020 as percent of 2000 Tree Cover	6.5%	7.7%
Primary Forest Cover, 2001, ha.	1,485,836	2,902,231
Primary Forest Area, 2001, % of Total Land Area	29.1%	39.0%
Primary Forest Area, 2020	1,459,666	2,823,803
Primary Forest Area, 2020, % of Total Land Area	28.6%	38.0%
Primary Forest Loss 2002–2020, ha.	26,170	78,428

Sources: https://rainforests.mongabay.com/deforestation/archive/Costa_Rica.htm; https://rainforests.mongabay.com/deforestation/archive/Panama.htm

as the Earth Summit. Both are active participants in annual United Nations COP meetings (UN, 2022). Perhaps most importantly, both countries participate in measuring carbon reduction, and in funding programs for forest conservation and reversal of deforestation provided by the UN's REDD+.

The REDD+ Forest Protection Program

REDD+ is the acronym for a forest protection program titled "Reducing Emissions from Deforestation and Forest Degradation." Its founding in 2007 by the UNFCCC Conference of the Parties (COP) was based on the logic that funding is needed for reforestation and forest management, given the negative roles that deforestation and degradation of forests in developing countries play in the buildup of carbon dioxide globally. The program additionally emphasizes the conservation of forests for carbon capture and storage (Herrera, 2022). Member countries include those of the Global South that have specific plans to reduce carbon emissions through halting deforestation and actively promoting reforestation and forest management (Carbon Credits, 2022).

Funding mechanisms under REDD+ are largely carried out through carbon credits that companies and governments can purchase, mostly on a voluntary basis, and primarily for individual projects. Services receiving funding include not only forest protection, but also monitoring of the performance of each partner country's forest management through a number of means, including remote sensing by satellites. Best practices are needed to reduce inaccurate reporting of deforestation and/or its mitigation. Each partner country is expected to take social results into account, including land tenure, employment in forestry, labor rights, and impacts on traditional agriculture (Carbon Credits, 2022). As elaborated below, Costa Rica and Panama are important, appropriate participants in REDD+ because of the continued importance of their rainforests to their respective economies and to global climate change mitigation. In short, the program enhances the two countries' abilities to not only contribute to global carbon reduction, but also to provide them invigorated revenue streams necessary for economic development.

Regional Forestry Agreements

In addition to their respective commitments to REDD+, both countries adhere to two regional agreements to boost reforestation. They include the Central American Regional Convention for Management and Conservation of the Natural Forest Ecosystems and the Development of Forest

Plantations (1999), and the Central American Commission on Environment and Development (CCAD). The first emphasizes reforestation of lands that previously experienced deforestation due largely to cattle ranching, and secondarily other agricultural activities. The agreement was designed to be regional, meaning that several adjoining countries will participate in reversing deforestation. The second, the CCAD, signed in 1994, promotes in an integrated manner between Central American countries the economic and social benefits of reforestation (FAO, 2002). Thus, in addition to the global-scale REDD+ initiative, Costa Rica and Panama participate in the two regional agreements to promote conservation and regulate land use where reforestation would be beneficial.

COSTA RICA

Costa Rica pursues a decarbonization policy that relies heavily, though not exclusively, on forestry, including conservation and reforestation. It is the world's first country to experience net reforestation, and has received global praise for its successful conservation of forests and biodiversity, such that rainforests in the heavily agricultural country account for more than half of the national territory. The country has been a leader in providing incentives for conservation and reforestation, which has occurred since the late 1980s. It is widely recognized as a leading country in successfully promoting such efforts in conservation and reversal of deforestation to the point that the national identity, and its governance, are based largely on these undertakings (World Bank, 2022). The halting of deforestation has resulted in forest cover that represents 52% of the national territory in 2019, and by that time carbon-neutral agricultural practices had already become a priority of the federal government under the administration of Carlos Alvarado Quesada (2018–2022). The administration stated a goal of 60% of national territory covered by forests by 2030. The country clearly sees the benefit of going beyond zero-carbon to become a model for other countries to follow (UN, 2019).

To become carbon neutral by 2050, in accordance with its commitment to the Paris Climate Agreement, Costa Rica has additionally set out to reduce emissions from transportation through more widespread use of electric vehicles and a continuation of its already-successful turn to renewable energy sources for the generation of electricity. The goal is to reach 70% electrification of buses and taxis, though a comparable goal for the use of personal vehicles has thus far been more difficult to achieve. As a result, ride sharing and personal electric vehicles will be more heavily promoted, with a goal of 100% electrification by 2050 (UN, 2019). Energy is another area of success thus far. As of 2019, 95% of electricity was derived from zero-carbon sources, primarily hydropower. All of these efforts, including energy (power) generation, transportation, agriculture, and forestry, are seen by Costa Rica's leadership to require corresponding social, fiscal and technological reforms to bring them about. Because of the country's successes thus far, and its ambitious plans to continue making progress, Costa Rica's initiatives have been heralded by UN officials, including General Secretary António Guterres, former High Commissioner on Human Rights Michelle Bachelet, and Executive Secretary for Climate Change Patricia Espinosa (UN, 2019).

Origins of Costa Rican Environmental Policy

Costa Rica's undertaking of programs to offer incentives for conservation and reforestation has a relatively lengthy history by the standards of the region. Such dedication to conservation and reforestation is based on the fundamental recognition that the country's economy is so strongly based on its natural environment. As a result of relatively high rates of deforestation in the 1950s and 1960s, due to expanded export-oriented beef production, the Costa Rican government began prioritizing a reversal of the trend by the mid-1960s. To illustrate the severity of the problem, between 1950 and 1960, during the rapid increase in global demand for beef, annual deforestation in Costa Rica typically varied between 42,000 and 52,000 hectares, and land in pasture increased from 680,000 to 1.56 million hectares that decade (Alfaro, 2020).

It was quite common for the small-scale Costa Rican family farm operator, an important component of the national culture, to cut trees down to facilitate farming, especially in the country's more recently settled frontier areas surrounding the more populated Central Valley. By the early 1990s, some two-thirds of the county's original forest had been cleared for crop and livestock farming, prompting the government to step up measures to promote reforestation. Particularly harmful was the rapid growth in the demand for beef in the U.S., prompting the Costa Rican government to curtail subsidies for basic grains and to redirect incentives toward export-oriented beef production on large and small farms alike (Yoder, 1994). However, throughout the 1960s, the damage to forests, ground water, and biodiversity had become abundantly clear to environmentalists, the state, and farmers alike. In short, the country had become keenly aware of the damages brought by forest loss and degradation, loss of biodiversity, and deterioration of water quality, and concluded that these needed to be reversed (Barrantes, 2000). The attention placed on forests prior to 1980 established a culture of caretaking of forests that ultimately proved useful when carbon reduction became an equally important (if not bigger) goal.

By 1969, well before global climate change conferences began, the federal government had already begun to incentivize conservation primarily to preserve biodiversity in the country's forests. This effort was based on socioeconomic and ethical concerns. The initiatives of that time period required a commitment by the government to create the institutional and legal frameworks to tackle such a challenge, and to invest in the necessary technical expertise. The focus of action was threefold: 1. Incentives to promote conservation; 2. funding for reforestation and identification of the most appropriate areas for reforestation; and 3. promotion of economic development with respect to the benefits derived from forestry, both in conservation and in reforestation (Barrantes 2000). The third action is the most robust socially, given the drive to produce livelihoods from deriving tangible goods from forests. It has garnered international recognition because of the steadfast commitment on the part of the government, including its implementation of laws. The World Bank, UN, and IMF have all lauded Costa Rica for its efforts, including the 1992 "Earth Summit," or The UN Conference on Environment and Development (UNCED), in Rio de Janeiro. These global institutions have recognized Costa Rica as a world leader in positive environmental action (Barrantes, 2000).

The legislation corresponding to the 1969 initiative was the Ley Forestal No. 4465 (Forestry Law Number 4465), created by the Dirección General Forestal (DGF), part of the Ministry of Agriculture (Ministerio de Agricultura y Ganadería, or MAG). The law ultimately laid the groundwork for the country's robust national park agenda, to reverse rapid deforestation in the country's remote frontier regions (Ramírez Cover, 2020). Two departments comprised the DGF: Protection and Research, and Forestry Promotion. The primary success at that time was to halt deforestation, while reforestation was a secondary priority at that point, given the lack of incentives offered. However, after 1973, the state offered incentives to landholders, who were able to deduct investments in reforestation from property taxes. By 1996, forestry laws providing incentives for reforestation more strongly reflected the recognition of economic benefits of forests. Such incentives were more direct, involving primarily deductions to income tax, rather than in the form of property tax abatements. Such direct incentives included discounting income taxes based on capital costs associated with the planting of trees (Barrantes, 2000).

The establishment of canton-level agricultural centers (centros agrícolas cantonales) and cooperatives became prioritized with Ley Forestal 7032 in 1988 (Yoder 1994). That law additionally included subsidies on a per-hectare basis for establishment of forest plantings for both small and mid-sized landholdings. Furthermore, in 1983, the federal government provided low interest loans to finance reforestation through the Banco Nacional de Costa Rica (National Bank of Costa Rica). For those landowners who possessed their own funds and could avoid loans, the government offered income tax deductions for the sale of forest products resulting from the planting of trees, and exemption from taxes on imported capital equipment used in reforestation. Finally, beginning in 1983, the state-owned Banco Nacional de Costa Rica offered low-interest (8%) credit for terms of up to 30 years (with 10 years of grace period) for larger-scale planters of trees (Barrantes, 2000).

To successfully carry out the halting of deforestation, the conservation of existing forests, and reforestation on degraded lands, the government realized the country needed a sizable group of forestry professionals. In 1975, the National University (UNA) began its program in forest engineering, and its first graduating class in 1978. Conservation and management of forest lands, including those in national parks, was the first priority. By 1979, the federal government began offering the first incentives for landholders to carry out reforestation. In 1986, the government, under Ley Forestal 7032, began to regulate land use in forested areas, and in 1988 it established a fund to subsidize the conservation of natural forests on private lands. These efforts resulted in an annual reduction on average of 18,000 hectares of deforestation by 1992. The policy was strengthened by legislation in the 1990s, including Ley Forestal 7174 (Forestry Law 7174) in 1990, which strengthened the regulation of natural forest land, and Law 7575 in 1996 that provided more funding for natural forest conservation. During the 1990s, UNA broadened its forestry engineering program beyond conservation to include agroforestry, reforestation, and the marketing of forest products. This broadening of focus reflected the country's modest turn toward neoliberal economic policy that included the signing of numerous trade agreements between Costa Rica and other countries of Latin America, North America, Europe, and East Asia, which ultimately opened up markets for wood products (Alfaro, 2020).

When all was said and done, between 1979 and 1995, about 74% of all direct and indirect incentives corresponded to reforestation and 24% to conservation. The dedicated tax deductions for sales of forest products (timber) and costs of tree plantings represented the top ranking incentive type by land area. The provision of incentives led to a steady increase in reforestation and forest conservation during that time period. From the mid-1990s onward, government incentive schemes became more diversified in terms of their objectives. Not only were the incentives oriented largely toward production of timber as an economic activity. The direct and indirect "services" that forests produce for society through biodiversity and water management, including groundwater quality, became valued to an even greater degree (Barrantes, 2000).

Administratively, the reforestation and conservation agenda laid out by law in the 1970s to mid-1980s came under the direction of the Ministry of Agriculture and Livestock (Ministerio de Agricultura y Ganadería, MAG).

More specifically, within MAG was the DGF (Dirección General Forestal). This organizational arrangement reflected the historical importance of the agricultural and livestock sector of the Costa Rican economy. Funding came from the general budget and from the Forestry Fund, which was fed by fees, taxes and aid from international organizations. Since 1986, the second Forestry Law laid out a more elaborate system of incentives for reforestation and forest conservation, and designated the creation of the Ministry of Natural Resources, Energy and Mines (MIRENEM, Ministerio de Recursos Naturales Energía y Minas). At that point, the DGF was reassigned from MAG to MIRENEM, reflecting the desire to bolster the latter ministry in the face of the country's increased commitment to conservation (Barrantes, 2000).

Carbon Reduction and Conservation Policy in Costa Rica Since 1996

The Costa Rican federal government implemented the Ley Forestal (Forestry Law) of 1996, also known as Ley 7575 (Law 7575), to focus primarily on conservation of forests and water. The law created the Forestry Financing Fund, or Fondo de Financiamiento Forestal (FONAFIFO). Key to Law 7575 is its explicit recognition of the social benefits of the environment, including biodiversity which provides raw materials for some pharmaceuticals, the basis for sustainable tourism, and protection of water resources, all of which are regarded as worthy targets of subsidies and credit. Furthermore, the law is viewed as crucial to another social objective: absorption and storage of carbon (Barrantes, 2000; Herrera, 2022).

All of these motivations underlying Law 7575 are viewed as working together to enhance not only quality of life for Costa Ricans, but economic development for the country as a whole. An impetus of the law was the idea that Costa Rica could successfully capitalize on its image as a leading country globally for environmental sustainability, to boost tourism and foreign earnings. Landholders would be paid for the services their forests provided, which normally occurred without compensation. In accordance with the law, between 1997 and 2000, 94% of government payments for forest services, which amounted to some $52 million (U.S.), involved conservation and management while the remainder was directed toward actual reforestation (Barrantes, 2000). This represents a shift in environmental spending since the 1980s, when reforestation was the primary target for incentives. The shift in fiscal priorities reflects the government's recognition of the importance of economic benefits in addition to carbon reduction provided by conservation, including products derived from enhanced biodiversity.

Costa Rica was among the earliest countries to embrace what would become a body of global ideas of pricing carbon and other greenhouse gasses through a carbon tax, cap-and-trade, and the market for carbon offsets. The corresponding policy, dating back to the late 1980s, recognized four primary sources of carbon dioxide emissions, including energy production, deforestation (the cutting of trees storing carbon), burning of wood, and industrial emissions. These factors underlie the portion of Law 7575 outlining payment for environmental services, which is to say, providing incentives and other payments in exchange for direct efforts to cut CO_2 emissions through reforestation and forest conservation, and for levying fines at some point for those who emit carbon. The policy additionally called for the creation of strategic alliances to facilitate payment for such services, as proposed by international institutions such as the World Bank and International Monetary Fund. For example, Norway agreed to pay a fixed fee of $10 (U.S.) per metric ton of carbon captured in Costa Rica, up to 200,000 metric tons. Meanwhile, the policy favored subsidies for hydroelectric power generation as a means of reversing carbon emissions in energy production (Barrantes, 2000).

Costa Rica is the first country in the region of Latin America and the Caribbean to receive funding under the REDD+ program. The payment came upon measured carbon reduction in 2018 and 2019. This reflects the country's commitment to (and success in) reversing deforestation and expanding its forest conservation programs. Participation in REDD+ has helped to invigorate the country's various incentive programs emanating from FONAFIFO (World Bank, 2022). FONAFIFO works with the World Bank's Forest Carbon Partnership Facility (FCPF), a part of REDD+ that provides funding and technical assistance to carry out the goal of funding emission reductions through forestry. Thus, the FCPF represents the ability for Costa Rica to continue its decades-long forest conservation and reforestation objectives (World Bank 2022). The REDD+ strategy fits well with the country's enduring attempts to boost tourism, a key sector of the national economy, and one that is based in part of the country's territory dedicated to national parkland (Ramírez Cover, 2020). Also included in the country's implementation of REDD+ are a funding mechanism for sustainable agroforestry on small farms operated by women, and one available to indigenous communities (Herrera, 2022).

Export-oriented agriculture is likewise an important sector of the Costa Rican economy, and one that the federal government prioritizes in terms of carbon neutrality. The government urges the agro-export industry writ large to take measures to reduce its carbon footprint, which can be considerable when it comes to grazing livestock, and sizable upon harvest (in the case of crops) and subsequent transport of both crops and livestock. To illustrate, the country ranks first in the world in pineapple production and first in exports of that tropical fruit. Del Monte, a large pineapple producer, has responded through a commitment, beginning in 2023, to offset the release of CO_2 through forest conservation that absorbs the gas naturally. Del Monte is promoting the idea heavily to attract the attention of consumers interested in the combatting of climate change. The company claims to possess 8,000 hectares of forest land that contains some five million trees (Portalfruticola, 2023). Similarly, agroforestry is an activity targeted by Costa Rican subsidies. Such activities include the planting of trees by livestock farmers to offset disturbances that accompany livestock grazing. Collectively, these agriculture- and livestock-focused activities enhance the country's ability to attract funding from companies seeking carbon offsets, non-governmental organizations focused on conservation, and individual (corporate) investment in carbon offsets (World Bank, 2022).

Costa Rica is making the effort to project to the world a positive image of the country's commitment to environmental stewardship. The carbon reduction goals of Costa Rica are not limited to reforestation, conservation and agriculture, but include transportation, energy production and tourism (Ramirez Cover, 2020). Elected officials talk of a robust transition to electric vehicles, and to the continuation of clean energy production. The country is increasingly embracing solar power and is already among the world's highest utilizers of hydroelectric power on a per-capita basis. Modeling of multiple scenarios by the University of Costa Rica and the government's Climate Change Directorate indicate that the country will reach not only net neutrality, but net negative carbon emissions by 2050. In short, the projections for reaching strict reforestation targets are seen as not only possible but likely. Costa Rica is increasingly seeing the success of its future economic development as based upon reduction of greenhouse gasses (Irving, 2021).

It is logical to assume that Costa Rican exceptionalism is expanded by virtue of the public sector's orientation toward mitigating climate change in such a way as to stand out among tropical developing countries. Costa Rican exceptionalism is a widely shared view of the country, based on its social democratic policy orientation since the latter 1940s, and its peaceful, stable society since that time, compared to its Central American neighbors to its north. The small-scale independent farmer, often described as middle class and associated with the production of coffee for export, forms the backbone of this image. The term "green exceptionalism" is invoked to reflect the revision of the country's identity, in tandem with agroforestry in small farm communities and in cattle ranching areas, and the heavy promotion of ecologically sustainable small-scale farming alongside ecotourism and agro-ecotourism. However, critics of this version of Costa Rica's reality argue that large-scale banana, sugar, and pineapple production, all of which present problems related to soil degradation and significant chemical fertilizer usage, are important components of the agro-export economy,

(Ramírez Cover, 2020). Such critiques should be questioned, given the country's successes in reforestation and conservation compared to the majority of other tropical developing countries. While green exceptionalism may be overstated, Costa Rica nonetheless is well positioned through its reforestation and conservation, and related governance, to meet its carbon reduction targets.

PANAMA

Panama has the distinction of being one of three countries in the world, along with Bhutan and Surinam, to have reached carbon-negative status, meaning that the country absorbs more carbon than it emits. This is due primarily to forestry and nature conservation, and the country boasts of forests covering approximately 65.4% of its land cover (República de Panamá, 2023). Panama's recent climate governance includes the promotion of its success in forestry, and highlighting that success as a model for nearby countries such as Ecuador, Colombia and Costa Rica (Burillo, 2021). It may seem remarkable that the country has achieved this status, given that the Panama Canal, a huge infrastructure project, has produced environmental challenges, especially in the realms of forest cover and water conservation, since its inception in 1903. Furthermore, the country lacked any serious legislation to frame policy related to the natural environment before 1966 (Wandiza, 2021). The country largely avoids many of the worst impacts of climate change such as drought, in large part because of the abundance of water. That said, there remain the threats of inundations from heavy rainfall, so the country remains committed to global climate initiatives (República de Panamá, 2023).

Today the Republic of Panama is a serious participant in REDD+ and in other multilateral, multinational environmental agreements, and has made sufficient progress as to legitimately count itself among tropical developing countries that might provide guidelines for other countries of the Global South to follow. Although the country in recent years has begun to address the reduction of carbon from transportation and energy production, its climate governance is primarily focused upon forest conservation and planting.

However, some two percent of forest cover was lost between 2012 and 2019, primarily due to logging that rural communities rely on economically. This has prompted efforts by the Ministry of the Environment (Ministerio del Ambiente) to temporarily stop its issuing of logging permits in 2021. The federal ministry has included the business community in the formulation of its plans to oversee a million hectares of reforestation by 2050 (Moloney, 2021). Furthermore, the federal government provides funding for monitoring forest cover to track rates of both deforestation and reforestation. Indigenous people are regarded as expert caretakers of rainforests, especially in the realms of protection and conservation. There are six indigenous reserves in the country, and they are regarded as important sites for forest conservation, given the cultural propensity for the groups to be proper caretakers of the lands they rely on for their livelihoods, which are often worked communally (Burillo, 2021; STRI, 2022). Thus, partnerships between the Ministry of the Environment and NGOs provide funding for indigenous groups to plant native species, which perform better than mono-cropped exotic species, in conjunction with the indigenous communities' agroforestry efforts (Sytsma, 2022; STRI, 2022).

Origins of Panamanian Environmental Policy

A decree in 1966 outlined the country's agenda for soil and water conservation, though it did not cover the natural environment in any comprehensive way. In 1977 the agreement between the U.S. and Panama that transferred ownership of the Panama Canal, signed by the Carter and Trujillo Administrations, respectively, included the outline of a bilateral commission to oversee the regulation of environmental deterioration resulting from the use of the canal. The 1977 policy, however, lacked vigor, despite the significant impact on the natural landscape that the canal brought about since its inception. But not until the mid-1990s did Panama implement consequential legislation: Law 1 of 1994, which established the National Institute of Renewable Natural Resources, and Law 24 (written in 1992 and implemented in1996), which introduced regulation of land use and incentives for reforestation. Law 41, the General Environmental Law (1998), created the National Authority of the Environment, or ANAM by its Spanish acronym. The latter became the governing authority of prior and subsequent forest-related laws, including those overseeing incentives for reforestation and establishment of protected areas (Wandiza, 2021). Thus, while Panama indeed developed a commitment to undertaking an array of environmental initiatives, including research and the outlining of regulations, the country was slower than Costa Rica in passing actual legislation to bring about reforestation and conservation of soils, water and forests.

The Ministry of the Environment was established by Ley 8 (Law 8) in 2015 to strengthen conservation, restoration and preservation of forests (UN, 2023; Wandiza, 2021). Likewise, agroforestry, or the incorporation of tree plantings within traditional agricultural systems, is a high priority of the Panamanian government as a way to offset the ill effects of deforestation and cattle ranching that dominated rural land-use change in the mid-20th century. The Ministry of Environment partners with the Smithsonian Tropical Research Institute, a non-governmental organization (NGO). At its Agua Salud facility 50 km. (35 miles) northwest of Panama City in the Panama Canal watershed, research is conducted on the regeneration of land degraded through overgrazing or deforestation. The result is that secondary forest, or natural forest regrowth, produces a landscape with good carbon capture and storage capabilities that the Panamanian government is actively promoting (Sytsma, 2022).

Recent Panamanian Climate Change Issues

The Panama Canal weighs heavily in discussions of Panama's record on climate policy. Because the canal enables ships to save considerable distances, fuel, and therefore, emissions, Panama regards the infrastructure as a contributor to carbon reduction (Burillo, 2021). Furthermore, the canal's watershed serves as an important source of clean water for the country. The government is targeting the transition of pasture to forests in the watershed, given that forests, even those categorized as mono cropping, have positive effects on water conservation and quality (Stefanski et al., 2015). The canal's watershed is quite large, and the water it provides the country is vulnerable to disruptions from drought, which is widely believed to result from climate change. In response to the reduced water flowing through the watershed and into the canal, the authorities have at times limited the sizes (drafts) of ships passing through the canal (Moloney, 2021).

Energy production is another area of interest on the part of the government in its efforts to combat climate change. The majority of energy production in Panama is renewable, including hydroelectric, solar, and wind power. The National Energy Plan 2015–2050 (Plan Energético Nacional 2015–2050) calls for a reduction by 30% of carbon dioxide emissions during that time period, but more recent studies estimate that the reduction could reach 67%. The adoption of an even greater quantity of energy sourced from renewable sources is estimated to leave fossil fuels accounting for a mere 13% of the country's energy grid by 2050 (Lasso, 2022). Hydro power, solar and wind are, in that order, the primary avenues toward the desired reductions (Moloney, 2021). In the realm of transportation, Panama is following the lead of neighboring Colombia in promoting electric vehicles. The objective is to cut the use of gasoline and diesel usage by 30% by 2030 (Moloney, 2021). Optimistic forecasts point to 89% of transportation energy coming from electrified cars, trucks and buses by 2050. In short, the Ministry of Energy expresses its commitment to maintaining and building on Panama's status as a carbon-negative country by reducing carbon emissions by energy usage and transportation, to dovetail with the policies related to the boosting carbon absorption, primarily through forestry (Lasso, 2022).

SIMILARITIES AND DIFFERENCES
IN THE TWO COUNTRIES' APPROACHES

Costa Rica and Panama are carrying out their respective neoliberal approaches to climate policy and governance. Both countries pursue environmental policies that are variations of the market-oriented paradigm of policymaking that is widely pursued throughout Latin America. Neoliberalism is a dominant economic development and fiscal management strategy that favors global trade and corresponding reduction of trade barriers. It also favors a reduction of state involvement in the economy and planning, to the extent possible, and thus, privatization of functions that are not essential for governments to do and are regarded as suitable for the private sector. To illustrate, throughout Latin America, forest conservation, reforestation, and related activities are most commonly treated as profitable ventures, because of the sale of lumber, wood chips, and other forest products. Private investment, including from foreign sources, is viewed as positive. Furthermore, forest management under neoliberalism in Latin America often includes the participation of NGOs (Liverman & Vilas, 2006).

Both countries similarly adhere to global agreements that rely on targets for carbon reduction, while targeting the private sectors for investment, and inserting the public sector's fiscal powers to accomplish their stated goals. In general, the public sector of each country provides incentives for the private sector to invest in reforestation and forest conservation, while regulating land use in and adjacent to forested areas. Incentives include tax relief, and in some cases the provision of legal residency for foreign investors. The two countries' hybrid approaches have produced impressive results by the standards of developing countries of the Global South, and both countries are taking into account rural groups who normally are regarded as disadvantaged. The latter include small-scale farmers who could benefit economically from the services that forestry provides (especially in conjunction with their diverse farming activities), and indigenous groups that historically and culturally work within sustainable rainforests through sustainable small-scale farming. In neoliberal fashion, incentives target more privileged stakeholders, but regulations and interventions are also designed to enable smaller-scale producers to benefit economically from carbon reduction.

One aspect of forestry policy that exhibits considerable similarity between the two countries relates to the promotion of forest plantations, particularly teak, though not only that species. In Costa Rica, Laws 4465 and 7575 provide the legal framework for incentives to plant trees on degraded lands. In addition to the direct incentives for planting, strict regulation guides logging of the mature trees, given that tree removal releases carbon and reduces the ability to absorb carbon, and trees in the ground represent collateral for the loans. In Panama, Forestry Law Number 1, enacted in the mid-1990s, calls for loans of up to 30 years in duration, depending on the species, to finance tree planting. Investments in tree plantations are deductible from income taxes, and import taxes and fees for equipment needed for forestry activities are waived. Panama in particular values the harvesting of logs, albeit selectively through thinning, such that teak logs, and those of other species, are recognized and promoted as important exports for the country (FAO, 2002).

Both countries offer residency status for investors in teak forests, given the favorable global market for that species, and its potential to fulfill the neoliberal objective of boosting exports. To illustrate, Panama will grant legal residency to any foreign national who invests at least $100,000.00 in a teak forest. Such investors are eligible to apply for citizenship after just five years of residency (Sytsma, 2022). Costa Rica's offerings are similar. In addition to monetary and residency incentives to attract investment in forestry, funding is provided for research on the relative advantages of different species and agroforestry arrangements, most often in partnerships with NGOs and other research institutions.

CLIMATE POLICY AND KNOWLEDGE PRODUCTION IN COSTA RICA AND PANAMA

Knowledge of sustainable development and mitigation of climate change is produced and reproduced through research promoted by the Ministry of the Environment and Energy, where data collection is centralized. The Ministry of Science, Technology and Telecommunications also directs and promotes such research. Climate change is integrated into curricula at all levels, from primary school to the university level. The Ministry of Public Education oversees the universal climate change curriculum. NGOs are included as well in research and training activities, and in dissemination of information about climate change and its mitigation. Among unique groups highlighted in the education and communication efforts, as well as in the mitigation efforts themselves, are indigenous groups, as laid out in the 2009 National Climate Change Strategy and the 2016 National Policy for the Adaptation to the Climate Change of Costa Rica. In these educational and communication plans, the term "environment" is commonly replaced by "sustainable development," which reflect the country's emphasis on forests and forestry services. Because of the existence of climate change in curricula from the elementary to the college levels, the consciousness of climate change is high in Costa Rica. According to a poll conducted in 2014 by the University of Costa Rica, as a result of these curricula, 79% of Costa Ricans were knowledgeable about climate change and 70% believed it is anthropomorphic (UNESCO, 2021). The National University in the city of Heredia currently offers engineering degrees in forestry sciences and in environmental management.

Panama has one of only six Regional Collaboration Centers (RCCs) in the world. Operated by the UN, their goals are to "support national climate action through capacity-building, technical assistance and strategic networking, sourcing know-how and resources to drive clean development." That is, the centers are established to incentivize sustainable development through assisting with projects that reduce carbon dioxide, and to help countries reach their goals of carbon emissions established through the Paris Climate Agreement in 2015. In the case of Panama, that means technical assistance in forest management, measuring deforestation and reforestation, and measuring and reporting carbon dioxide emissions. Panama was selected as a site by the UN for a RCC because of its policies favoring carbon reduction through forest management, and the belief that Panama can be held up as a model for other countries to follow such policies (UN, 2020).

Panama's Ministry of the Environment sets up workshops under a project titled Systems of Sustainable Production, Conservation and Biodiversity (SPSCB by its Spanish acronym), with support from the World Bank, to bring together the public and private sectors to promote sustainable rural development, including for indigenous groups. This is an example of dissemination of information that brings together a number of stakeholders. Amont the twenty participating organiztions are agricultural cooperatives, forestry cooperatives, NGOs, and agricultural trade association. The three main objectives of the SPSCB are conservation of biodiversity, maintaining sustainable productive landscapes, and production of knowledge (República de Panamá, 2019). Furthermore, la Universidad Metropolitana de Educación, Ciencia y Tecnología (Metropolitan University of Education, Science and Technology) offers a bachelors degree in environmental management, and the Universidad de Panamá (University of Panama) offers a bachelors degree in environmental economics, and one in ecological geographical tourism.

LIMITATIONS OF REFORESTATION
IN COSTA RICA AND PANAMA

Despite the positive media coverage and promotional narratives surrounding the efforts by the two countries to reverse reforestation as the strongest basis of the two countries' respective climate change efforts, criticisms nonetheless exist. Small farm forestry in the canton of Hojancha in Costa Rica's Nicoya Peninsula was found to have failed at creating equality in villages and towns in that rural area. Producers of tree seedlings varied in terms of the scale of their nursery operations: the smaller producers relied on the larger production units to market seedlings for them. Most small-scale seedling nurseries were operated on rented land while larger-scale seedling producers owned the land. Only the large farms possessed sufficient land to plant large parcels of teak and other hardwoods for the production of timber. On the other hand, the DGF provided technical assistance to seedling producers of all sizes, given the priority nationally and locally (Yoder, 1994). In eastern Panama, critics of teak farms claim that such mono cropping fails to conserve water as effectively as native species in secondary forest regrowth (Sytsma, 2022).

Furthermore, there are difficulties in convincing cattle ranchers that mono cropping of teak and other hardwood species can generate as much income per hectare as does livestock production. This is because incentives per hectare would need to be longer in term than the policy involves. Risk also factors into the decision-making process of land managers, such that the question of timber prices is taken into account. Cattle production is much more familiar in that regard, which further complicates the process of convincing ranchers to make the switch to planting trees (Stefanski, 2015). This is especially the case given that Panama's and Costa Rica's incentives for planting trees last only five years. The process of conversion to tree plantations is, therefore, slower than the public sector has hoped. Teak mono cropping is not as beneficial from the standpoint of carbon capture and storage as secondary (natural) regrowth; however, the exotic hardwoods produce better economic yields, and are, therefore, more attractive to land managers. Finally, lending by the National Bank of Panama for livestock and agricultural activities complicates the policy of providing incentives for reforestation (Stefanski, 2015).

Despite successes in approaching carbon neutrality through forestry in the 1970s through the 1990s, Costa Rica nonetheless discovered some challenges that required subsequent policy revisions. The main issues underlying such revisions included a lack of absolute certainty in funding conservation and reforestation schemes, and some weaknesses in monitoring progress in these endeavors. Some criticisms of bureaucratic centralization arose during that time period (Barrantes, 2000). Despite imperfection, the spirit of the policy and related environmental laws remain to the present, as do the administrative and funding arrangements implemented in 1996.

One critique of Panama's environmental policy is that the country at times lacked the legal and judicial capacity, as well as the research infrastructure, to successfully bring about its ambitious, fairly complete decarbonization policies. This particular criticism is based on the country's long-time dependence on the natural environment for economic development, and the need for more effective and decentralized management of the natural environment, including exploitable forests and forest products. Executive decrees have been ambitious, but a criticism emerged that the mechanisms for carrying them out have often been lacking (La Estrella de Panamá, 2013; Wandiza, 2021). A counter argument in defense of Panama's climate governance emphasizes that the country absorbs more carbon than it emits, because of its commitments to funding decarbonization in accordance with global climate agreements (Goering, 2021).

Finally, despite the two countries' favorable reputations for effective forest conservation through inclusion of indigenous groups, the latter have not fully acquired rights to occupy the land they have been promised. To illustrate, in Costa Rica, the 1977 Indigenous Law outlined the return of lands to indigenous communities previously occupied by them but appropriated by ranchers during the beef export years of the 1950s–70s. Because of inaction, the indigenous communities have begun to actively reclaim their lands, resulting in conflicts that at times have been violent. Thus, the country's neoliberal approach to forest conservation has largely been a failure to date, despite the visible efforts by the different governments since the early 1990s to provide payment for environmental services to both indigenous communities and large-scale ranchers, largely financed by a fuel tax. Underlying the payments for environmental services policy is the belief that the ecotourism it stimulates will ultimately be an economic source for funding the policy (Pearce, 2023).

Currently, more than half the land identified as rightfully belonging to indigenous groups through title remains in the hands of cattle ranchers; in some areas, less than a fourth of such land has been handed over to the rightful owners. Much of this is due to bureaucratic inefficiency, including the determination of who is legally indigenous and therefore legally entitled. One result has been some violent incidents, and often the only people to work on behalf of the indigenous people affected are NGOs (Pearce, 2023). Similarly, in the indigenous farming community of Majé, in Panama's Pacific Coastal region some 100 km. east of Panama City, indigenous farmers have continued to be displaced by loggers and ranchers. Illegal logging remains an activity in such indigenous areas that the government has been incapable of halting. The government is often accused of siding with ranchers and loggers by selectively carrying out conservation policies that prioritize indigenous care taking of forest and farmland (O'Donnell, J., & Heater, C, 2022).

SUMMARY AND CONCLUSIONS

This chapter provides an overview of climate policies of Costa Rica and Panama, two countries endowed with abundant rainforests that emphasize the interrelationships between forest management and carbon reduction.

Both have been relatively successful in reducing carbon emissions and absorption of carbon through a combination of public-sector initiatives such as legislation that encourages conservation, and provision of incentives to the private sector for reforestation and afforestation. Furthermore, their promotion of forestry emphasizes greater reliance on markets for forest products as key elements of economic development.

The carbon absorption efforts of both countries are worthy of study, given that their respective policies might be useful to other tropical countries that seek to carry out Paris Climate Agreement carbon reduction objectives. They arguably have developed a balanced version of neoliberal policy, including a positive contribution to economic development, while inserting enough state support of forest management and corresponding legal framework to bring out the non-market benefits of climate change mitigation. However, other tropical countries would need to take into account the unique circumstances of Costa Rica and Panama that may or may not be transferable. Both countries have lengthy histories regarding forest conservation as a public benefit. This is especially the case in Costa Rica, whose "exceptionalism" is based not only on its strong social democratic orientation, free of civil wars, but also its tendency to turn its natural environment into a basis for tourism and a source of forestry products. For its part, Panama has a longstanding tendency to value forest conservation and to manage the basin of the canal, which has resulted in its status as a carbon-negative country. Both countries have favorable democratic institutions and the rule of law necessary for successful conservation efforts and sustainable forestry, including harvesting of trees and other features that its forests offer. Both countries value small-scale producers as the backbone of society and indigenous groups as experienced stewards of forests. However, they would be well advised to ensure that smallholders and indigenous groups truly participate on a level playing field, to transcend the well-cited problems of inequality under neoliberal economic development in the two countries, as well as Latin America as a whole. That is, social and cultural concerns ought to be more deliberately front and center in the forest management side of policy with respect to carbon reduction.

Costa Rica and Panama both enjoy favorable relationships with the UN, World Bank, and other such international organizations. Such organizations, through COPs meetings and REDD+, offer funding for participating countries, along with requirements for assessment and measurement of results. Likewise, their participation in regional environmental agreements reinforces their commitment to practice public-private partnership in forest management, a departure from pure neoliberal policy. Both countries, while ahead of most others in this regard, still could benefit from more robust assessment than the current practice of relying heavily on NGOs. This point could prove useful to other tropical countries that seek to meet climate-related goals in large part through forest management. Finally, both countries rely on their forest conservation as a magnet to attract foreign expatriates as residents, an activity they see as a key feature of economic development. Their respective peaceful, safe and sustainable societies, with well-developed legal institutions, might be difficult for many other tropical countries to replicate. Policymakers in other tropical countries seeking to rely on natural resources to attract residents and visitors, to boost economic development, and to reduce carbon emissions would do well to examine the experiences of the two countries examined in this study but should also take note of their own assets and other unique circumstances and challenges.

REFERENCES

Alfaro, Marielos 2020. Evolución del sector forestal de Costa Rica entre 1969 y 2020: Reflexiones sobre la formación de profesionales forestales [Evolution of the forestry sector of Costa Rica between 1969 and 2020: Reflections on the formation of forestry professionals]. *Revista Trimestral sobre la Actualidad Ambiental, 275,* 12–17.

Barrantes, G. (2000, November). *Aplicación de incentivos a la conservación de la biodiversidad en Costa Rica* [Application of incentives for the conservation of biodiversity in Costa Rica]. Sistema Nacional de Areas de Conservación (SINAC) and Ministerio de Ambiente y Energía de Costa Rica (MINAE). https:www.fb-sgo.cr/sites/default/files/; bibliotecabiodiversidad_psa_estudio_caso_cr.pdf

Burillo, Y. (2021, December 14). Panama leading by example on climate change. *The Diplomatist.* https://diplomatist.com/2021/12/14/panama-leading-by-example-on-climate-change/

Carbon Credits. (2022). What does REDD+ mean? Everything you need to know. *Carbon Credits.* https://carboncredits.com/what-does-redd-plus-mean-everything-you-need-to-know/

FAO. (2002). *Teak (tectonic grandis) in Central America.* Forest Plantations Working Paper 19. Forest Resources Development Service, Forest Resources Division. FAO (Food and Agriculture Organization), Rome. https://fao.org/3/y7205e/y7205e08.htm

La Estrella de Panamá. (2013, January 25). ¿Hace falta una nueva Política Ambiental en Panamá? [Is a new Environmental Policy needed in Panama?]. *La Estrella de Panamá.* https://www.laestrella.com.pa/opinion/redaccion-digital-la-estrella/130125/falta-nueva-panama-politica-ambiental

Forest Carbon Partnership. (2022). *What is REDD+?* https://www.forestcarbonpartnership.org/what-redd

Goering, L. (2021, November 3). Forget net-zero: Meet the small-nation, carbon-negative club, *Reuters.* https://www.reuters.com/business/cop/forget-net-zero-meet-small-nation-carbon-negative-club-2021-11-03/

Herrera U., M. E. (2022, January-March). Estrategia Nacional REDD+: Una serie de políticas ambientales con intervenciones inclusivas que promueven soluciones basadas en la naturaleza en el corto y mediano plazo en Costa Rica [A series of environmental policies with inclusive interventions that promote solutions based on nature in the short and medium term in Costa Rica] *Revista Trimestral sobre la Actualidad Ambiental, 281* (Enero-Marzo 2022), Artículo 4 (28–35). https://www.ambientico.una.ac.cr/wp-content/uploads/tainacan -items/5/37677/005-Herrera.pdf

Irving, D. (2021, August 30). Costa Rica leads the way in cutting carbon emissions. *The Rand Blog.* https://wwwl.rand.org/blog/rand-review/2021/08/costa -rica-leads-the-way-in-cutting-carbon-emissions.html

Lasso, M. (2022, October 4). Proyectan que en 2050 Panamá reducirán en 67% la emisión denCO^2 [Projections that in 2050 Panama will reduce CO^2 emissions by 67%]. *La Estrella de Panamá,* 4 October. https://www.laestrella .com.pa/economia/221004/proyectan-2050-panama-reduciran-67-emision-co

Liverman, D. M., & Vilas, S. (2006, November). Neoliberalism and the Environment in Latin America, *Annual Review of Environment and Resources, 31*, 327–363. https://doi.org/10.1146/annurev.energy.29.102403.140729

Moloney, A. (2021, April 26). Panama's plan to go green and reforest 1 million hectares by 2050. *World Economic Forum..* https://www.weforum.org/agenda/ 2021/04/panama-forests-environment-nature-climate-change/

O'Donnell, J. and Heater, C. (2022, May 5). Panama's indigenous groups wage high-tech fight for their lands, *Yale Environment 360.* https://e360.yale.ude/ features/panamas-indigenous-groups-wage-a-high-tech-fight-for-their-lands

Pearce, F. (2023, March 21). Lauded as green model, Costa Rica faces unrest in its forests. *Yale Environment 360.* https://e360.yale.edu/features/costa-rica -deforestation-indigenous-lands

Portalfruticola (2023, January 16). Este año Costa Rica exportará piña cuya huella de carbono es igual a 0 [This year Costa Rica will export pineapples whose carbon footprint is equal to 0]. *Portalfruticola.* https://www.portalfruticola .com/noticias/2023/01/16/este-ano-costa-rica-exportara-pina-cuya-huella -de-carbono-es-igual-a-0/

Ramirez Cover, A. (2020). Excepcionalismo verde y desarrollo sostenible en Costa Rica [Green exceptionalism and sustainable development in Costa Rica]. *Anuario del Centro de Investigación y Estudios Políticos, 11*(1–21). https://doi .org/10.15517/aciep.v0i11.44774

República de Panamá. (2019, December 11). *Proyecto de sistemas de producción sotenible y conservación de la biodiversidad realiza taller de transición* [Sustainable production systems and biodiversity conservation project holds transition workshop]. Ministerio de Ambiente. https://wwwmiambiente.gob.pa/proyecto-d-sistemas -de-produccion-sostenible-y-conservacion-de-la-biodiversidad-realiza-taller-de -transicion/

República de Panamá. (2023). Principales Impactos del Cambio Climático [Principal Impacts of Climate Change]. https://dcc.miambiente.gob.pa/impacto -del-cambio-climatico-en-panama/

Smithsonian Tropical Research Institute. (2022, October 22). *Indigenous reforestation: Seeking win–win solutions to combat climate change and improve livelihoods in Panama's indigenous Ngäbe-Buiglé Comarca.* https://stri.si.edu/story/indigenous -reforestation

Stefanski, S. F., Shi, X., Hall, J. S., Hernandez, A., & Fenichel, E. (2015). Teak–cattle production tradeoffs for Panama Canal Watershed small scale producers. *Forest Policy and Economics, 56,* 48–56. http://dx.doi.org/10.1016/j.forpol.2015.04.0011389-9341

Steinmetz, J. (2021). *Politics, power, and purpose: An orientation to political science.* Fort Hays State University Scholars Repository, Open Educational Resources. https://fhsu.pressbooks.pub/orientationpolsci/chapter/chapter-9-public-law-and-pre-law-training/

Sytsma, C. (2022, March 10). From teak farms to agrofestry: Panama tests reforestation strategies. *Mongabay Conservation News.* https://news.mongabay.com/2022.03/from-teak-farm-to-agroforestry-panama-tests-reforestation-strategies/amp/?print

United Nations. (2019, March 4). Costa Rica commits to fully decarbonize by 2050, *United Nations Climate Change News.* https://unfccc.int/news/costa-rica-commits-to-fully-decarbonize-by-2050

United Nations. (2020, June 9). Panama launces new programme towards implementing its NDC, *UN Climate Change News.* https://unfccc.int/news/panama-launches-new-programme-towards-implementing-its-ndc

United Nations. (2023). Ley N° 8–Crea el Ministerio de Ambiente [The Ministry of Environment Creates Law No. 8]. *United Nations Environment Program.* https://leap.unep.org/countries/pa/national-legislation/ley-no-8-crea-el-ministerio-de-ambiente

UNESCO. (2021). *Costa Rica climate change communication and education.* https://education-profiles.org/latin-america-and-the-caribbean/costa-rica/~climate-change-communication-and-education

Wandiza, G. J. (2021, March 11). Leyes ambientales en Panamá y su historíe, Te hecho el cuento [Environmental laws in Panama and their history: I will tell you the story]. *LinkedIn.* https://www.linkedin.com/pulse/leyes-ambientales-en-panama-y-su-historiate-hecho-el-gary-pinedo/?originalSubdomain=es

World Bank. (2022, November 18). Costa Rica's forest conservation pays off. *World Bank News.* https://www.worldbank.org/en/news/feature/2022/11/16/costa-rica-s-forest-conservation-pays-off?fbclid=IwAR2m-Yq9b3oqztoAoX_AW8uzdunhsDPTD3d6uMBKaHitznFij4eqMTCHJzQ

Yoder, M. (1994). *Critical chorology and peasant production: Small farm forestry in Hojancha, Guanacaste, Costa Rica.* [Unpublished doctoral dissertation]. Louisiana State University.

CHAPTER 6

CLIMATE GOVERNANCE IN THE CENTRAL AMERICAN NORTHERN TRIANGLE

Opportunities and Challenges of the Public Transport System

Hugo Renderos
University of Akron

Global warming is an issue that increasingly needs attention as time passes due to the severity of damage it is causing the planet. It is extremely important to increase efforts to reduce black carbon emissions at the local, regional, and international levels as global warming affects human life equally regardless of geographical location (Climate and Clean Air Coalition, 2014). Recent studies show that black carbon is the second most powerful climate warming pollutant after carbon dioxide (Bond, et al., 2013). For the above-mentioned reason, this chapter shall focus on climate governance and what the Central American Triangle is doing to decrease black carbon emissions. There are four major climate warming pollutants affecting

Climate Governance in International and Comparative Perspective, pages 105–128
Copyright © 2024 by Information Age Publishing
www.infoagepub.com

the environment; methane, tropospheric ozone, hydrofluorocarbons, and black carbon. All four pollutants are short lived, meaning they live for a short period of time in the atmosphere. Reducing black carbon, being a short lived climate pollutant, would bring about rapid climate benefits.

Black carbon has the shortest life span out of all four short lived climate pollutants, staying in the environment for a few days or even weeks. Therefore, reducing black carbon emissions would provide immediate benefits to the environment (Climate and Clean Air Coalition, 2014). But what is black carbon and what causes it? Black carbon is produced both naturally and by human activities as a result of the incomplete combustion of fossil fuels, biofuels, and biomass (World Bank and International Cryosphere Climate Initiative, 2013). Primary sources include emissions from diesel engines, cook stoves, wood burning, and forest fires. Emitting black carbon in the atmosphere contributes to climate warming by converting incoming solar radiation to heat. It also influences cloud formation and impacts regional circulation and rainfall patterns (Rabatel et al., 2013). Black carbon particles fall on snow, glaciers, and ice thereby darkening light surfaces, causing them to absorb more sunlight and melt faster. This significantly affects water supplies, which rely heavily on snowmelt runoff.

Although black carbon is emitted from a wide variety of combustion sources such as biomass burning, domestic cookstoves, and industry, the transport sector is the largest source of black carbon emissions in Central America's Northern Triangle. The reasons behind why black carbon emission is high in Central America are several. One, yellow school buses circulate the streets of Guatemala, Tegucigalpa, and San Salvador without regard of the dangerous black carbon emitted in the environment. Two, the Northern Triangle states' authorities arbitrarily enforce emissions control laws. And third, importing yellow outdated school buses are big business in the Northern Triangle, generating millions of dollars for the private sector by offering affordable public transport, and kickbacks received by politicians (Interviewee 1, 2023).

ISSUE AND QUESTIONS

Central America is an isthmus connecting the Continents of North and South America. The Isthmus is made up of seven small states. What is known as the Central American Northern Triangle, consisting of Guatemala, Honduras, and El Salvador, is also known as the states having highly polluted capitals. All three states have a public transport system run by the private sector but regulated by central governments (Ardila, 2006). El Salvador, the smallest of all seven states in Central America, is one of three capitals in Latin America with the most air pollution. Guatemala, Honduras, and El

Salvador's transport system offers inter and intra departmental transport to its public (Ardila, 2006). Almost 90 percent of the population employ public transport as its method of transport.

The private sector offers public transport services to its citizens by circulating yellow school buses, imported from the USA, throughout the country. These school buses are used for decades with little maintenance, if any (Centre for Clean Air, 2022; Interviewee 1, 2023). As a result of the poor or lack of mechanical maintenance, the buses emit highly contaminating black carbon, thereby polluting the environment. Consequences of a polluted environment have costly effects for the people, environment, and governments. Can newer buses be replaced by the old ones? They can be but bus line owners raise prices. Can government officials regulate public transport prices? They can but are often bought out by bus line owners in order to legislate favouring the public transport industry.

The Northern Triangle countries have been governed by both left- and right-wing regimes. Both have governed the same regarding climate change. The question this chapter endeavours to seek answers to is "Do the governments of Guatemala, Honduras, and El Salvador have the willingness of protecting its populations by regulating air pollution caused by its public transport"? "What are there challenges these governments face impeding them from regulating black carbon pollution in their capitals"?

LITERATURE REVIEW

Urbanization coupled with urban transport go hand in hand in the Northern Triangle. Urbanization has risen to a major world topic in global politics. With urbanization, public transport is affected by the increased use of it (Mcgranahan, Balk, & Anderson, 2007). Because major cities in the Northern Triangle attract thousands of people for various reasons, use of major resources become essential and necessary for survival. One of these major resources is clean air (McDonald et al., 2011). With public transport being the major source of transport in the Northern Triangle, it is imperative that this issue gets addressed due to high emissions of black carbon in the environment. The use of public transport system in the Northern Triangle causes concern for the environment because the system of transport employs old yellow school buses imported from the United States. These buses are dated and therefore, removed from school bus fleets. However, once transported to Central America, these buses are refurbished and usually can run for an additional twenty to twenty five years more (Mosher & Ekstrom, 2010).

Black carbon emissions have been studied both in laboratory and field research. Whilst laboratory studies have typically produced detailed specific emission information, the field studies have included measurements using portable emission measurement systems, mobile measurements, and next-to-source experiments (Dallmann et al 2013, Liu et al 2021, Saarikoski et al 2021). Results of many field studies do not represent a single black carbon source, but the entire traffic fleet. It is evident that black carbon is determined by different methods and in different research environments may not be fully comparable. Their use in emission climate and air quality models requires careful and expert evaluation. This is something that the Northern Triangle lacks, expertise, systems for measuring emissions, and the tools to carry out the testing.

Approximately 8,000 kilotons of black carbon are emitted globally each year (Bond et al., 2004). These emissions, the black carbon's contribution to global warming, come predominantly from the northern hemispheres (North America and Europe). This fact is particularly important given the fact that black carbon emissions north of 40°N are most likely to be transported to the North Pole region, where their warming impact is amplified (Ramanathan & Carmichael, 2008). A regional breakdown shows that the developing world, Africa, Latin America, and Asia, emits roughly 80% of total black carbon emissions, whilst Europe and North America represent a combined 13% (Bond et al., 2004). However, Europe, North America, and Russia emit black carbon primarily from fossil fuel combustion and largely above 40°N, thereby contributing disproportionately to warming due to black carbon ratios and the impact of latitude on arctic melting. China and India account for 39% of global black carbon emissions, and about 54% of warming black carbon emissions. In Central and South America, contained combustion, forest burning and savanna burning each contribute about 5% of global black carbon (Bond et al., 2004).

Throughout history, most climate warming agents such as black carbon have been emitted by developed countries where most of the world's energy and resources are consumed. By contrast, in this chapter, it is shown that not only are Asian countries affected by this problem as Central and South America experience the same environmental problems as well (Bond et al., 2004). The Northern Triangle has the issue of black carbon emitted in to the air since concentrations are extremely high in its urban areas with very high population density (World Bank, 2020). Furthermore, these countries have trouble with administrative capacity, as indicated below. Consequently, where benefits from black carbon emissions reduction are greatest, constraints on funding and institutional capacity are the greatest as well.

The problem of mitigating black carbon emissions in the Northern Triangle is amplified by the expected exponential growth in transport means. These trajectories suggest that these countries urgently need to take steps toward reducing black carbon emissions because it affects the entire region's well being, economically, health wise, and its productivity. Examples of this include the introduction and use of fuel-efficient vehicles and buses, black carbon measuring systems that accurately measure emissions, willingness from the governments to introduce and implement emissions control systems, and the citizenry doing their part in contributing to the overall benefits of the cities by adhering to the regulations and policies controlling black carbon emissions.

There are existing studies showing how black carbon emissions impact current climate change. Although there are continuing studies, those existing studies overwhelmingly indicate a consensus showing how black carbon emissions accelerate the melting of snow and ice, glaciers, and affect temperatures and weather patterns in both North and South poles (Bond et al., 2013). The latest research shows that black carbon emissions also affect photosynthesis when it is deposited on leaves, reduce the quantity and or quality of water available for agriculture, and decrease visibility due to suspended particles in the air. It is important to emphasise that despite climatic impacts from black carbon emissions, certain trends being already established, more research and data needs to be carried out in Latin America. There is a dire need to focus and conduct regional and local studies discerning the full extent and implications of how black carbon emissions affects this area of the world (Centre for Clean Air, 2022).

Existing research results show it is imperative that countries in the Northern Triangle do whatever they can to reduce black carbon emissions as soon as possible because the health of its populations are directly affected. Black carbon is a component of particulate matter, it is a complex mixture of very small particles and liquid droplets in the atmosphere (Yang et al., 2021). The most common source of transportation related particulate matter is incomplete combustion of diesel engines (Yang, et al., 2021). The most common type of public transport in the Northern Triangle is offered by the private sector through yellow school buses circulating each countries' capitals.

Health experts have travelled to the Northern Triangle countries to conduct research for themselves on health issues affecting the citizens. The studies' results show how people in all three countries' capitals suffer from respiratory and vascular illnesses (Yang et al., 2021). The studies' results found clear links between black carbon exposure and a decrease in breathing and cardiovascular functioning. These studies found additional evidence demonstrating how people were affected with black carbon emissions by thrombosis, acute respiratory symptoms, aggravated asthma symptoms, decreased lung functioning, and lung inflammation (Farzad et al., 2020; Yang, 2021). Furthermore, these same studies' results revealed that outdoor pollution, diesel exhaust, and particulate matter can cause cancer if exposed to them for a long time period.

The aforementioned studies' results ought to be alarming for leaders of the Northern Triangle countries and must prioritise effective and efficient policies toward reducing black carbon pollution. Policies toward reducing black carbon need to be implemented because expanding the use of diesel fuels and vehicles in all three countries means that more and more people will continue being exposed to harmful fumes damaging their health and the environment. By prioritising and placing great effort in producing policies that effectively reduce the emission of black carbon to low levels, combating climate change and preventing and or detecting early signs of pulmonary and cardiovascular diseases would greatly improve the health and the overall lives of all affected people.

POLICY REVIEW

This study was conducted using the case study tradition. The author currently lives in Central America, making it easy for travel throughout the region. All of the information obtained for the chapter was via personal interviews for El Salvador and Honduras. The interviews for the Salvadorean deputies were scheduled according to the deputies' schedule availability. The author called each deputies' office and requested time for an interview. The deputies were happy to be of service and explain their views on the issues and what the current government is doing to counter black carbon emissions.

For the Honduras information, the author travelled to Tegucigalpa, Honduras. Prior to travelling to Tegucigalpa, the author called the office of the Honduran deputy and arranged to meet at the deputy's office on the time, day, and date specified. Only one deputy was interviewed for Honduras as other deputies' schedules were not open during the times the author was able to travel and speak to each. Both Salvadorean and Honduran deputies were requested to answer open ended questions. Both countries' deputies sat down and explained what the issues were that affected their cities, what the government was doing to counter it, and what could be expected once the laws were put into force. For Guatemala, the case was different because the deputies which were contacted were not available for giving personal interviews. As a result, there were no interviews conducted for Guatemala.

GUATEMALA

The Republic of Guatemala has an area of 108,889 km^2. It borders with Mexico, El Salvador, Honduras, and Belize, as well as with the Caribbean and the Pacific Ocean. Guatemala is a country with a great climatic diversity varying markedly with geography due to its mountainous terrain (Central Intelligence Agency, 2023a). Approximately two thirds of Guatemala is mountainous, and thirty four per cent is covered by woodland. Located in the Central American Isthmus, Guatemala's climate is tropical, with high mountainous areas being cooler than the lower sea level areas. The average temperature is between 18 and 22° C. Guatemala has a population of more than 17 million (World Bank, 2023). Guatemala City's population is more than 3.5 million (OHCHR, 2023a). It is the largest capital in the Central American Isthmus, and in the Caribbean Basin.

Like many cities in Latin America, Guatemala City's population has grown rapidly since the 1960's. Since its civil war started in the late 1950's and concentrated in the jungles, people migrated to the cities, making Guatemala City their destination of choice. With its massive population, the city is the most crowded one in Central America. At the same time, the city accommodates over 2 million vehicles, of these, 3,000 are public transport buses (Ardila, 2008; Cutz et al., 2020). Most of the city's inhabitants employ public transport for travelling around the city. These same buses are responsible for the emission of black carbon in the city. The city's 3,000 strong bus fleet is ageing. The reason why the bus fleet is ageing is because when cars and school buses in the United States are no longer able to pass emissions tests, they are sold in auctions (The Guardian, 2019; Kroger, 2022). When these vehicles go up for sale, they are not bought by local residents. They are sought after by enterprising operators from Guatemala, Honduras, and El Salvador. Public transport providers are business people providing public transport to their countries' capitals.

Pollution in Guatemala

Purchasing buses and vehicles at public auctions in the United States is big business for these entrepreneurs as they buy cheap, old, and outdated vehicles and import them back to the south. These same buses and vehicles are banned from circulating the USA but for a few minor repairs and they can last for 20 or more years in the Northern Triangle (Ardila, 2006, 2008). Often times, vehicles bought at auctions in the USA are totally damaged, what in the USA is known as totally destroyed, these vehicles are sold to Central American vehicle resellers and after giving the vehicles a make over, they are put on the market for sale. The yellow school buses are refurbished, engines are modified, a fresh coat of paint is applied to all buses, and more powerful motors replace the original ones to transport the public (Cutz et al., 2020; Fridolin, 2019). Mechanics may remove the catalytic converters, even though the devices reduce pollution and are legally required in the city.

The result of having old, refurbished, freshly painted school buses circulating in Guatemala City is heavy pollution as black carbon gets emitted along the roads. As a consequence, the heavily polluted air becomes inhaled by the city's inhabitants (Fridolin, 2019). The capital city has the most polluted air than any other area in the country. Bear in mind that the metropolitan area of Guatemala City is home to more than 3.5 million people, making it the largest urban centre in Central America. With the number of private vehicles circulating in the city, it is the sixth most polluted city in the Americas (Grutter & Grutter, 2014). Guatemalan authorities, far from regulating air pollution, struggle even to measure it. Unfortunately, Guatemala is one of many developing nations where air pollution is driven by an influx of old, used vehicles from countries with strict environmental standards.

According to the World Health Organisation (WHO), there are approximately 4.2 million deaths per year due to outdoor pollution inhalation (Shaddick et al., 2020). The number of inhabitants living in Guatemala City is 3.5. This number is just under the number of deaths world wide from black carbon inhalation. It is important to keep in mind that pollution studies in the Northern Triangle remain understudied. This understudy causes there to be unaccounted deaths from air pollution in the region. So, it may well be that pollution deaths are much higher than reported by WHO. The Guatemalan Health Ministry released data showing that respiratory illnesses are the most common cause of death in children under 14 years of age (Tomczyk et al., 2019). A Guatemalan pulmonologist expert indicated that young children under the age of 14 and older citizens over the age of 60 are the most vulnerable and most affected by black carbon inhalation. These two groups commonly experience long lasting coughs, sometimes lasting more than two weeks. If coughs persist longer than two weeks, it is a sign of a more serious illness (Balakrishnan et al., 2023).

The Guatemalan Health Ministry, with what little economic resources it is provided and scarce availability of experts on which to rely, carried out an air pollution study on the severity of the polluted air circulating the streets of Guatemala City. The study results showed that there were several different kinds of air pollution particles roaming the air. The particle having the greatest impact on the respiratory health of Guatemalans is black carbon consisting of a tiny particle approximately the size of red blood cells inside the body (Wang et al., 2022). The size of these particles allows them to penetrate deep into the lungs and enter the blood stream. Furthermore, the study showed that they can cause numerous short and long term respiratory problems such as coughing, asthma, laryngitis, sinusitis, and even lung cancer.

Guatemala's Response to Black Carbon Pollution

Guatemala is a developing nation, where corruption runs rampant. Self interests and those of the elite drive the Guatemalan politicians. The people in Guatemala know very well that politicians have rarely served their constituents' general interests. Like in many developing nations, politicians are vulnerable to the highest bidders. Many times, politicians are bought by private interests and legislatively favouring their donors (Flores & Rivers, 2020). As mentioned earlier, importing old and damaged vehicles for refurbishment is big business in Guatemala. There is a union that dictates what tariffs importers pay and how old cars may be for importing into the country (Rodriguez, 2018). There have been several occasions when different Guatemalan administrations passed laws to regulate black carbon emissions but failed for reasons all too common the developing nations. For example, in 1997 a law passed regulating vehicle emissions. However, the law was overturned in 1998. In 2012, a new tax law was introduced establishing an age limit on imported vehicles. Nevertheless, the Union of Importers of Used Vehicles and Buses challenged the law claiming it was unconstitutional to regulate citizens' livelihoods and means of sustaining themselves (Flores & Rivers, 2020; Rodriguez, 2018). The Constitutional Court of Guatemala ruled in favour of the union declaring that law as written and passed regulated citizens' means of supporting themselves. It is highly probable that the judge was paid to rule in favour of the union. Evidently, judges are influenced by the wealthy businesses that are able to pay their way and continue doing business as it benefits them.

Even with existing laws regulating black carbon emissions in place in Guatemala, the Ministry of the Environment and Natural Resources (MARN) has no way of enforcing them as the enforcers are not willing to do so or refuse doing it. Many times, these environmental enforcers are bribed into accepting a few dollars by bus owners and look the other way as if there were no laws broken (Cohn-Berger & Quezada, 2016). More recently, with bus fleets being replaced with newer models, the Guatemalan government has encouraged bus owners to invest on electric buses or on buses running on biodiesel fuel. However, due to the costs of purchasing and replacing newer model fleets with electric ones or running on biodiesel fuel, many bus owners refuse to do so. The government has even offered subsidies for bus owners to buy on large scale and still they claim they are not able to afford them (Dotson, 2014). Bus owners claim that even if they did replace their current models with electric one, the majority of the public would not be able to afford them because fares would increase in order to make a profit. The Transurbano Initiative was a government response to reducing black carbon emissions. It was a programme designed so that bus owners could trade out their fleet of buses and replace them with newer models which would utilise pre paid fare cards instead of cash (Dotson, 2014). The programme was not successful at all because bus owners continued circulating their older fleet as usual.

If Guatemalan authorities really want to modernise their transport system and reduce as much as possible black carbon emissions, it will take a lot of political will. Additionally, it may even have to offer many incentives in the short term, perhaps overly priced ones, in order to convince bus owners to invest in buses that do not pollute as much as the current fleet does. While several international development banks offer government loans for infrastructure projects in developing nations, the Guatemalan government will need to renew its own commitment to modernising the transport sector if the country is going to change its air from a densely polluted one to a cleaner one.

Guatemala's current administration, under the supervision of the Ministry of Environment and Natural Resources organised a commission to draft air pollution regulations (Riojas-Rodriguez et al., 2016). However, the Ministry is encountering issues due to the fact that MARN has been plagued with Guatemala's pervasive political corruption. In 2015, Otto Pérez Molina's administration was found to be involved in corrupt political activities (Beltran, 2016). Pérez Molina was forced to resign as the findings were irrefutable and convincing. Jimmy Morales, Pérez Molina's successor is currently under investigation for receiving illicit funds to finance his presidential campaign. Whilst in office, Morales expelled the United Nations appointed anti corruption commission that uncovered several of his politicians' illegal activities (Malkin, 2017). During Jimmy Morales' administration, his environmental minister declared that vehicle emissions regulations would go into force, but resigned from his post in 2018 January. The current administration under President Giamattei, has indicated that MARN plans on carrying out an analysis establishing the country's baseline for vehicular pollution (Sánchez Girón & Urrutia Campo, 2021). Giamattei's goal is to improve Guatemala's technical capacity because at the governmental level it does not have much experience, expertise, or know how measuring emissions.

HONDURAS

Honduras is the second nation making up what is known as the Northern Triangle. It has a square area of 112,000 kilometres. Its most current population is estimated to be a little more than 10 million people, of which it is calculated that 45% live in urban areas, mainly concentrated in Tegucigalpa and San Pedro Sula. Honduras has two major cities, Tegucigalpa, its capital city, and San Pedro Sula, its main economic activity centre (U.S. Dept. State, 2023). San Pedro Sula has an important base of light industry and commercial production of tobacco, and wood products such as lumber. It is the second largest nation after Nicaragua, in land size, of all Central American states. More than three-fourths of Honduras is mountainous and wooded. The eastern lowlands include part of the Mosquito Coast. The economy is primarily agricultural; bananas, coffee, and sugar are the main export crops, and corn, rice and beans are the chief domestic staples (OHCHR, 2023b).

Honduras is a multiparty republic with one legislative house, and the head of state and government is the president. In the 20th century, under military rule, there was nearly constant civil war. A civilian government was elected in 1981. The military remained influential as the activity of leftist guerrillas increased in neighbouring El Salvador, Guatemala, and Nicaragua (Central Intelligence Agency, 2023b). Honduras is very susceptible to suffer from natural disasters. Flooding caused by a hurricane in 1998 devastated the country, killing thousands of people and leaving hundreds of thousands homeless. In 2001, Honduras was hit by a severe drought. Recovery and rebuilding efforts followed for the next several years. Throughout its history, it has experienced devastating earthquakes due to its geographical position. A total of 2,516 tremors with a magnitude of four or above on the Richter Scale occurred within 300 kilometres of Honduras in the past 10 years (USAID, 2023). This comes down to a yearly average of 251 tremors per year, or 20 per month. In 2009 President Manuel Zelaya was ousted in a coup, the first military coup in Central America since the end of the Cold War (Gordon & Weber, 2013) A military-supported interim regime held power only until January 2010, when an elected president took office.

Pollution in Honduras

Like Guatemala, Honduras is another nation plagued with black carbon emission issues. Also, for the same reasons it is having problems containing, reducing, and mediating it, political corruption, lack of political will, and a strong union of buses and old, used vehicle importers opposing any legislation regulating vehicle imports. The most vulnerable populations from black carbon inhalation tend to be the elderly, children, and women (Rivas, Suarez-Aleman, & Serebrisky, 2019). As in Guatemala, black carbon emission pollution s a serious health problem in Honduras. Honduras, like Guatemala, lacks air quality monitoring systems and appropriate experts to rely on for advice on proposing mitigating factors for reducing and controlling black carbon emissions in the country. The Honduran Health Ministry identified acute respiratory infections, cardiovascular and cardiopulmonary and eye diseases as the main health issues in children under the age of five, people older than 60 years, and women of child bearing age (Interactive Country Fiches, 2023). An important risk factor for these diseases is being constantly and heavily exposed to black carbon from older vehicles circulating the streets of Tegucigalpa and San Pedro Sula.

In 2006, the Health Ministry reported more than a million cases of medical care stemming from respiratory diseases. Of the one million reported cases, 27% of these were recorded in Tegucigalpa affecting children under age five (Ostro & WHO, 2004; Timilsina & Shrestha, 2009). According to the World Bank, there are approximately 500 premature deaths occurring a year that are easily traceable to polluted air in the two major Honduran cities (Awe, Nygard, Larssen, Lee, Dulal, & Kanakia, 2015; Lelieveld, Pozzer, Poschl,Fnais, Haines, & Munzel, 2020). The ones most adversely affected by black carbon emissions in the study were older adults whose lives were shortened by inhaling black carbon. This trend not only affects the victims, but also their families, the nation's economy and the overall lively hoods of Honduran citizens. The high costs related to medical care, low production or under production of the nation's productivity as well as the losses this entails, factor in the effects of health issues from air pollution. It is reported by the Health Ministry that there is approximately a loss totalling $51.6 million per year from air pollution health related problems (Organisation for Economic Cooperation and Development, 2023).

One of the main causes of contaminated air in Honduran cities is the increasingly and heavy concentration of polluted air from buses and vehicles' black carbon emissions. According to the Health Ministry, approximately 63% of the vehicle and bus fleets combined are concentrated in Tegucigalpa and San Pedro Sula (Interviewee 2, 2023) It is estimated that the vehicle and bus fleets could double in the next ten years because demand in public transport has increased due to COVID-19 aftermath (Interviewee 2, 2023). Also, due to the present economic downturn worldwide and people migrating to the cities in search for employment, urban sprawl is making longer travel distances, making way for increased demand for public transport. Honduras faces the same issue Guatemala has with controlling black carbon emissions. Additionally, it does not count on emissions control equipment to measure amount of the air's contamination. Furthermore, it lacks ways of controlling the high number of imported vehicles from circulating without proper emission control systems (Interviewee 2, 2023).

Honduran buses and autos are imported in large numbers annually from the USA. On average, the autos and buses average at least ten years of use prior to being imported in Honduras. Some buses are as old as fifteen years and have no working emissions control systems. Auto and bus maintenance standards are poorly monitored, enforced, and demanded from the government (Interviewee 2, 2023). The practices of maintaining and up-keeping an auto or a bus in Honduras is not like the developed world where oil changes and tire rotations or replacements thereof are performed every five thousand miles or six months, whichever comes first. The reader must keep in mind that Honduras is a developing nation and people do not earn enough to care for secondary necessities. Uber, taxi and bus drivers do the least necessary maintenance to their vehicles because they have other pressing priorities than changing oil or tires on their vehicles.

A recent study conducted by university students on the efficiency and effectiveness of the Honduran public transport system found that bus owners providing public transport to the public do not perform any maintenance on their units (Interviewee 2, 2023). For example, the study concluded that no maintenance had been performed on the units for such a long time due to the fact that the tires on almost all buses were completely bald. Some of the tires even displayed some wires poking through. What is even more shocking is the fact that some of these buses' engines were held together with wires or strings. Brakes squeaking were normal for many of the observed buses. When riding on the buses, buckets were placed as seats when the appropriate repairs or replacements were not performed. It is because of the aforementioned reasons that these buses have earned the name "rolling coffins" (Interviewee 2, 2023).

Honduran Response to Black Carbon Pollution

The government has tried several times to control and force changes in the transport system, making it safer and securer to travel in them. The legislative assembly has passed several laws imposing heavy penalties, mandating safety inspections, and even sanctioning bus owners. Unfortunately, sanctions, fines, impounding, nothing has worked. One of the main concerns regarding the improvement of the transport system was sanctioning buses polluting the air with black carbon. A new law established that buses had to have smoke stacks in order to avoid black carbon emission from being inhaled by people driving behind them (Almeida, 2021). This government response turned out a failure.

Honduras has a need for affordable transport in urban areas. Providers of public transport argue that if they are forced to make maintenance repairs on their buses, the cost of the bus fare will increase. If bus fares increase, people will protest and demand that the government force bus owners provide public transport. When the government obligates these bus owners in providing transport at lower prices, they threat with striking until there is an agreeable accord with government officials (Secretaría de Finanzas, 2015). This is a vicious cycle that several Honduran administrations have been through. Since the mid 1990's, every presidential candidate made promises of reforming the public transport system making it safer, securer, and affordable but have all failed to deliver. The opposition to such reforms is powerful making it difficult for governments from improving it.

Another intent at reforming and improving air quality in Honduras, but a failed effort, was the introduction of the Pure Air programme in the late 1990's (Cision PR Newswire, 2023; Secretaria de Recursos Naturales y Ambientales, 2023). The programme was introduced and implemented by the Reina administration. The programme was designed to support the newly established legal framework of regulating motor vehicle emissions, fuel quality, and the implementation of a vehicular technical inspection. The programme put four monitoring stations in place in the two major cities (three in Tegucigalpa and one in San Pedro Sula) of the country in order to monitor air quality (Secretaria de Finanzas, 2019). Unfortunately, this governmental programme created for reducing the air pollution failed because the stations were not operating as designed, they did not receive appropriate maintenance, and those operators assigned to monitor the stations were not appropriately trained to operate them (Secretaria de Recursos Naturales y Ambientales, 2023). Additionally, there were no air quality standards to allow setting limits for the protection of the population from the harmful effects of air pollutants.

In 2010, during the Lobo administration and under the direction of the Ministry of the Environment and Natural Resources (SERNA), the National Plan for Air Quality Management was implemented to reduce air pollution (Secretaria de Recursos Naturales y Ambientales, 2023). The Plan's purpose was to help improve air quality and preventing it from deteriorating, promoting sustainable human development, and protecting the citizenry's health. The Plan's designed was to reduce air pollutant concentrations in the environment, contribute with the reduction of greenhouse gas growth rate, strengthen Honduras' air quality management system, provide sustainable transportation that does not pollute the environment, and provide clean and efficient energy (Secretaria de Recursos Naturales y Ambientales, 2023). By implementing the Plan, the government wanted to focus on the aforementioned factors because it would bring national, health, economic, and productivity benefits for all. Furthermore, the National Plan was a way of showing bus owners that the government was making every effort for dialogue and building political commitment of the institutions involved (Secretaria de Recursos Naturales y Ambientales, 2023).

Lately, the Hernandez administration (2014–2022), under the SERNA, put together and assigned a group to study and address black carbon emissions and its mitigation in Tegucigalpa and San Pedro Sula. The group included members from government entities, civil society, businesses, national, and international non governmental organisations (Secretaria de Recursos Naturales y Ambientales, 2023). The group was assigned the task of determining historical and current conditions arising from black carbon and air contamination emanating from it. The group collected, reviewed, and evaluated information from previous administrations studies, and from private organisations having undertaken studies investigating the same issue. It looked at previously conducted assessments, measurements, and formerly implemented programmes and projects on air quality matters (Open Knowledge Report, 2023). Up to this point everything was conducted fairly easy, simple, and in the best interests of the nation. However, when it came time to propose legislation, this is where all discord erupted. Bus owners did not want to propose any legislation proposing black carbon emission controls because it would cause the bus fares to increase, which they argued, the public would protest and not pay the increase as needed for bus owners to make a profit. (Open Knowledge Report, 2023). A stalemate such as the one above is not uncommon when dealing with air quality controls in Honduras. It is very difficult, almost impossible for the government to control air quality due to heavy and strong opposition from those groups providing an essential service to the public.

EL SALVADOR

El Salvador is the smallest country in the Central American Isthmus. It is a little smaller than the USA State of Massachusetts with over 21,000 square kilometres (Banco Mundial, 2023). This mountainous country is bordered by the Pacific Ocean, Guatemala, and Honduras. Known as the Valley of the Hammocks, El Salvador has frequent earthquakes and volcanic activity. It is the only country in Central America that does not have a coastline on the Atlantic Ocean. About half of the a little under 7 million Salvadoreans live in urban areas. Salvadoreans are poor and face difficulties keeping electricity or running water in their homes (Datos Mundial, 2023). Most wealthy Salvadoreans reside in San Salvador's suburbs. The government offers its students free education up through the ninth grade, but many families struggle covering the cost of supplies and transporting their children to and from schools. There are approximately three million Salvadoreans living in the United States. About 95 percent of Salvadoreans are mestizos, a mixed race of Aboriginal and Spanish descendants. The Salvadorean food staples are rice, beans, and corn. Economic inequality led to the bloody, civil war in 1980. About 3.5 million Salvadoreans fled to the United States avoiding the war. The civil war lasted for 12 years, costing the lives of about 75,000 people. It ended in 1992 when the government and leftist rebels signed a peace accord providing for military and political reforms (UNESCO, 2023).

Pollution in El Salvador

Vehicular emissions are a major source of air pollution in El Salvador. Gas and particulate emission levels from the fleets of buses and autos in El Salvador are very high, when compared with international standards defining acceptable levels. These levels are high because the fleets of buses and vehicles consist predominately of very old autos with little, if any, maintenance. The consequences of this situation are varied and serious. The population in the major Salvadorean cities are frequently dealing with respiratory and eye problems (Interviewee 3, 2023). Income levels of most auto owners in El Salvador do not allow them to acquire better or newer vehicles. This is a serious limitation for reduction of current emission levels. However, experiences in other developing nations have shown that a considerable decrease of black carbon emissions is within reach in El Salvador, if it only allows inspection and maintenance programmes to be introduced and passed in Congress (Interviewee 3). Passing and implementing an emissions control programme in El Salvador would generate positive effects on the health and quality of life of the population, as well as additional savings in vehicle maintenance costs and health services (Interviewee 4). Something that is worth emphasizing is the fact that this type of programme provokes significant reductions in greenhouse gas emissions as well.

One of the vital factors of heavy pollution found currently in El Salvador is auto black carbon. The lack of auto maintenance, be it for pure neglect or lack of economic resources, vehicle's age, and second-hand auto imports (often times autos wrecked and or already retired from the vehicle fleet from the USA), taken together with the poor quality of the fuel sold, all represent an extremely and serious health, environmental, and economic problem for the Salvadorean people. The exceedingly high rates of black carbon emissions, the resulting respiratory problems among the populace, continuous expenditures on patching instead of real and effective repairs, and high probabilities of both accidents and breakdowns on the road provide clear evidence of this situation (Interviewee 3 & 4, 2023). Although it is true that it is less expensive to have a vehicle in good repair, it is also true that the low levels of income do not allow the purchase of the type of automobile which could prevent, or at least ease, these problems. However, not all of the problems are due to the autos' old age. In a large number of cases, the disrepair of the autos is due more to a lack of awareness than a lack of knowledge of the need for correct engine functioning (Interviewee 3). Therefore, the need arises for a type of public control that would not only reduce or prevent the problems aforementioned, but would also attempt to inform and raise the levels of awareness of auto owners on the impact of their own economic resources, the public's health, and their fellow citizens caused by driving an auto in deterioration (Interviewee 4). This is why there is a need introduce and implement a control system regulating vehicular emissions.

In El Salvador, the emissions of black carbon are not so much from old yellow school buses but from the importing of old wrecked automobiles. Similar to Guatemala and Honduras, importing old and wrecked autos in

El Salvador and repairing them in a fully refurbished state is big business as well (Interviewee 5, 2023). It is estimated that 44% of autos circulating in El Salvador were produced during the 1980's. There is a smaller percentage, 27%, were produced in the 1970's, whilst another 6% were produced before the 1970's. In contrast, only 23% were produced during the 1990's forward. According to the Salvadorean Ministry of the Environment and Natural Resources, MARN, autos are the most extensive contributors of black carbon in the environment (Interviewee 5, 2023). This is not to say that old yellow buses are not imported, but are to a lesser extent responsible for all the smog found the air. A study conducted by MARN in 2021 found that a high percentage of pollution in El Salvador is produced mostly by the smoke emitted by private vehicles and some comes from the public transport system (Interviewee 5, 2023). The country's governments several times alluded to implementing an emissions control programme but have failed to carry it through. Often times, former administrations have only briefly discussed it whilst campaigning, but once in office have never executed anything that would address air pollution in its cities (Interviewee 5, 2023).

Salvadorean Response to Black Carbon Pollution

El Salvador, under the Bukele Administration, the National Council of Environmental Sustainability and Vulnerability (CONASAV) presented a series of reforms to the Land Transport Law. In essence, the proposed laws essentially aim at reforms to reduce atmospheric black carbon pollution (Interviewee 6). Former Salvadorean Administrations have been reluctant to pass any emissions regulating legislation because they fear that those mostly affected such as the transport providers will go on strike like they have in the past and the public will be affected as most capital city residents rely on public transport for getting to their destinations, be it via Uber, Lyft, buses, or taxis.

El Salvador has four black carbon measuring stations for monitoring the levels of polluted air. There are two stations in San Salvador, one in San Miguel, and the other in Santa Ana. Despite having stations available to monitor levels of pollution in the country, it still struggles to contain it because the nation does not have in place laws regulating black carbon emissions (Interviewee 6). After establishing that current levels of air pollution have reached dangerous levels, the CONASAV reforms focus particularly on the reduction of emissions by public transport and allow the police force to confiscate vehicles which exceed specific levels of pollutant emissions. The reforms also recommend the replacement of public service vehicles that are over 20 years old and that this new measure should be implemented within three years (Interviewee 6). The World Health Organisation has placed El Salvador among the Latin American countries with the highest levels of atmospheric pollution and relates these levels specially with public transport.

Just this year, the Salvadorean Congress proposed and passed a bill proposing a fine of $57 to be imposed on drivers of vehicles emitting excessive black carbon. The new proposed law imposing stricter contamination regulations for autos will be voted on later this year. The proposed law submitted for approval by the Salvadorean Legislative Assembly targets every mode of transportation, private autos, public transport buses, and cargo vehicles (Interviewee 6, 2023). Several anti reform groups have voiced that the proposed measure will increase the economic burden on individuals who are unable to afford the costs associated with meeting the new environmental standards. However, a large sector of public opinion has expressed that the policy proposal is an important step in enforcing the public health, specially for the most vulnerable such as children, elderly, and those with chronic respiratory illnesses (Interviewee 6, 2023). It appears that there is demand among the public to take measures in order to address the high levels of contamination and the high costs on the people suffering and living with respiratory related issues.

FINDINGS SECTION

Reducing short-lived climate pollutants like black carbon is becoming an increasingly important component of national interest for all three nations in the Northern Triangle's efforts to fight air pollution. Because of black carbons' short lifetimes, reducing these emissions can have almost immediate climate benefits. Recent studies have concluded that black carbon is the second most powerful climate warming pollutant after carbon dioxide. Transportation, especially diesel powered vehicles, is one of the largest sources of black carbon emissions in Guatemala, Honduras, and El Salvador. The major factor contributing to the black carbon problem is its high imports of USA used buses and older automobiles to the Northern Triangle.

Given the dominant characteristics of the vehicular fleets in Guatemala, Honduras, and El Salvador, particularly the age and lack of maintenance, and the problems of environmental contamination, an emissions control programme is in order. This type of programme would generate various benefits. The benefits can have a domino effect. For example, when the health of the residents of the Northern Triangle is cared for, more productivity produces more economic activity. When the health of the people is taken care of, they will produce, have more resources, and perhaps have enough for providing appropriate maintenance for their vehicles. With well maintained vehicles circulating in the Northern Triangle, there would be less black carbon emissions affecting the environment and the health of the people. Given these benefits and costs, the emissions control programme would be expedient, as long as the costs are affordable to all vehicle owners. All three countries have experienced heavy opposition to any legislation regulating black carbon emissions, it would be necessary to include, from the point of view of the vehicle owners or the firms that

would offer inspection services, their input to arrive at a mutually agreeable emissions control programme. Including all parties affected by the emissions control programme is appropriate because it would educate, inform, and transmit the information necessary that vehicular inspection services is not an expense but an investment with a series of associated benefits.

CONCLUSION

In addition to the recommendations above, the Northern Triangle will benefit from more extensive air quality monitoring. Currently, Guatemala and Honduras have shown the intent of trying to mitigate the black carbon problem existing in their countries but fallen short due to interests are being focused on other issues, their legislative assemblies have not shown the willingness to address the issues adequately, and pressure from interest groups seems to be overwhelming for the legislative and executive branches of government.

With improved air quality monitoring, public education, and communications campaigns can only increase the quality of life for all living in the Northern Triangle. Pollution, public health, and climate change need to be seriously be addressed and dealt with in order to accelerate progress toward better government policies. If policymakers in Guatemala and Honduras follow the steps the Salvadoreans are taking to mitigate air pollution, black carbon emissions will be significantly reduced in the Northern Triangle, providing important public health and environmental benefits to hundreds of Guatemalans, Hondurans, and Salvadoreans as well as throughout the region and globally. Simply put, reducing black carbon in the transport sector is a win-win strategy to address both climate change and public health concerns.

RECOMMENDATIONS

Studies prove that black carbon has widespread and profound impact on the environment as it can lead to anything from premature deaths, heart, lung, and respiratory problems, eye and vision infections to lung cancer and death as a result from outdoor air pollution. Despite black carbon emissions and climate research being limited in the Northern Triangle, enough is known to strongly recommend that governments need to urgently act.

Because black carbon is emitted as a component of particulate matter, one strategy to reduce it the Northern Triangle that can effectively decrease it is implementing a strict emissions control policy. Stricter emissions regulations for newer vehicles once in place, the populace can adopt vehicle emission standards for newer vehicles that will require passing the emissions test. This will require educating the people that this investment (fines or penalties for not meeting emissions standards) is required in order to protect all of the people, their health, and overall environment. Because so many older, high-emitting vehicles will remain on the road for years to come, all three countries should consider adopting complementary measures that will reduce emissions from their existing bus fleets, specifically for Guatemala and Honduras.

Of all three countries in the Northern Triangle, currently, El Salvador's government under the Bukele Administration seems the one to be willing to do something about the air pollution affecting its cities and people. Therefore, another recommendation is to follow what the government of El Salvador is doing to remedy the black carbon issue. It is urging the legislative assembly deputies to introduce and pass legislation that will phase out all buses older than 25 years and provide incentives to invest in emissions control systems and install it onto their vehicles. The current Bukele Administration has been in power since 2019 and is making visible, long term, and profound changes in the economy, educational, security, and foreign policy. Of course, with this comes also the willingness of the politicians to want to make a difference in their countries and not just look out for their own personal interests and of those financing their campaigns.

APPENDIX

	Guatemala	Honduras	El Salvador
Type of Government	Democratically Elected	Democratically Elected	Democratically Elected
Population	17.11 Million	10.28 Million	6.314 Million
GDP per capita	5,025.54 U$D	2771.72 U$D	4551.18 U$D
Land Area	108,890 km2	112,490 km2	21,040km2
Approach to Mitigating Black Carbon	Legislation	Legislation	Legislation

Source: CIA, 2023; World Bank, 2023.

REFERENCES

Almeida, M.D. (2021). Experiencias de política fiscal con contenido ambiental en países del Sistema de la Integración Centroamericana (SICA)/COSEFIN y recomendaciones de política pública [Fiscal policy experiences with environmental content in countries of the Central American Integration System (SICA)/COSEFIN and public policy recommendations. Project Documents (LC/TS.2021/9), Santiago, Economic Commission for Latin America and the Caribbean (ECLAC)]. *Documentos de Proyectos* (LC/TS.2021/9), Santiago, Comisión Económica para América Latina y el Caribe (CEPAL).

Ardila, A. (2006). Resumen técnico del sector transporte urbano en Centroamérica y Panamá [Technical summary of the urban transportation sector in Central America and Panama]. *Inter American Development Bank.*

Ardila, A. (2008). Limitation of competition in and for the public transportation market in developing countries: Lessons from latin american cities. *Transportation Research Record, 2048*(1), 8–15. https://doi.org/10.3141/2048-02

Awe, Y., Nygard, J., Larssen, S., Lee, H., Dulal, H., & Kanakia, R. (2015). *Clean air and healthy lungs: enhancing the World Bank's approach to air quality management.*

Balakrishnan, K., Steenland, K., Clasen, T., Chang, H., Johnson, M., Pillarisetti, A., & Peel, J. L. (2023). Exposure–response relationships for personal exposure to fine particulate matter (PM2· 5), carbon monoxide, and black carbon and birthweight: an observational analysis of the multicountry Household Air Pollution Intervention Network (HAPIN) trial. *The Lancet Planetary Health, 7*(5), e387-e396.

Banco Mundial. (2023). https://www.bancomundial.org/es/country/elsalvador

Beltrán, A. (2016). A new era of accountability in Guatemala? *Current History, 115*(778), 63–67.

Bond, T., Venkataraman, C., & Masera, O. (2004). Global atmospheric impacts of residential fuels. *Energy for Sustainable Development, 8*(3), 20–32.

Bond, T. C., Doherty, S. J., Fahey, D. W., Forster, P. M., Bernsten, T., DeAngelo, B. J., Flanner, M. G., Karcher, S., Koch, D., Kinne, S., Kondo, Y., Quinn, P. K., Sarofin, M. C., Schultz, M. G., Schulz, M., Venkataraman, C., Zhang, H., Zhang, S., Bellouin, N., ... Zender, C. S. (2013). Bounding the role of black carbon in the climate system: A scientific assessment. *Journal of Geophysical Research: Atmospheres, 118*, 5380– 5552. https://doi.org/10.1002/jgrd.50171

Central Intelligence Agency. (2023a). https://www.cia.gov/the-world-factbook/countries/guatemala/

Central Intelligence Agency. (2023b). https://www.cia.gov/the-world-factbook/countries/honduras/

Centre for Clean Air. (2022). https://doi.org/http://cleanairinstitute.org/Proyecto_tegucigalpa_honduras.php

Climate and Clean Air Coalition. (2014). *Time to act to reduce short-lived climate pollutants.* https://www.ccacoalition.org/en/content/short-lived-climate-pollutants-slcps

Cision PR Newswire. (2023). https://www.prnewswire.com/news-releases/gobierno-de-honduras-ejecuta-acciones-efectivas-contra-el-cambio-climatico-899180538.html

Cohn-Berger, G., & Quezada, M. (2016). Líquenes como bioindicadores de con-taminación aérea en el corredor metropolitano de la ciudad de Guatemala [Lichens as pollution bioindicators aerial in the metropolitan corridor of Guatemala City]. *Revista Científica, 26*(1), 20–39.

Cutz, L., Tomei, J., & Nogueira, L. A. H. (2020). Understanding the failures in developing domestic ethanol markets: Unpacking the ethanol paradox in Guatemala. *Energy Policy, 145*, 111769.

Dallmann, A., Simon, B., Duszczyk, M. M., Kooshapur, H., Pardi, A., Bermel, W., & Sattler, M. (2013). Efficient detection of hydrogen bonds in dynamic regions of RNA by sensitivity-optimized NMR pulse sequences. *Angewandte Chemie International Edition, 52*(40), 10487–10490.

Datos Mundial. (2023). https://www.datosmundial.com/america/el-salvador/volcanes.php

Dotson, R. (2014). *Mobility in crisis: Security, rights and responsibility on Guatemala city buses.*

Farzad, K., Khorsandi, B., Khorsandi, M., Bouamra, O., & Maknoon, R. (2020). A study of cardiorespiratory related mortality as a result of exposure to black carbon. *The Science of the Total Environment, 725*, 138422. https://doi.org/10.1016/j.scitotenv.2020.138422

Flores, W., & Rivers, M. (2020). *Curbing corruption after conflict: Anticorruption mobilization in Guatemala.* United States Institute of Peace.

Fridolin, B. R. (2019). Enabling underground transport construction Guatemala City, Central America. In D. Peila, G. Viggiani, & T. Celestino (Eds.), *Tunnels and underground cities: Engineering and innovation meet archaeology, architecture and art* (pp. 5100–5109). CRC Press.

Girón, A. I. S., & Campo, M. A. U. (2021). Modelación del ruido producido por tránsito vehicular en la ciudad de Guatemala [Modeling of the noise produced by vehicular traffic in Guatemala City]. *Revista Naturaleza, Sociedad y Ambiente, 8*(1), 41–53.

Gordon, T., & Webber, J. R. (2013). Post-coup Honduras: Latin America's corridor of reaction. *Historical Materialism, 21*(3), 16–56.

Grütter, J., & Grütter Consulting, A. G. (2014). Real world performance of hybrid and electric buses. *Renewable energy & energy efficiency promotion in international cooperation.*

The Guardian. (2019, January 24). *Zombie clunkers: Has your local bus been resurrected in Guatemala?* https://www.theguardian.com/cities/2019/jan/24/zombie-clunkers-has-your-local- bus-been-resurrected- in-guatemala

Interactive Country Fiches. (2023). https://dicf.unepgrid.ch/honduras/pollution

Kroeger, T. (2022). *The famous chicken buses of Guatemala.* https://www.guatemalatransportservice.com/chicken-bus-guatemala/

Lelieveld, J., Pozzer, A., Pöschl, U., Fnais, M., Haines, A., & Münzel, T. (2020). Loss of life expectancy from air pollution compared to other risk factors: A worldwide perspective. *Cardiovascular Research, 116*(11), 1910–1917.

Liu, S., Xing, J., Westervelt, D. M., Liu, S., Ding, D., Fiore, A. M., Kinney, P. L., Zhang, Y., He, M. Z., Zhang, H., Sahu, S. K., Zhang, F., Zhao, B., & Wang, S. (2021). Role of emission controls in reducing the 2050 climate change penalty for $PM_{2.5}$ in China. *Science Total Environment, 765,* 144338, https://doi .org/10.1016/j.scitotenv.2020.144338

Malkin, E. (2017). Guatemala president who championed honesty orders anticorruption panel chief out. *New York Times, 27.*

McDonald, R., Green, P., Balk, D., Fekete, B. M., Revenga, C., Todd, M., & Montgomery, M. (2011). Urban growth, climate change, and freshwater availability. *Proceedings of the National Academy of Sciences, 108*(15), 6312–6317.

Mcgranahan, G., Balk, D., & Anderson, B. (2007). The rising tide: Assessing the risks of climate change and human settlements in low elevation coastal zones. *Environment and Urbanization, 19*(1), 17–37. 10.1177/0956247807076960.

Office of the High Commissioner Human Rights. (2023a). https://www.ohchr.org/ en/countries/guatemala

Office of the High Commissioner of Human Rights. (2023b). https://www.ohchr .org/en/countries/honduras

Open Knowledge Report. (2023). *Country climate and development report.* https:// openknowledge.worldbank.org/entities/publication/cfe562d2-1a82-4e00 -9b1f-d925c33668c6/full

Organisation for Economic Cooperation and Development. (2023). https://www .oecd.org/environment/the-cost-of-air-pollution-9789264210448- en.htm

Ostro, B., & World Health Organization. (2004). *Outdoor air pollution: assessing the environmental burden of disease at national and local levels.* World Health Organization.

Rabatel, A., Francou, B., Soruco, A., Gomez, J., Cáceres, B., Ceballos, J., Basantes, R., Vuille, M., Sicart, J., Huggel, C., Scheel, M., Lejeune, Y., Arnaud, Y., Collet, M., Condom, T., Consoli, G., Favier, V., Jomelli, V., Galárraga-Sánchez, R., Wagnon, P. (2013). Current state of glaciers in the tropical Andes: A multi-century perspective on glacier evolution and climate change. *The Cryosphere, 7,* 81–102. https://doi.org/10.5194/tc-7-81-2013

Ramanathan, V., & Carmichael, G. (2008). Global and regional climate changes due to black carbon. *Nature geoscience, 1*(4), 221–227.

Riojas-Rodríguez, H., da Silva, A. S., Texcalac-Sangrador, J. L., & Moreno-Banda, G. L. (2016). Air pollution management and control in Latin America and the Caribbean: Implications for climate change. *Revista Panamericana de Salud Pública, 40,* 150–159.

Rivas, M. E., Suarez-Aleman, A., & Serebrisky, T. (2019). *Stylized urban transportation facts in Latin America and the Caribbean.* Inter-American Development Bank.

Rodríguez, G. (2018). *Fifth former president arrested on corruption charges in Guatemala.* https://digitalrepository.unm.edu/noticen/10507

Moser, S., & Ekstrom, J. (2010). A framework to diagnose barriers to climate change adaptation. *National Academy of Sciences of the United States of America, 10*(107). 22026-31. https://doi.org/10.1073/pnas.1007887107

Saarikoski, S., Niemi, J. V., Aurela, M., Pirjola, L., Kousa, A., Rönkkö, T., & Timonen, H. (2021). Sources of black carbon at residential and traffic environments obtained by two source apportionment methods. *Atmospheric Chemistry and Physics, 21*(19), 14851–14869.

Sánchez Girón, A. I., & Urrutia Campo, M. A. (2021). Modelación del ruido producido por tránsito vehicular en la ciudad de Guatemala [Modeling of the noise produced by vehicular traffic in Guatemala City]. *Revista Naturaleza, Sociedad Y Ambiente, 8*(1), 41–53. https://doi.org/10.37533/cunsurori.v8i1.6

Secretaria de Finanzas. (2015). *Guía Metodológica General para la Formulación y Evaluación de Programas y Proyectos de Inversión Pública* [General methodological guide for formulation and evaluation of public investment programs and projects]. Honduras.

Secretaria de Finanzas. (2019). *Análisis del marcaje de Presupuesto General de la República, Cambio Climático* [Analysis of the general budget marking of the republic, climate change]. Honduras.

Secretaria de Recursos Naturales y Ambientales. (2023). *Ley General del Ambiente* [General environmental law]. https://www.google.com/url?sa=t&rct=j&q=&esrc= s&source=web&cd=&ved=2ahU KEwi2ud3l_tr_AhVRjIkEHUfaCDo4HhAWe gQIDxAB&url=https%3A%2F%2Fportalunico.iaip.gob.hn%2Fportal%2Fver_ documento.php%3Fuid%3DNTk1NjM4OTM0NzYzNDg3MTI04NjE5ODcyMz Qy&usg=AOvVaw3KUl7-SXlSS- vPLu7MlfWs&opi=89978449

Shaddick, G., Thomas, M. L., Mudu, P., Ruggeri, G., & Gumy, S. (2020). Half the world's population are exposed to increasing air pollution. *NPJ Climate and Atmospheric Science, 3*(1), 23.

Timilsina, G. R., & Shrestha, A. (2009). Factors affecting transport sector CO_2 emissions growth in Latin American and Caribbean countries: An LMDI decomposition analysis. *International Journal of Energy Research, 33*(4), 396–414.

Tomczyk, S., McCracken, J. P., Contreras, C. L., Lopez, M. R., Bernart, C., Moir, J. C., & Verani, J. R. (2019). Factors associated with fatal cases of acute respiratory infection (ARI) among hospitalized patients in Guatemala. *BMC Public Health, 19*(1), 1–11.

UNESCO. (2023). https://siteal.iiep.unesco.org/pais/el_salvador#perfil-educacion

United States Agency on International Development. (2023). https://www.usaid. gov/honduras

United States Department of State. (2023). https://www.state.gov/countries-areas/ honduras/

Wang, Y., Shupler, M., Birch, A., Chu, Y. L., Jeronimo, M., Rangarajan, S., & PURE-AIR Study Investigators. (2022). Personal and household PM2.5 and black carbon exposure measures and respiratory symptoms in 8 low-and middle-income countries. *Environmental Research, 212*, 113430.

World Bank. (2020). https://www.worldbank.org/en/search?q=central+america&currentTab=1&label=1bc6 0e89-879d-4ff4-acae-deadd67919a8

World Bank. (2023). https://www.worldbank.org/en/country/guatemala/overview

The World Bank and the International Cryosphere Climate Initiative. (2013). *On thin ice: How cutting pollution can slow warming and save lives.* The World Bank Group. https://documents.worldbank.org/en/publication/documentsreports/ documentdetail /146561468180271158/main-report

Yang, J., Hoogh, M., Vienneau, D., Siemiatyck, J., Zins, M., Goldberg, M., Chen, J., Lequy, E., Jacquemin, B. (2021). Long-term exposure to black carbon and mortality: A 28-year follow-up of the GAZEL cohort. *Environmental International, 157.* https://doi.org/10.1016/j.envint.2021.106805

PART III

NATIONAL PERSPECTIVES

CHAPTER 7

CLIMATE GOVERNANCE IN DEVELOPMENT PERSPECTIVE

A Comparative Study of Ghana and Nicaragua

Peter F. Haruna
Texas A&M International University

Hugo Renderos
University of El Salvador-San Salvador

ABSTRACT

The UN's SDG13 urges governments to take action that combats climate change impacts. To do so, governments must pay attention to policy design and implementation. The chapter raises concerns about the political will and capacity of governments to act. Using Ghana and Nicaragua as case studies, public value governance theory is leveraged to examine their experiences with land, forest, and wildlife policy. The chapter questions the ability of 'post-colonial' states to manage natural resources as part of climate governance: Their climate policies align poorly with the structure of their respective societies, beliefs about nature, understandings of conservation and pres-

Climate Governance in International and Comparative Perspective, pages 131–155
Copyright © 2024 by Information Age Publishing
www.infoagepub.com
All rights of reproduction in any form reserved.

ervation practices, and their philosophy of work. We argue that the issue of political legitimacy requires that governance must transcend not only short-term materialistic motivations but also consider indigenous knowledge in designing climate governance. After analyzing and comparing national policies and programs in both countries, we discuss lessons that other countries in the Global South might find useful as they address the requirements of SDG13.

Climate change—global warming—is forecast to worsen and do irreversible damage to Planet Earth, if the world does nothing (Intergovernmental Panel on Climate Change—IPCC, 2022). At 1.5° C, the effect of global warming is real, including extreme weather events such as flooding, drought, deforestation, and degradation that sink millions of people into poverty and hunger. As part of global climate governance, the UN's SDG13 urges governments to combat climate change impact on ecological, social, and economic systems. Sadly, the climate discourse is narrowly tailored even as the World Meteorological Organization predicts with certainty that the annual mean temperature will exceed the 1.5° C threshold in 2023 (Tetrault-Farber, 2023).

The questions of what governments should do and how they should do it to fend off climate change impacts lies at the core of climate governance in Global South countries that, ironically, are the poorest in the world and most severely affected. For example, deforestation, degradation, and wild-fires in the Amazon and the Congo Basin raise questions whether governments in these regions have political will and ability to confront the loss of livelihoods and biodiversity and achieve sustainable development. Sadly, mainstream administration and governance literature barely covers the realities of life in Global South regions such as Africa, Latin America, and the Caribbean. As Sapiains et al (2020) argue, "Southern" perspectives are underrepresented and reflected in climate governance literature. As a result, what is known about climate governance is through the prism of Western notions about nature and environment without considering alternative non-Western ways of knowing and acting on public problems.

To shed light on climate governance in Africa and Latin America, the chapter raises questions about the UN's SDG13. The question we address is whether the post-colonial states in these regions, using liberal democracy and bureaucracy as theories for governing, can build climate adaptation and mitigation capacities to minimize vulnerabilities. Using Ghana and Nicaragua as case studies, we examine their experience with land, forest, and wildlife policies by leveraging public value governance theory (Bryson et al., 2014). The chapter questions the ability of the post-colonial states to manage natural resources equitably as part of climate governance. Based on inherited colonial knowledge, their policies align poorly with the structures of their respective societies, beliefs about nature, understandings of conservation and preservation practices, and philosophy of work (Gyekye, 1997). We argue that the issue of political and administrative legitimacy requires that climate governance transcend short-term materialistic motivations and consider embedding indigenous knowledge in designing climate governance.

To make the argument, the chapter is structured in four parts. The next section presents literature by exploring public value governance theory while the third analyzes and compares policies in both countries before extrapolating and discussing lessons learned. We conclude by highlighting implications for climate governance theory and practice in development settings.

LITERATURE REVIEW

Whether viewed through the prism of policy (Thobani, 2022), or climate governance (Sapiains et al., 2020), or sustainability (Banerjee, 2003), or decoloniality (Yacob-Haliso et al., 2021), values concerns are at the core of the debate on what governance should look like. As a concept, governance contrasts with the fading state-centered governing approach, signaling "a new paradigm in which citizens and states, governments and private sectors, organizations and citizens, form a web of relations" (Kim et al., 2005, p. 647). Challenging governments to be effective, responsive, transparent, and accountable, governance emphasizes the commitment to the rule of law. Over time, it has morphed into a multifaceted function and expectations with normative values including equity, participation, inclusiveness, decentralization, human rights, gender and racial equality, collaboration, growth, and performance (Grindle, 2004). The inherent values conflict and tensions apart, the demands of governance contained in the foregoing list are challenging to any Global North government (industrialized and high-income countries), not to talk about the Global South (non-industrialized, middle, and low-income countries).

In development settings, the governance challenge is even more daunting. Under the rubric of governance reform, expectations have multiplied, including "delivering quality services with fewer resources to diverse populations of users, partnering effectively with private and nonprofit sectors, responding flexibly and rapidly to shifts in demands and needs, assuring citizens safety and security, stimulating widespread and equitable economic growth and opportunity" (Brinkerhoff & Brinkerhoff, 2015, p. 22). Other scholars are concerned that governance principles provide little guidance in practice as they do not specify "what's essential and what's not, what should come first and what should follow, what can be achieved in the short term and what can only be achieved over the longer term, what is feasible and what is not" (Grindle, 2004, p. 525). In sum, governance-based reform tends to focus on form and rhetoric of reform rather than ground-up public management functions in the context of development.

Nonetheless, the governance literature remains values driven. Historically, pioneering scholars such as Wilson (1887) championed values that continue to define current research and scholarship. From his perspective, fact-value separation is important to assure that administration and governance are scientific, professional, and efficient. As he argues, "there should be a science of administration. . . . to make its business less unbusiness like. To strengthen and purify its organization, and to crown its duties with dutifulness" (p. 201). Rooted in efficiency as a key administrative value, Wilson's perspective provides a template to guide government operations and performance management, as well as a focus for research and scholarship in administration and governance (Light, 1997; Behn, 2003).

In view of the sustained focus on values, a body of work is emerging in mainstream administration and governance that Bryson et al. (2014) describe as "public value governance." This approach purports to move administration and governance beyond the traditional values of economy, efficiency, and effectiveness to include a diverse set of public values to address "current challenges and old shortcomings" (Bryson et al., 2014, p. 445). Comprising varied strands of literature, this perspective recognizes the role of government as guarantor, clarifies the meaning, and emphasizes the importance of concepts such as value, public value, values, and public values to governance. For example, Bozeman (2007, p. 13) conceptualizes public values as, "normative consensus about a) the rights, benefits and prerogatives to which citizens should (and should not) be entitled; b) the obligations of citizens to society, the state, and one another; and c) the principles on which governments and policies should be based."

On her part, Nabatchi (2017) considers public values as normative citizens' rights and obligations, as well as policy principles. Nabatchi argues that public values are foundational to the study and practice of administration and governance, and that the task of administration and governance involves resolving public values conflict and avoiding public values failure. Structuring public values away from familiar bureaucratic and democratic ethos, Nabatchi offers multiple public values frames, specifying liberty, equality, rights, equity, neutrality, merit, efficiency, innovation, and cost-effectiveness, among several others. Along similar lines, Rosenbloom and Piotrowski (2005) emphasize the protection of democratic values such as individual rights, citizen participation, and transparency from being overlooked in favor of bureaucratic and professional values.

At the center of scholarly debates about value, public value, values, and public values are questions about public value failure when both public and private sectors fail to provide goods and services needed to attain public value (Bozeman, 2007) and public sphere or democratic space encompassing "a web of values, places, organizations, rules, knowledge, and other cultural resources (Benington, 2011, p. 31). Alongside these issues, Bryson et al (2014, p. 448) raise multiple questions relevant to public value governance: "whether the objects of value are subjective psychological states or objective states of the world; whether value is intrinsic, extrinsic, or relational; whether something is valuable for its own sake or to something else; whether there are hierarchies of values; who does the valuing; how the valuing is done; and against what criteria the object of value is measured."

Consistent with the above critique of public value governance, scholars question the lack of attention to social equity, which Gooden (2015) describes as the nervous area of government. This is more so because the American Society for Public Administration's Code of Ethics identifies social equity as a core societal value. Principle 4 directs members to "treat all persons with fairness, justice, and equality and respect individual differences, rights, and freedoms. Promote affirmative action and other initiatives to reduce unfairness, injustice, and inequality in society." Seen through a variety of identities, Blessett et al (2019) argue that social equity lags in administration and governance practice, research, and teaching. They propose a social equity manifesto "to assist scholars and practitioners move beyond rhetorical acknowledgement toward a meaningful action to secure equitable outcomes for all people" (p. 296).

The issue of social equity frames the debate on what sustainable development should look like and imply for administration and governance. The UN's global agenda advocates "inclusive and sustainable economic growth in particular for developing states" (UN, 2015, para 21). Urging governments to take climate action, the agenda reiterates the need to end poverty and hunger, ensure healthy lives and promote well-being, pursue universal gender equality, create decent work, and reduce inequalities. The UN explains that inclusive development entails that every person has fair share and equitable access to resources and opportunities to earn a dignified living. Thus, at issue is how to adopt a development paradigm that prioritizes social equity and quality of life beyond only economic growth.

To address equity and the quality of life, the World Bank (2023, p. 30) explores the concept of social sustainability, which "increases when more people feel part of the development process and believe that they and their descendants will benefit from it." The Bank elaborates on social sustainability, highlighting cohesion, inclusion, resilience, and process legitimacy as the key components of sustainable communities and societies. Drawing on successful interventions in the Global South countries, the Bank offers evidence of the positive effects of the process legitimacy index (rule of law, control of corruption, accountability, and justice system). The evidence suggests strong correlations with "low poverty levels, higher GDP per capita, lower inequality, stronger human capital accumulation, and greater human development" (pp. 47–49). While arguing that process legitimacy contributes to building sustainable societies, the Bank recognizes the tensions that it entails. Reconciling competing interests and claims in making public decisions seen as fair poses challenges to governments, groups, and individuals.

While the focus on human wellbeing, equity, fair income, inclusion, environmental justice, and empowerment in administration and governance is clear in literature, there is no agreement on how to achieve them. As indicated above, multinational institutions such as the UN and World Bank advocate policy reform that ensures economic growth and neoliberal strategies. Much of the academic literature supports this approach, arguing in favor of green growth or a shift toward renewable energy and circular economy. For example, Hartwick and Peet (2003, p. 209) consider sustainability as the "effects most people can be persuaded to find tolerable, as the necessary environmental consequences of an even more necessary growth process." And Cohen (2023, p. iv) emphasizes that "we need to recognize what we know how to do and utilize our experience and brainpower to end poverty while protecting the Planet." Thus, this narrative entertains the possibility of bridging the divide between economic growth, human wellbeing, and environmental degradation (Adams, 2009).

However, the continued focus on growth as the means to achieve social equity and equality draws criticism from scholars emphasizing historical and structural inequities (Gooden, 2015). They agree that growth yields benefit but fails to distribute it fairly. Adams (2009) argues that growth "creates winners and losers and allows winners to carry their benefits between generations." He contends that wherever growth occurs, inequality increases alongside, perpetuating thereby the divide between economic growth and human wellbeing. Raworth (2017) argues the need to break from economic growth and create regenerative and distributive economies. Ecological economists focus on degrowth, arguing that it entails a planned contraction of economic activity aimed at enhancing human wellbeing and equality (Schmelzer, 2023). From the perspective of the Global South, Banerjee (2003) argues that sustainable development pursues an economic paradigm that impoverishes rural populations.

Nonetheless, growth and efficiency are the underlying themes of administration and governance as they have evolved since the twentieth century. Premised on the neoliberal economic model, the World Bank and International Monetary Fund-sponsored policies target efficiency based on economic growth theory (Williamson, 2004). Such policies emphasize competition, growth, and efficiency in administration and governance. Even *Poverty Reduction Strategy Paper Approach* that both the World Bank and International Monetary Fund adopt stipulates growing the economy (Levinsohn, 2002). But to the extent that they are implemented, such policies are harsh and anti-poor because of cuts in social welfare spending and the removal of subsidies designed to control prices of food and agricultural products (van de Walle, 2004). While good governance literature encompasses equity, participation, inclusiveness, decentralization, human rights, equality, collaboration, growth, and performance (Grindle, 2004), the focus remains on efficiency. The next section explores and evaluates Ghanaian and Nicaraguan experiences with natural resources management as climate governance.

COMPARING GHANA AND NICARAGUA

In this section, we present Ghana and Nicaragua as case studies, focusing on comparing both countries on dimensions of society, governance, and land, forest, and wildlife policy. In public administration, comparative studies provide insights and understanding of issues that are of concern in different countries and the profession. For example, Guljarani and Moloney (2012, p. 78) uncover "a false North-South administrative dichotomy and advance a vision for public administration as a global social science." In that study, theory is in the Global North with study subjects located in the Global South. In contrast, the current two-country study applies a comparative approach to academic engagement by locating theory in the Global South. Based on the issues specified in literature, we leverage research experiences to synthesize information about where both countries diverge, converge, and overlap, what they can learn from each other about climate governance, and how their experience can be of benefit to other developing countries and the fields of public administration and governance. We begin by highlighting the national contexts and cultures before launching the comparative analysis.

Climate Governance Contexts and Cultures

Nested in unique geographic regions within a globalizing world, Ghana and Nicaragua share development experiences that make comparison intellectually exciting and professionally insightful. What do they strive to achieve and how do they strive to achieve it? In sum, what values systems frame their development effort? While it is true that layers of differences including cultures, demographics, economics, and politics exist between them, we argue that despite those differences both countries face shared development, theory, technology, and climate change challenges. For, if governance is a human enterprise as Ford and Ihrke (2019) argue, Ghana and Nicaragua have more deep-rooted connections in common than meets the eye. Looking at both countries clinically enables the authors to glean information about who governs, with what purpose, and through what mechanisms, which is the necessary and starting point of comparing and analyzing climate governance in the varied contexts included in this book.

On the surface, Nicaragua is smaller in population size (6.6m) and land mass area (129,494 sq. km) but has a longer political history, having gained political independence from Spain in 1821. In addition to natural resources wealth, Nicaragua's influence revolves around the possible canal route through Central America (https://americanarchive.org/exhibits/newshour-cold-war/nicaragua). In contrast, Ghana is larger in population size (32.1m) and land mass area (238,537 sq. km), more demographically diverse, and much younger, having gained political independence from Britain in 1957. Its influence lies in having been a model British colony and having led Africa's liberation struggle (Young, 2012). But at the deeper level, while their colonial heritages are different, predominantly Spanish and British, those "ruptures" (Olivier de Sardan, 2009; Tignor, 1999) create marks in their histories and systems of governing. The "ruptures" reflect in governance and in "colonial epistemology" (Grosfoguel, 2007).

Nonetheless, both countries maintain connections to and practices of Indigenous political, social, and cultural belief systems that ground their societies and inform their ways of living, being, knowing, and acting on public problems. The Nicaraguan Constitution (1987) recognizes and protects diverse social and cultural practices among Indigenous groups including Mestizos, Chorotega, Miskitu, Cacaopera, Sutiaba, Mayangna, and Afro-descendant groups such as the Kriol and Garifuna. Likewise, the Ghanaian Constitution (1992) recognizes a multi-cultural society consisting of Akan, Ewe, Ga-Adangbe, and Mole-Dagbani speaking groups, among others. Knowledge of these groups along with their beliefs and practices is well documented and taught. For example, Ghana's political development and social progress include knowledge and teaching of its social structure from precolonial through colonial to postcolonial eras (Adu, 1965; Gyekye, 1997). Such knowledge systems as Grosfoguel (2007) argues, have a lot to contribute to decolonial theory beyond the Western canon of reproducing coloniality of knowledge.

To elaborate on the understanding of Indigenous knowledge, Table 7.1 summarizes information about Ghana and Nicaragua as rural and agriculturally based societies even though manufacturing, banking, and service industries are creeping in. In these kinds of societies, social relations—the sense that one is not alone—defines knowing and becoming with implications for how to organize and respond to public problems such as climate change. As theorists argue, in small-scale societies such as those of Ghana and Nicaragua, social relations discourage competition, are primary, originate from work, mutual orientations, and self-enforcing (Hummel, 1994; Suzman, 2020). In short, the texture of life is different and is based on

TABLE 7.1 Comparative Philosophical, Political, Economic, and Administrative Framework

Dimension	Ghana	Nicaragua
Ontology	Human Beingness	Human Beingness
Epistemology	Relational Humanism	Rational Existentialism
Human–Nature Relationship	Mutual Caring	Competition
Belief System	Supreme Being	Judeo Christian
Historical Background	Former British Colony	Former Spanish Colony
Nature of Society	Decentralized, rural (65%)	Centralized, urban (52%)
Nature of Modern Politics	Multiparty, Liberal, Democracy	Authoritarian, Presidential Republic, Multiparty
Nature of Economy	Mostly Agricultural	Agricultural, Light Manufacturing
Indigenous Economic Decision-Making	Cooperative Endeavor	Autonomous Regional
Indigenous Political Decision-Making	Consensus	Social Equality, Participatory
Administrative Decision-Making	Developmental Collaboration	Centralised, Hierarchical

Source: Compiled by authors (May–June 2023) based on review of literature, e.g., see Kwame Gyekye's (1997) conceptualization of African society in *Tradition and Modernity: Philosophical Reflections on the African Experience*, Oxford University Press.

human beingness, while political decision making is via collaboration and consensus.

In contrast, in urbanized and bureaucratized societies such as those of Europe, social relations are secondary, derive from top down, are available in job descriptions, and need enforcing. Thus, in this latter model of society, centralized and hierarchical systems dominate. The difference between both models of society is value-based: bottom up social relations v. top down rational-functional values (Schon, 1983). Over the period of self-governance, Ghana and Nicaragua alongside others in the Global South transplanted "universal bureaucratization" to their societies, especially in the recent wholesale adoption of liberal democracy as a theory of governing. The result of the imposition of that model, predictably, is the mismatch with their societies reflected in what one can describe as an excessive degree of authoritative centralization, careerism, and governance control (Cox, 2016). Under this circumstance, and as pathology of command-and-control theorists argue, governance entails objective, technical, and professional constructs with misalignment to and engagement with the world of experience. The outcomes in administration and governance include degradation and loss of local resilience (Scott, 1998).

To be sure, no one denies that Ghana and Nicaragua are making advances in science, technology, and medicine through modern civilization with bureaucratization at its spearhead. Nor is the issue that there are no hard working professionals dedicated to the bureaucratic ideals of rationality, neutrality, and predictability (Wilson, 1887). However, among colonialism's bequests to these countries—including the arbitrary demarcation of lands and forests, the creation of exploitative agro-export economies, and failure to nurture democratic traditions—is bureaucratized and disconnected administration and governance. For example, the agro-export economies in both countries are based on raw materials production for European markets, perpetuating cycles of inequality in land ownership and access to natural resources. The excessive growth of the administrative state (Waldo, 1948) in these countries consisting of the regulatory power of their national bureaucracies churns out and enforces thousands of rules that often exceed statutory authority. Hummel's (1994, p. 5) incisive critique of bureaucracy is relevant here: "If bureaucracy enables, what are the kind of social life, the kind of values, the kind of psyche, the kind of speaking and thinking, the kind of politics that are enabled?"

Vulnerability and Climate Governance

Given the above context and background, how do Ghana and Nicaragua approach climate governance, the concept that researchers, scholars, and policy makers mobilize to illuminate development thinking and practice in the era of climate change. To address this question, it is reasonable to briefly explore climate change risks and vulnerabilities, as well as national actions and outcomes before assessing their approach to climate governance. Located in regions with vulnerability to climate change impacts, both countries share challenges ranging from socio-economic volatility through institutional weaknesses to political conflicts. The regions face risks of climate and weather extremes with exacerbated vulnerability due to low finance, low adaptive capacity, dependence on rain-fed agriculture, and shrinking productivity (Intergovernmental Panel on Climate Change–IPCC, 2023). However, adaptation, especially ecosystem measures are beneficial in the regions but not new. As discussed below, climate governance in both countries follows a familiar pattern with slight variations in outcomes.

Ghana's vulnerability to climate change is multi-dimensional with economic, political, social, and ecological factors intersecting and mutually reinforcing. With an economy dependent on commodity markets, Ghana's fiscal environment is unstable. Plunging from 6.5% in 2019 to 1.7% growth in 2020, Ghana remains at high fiscal and debt risk (African Development Bank, 2021), making it difficult to meet the UN's SDG targets. Sadly, delays in reaching an agreement with creditors on restructuring external debt increase vulnerabilities and uncertainties, putting Ghana in a delicate fiscal predicament in the short-to-medium term (https://www.worldbank.org/en/country/ghana/overview). Like Ghana, Nicaragua has an externally dependent economy—over 80% of its GDP originates from outside. With its economic growth forecast to be 1.8% in 2023 along with increasing debt service, Nicaragua faces a challenging task of funding climate adaptation and mitigation initiatives even as ecosystem-based solutions are relatively low cost (https://www.thedialogue.org/wp-content/uploads/2023/02/Economic-Growth-Prospects-2023_FINAL.pdf; IPCC, 2023).

The political and social environments in both countries pose externally generated risks as well. While political power alternates between two political parties in Ghana (both of which recognize the need for climate action), climate change is ironically not prominent in their political agendas. Both parties support fossil fuel extraction and large scale mining driven by the global market without clarifying a policy direction for achieving a decarbonized economy (Ayelzuno & Yevugah, 2019). As Ghana prepares for US-style parliamentary and presidential elections in 2024 and as the political temperature rises, the risks involve hyping large-scale infrastructure projects rather than long-term climate policy planning. As far as the general mood of the country, there is simmering disquiet about living conditions as measured by European standards. For example, Afro barometer surveys (2019) indicate that majorities of Ghanaians believe the government is performing "fairly badly" or "very badly" in narrowing income gaps (66%), improving living standards of the poor (56%), and creating jobs (54%).

On the other hand, Nicaragua faces political tensions and conflicts of its own that pose risks to climate action. Plagued by the Somoza dictatorship and the Ortega-Murillo repressive regime, Nicaraguan democracy and economic policy appear criminalized based on extortion, expulsion and tax penalty (https://www.thedialogue.org/wp-content/uploads/2023/02/Economic-Growth-Prospects-2023_FINAL.pdf). As a result, 8.7% of the population left the country between 2019 and 2022. In fact, 15% of individuals that left the country between 2021 and 2022 are people with tertiary education in a country where less than 20% of the labor force has a tertiary education (InterAmerican Dialogue Survey of Migrants, 2022). While these challenges have different genealogies, they raise related questions such as whether in Ghana and Nicaragua the administrative state has the capacity to implement climate policy required by SDG13 and fend off climate impacts, especially on the most vulnerable in society. In the next section, we explore climate governance architecture, examine opportunities, and discuss natural resources management amidst evolving complex socioeconomic and political challenges.

In reviewing literature, the conception of climate governance in Ghana and Nicaragua derive from and/or are informed by Westernized notions of nature, environment, conservation, and preservation. This is not surprising because of the complexity of international and global relationships with multinational institutions in which the Global North dominates in writing the rules. The nature and influence of these relationships is the subject of literature emphasizing global public policy and transnational administration (Stone & Moloney, 2019). For example, within the environmental space alone, there are over five hundred conventions to most of which Ghana and Nicaragua are signatories. As Stone and Moloney argue, these conventions do not only inform governance paradigms but also broaden the kaleidoscopes of governance actors and scales beyond nation-states as social forms for the contemporary world.

As indicated above, the influence of multinational institutions dominated by the Global North is self-evident in Ghana, especially as it relates to institutional structure and function. For example, informed by the IPCC Assessment Reports, climate governance in Ghana focuses on evidence-based adaptation and mitigation strategies. Framed as a challenge to adapt by reducing vulnerability, Ghana's adaptation strategy is part of the broader national development policy framework as it evolved since the 1990s. The priority areas include sustainable development, poverty reduction, gender sensitivity, and inclusion. With support from multiple partners (UNEP, UNDP, African Union, African Capacity building Foundation, central government, sub-national governments, and civic and private sectors), the strategy aims to build institutional capacity, create public awareness, improve livelihoods, promote off-grid energy use, increase agricultural productivity, and protect water resources, to mention but a few (https://www.adaptation-undp.org/sites/default/files/downloads/ghana_national_climate_change_adaptation_strategy_nccas.pdf).

Ghana combines both a hierarchical and multi-level/multi-actor climate governance approach. Derived from European Union studies, this model emphasizes decentralized decision making as actors other than the state itself are engaged in tackling the complexity of climate change (Sattler et al., 2016). While recognizing the need for multi-actors at multi-levels, Ghana's climate governance approach is centralized and incorporated into administration and governance (see p. 15). Among climate lead agencies are the Ministry of Environment, Science, Technology, and Innovation and the Environmental Protection Agency with support from the Ghana National Climate Change Committee. Table 7.2 summarizes information about the focus of Ghana's climate governance. The Climate Action Tracker (2021) report indicates that the structures for vertical and horizontal coordination are in place

TABLE 7.2 Comparative Adaptation and Mitigation Governance Framework

Dimension	Ghana	Nicaragua
Governance Perspective	Purposive activities of actors	Subordinate-Institutional Maintenance
Climate Governance Approach	Multi-Actor and Multi-Level	Single actor, Multi-Level
Climate Governance Theory	Liberal Democratic Theory	Authoritarianism
Climate Governance Values	Efficiency, Effectiveness, Equality	Humanistic, Prosperity, Well Being
Adaptation Definition	Generic (e.g., Capacity, Resilience)	Specialized, Focused Based
Adaptation Approach	Infrastructure and Ecosystem-Based	Ecosystem Based
Mitigation Approach	Specific and Global (e.g., CO_2 Emission)	Local, Regional, National
Climate Governance Lead Agency	Environmental Protection Agency	Ministry of Natural Resources and the Environment
Mode of Climate Governance	Centralization, Top-Down	Centralized Decision Making
Implementation	Centralized and Mainstreamed	Decentralized, Conventional
Norms and Principles	Laws, Rules, Plans, Protocols, Participation	Laws, Executive Orders, Decrees, Mandates, Participatory

Source: Compiled by authors (May–June 2023) based on review of literature, e.g., Sapiains et al. (2021), Huitema et al. (2016), and Baninla et al. (2022).

and that the government provides sufficient allocations for them to perform the statutory tasks. In sum, climate governance involves multi-actors at multi-levels, but it is top down, emphasizing time-honored organizational functions such as planning, organizing, directing, and coordinating.

The Climate Action Tracker (2021) critiques Ghana based on political, institutional, policy process, and stakeholder engagement assessment of progress being made toward achieving targets under the Paris Agreement (2015). In addition to insufficient political commitment, Ghana lacks long term climate policy direction and comprehensive climate legislation covering all economic sectors and embedding all emission targets. Ghana also lacks a long-term 1.5° C compatible emission reduction target and relevant policies. Institutionally, ministries lack alignment and climate policies lack mainstreaming and sufficient distribution (https://climateactiontracker .org/documents/864/2021_08_CAT_Governance_Report_Ghana.pdf).

Underlying all these systemic weaknesses is resource inadequacy which Ghana must address by identifying barriers to accessing international funds and developing a plan to tackle them.

While the Climate Action Tracker's critique of Ghana has merit, it is not surprising and it is in fact predictable, especially in regard to the inability to access international climate funds. The critique is not unique to Ghana because it relates mostly to the problem of lack of human and technical capacity facing Global South countries, especially when it comes to the application of project-based models. The IPCC (2023) notes that annual finance flows targeting adaptation for Africa are billions of US dollars less than the lowest adaptation cost estimates for near-term climate change. And the UN recommends a shift away from the current project-based technical approach toward considering investment in sustainable long-term capacity building through program approaches. A complex and sensitive global issue, climate financing is the continuing subject of creating "Loss and Damage" fund discussed at length at *COP27* without a concrete resolution. It is also the focus of the recent summit for a *New Global Financing Pact* exploring financial opportunities to help low-income countries (Bosworth, 2023).

As well, while the institutional misalignment aspect of the critique is correct, it is a classic manifestation of bureau-pathology, which is not unique to Ghana either. While vertical and lateral coordination is needed for multi-actors to address positive and negative spillovers across jurisdictions, it remains an administrative dilemma (Hooghe & Mark, 2003). In fact, literature suggests coordination costs increase as more jurisdictions get involved in decision making (Benz et al., 2009) such as in Ghana's climate governance. But the Climate Action Tracker critique of Ghana misses broader and deeper societal challenges that we explore in the next section on natural resources policy and management as they intersect with livelihoods.

NATURAL RESOURCES POLICY AND MANAGEMENT

This section explores natural resources policy and management in Ghana and Nicaragua to compare experiences with and outcomes in the land, forest, and wildlife sector and amplify the nature of climate governance. This policy domain is receiving research attention as Baninla et al (2022) argue, highlighting vulnerability associated with agriculture, forest management, and energy sustainability. Both countries have a long history of natural resources policy, dating back to the colonial era but the policy and management frameworks remain unchanged—centralized, top down, and state-controlled with minimal indigenous knowledge input. In Ghana, literature reverberates with the themes of centralized control of land, forest, and wildlife policy and management under British colonial rule (Attuguayefio

& Fobil, 2005). Motivated by rising population growth in the country, the colonial government passed legislation promoting conservation, regulating forest sequestration, and creating Ghana's Forestry Department.

Although limited, the colonial Forest Act (1927) in Ghana provides conditions and procedures for creating forest reserves and clarifying land ownership within reserved areas. In addition, the colonial government took advantage of the *Convention for the Preservation of Wild Animals, Birds, and Fish in Africa* to regulate wildlife management. While the colonial Forestry Act (1948) broadened conservation by protecting river drainage zones, controlling woodland, and advancing forestry studies, its real objective focused on maximizing the harvest and export of forest products as a long-term exploitative economy (Oduro et al., 2011).

While Nicaragua has a different historical experience with land, forest, and wildlife policy and management, the outcome is like that of Ghana in terms of lack of consultation, public participation, and community engagement. Under the Somoza dynasty, land and forest policy and management focused narrowly on cotton production and export at the expense of

food crops production. Maldidier (2004) argues that between 1950 and 1980, cotton production expanded from 4,000 hectares to 85,000 hectares, making Nicaragua the fifteenth largest cotton producer in the world by 1975. The introduction of mono-crop agriculture had two negative consequences: On the one hand, small-holder crop farmers lost their land to large-scale cotton producers, unjustly depriving them of their livelihoods. On the other hand, arable land reduced by 50% (Marti, 2020), and deforestation hit due to large-scale cotton cultivation in Nicaragua's agricultural landscape (Matamoros-Chavez, 2014).

Like Nicaragua, centralized colonial land, forest, and wildlife policy and management had negative consequences in Ghana. While colonial land, forest, and wildlife policy formulation and management followed British civil service and governing tradition—emphasizing the values of loyalty, centralization, control, objectivity, and efficiency—its exclusive and centralized focus with minimal engagement with the public and communities raised questions for administration and governance. For example, Agyeman et al. (2007) argue that communities interpreted the forest reserve and conservation policy as land and forest resources confiscation tactic by the colonial government, resulting in resistance by the Aborigines' Rights Protection Society (ARPS). A local anti-colonialist organization, the ARPS protested colonial land legislation which threatened indigenous land tenure systems (Nti, 2013).

It is true that in the postcolonial era Ghana adopted policies with a broader scope in the search for sustainable land, forest, and natural resources policy and management, but such policies remain centralized and top down with minimum impact on deforestation and degradation. Largely state centered, policies and legislative frameworks, and natural resources management remain hierarchically focused on conservation. In fact, Ghana's Constitution (1992, Art. 36, 9) requires that "the State shall take appropriate measures needed to protect and safeguard the national environment for posterity; and shall seek cooperation with other states and bodies for purposes of protecting the wider international environment for mankind." But due to geo-political changes, Ghana now attempts to pursue a consultative and participatory policy process through devolution, and yet much remains to be done as discussed below.

Over the past three decades, Ghana has made steady shifts, fostering a new climate in the development of land, forest, and wildlife policy by authorizing citizen involvement in devolved flora and fauna governance. For example, both the 1994 and 2012 policy instruments emphasize multi-sectoral approaches to administering flora and fauna reserves, recognize the concerns of multi-forestry actors, promote partnered governance, and devolve forest functions (Government of Ghana, 2012). In addition, the policies authorize the creation of a fund to support community-based natural resources management through skill development, organizational capacity building, and motivating individual behavioral changes. This is in line with research findings indicating that natural resources managed by and/or together with communities show low levels of degradation (Addison et al., 2019). But uncertainties persist with narrowly tailored policies.

While pursuing a different approach, Nicaragua similarly witnessed land, forest, and wildlife policy shifts since Ortega gained power in 1979. At the core of the shifts is the creation of more environmentally friendly policies—banning the use of soil damaging pesticides and agrochemicals, creating an integrated pest management program, offering incentives to small-scale farmers, and collaborating with universities to conduct research on alternatives to harmful pesticides (Van Hecken et al., 2021; Maldidier, 2004). In addition, Ortega's policies authorized the creation of institutions and organizations responsible for restoring damaged lands, including the Centre on Cotton Experimentation, Institute for Natural Resources and the Environment, and the Tropical Agronomy Centre for Research and Training (Gonda, 2019). To promote workers' wellbeing, Ortega enacted legislation requiring plantation owners to provide protective gear to their workers and educate them about safe working conditions (Sylvander, 2021). Among others, the result of these initiatives is re-afforestation and the shift away from coffee and cotton production to soybean and sesame cultivation (Ministry of Finance and Public Credit, 2019).

It is true that both countries are striving toward improving land, forest, and wildlife policy formulation and implementation, but challenges remain. For example, Ghana's policy instruments lag deforestation and degradation effects. For example, Marfo (2010) and Osei-Mainoo (2012) assert that Ghana lost 1.6 million hectares of forest cover between 2000 and 2010. And the UN's Food and Agricultural Organization (2016a) reports a forest degradation rate of 45,931 ha per annum. In both countries even as policies improve afforestation, preservation, conservation, and the management of land, forest and wildlife, natural resource management programs deemed to be successful do not address climate vulnerability, unequal opportunities, and marginalization. As Baddianaah and Baaweh (2021) argue, this evaluation raises questions about natural resources management and sustainable development.

LESSONS LEARNED AND LEGITIMACY

While Ghana and Nicaragua have different histories, cultures, and governing systems, their approaches to climate governance are similar mainly in terms of how they domesticate rather than indigenize Western institutional framework and function. The focus is on multi-actors and multi-level of climate governance with centralization based on state control and the influence of multinational institutions and networks. As a result, climate governance is not indigenized as part of a long-term policy direction or strategy to exercise state authority in pursuing adaptation and mitigation measures to which natural resources policy and management can be linked. The policy institutional frameworks—inherited and transplanted from colonial governments—do not align well to the structure and philosophical basis of their diverse, decentralized, mostly rural and agricultural societies. Despite reform efforts, for example, to devolve natural resources management to communities in Ghana and to revamp and rebuild institutional capacity in Nicaragua, programs remain ungrounded in real life experience. What lessons can one extrapolate about climate governance and legitimacy from this analysis?

First, both countries do not promote the engagement of actors as co-equals in the public policy space, especially subnational jurisdictions and communities. By applying familiar but centralized and top down climate governance, people suffer from the pathology of command and control approach. To be sure, command and control enhances the values of technical efficiency and economy as literature suggests. However, and as Cox (2016) argues, command-and-control is also associated with several deficiencies—lack of fit with local social conditions, suppression of local social and ecological function, loss of resilience, and system degradation. Where public engagement occurs, if anything, it is through a centralized authoritative framework of technical professionals, using a government know-how approach that stifles learning about public values conflict from the ground up. Because of the lack of engagement, indigenous knowledge of and experience with environmental governance is not well understood and tapped into. As Ntiamoa-Baidu (1995) argues in the case of Ghana, cultural beliefs and the superstitions attached to them compel communities to live in harmony, not in competition, with nature. Such beliefs strengthen the links between the social and natural realms and facilitate community cooperation in natural resources management. But Ntiamoa-Baidu also argues that such beliefs are at odds with westernized thought, often perpetuating the taken-for-granted notion that knowledge only flows unidirectionally from the Global North to the Global South.

Second, the policy frameworks and programs exemplified in natural resources policy and management indicate that the administrative state in these countries lacks the political will to develop relational governance grounded in democratic principle and public service ethos that enhances compassionate treatment of the most vulnerable and underprivileged in society. Such a relational governance model is the subject of transformative public service (King & Stivers, 1998; Stivers, 2005). For example, King and Stivers (1998) suggest ways in which administrators and citizens should and must connect at the basic level of human existence to enhance legitimacy. Likewise, King and Zanetti (2005, p. xi) argue that public service can transform "institutions, practices, people's lives and experiences in a manner that serves democracy, engagement, and social and economic justice."

Third, both countries do not focus as much attention on exploring the trade-offs, synergies, and dilemmas entailed in managing competing demands in climate governance and development goals. For example, with its rapid population growth, increasing urbanization, high demand for forest products, farmland, mining, oil, and gas, Ghana needs a policy reappraisal. If the world decarbonizes as expected and fossil fuel prices decline, the pressure on its land use, agriculture, and forest products will worsen current deforestation and degradation levels. In short, reconciling inherent conflicts between ending poverty and hunger (SDG1 & 2) on the one hand and maintaining environmental sustainability (SDG11) on the other puts public values conflict at front and center of climate governance. Likewise, at the current levels of migration, Nicaragua must seek to reconcile creating decent work and maintaining economic growth (SDG 8) on the one hand and building peace, justice, and strong institutions (SDG16) on the other.

Finally, while both countries are scaling up social protection interventions to mitigate climate change impacts on their most vulnerable populations, maladaptation and mitigation deficits remain. For example, Table 7.3 shows the 2022 Bertelsmann Transformation Index, an objective measure for comparing vulnerability in Ghana and Nicaragua. The data indicate that on a scale of 1–10 both countries score below the mean on equal opportunity (4) and safety nets (4). They also score below the mean (4) on environmental policy, which lies at the heart of climate governance.

TABLE 7.3 Comparing Ghana and Nicaragua on Select 2022 Bertelsmann Transformation Indices (Scale: 1–10)						
Dimension	Policy Coordination	Structural Constraints	Governance Index	Environmental Policy	Equal Opportunity	Safety Nets	Conflict Intensity
Ghana	8	6	6.23	4	6	4	4
Nicaragua	5	7	2.45	4	5	4	5

Source: Compiled by authors and based on data available at: https://bti-project.org/en/reports/country-report/

These indicators, which capture social protection responses and gaps, cut across the UN's SDGs including no poverty (SDG1), zero hunger SDG2), good health (SDG3), quality education (SDG4), and reduced inequalities (SDG10). To be sure, the reality will differ within and between the countries due to finance availability and implementation capacity. But the majority of poor and food insecure people are likely to be small-scale farmers who face the combined effects of climate impacts, social conflicts, post-season losses, as well as market failures on agricultural production (Gentilini et al., 2020). In sum, both countries do not collaborate with communities to maintain lands and forests as the IPCC (2023) suggests.

SUMMARY AND CONCLUSION

The chapter focuses attention on Ghanaian and Nicaraguan experience with designing and implementing climate governance based on commitment to the Global South's agendas for inclusive development. Both countries are trapped in a paradox of climate action: They are required to address a climate problem they did not create, use resources they do not have, and apply external knowledge incompatible with their lived experience. And they are required to forgo the most cost effective way to development to solve a problem they have not created. It is equivalent to tying their hands behind their backs and demanding that they act on climate change under SDG13. The chapter contests this inequitable approach by arguing for the need to engage with communities based on insights from relational and values-based governance instead of managerialism and technicism. As Francis et al. (2020) argue, technocratic solutions are good, but they are unfeasible unless scholars address the contextual, social, and political forces.

The comparison between Ghana and Nicaragua suggests how to improve climate governance and strengthen legitimacy, which is at the core of the re-founding movement in public administration (Wamsley, 1990). Responding to and fighting climate change not only juxtaposes social with economic and ecological values, but also presents the opportunity to learn about time-tested indigenous ways of knowing and becoming. To do so requires bridging society and climate governance through community-based policy making that bolsters democracy and recognizes what Ajei (2018, p. 12) sees as the imperative of "trans-modernity." To date, there are no homegrown initiatives in either country to address climate change. From this perspective, the questions remaining are clear—how to adapt hegemonic objective and rational constructs to aspirations and actions anchored to ecology and well-being and how humanity can begin to rethink its relationship with nature (Raworth, 2017; Schmelzer at al., 2022).

CONFLICTS OF INTEREST

There are no competing interests to declare.

REFERENCES

Adams, W. (2009). *Green development: Environment and sustainability in a developing world* (3rd ed.). Routledge.

Addison, J., Stoeckl, N., Larson, S., Jarvis, D., & Aboriginal, B. (2019). The ability of community-based natural resource management to contribute to development as freedom and the role of access. *World Development, 120*, 91–104.

Adu, A. L. (1965). *The civil service in new African states.* Allen & Unwin.

African Capacity Building Foundation. (2021). 2021 *Annual report: Securing Africa's future through capacity development.* Retrieved April 21, 2023, from https://www.acbf-pact.org/sites/default/files/Annual%20Report%202021%20English.pdf

African Development Bank. (2021). *Annual report 2021.* Retrieved July 16, 2023, from https://www.afdb.org/en/documents/annual-report-2021

Africa Progress Panel. (2015). *Why climate challenges are an opportunity for Africa.* Retrieved July 15, 2021, from http://africaprogresspanel.org/why-climate-challenges-are-an-opportunity-for-africa/

African Union. (2013). *Africa agenda 2063 for transformation and change.* Retrieved April 21, 2023 from https://au.int/en/agenda2063/overview

African Union. (n.d.). *Draft Africa climate change strategy 2020–2030.* Retrieved April 21, 2023, from https://archive.uneca.org/sites/default/files/uploaded -documents/ACPC/2020/africa_climate_change_strategy_-_revised_draft_16 .10.2020.pdf

Afrobarometer Survey. (2019). Retrieved July 16, 2023, from https://www.afrobaro meter.org/countries/ghana/

Agyeman, V. K., Oduro, K. A., & Gyan, K. (2007). *Review of existing policy and legislative documents on definition of timber legality in Ghana.* VLTP Background Paper

No. 4. Validation of Legal Timber Program, Forestry Commission, Accra, Ghana.

Ajei, M. O. (2018). Trans-modernism and a Legon tradition of African philosophy. *Legon Journal of Humanities, 29*(2), 1–25.

Attuquayefio, D. K., & Fobil, J. N. (2005). an overview of biodiversity conservation in Ghana: Challenges and prospects. *West African Journal of Applied Ecology, 7*, 1–18.

Ayelazuno, J., & Yevugah, M. (2019). Large-scale mining and ecological imperialism in Africa: The politics of mining and conservation of the ecology in Ghana. *Journal of Political Ecology, 26*(1), 243–262.

Baddianaah, I., & Baaweh, L. (2021). The prospects of community-based natural resource management in Ghana: A case study of Zukpiri community resource management area. *Heliyon, 7*(10), 1–11.

Baninla, Y., Sharifi, A., Allam, Z., Tume, S., Gangtar, N., & George, N. (2022). An overview of climate change adaptation and mitigation research in Africa. *Frontiers in Climate.* https://doi.org/10.3389/fclim.2022.976427

Banerjee, S. B. (2003). Who sustains whose development? Sustainable development and the reinvention of nature. *Organization Studies, 4*(1), 143–180.

Benz, A., Breitmeier, H., Uchimank, U., & Simonis, G. (2009). *Politik in Mehrebenen-systemen* [Politics in Multi-level Systems]. Wiesbaden.

Behn, R. (2003). Why measure performance? Different purposes require different measures. *Public Administration Review, 63*(5), 586–606.

Benington, J. (2011). From private choice to public value? In J. Benington & M. Moore (Eds.), *Public value: Theory and practice* (pp. 31–51). Palgrave McMillan.

Blessett, B., Dodge, J., Edmond, B., Goerdel, H., Gooden, S., Headley, A., Riccucci, N., & Williams, B. (2019). Social equity in public administration: A call to action. *Perspectives on Public Management and Governance, 2*(4), 283–299.

Blewitt, J. (2018). *Understanding sustainable development.* Routledge.

Bozeman, B. (2007). *Public values and public interest: Counterbalancing economic individualism.* Georgetown University Press.

Brinkerhoff, D., & Brinkerhoff, J. (2015). Public sector management reform in developing countries: Perspectives beyond NPM orthodoxy. *Public Administration and Development, 35*(4), 222–237.

Bruere, H. (1912). *The new city government.* D. Appleton.

Bryson, J. M., Crosby, B., & Bloomberg, L. (2014). Public value governance: Moving beyond tradition public administration and the new public management. *Public Administration Review, 74*(4), 445–456.

Bosworth. (2023). *Motley's climate advocacy offers a new model of leadership.* https://www.worldpoliticsreview.com/mia-mottley-barbados-climate-change-financing-mitigation/?utm_source=Active+Subscribers&utm_campaign=282186dfea-daily-review-071023&utm_medium=email&utm_term=0_35c49cbd51-282186dfea-%5BLIST_EMAIL_ID%5D&mc_cid=282186dfea&mc_eid=6c86181c79

CABEI. (2019, March 18). *Bio-CLIMA Nicaragua: Integrated climate action for reduced deforestation and strengthened resilience in the BOSAWÁS and Rio San Juan Biosphere Reserves.* Central American Bank for Economic Integration, (Green Climate Fund).

Casolo, J., Flores Cruz, S., Gonda, N., & Nightingale, J. (2022). Choosing to "stay with the trouble": A gesture towards decolonial research praxis. *Undisciplined Environments, 50*(1), 1–27.

Climate Action Tracker. (2021). Retrieved July 16, 2023, from https://climateaction tracker.org/documents/864/2021_08_CAT_Governance_Report_Ghana .pdf

Cumes, A. (2012). Mujeres indígenas patriarcado y colonialismo: un desafío a la seg-regación comprensiva de las formas de dominio [Indigenous women patriar-chy and colonialism: a challenge to comprehensive segregation of domain forms]. *Anuario de Hojas de WARMI 17,* 1–16.

Cohen, S. (2023). *Environmentally sustainable growth: A pragmatic approach.* Columbia.

Cox, M. (2016). The pathology of command and control: A formal synthesis. *Ecology and Society, 21*(3), 33.

Economic Commission for Africa. (2015). Retrieved July 15, 2023, from https://www .nmun.org/assets/documents/conference-archives/new-york/2015/NY15_ BGG_ECA_Update.pdf

Economic Commission for Latin America. (2016). *Latin American economic outlook.* Retrieved July 15, 2023, from https://www.oecd-ilibrary.org/development/ latin-american-economic-outlook-2016_9789264246218-en

Food and Agricultural Organization. (2016a). *Ghana case study prepared for FAO as part of the State of the World's Forests 2016 (SOFO).* Retrieved May 26, 2023 from http://www.fao.org/documents/card/en/c/881a9917-a549-45b2-93f1 -f26ec9482e03/

Ford, M., & Ihrke, D. (2019). Perceptions are reality: A framework for understand-ing governance. *Administrative Theory & Praxis, 41*(2), 129–147.

Gentilini, U., Almenfi, M., Orton, I., & Dale, P. (2020). *Social protection and jobs re-sponses to COVID-19: A real-time review of country measures.* World Bank.

Gonda, N. (2019). Re-politicizing the gender and climate change debate: the poten-tial of feminist political ecology to engage with power in action in adaptation policies and projects in Nicaragua. *Geoforum, 106,* 87–96.

Gooden, S. (2015). *Race and social equity: A nervous area of government.* Routledge.

Government of Ghana. (1992). *The 1992 Constitution.* Government Printing Press.

Government of Ghana. (1993a). *Civil service law, PNDC Law 327.*

Government of Ghana. (1994). *Forest and wildlife policy.* Ministry of Lands and Forestry.

Government of Ghana. (1996). *Forestry development master plan 1996–2020.* Ministry of Lands and Forestry.

Government of Ghana. (2012). *Forest and wildlife policy.* Ministry of Lands and Forestry.

Government of Ghana. (2012). *National climate change adaptation strategy.* Retrieved July 16, 2023 from https://www.adaptation-undp.org/sites/default/files/ downloads/ghana_national_climate_change_adaptation_strategy_nccas.pdf

Government of Nicaragua. *The Constitution of Nicaragua.* U.S. Library of Congress.

Grindle, M. S. (2004). Good governance, R.I.P.: A critique and an alternative. *Gov-ernance, 30*(1), 17–22.

Grosfoguel, R. (2007). The epistemic decolonial turn: Beyond political economy paradigms. *Cultural Studies, 21*(2–3), 211–223.

Guljarani, N., & Moloney, K. (2012). Globalizing public administration: Today's research and tomorrow's agenda. *Public Administration Review, 72*(1), 78–86.

Gyekye, K. (1997). *Tradition and modernity: Philosophical reflections on the African experience.* Oxford University Press.

Hartwick, E., & Peet, R. (2003). Neoliberalism and nature: The case of the WTO. *The annals of the American academy of political and social science, 590*, pp. 188–211.

Holzer, M., & Ballard, A. (2021). *The public productivity and performance handbook* (3rd ed.). Taylor & Francis.

Hummel, R. (1994). *The bureaucratic experience: A critique of life in the modern organization.* St. Martin's Press.

InterAmerican Dialogue Survey of Migrants. (2022). Retrieved July 16, 2013, from https://www.thedialogue.org/analysis/?s-region=3644

IPCC. (2022). *Climate change 2022: Impacts, adaptation, and vulnerability, contribution of Working Group II to the Sixth Assessment Report of the Intergovernmental Panel on Climate Change.* Cambridge University Press.

IPCC. (2023). *Climate change 2023: Synthesis report.* Retrieved July 13, 2023, from https://www.ipcc.ch/report/ar6/syr/downloads/report/IPCC_AR6_SYR_SPM.pdf

Levinsohn, J. (2002). *The World Bank's poverty reduction strategy paper approach: Good marketing of good policy.* Paper prepared for the Group of 24 research program.

Maldidier, C. (2004). Agricultural pioneer fronts, the crest of a far-reaching wave: The social and spatial dimension of lowland colonization in Nicaragua. In D. Babin, (Ed.). *Beyond tropical deforestation: From tropical deforestation to forest cover dynamics and forest development* (pp. 185–199). UNESCO.

Marfo, E. (2010). *Chainsaw milling in Ghana: Context, drivers, and impacts*: Tropenbos International.

Martí Puig, S., & Serra, M. (2020). Nicaragua: De-democratization and regime crisis. *Latin American Politics and Society, 62*(2), 117–136.

Matamoros-Chávez, E. (2014). Micropolíticas de campesinos colonos en territorios indígenas de Nicaragua [Micropolitics of peasant settlers in indigenous territories of Nicaragua]. [Doctoral dissertation, The University of Texas at Austin]. The University of Texas.

Moloney, K., & Stone, D. (2019). Beyond the state: Global policy and transnational administration. *International Review of Public Policy, 1*(1), 104–118.

Nti, K. (2013). This is our land: Land, policy, resistance, and everyday life in colonial Southern Ghana, 1894-7. *Journal of Asian and African Studies, 48*(1), 3–15.

Olivier de Sardan, J. (2009). State bureaucracy and governance in francophone West Africa: An empirical diagnosis and historical perspective. In G. Blundo & P. Y. Le Meur (Eds.), *The governance daily life in Africa* (pp. 39–72). Brill.

Osei-Mainoo, D. (2012). *Assessing the contribution of collaborative forest management to the livelihoods of households in the Ashanti region.* Thesis. Kwame Nkrumah University of Science and Technology, Kumasi Ghana.

Nabatchi, T. (2018). Public values frames in administration and governance. *Perspectives on Public Management and Governance, 1*(1), 59–72.

Nicaraguan Ministry of Finance and Public Credit. (2019). *Bio-CLIMA Nicaragua: Integrated climate action for reduced deforestation and strengthened resilience in the BOSAWÁS and Rio San Juan Biosphere Reserves.* (Managua, Nicaragua).

Oduro, K., Acquah, B., Adu-Gyamfi, A., & Agyeman, V. (2012). *Ghana forest and wildlife handbook: A compendium of information about forests and wildlife resources, forestry-related issues and wood processing in Ghana.* Forest Commission. https://www.researchgate.net/publication/275967948_Ghana_Forest_and_Wildlife_Handbook_A_compendium_of_information_about_forests_and_wildlife_resources_fore stry-related_issues_and_wood_processing_in_Ghana.

Owusu-Ansah, N. (2020). Leading sustainability: Understanding leadership emergence in community resource management areas in Ghana. *Qual. Rep., 25*(7), 1766–1779.

Orosco, M. (2023). *Dimmed economic growth prospects in Nicaragua.* Retrieved July 13, 2023 from https://www.thedialogue.org/wp-content/uploads/2023/02/Economic-Growth-Prospects-2023_FINAL.pdf

Raworth, K. (2018). *Doughnut economics: Seven ways to think like a 21st century economist.* Amazon.

Sattler, C., Schroter, B., Meyer, A., Giersch, G., & Matzdorf, B. (2016). Multilevel governance in community-based environmental engagement: A case study comparison from Latin America. *Ecology and Society, 21*(4), 24.

Sapiains, R., Ibarra, C., Jimenez, G., O'Ryan, R., Blanco G., Moraga, P., & Rojas, M. (2021). Exploring the contours of climate governance: An interdisciplinary systematic literature review from a southern perspective. *Environmental Policy and Governance, 31*(1), 46–59.

Schon, D. (1983). *The reflective practitioner: How professionals think in action.* Basic Books.

Schmelzer, M. (2023). *"Green growth" won't prevent climate collapse: Degrowth might.* Retrieved May 7, 2023 from https://www.worldpoliticsreview.com/degrowth-movement-policies-climate-change-policies/

Scott, C. (1998). *Seeing like a state: How certain schemes to improve the human condition have failed.* Yale University Press.

Stone, D., & Moloney, K. (2019). The rise of global policy and transnational administration. In D. Stone & K. Moloney (Eds.), *Oxford handbook of global public policy and transnational administration.* https://doi.org/10.1093/oxfordhb/9780198758648.013.27

Suzman, J. (2020). *Work: A deep history, from the stone age to the age of robots.* Penguin.

Sylvander, N. (2021). Territorial cleansing for whom? Indigenous rights, conservation, and state territorialization in the Bosawas Biosphere Reserve, Nicaragua. *Journal of Physical, Human, and Regional Geosciences, 121*(3), 23–32.

Thobani, S. (2022). *The deadly intersections of COVID-19: Race, states, inequalities and global society.* Bristol University Press.

Tetrault-Farber, G. (2023). *Temperatures seen surging as El Nino weather pattern returns.* Retrieved July 4, 2023, from https://www.reuters.com/business/environment/temperatures-seen-surging-el-nino-weather-pattern-returns-wmo-2023-07-04/

Tignor, L. (1999). Colonial Africa through the lens of colonial Latin America. In J. Adelman (Ed.), *Colonial legacies: The problem of persistence in Latin American history.* Routledge.

United Nations. (n.d.). *Transforming our world: The 2030 agenda for sustainable development.* Retrieved May 5, 2023, from https://sdgs.un.org/2030agenda

United Nations. (2022). *Accessing climate finance: Challenges and opportunities for small island developing states.* Retrieved July 13, 2023, from https://www.un.org/ohrlls/sites/www.un.org.ohrlls/files/accessing_climate_finance_challenges_sids_report.pdf

Van De Walle, N. (2004). Economic reform: Patterns and constraints. In E. Gyimah-Boadi (Ed.), *Democratic reform in Africa: The quality of progress* (pp. 29–63). Lynne Reiner.

Van Hecken, G., Kolinjivadi, V., Huybrechs, F., Bastiaensen, J., & Merlet, P. (2021). Playing into the hands of the powerful: extracting "success" by mining for evidence in a payments for environmental services project in matiguás-río blanco, nicaragua. *Tropical Conservation Science, 14*(1), 1–8.

Waldo, D. (1948). *The administrative state: A study of the political theory of American public administration.* Ronald Press.

Wamsley, G. (1990). *Re-founding public administration.* Sage Publications.

Williamson, J. (2004). *A short history of the Washington consensus.* Paper presented at the conference on, From the Washington Consensus towards a new Global Governance, September 24–25, Barcelona, Spain.

Wilson, W. (1887). The study of administration. *Political Science Quarterly, 2*(2), 197–222.

World Bank. (2023). *Social sustainability in development: Meeting the challenges of the 21st century.* World Bank.

Yacob-Haliso, O., Nwogwugwu, N., & Ntiwunka, G. (2021). *African indigenous knowledges in a postcolonial world: Essays in honor of Toyin Falola.* Routledge.

Young, C. (2012). *The postcolonial state in Africa: Fifty years of independence, 1960–2010.* University of Wisconsin Press.

CHAPTER 8

PUBLIC–PRIVATE–COMMUNITY PARTNERSHIPS AS PATHWAYS FOR CLIMATE GOVERNANCE IN ZIMBABWE

Brighton M. Shoniwa
Women's University in Africa

ABSTRACT

This study advances the narrative that active participation of local communities is essential for sound climate governance and aligns with the paradigmatic shift from New Public Management (NPM) to New Public Governance (NPG). A qualitative methodological approach, which is an explorative and multi-district case study, examines the stakeholder perspectives on climate governance in Zimbabwe. The architecture for climate governance in Zimbabwe includes a myriad of policy and institutional frameworks, cascading from the national to the village level. Public-Private-Community Partnerships (PPCPs), which entice collaborative action and input of local communities, are relevant in the wake of complex and unprecedented climate extremes. Informed by literature, the Governance Network Theory (GNT), and experiences in selected Zimbabwean Districts, this study proposes a PPCP frame-

Climate Governance in International and Comparative Perspective, pages 157–183
Copyright © 2024 by Information Age Publishing
www.infoagepub.com
157

work for sound climate governance. In line with the NPG paradigm, commitment towards citizen engagement and society-centred policymaking in climate governance is essential.

According to Ross (2010), humanity's ecological footprint exceeded the earth's natural environmental limits in the 1980s and continued to rise. The planets' regenerative capacity is much lower than demand and the effects of this failure to protect and maintain the earth's resources are numerous, including climate change, loss of biodiversity, and failure to meet basic human needs. Because of this predicament, in 1987, the World Commission on Sustainable Development produced a report, titled 'Our Common Future', which offered an alternative paradigm for humanity, and brought, into the central focus, the issue of sustainable development. Sustainable development refers to the process of meeting the needs of the present without compromising the ability of future generations to meet their own needs (World Commission on Environment and Development, 1987). Similarly, Mensah (2019) argues that sustainable development is an approach to development, which uses resources in a way that allows them (the resources) to continue to exist for others in the future. In sustainable development, there is imposition of limits to the present generations so that there is preservation of natural resources. However, human action has led to climate change with complex, and unprecedented effects.

Mugambiwa and Rukema (2019) argue that climate change is one of the most complex problems facing the globe today. Climate change at international, regional, and local levels has posed as a major threat both to the environment and to the continued existence of humanity. Accordingly, there is a strong case for sound climate governance. In the same vein, and in line with the Sustainable Development Goals (SDGs), whose five themes are people, planet, prosperity, peace, and partnership, it is essential to ensure that 'no one should be left behind' in climate governance. More specifically, SDG number 13 seeks to protect the planet from degradation and take urgent action on climate change. The focus extends to ensuring that all people can enjoy prosperous and fulfilling lives, and that progress takes place in harmony with nature (United Nations, 2022).

According to Lipper et al. (2018), eradicating poverty, ending hunger, and taking urgent action to combat climate change and its impacts are three objectives the global community has committed to achieving by 2030 by adopting the sustainable development goals. However, as argued by the United Nations (2023), the world is on the brink of a climate catastrophe and current actions and plans to address the crisis are insufficient. In Africa, Latin America, and Caribbean (ALAC) extreme weather events make it difficult to sustain livelihoods. Accordingly, global developmental aspirations underline need for action, with the Sustainable Development Goals seeking to ensure prosperity in harmony with nature. In particular, Africa will be the most affected by the climate catastrophe if there is no immediate action. Africa, which is also the world's poorest region, is the continent most vulnerable to the impacts of projected change because of widespread poverty limits and adaptation capabilities (Post et al., 2021). More so, because most African countries have agro-based economies, they are more vulnerable to climate change impacts compared to the developed world (Mugambiwa & Rukema, 2019). This requires urgent attention in the discourse of climate governance. Worth noting is that African countries value environmental protection, and Aspiration Number 1 of African Union's Agenda 2063 envisions prosperity and protection of natural environment. Focus is also on the protection of natural endowments, environment and ecosystems. Africa must ensure equitable and sustainable use and management of water resources for socio-economic development, regional cooperation and the environment (African Union Commission, 2015).

The impacts of climate change are significantly adding to the development challenges of ensuring food security and poverty reduction in in Zimbabwe (Mugambiwa & Rukema, 2019). Therefore, Zimbabwe prioritises climate action, and Section 73 of the Constitution underscores the right to a safe environment. There is a cocktail of legal and institutional frameworks, with structures at national, provincial, and district levels. The multi-level approach ensures multi-stakeholder participation as the Zimbabwean government must collaborate with private sector and local communities. Active participation of local communities is essential for sound climate governance and aligns with the paradigmatic shift from New Public Management to New Public Governance. Accordingly, this study examines Public-Private-Community Partnerships (PPCPs) as pathways for climate change resilience and answers the following three research questions.

1. Is collective climate governance possible in Zimbabwe without hierarchies/inequities?
2. How do democratic and participative processes facilitate climate self-governance?
3. How do communities act as levelling mechanisms to assure equity in climate governance?

LITERATURE REVIEW AND THEORETICAL FRAMEWORK

There is review of related literature on concept of collective climate governance, and the relevance of PPCPs in climate governance. Focus is also on the role of the local community in climate governance. The Governance

Network Theory (GNT), which is part of the New Public Governance (NPG) Paradigm, is the guiding theory in this study.

THE CONCEPT OF COLLECTIVE CLIMATE GOVERNANCE

The meaning of governance is as contested as the meaning of Climate change itself. There is no general scholarly agreement as to what governance of climate change means. However, Mutambisi et al. (2021) assert that climate change governance is a concept that embraces inclusivity in designing mitigation and adaptation strategies by all climate stakeholders, including indigenous communities affected by climate change and it is this conceptual definition, which this study applies. Governance of climate change is multi-faceted and multi-level in nature due to the wide range of causes and consequences of climate change.

According to Post et al. (2021), tackling climate change is an extensive, time-consuming and costly task, which is multi-disciplinary in nature, and hence, is not possible to achieve solely through the policy implementation and regulation from central governments, and bodies alone. This therefore, warrants the extreme importance of engaging institutions and organisations from the national, sub-national to the local level. It is therefore, critical to enhance institutional architecture and the arrangements under which such institutions operate in climate change response. The United Nations Framework Convention on Climate Change, as cited by Whaley and Cleaver (2017), provides that all parties must formulate and implement national and regional programmes containing measures to facilitate adequate adaptation and mitigation to climate change. This formulation can only be possible when there is a clear understanding and appreciation of indigenous people's perceptions on climate change and the strategies for addressing it. Lankford et al. (2016) agitated that culture and development link in a number of different ways that influence sustainability. Furthermore, Post et al. (2021) stated that the local communities' ability to perceive climate change is a key precondition for their choice to adapt. It is essential to gain an understanding of traditional African perspectives on climate change.

As argued by Kashwan et al. (2020, p. 1), "We are in the middle of a planetary crisis that urgently requires stronger modes of earth system governance." Mutambisi et al. (2021) point out that climate change adaptation is a complex process that requires sound governance and efforts from different stakeholders. Stranadko (2022) states that climate change is a global environmental problem that involves multi-level scale, multi-actor involvement, multi-sector binding and vertical and horizontal dimensions of interactions in the global governance system. However, the various problems that currently overwhelm Africa, for example, poor governance, lack of technology, and prevalence of poverty, make it difficult to establish the best approaches to achieve climate change adaptation (Mutambisi et al., 2021). This study seeks to enhance climate governance practices through PPCPs, which entail active interaction among the public, private, and community actors. Complementary action by these tripartite actors could help address the problems isolated by Mutambisi et al (2021) that overwhelm African countries.

PUBLIC–PRIVATE–COMMUNITY PARTNERSHIPS AND CLIMATE SELF-GOVERNANCE

A partnership is a voluntary and collaborative relationship between various parties, both public and non-public, in which all participants agree to work together. Partnerships aim to achieve a common purpose or undertake a specific task and, as mutually agreed, to share risks and responsibilities, resources and benefits (Perez, 2015). Sathiyah (2013) states that a PPCP is a joint collaboration between different stakeholders from the governmental, commercial and community sectors in order to achieve a common goal. PPCPs are multi-stakeholder synergies, which are binding contractual relationships, either formal or informal, between the public sector, the private sector, and community-rooted initiatives (Perez, 2015). PPCPs delve on the democratisation of collaboration, communication, and decision-making, in which the relevant actors from the three sectors involved are included.

Multi-stakeholder partnerships initiate a dialogue aiming at building trust among different stakeholders that, by sharing resources, responsibilities, risks and benefits, become partners for the realisation of common objectives. In this regard, PPCPs seem to be gaining popularity in the field of sustainable development (Adnyana et al., 2015). The PPCPs are an extension of the PPPs, which did not include the community. The collaborative relationships between various actors in PPCPs are essential in the resolution of complex problems like climate change (Perez, 2015). Sound climate governance requires active interaction by the tripartite actors, that is, the government, private sector, and community.

PPCPs are also networks and are new forms of cooperation between public and private actors as well as platforms for citizen engagement and stakeholder participation (Koppenjan, 2015). In light of the significance of PPCPs in the resolution of contemporary challenges, they could be the answer to the threat of climate change. Contemporarily, networks are a fundamental characteristic of modern society and are in line with the multi-actor perspective of managing (Osborne et al., 2012; Opolski et al., 2013;

& Torfing et al., 2013). In a similar vein, Koppenjan (2015) argues that in networks the actors interact because actors are dependent upon each other. According to Dickinson (2016), the multiple actors or partners require each other's resources, for example, money, production means (land, capacity of personnel, and machinery), competences/authorities, information, knowledge/expertise, legitimacy/support, relations, and media access.

There is recognition of multi-stakeholder partnerships as vehicles for addressing complex problems and facilitating development. According to AtKisson (2015), partnerships have always been part of the global sustainable development movement, starting from the first United Nations Conference on Environment and Development, or 'Earth Summit,' in Rio de Janeiro in the year 1992. Nevertheless, the concept of partnerships as a recognised element of the global process of implementing sustainable development had its formal starting point in the year 2002, at the World Summit on Sustainable Development (WSSD), in Johannesburg. The voluntary multi-stakeholder partnerships are an important mechanism for the promotion and implementation of sustainable development (Neely et al., 2017). "The need to implement these ambitious, integrated goals simultaneously requires a fundamental shift in the way policies are developed and implemented" (Food and Agriculture Organisation, 2017, p. 37). Collaboration means working in strategic partnerships and developing funding mechanisms that will support harmonisation and synergies across sectors (Koppenjan, 2015). There was recognition of multi-stakeholder partnerships as key mechanisms for supporting effective policy planning, prioritisation, implementation, and outcomes (AtKisson, 2015).

The multi-stakeholder partnerships could be useful vehicle for enhancing sound climate governance and are in line with the Paris Agreement's forward-oriented and universal framework for enabling collective action (Sharma, 2017). In his presentation, the U.S. President Barack Obama, highlighted the vision for the Paris Agreement when he argued that, "here, in Paris, we can show the world what is possible when we come together, united in common effort and by a common purpose" (Obama, 2015, p. 1). In addition, the Paris Agreement stated that interdependencies within the climate context extend beyond country boundaries. No country can address climate change alone and sufficient mitigation will necessarily involve more actors than ever, including developing countries. Simultaneously, climate impacts are likely to cut across jurisdictions either directly or indirectly through other systems, such as migration and economic globalisation (Klinsky, 2018). Therefore, multi-stakeholder cooperation is necessary for sound climate governance, globally, and in Zimbabwe. The PPCPs draw their relevance for the fact that unity of purpose is key in addressing the complexities arising from climate change.

ROLE OF THE LOCAL COMMUNITY
IN CLIMATE GOVERNANCE

There are two contrasting approaches for stimulating sustainable development. The first approach is the 'blueprint' method, whereby the government takes a leading role in the quest to promote sustainable development. There was use of the 'blueprint' approach during the Classical Public Administration (CPA) era. The second method is the 'participatory' approach, in which there is the argument that 'development should be woven around the people and not people around development' (Chambers, 1995). The PPCPs are part of the participatory method because there is the involvement of three parties, that is, government, private sector (development partners) and the community. Government's main role is to establish the legal and institutional frameworks for the implementation of the PPCPs. The private partners provide capital and professional expertise while the community provide labour and indigenous knowledge (Maung, 2013).

Participation ensures mutual understanding, empowerment, transparency, accountability, and leads to inclusive solutions. According to Hodge and Greve (2007, p. 256), "the concept of participation in development activities is certainly not a new one, and in rural development, community participation has been recognised as an essential component since the early 1950s." Government's main role is to establish the legal and institutional frameworks for the implementation of the PPCPs. The private partners provide capital and professional expertise while the community provide labour and indigenous knowledge (Maung, 2013). Active interaction between the three parties (government, private sector, and community) could result in the political, economic, social, and environmental sustainability. In addition, the PPCPs base on the principle of collective responsibility and the idea that governments alone cannot facilitate development. Therefore, this study attempts to understand how active participation of local communities is essential for good climate governance in Zimbabwe.

THEORETICAL FRAMEWORK:
GOVERNANCE NETWORK THEORY

The feasibility of the PPCPs in climate change governance cannot be fully analysed outside the theoretical realm of the Governance Network Theory (GNT), which is part of the New Public Governance (NPG). According to Kanniainen (2017), the NPG emerged in 2000s and its global attention was by Osborne (2010), who presented it in contradiction with NPM and its predecessor, the Classical Public Administration (CPA).

The GNT does not seek to replace the previous paradigms like the NPM and CPA, but embeds them in a new context (Osborne et al., 2012). Hinging upon the fundamental principles of the NPG, the GNT underscores the need for emphasis on citizen and multi-stakeholder participation in the development processes of nation-states. The 21st century nation-state ought to forge networks and partnerships with the private entities in strategic sectors of the economy that include climate governance. Notions of governments as the sole decision makers and providers of public services are no longer valid as nation-states experience diminishing external and domestic financial support. In line with this argument, the GNT emphasises the resolution of complex problems like climate change through a multi-stakeholder approach.

Klijn and Koppenjan (2012) argue that the resolutions of complex problems confronting contemporary societies involve multiple actors. Operating in isolation, governments, businesses, and the community are often not able to tackle the complex societal challenges as they lack the resources or problem-solving capacities to do so (Osborne, 2010). The complexity of the contemporary challenges and interdependencies between actors result in intensive interactions between actors (Dickinson, 2016). As a result, governance networks emerge, which entail interrelated systems of enduring patterns of social relations between actors involved in dealing with a problem, policy, or public service (McQuaid, 2010).

Additionally, Klijn and Koppenjan (2012) argue that the GNT could deal with the complexities, interdependencies and dynamics of public problem solving and service delivery, which NPM failed to address. The application of the GNT ensures that there is multi-stakeholder collaboration in developmental efforts. Therefore, attaining food security, which is the central objective of this study, could be possible through governance networks. According to Neely et al. (2017), multi-stakeholder collaboration can contribute towards cross-sectorial coordination by bringing diverse actors from civil society, private sector and government (including multiple ministries in some cases). The partnerships aim to enhance information exchange, increase possibilities for negotiation and advocacy and inclusive decision-making across sectors, all of which can enhance governance. It is worthy to note that development is about people and there is need to ensure participation of all stakeholders in any intervention.

The GNT is one of the alternatives for ensuring that there is effective participation of community in developmental activities. According to Vangen and Huxham (2010), participation is more of a facilitator than a hindrance to development. Community participation is ideal because development must be 'woven' around people and not people around development. Therefore, participation has emerged as a concept that will not only change the nature and direction of development interventions but lead to a type of development, which is more respectful of poor people's position and interests (Pestoff and Brandsen, 2010). Participation, enhanced through PPCPs, has now become the hallmark of sustainable development, with a general shift from prescriptive 'top-down' to participatory 'bottom-up' approaches to development (Vangen and Huxham, 2010). Development initiatives should be from bottom up, to ensure ownership, and collaboration with new groups hitherto neglected in the development process. Putting people at the centre, through PPCPs, means letting the beneficiaries of development projects and programmes identify their needs, initiate and plan interventions, and seek assistance from interested partners.

METHODOLOGY

This study applied qualitative and explorative research because climate change governance has and continues to be an area of debate as to how best to implement it. Issues pertaining climate change mitigation and adaptation always ignite debates and variations of viewpoints. Accordingly, an examination of public–private–community partnerships as pathways for sound climate governance in Zimbabwe was not possible without reference to qualitative research, which is a flexible approach. In line with Figure 8.1, the research was multi-district case study, across three of Zimbabwe's Provinces that have a high potential for agricultural production. The population comprised multiple stakeholders in the area of climate change governance in Zimbabwe.

There was use of multiple sampling techniques, namely quota (there are categories of stakeholders), purposive, and snowball sampling techniques. Sampling was in phases, and at each stage, there was use of an appropriate method. In the first instance, there was use of purposive sampling to identify the stakeholder institutions involved in climate change governance. There was also use of snowball or chain sampling when identifying some of the stakeholders, for instance, in some districts the Agricultural Advisory Services (ADS) and the Environment Management Agency (EMA) helped to identify climate change champions.

The purposive sampling method (expert-based) was used to select the knowledgeable resource persons within the stakeholder entities. For instance, in the Ministry of Lands, Agriculture, Fisheries, Water, and Rural Development (MLAFWRD), there was selection of the ADS, which have close interactions with the farmers. In the Ministry of Environment, Tourism, and Natural Resource Management (METNRM), there was selection of EMA, which has a mandate for environment protection in Zimbabwe. The choice of the four provinces (Mashonaland East, Mashonaland Central, Mashonaland West, and Manicaland), as well as the 12 districts was

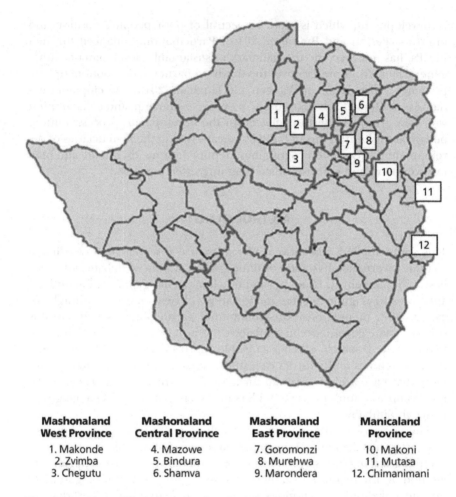

Mashonaland West Province	Mashonaland Central Province	Mashonaland East Province	Manicaland Province
1. Makonde	4. Mazowe	7. Goromonzi	10. Makoni
2. Zvimba	5. Bindura	8. Murehwa	11. Mutasa
3. Chegutu	6. Shamva	9. Marondera	12. Chimanimani

Figure 8.1   Map of Zimbabwe showing the selected districts.

based on their high potential for rain-fed agricultural production, as guided by the ADS. The selected government ministries also have structures, devolved down to the ward level, which guided the selection of the key informants. There was also selection of highly active community-based organisations who partake in climate action, for example, Chimanimani District provided two of such entities, namely Towards Sustainable Use of Resources Organisation (TSURO), and Chikukwa Ecological Land Use Community Trust (CELUCT).

Using multiple tools, the data collection extended over a period of six months, from October 2022 to March 2023, and there was use of multiple data collection tools. There was purposive selection of knowledgeable key informants drawn from the multi-entities in the climate governance sector and domiciled at the national, provincial, and district levels. The saturation principle helped to determine the number of key informant interviews, implying that there was discontinuation of data collection when no new insights were coming from the participants. In addition, in each of the selected districts, as indicated in Figure 8.1, the ADS and EMA helped to identify a team of 12 climate change champions, who participated in a community focus group discussion (FGD). The study ensured gender balance and each community FGD comprised an equal number of males and females.

There was the adoption of a flexible model for analysing qualitative data. There was basic analysis during data collection (documentary review, interviews, and FGDs). After data collection, qualitative data analysis was thematic, in which there was grouping of the related findings. The method is a qualitative analytical method for identifying, analysing, and reporting patterns (themes) within the data. Thematic analysis organises and describes the data into sets.

FINDINGS AND DISCUSSIONS

There is presentation and discussion of findings in line with three themes. First, focus is on the architecture of collective climate governance in Zimbabwe. Second, there is discussion on the relevance of participative processes in climate self-governance in Zimbabwe. Finally, there is presentation on Zimbabwean communities as levelling mechanisms for ensuring equity in climate governance.

COLLECTIVE CLIMATE GOVERNANCE IN ZIMBABWE

This study showed that the governance of climate change in Zimbabwe is informed by the principles of decentralisation and autonomy, accountability and transparency, responsiveness, flexibility, participation, and inclusion. The institutional architecture for climate change governance in Zimbabwe includes the policy and legislative framework as well as various organisations. The Constitution of Zimbabwe (2013) is the supreme Law of the Land and it gives every person environmental rights. The Constitution of Zimbabwe therefore, gives the legal basis for all environmental legislation and climate governance. Policies include the Climate Change Response Strategy, the National Climate Change Policy, the Interim Poverty Reduction Strategy Paper, the Intended Nationally Determined Contribution and the National Adaptation Plan among others.

One of the key informants from the Agricultural Advisory Services pointed out that Zimbabwe has a multi-stakeholder committee that deals with climate change issues at the national level. It is comprises members from various government ministries, civil society actors, private sector and academia. Represented ministries include the Ministry of Environment, Tourism, and Natural Resource Management (METNRM), Ministry of Lands, Agriculture, Fisheries, Water, and Rural Development (MLAFWRD), and the Ministry of Energy and Power Development. The National Climate Change Committee coordinates all climate change related activities at the national level. Further to that, the committee provides a forum for exchange of views on climate change issues, consensus building on national positions and advice to the government ministries when required. Other national institutions, universities, research organisations, industry associations and non-governmental organisations can also provide technical inputs when required. The transformation of the National Climate Change Office into a full-fledged National Climate Change Department in the year 2013 demonstrates the commitment of the government of Zimbabwe to set up a robust institutional framework for climate change governance. The primary role of the Department is to assist the government in designing climate change policies and coordination of specific national climate change projects such as the compilation of national inventories of greenhouse gases. The Department also coordinates other climate change activities between various ministries and organisations, including the private sector, and functions as the secretariat for the Clean Development Mechanism (CDM) in Zimbabwe.

Moreover, the agricultural institutions are an integral part of the institutional architecture for climate change governance. The ADS is the lead department primarily responsible for increased and sustainable agricultural production and the provision of appropriate technical, professional and other support services to the agricultural industry. Further to that, ADS provides regulatory, advisory and technical services, farmer training, food technology and dissemination of technologies. In the process of executing their mandate, ADS staff interacts with various facets of climate change governance including regulation, communication and training on climate smart technologies.

Various forms of non-governmental organisations including local and international organizations, participate in climate change governance in a variety of ways. Climate change activities in Zimbabwe are also carried out in close cooperation with a number of research organisations and non-governmental organisations, among them the United Nations (UN) agencies. Besides the UN agencies, some of the prominent non-governmental organisations (NGOs) in the climate governance fraternity include Zimbabwe Environmental Law Association (ZELA), Care International, World Vision, and SNV among others. These organisations are usually active in the formulation and implementation of climate change adaptation although they participate in policy-making processes through advocacy initiatives and provision of funding. There has been effective participation of NGOs in the development of climate change policy, and in defining the country's position on climate change in international negotiations. There are also community-based organisations, and traditional leaders who play an active role in climate governance. Thus, Zimbabwe's climate governance architecture has diverse institutions that contribute differently to climate governance, as shown in Figure 8.2.

Understanding the political economy of multi-level governance approaches helps to break down the state centeredness of institutions and

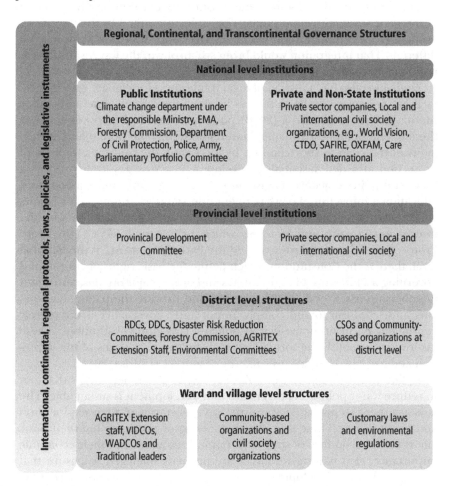

Figure 8.2   Multi-level governance approach to climate change. *Source:* Author improvements, Adapted from Chinyanga (2017) and Musarandega (2019)

improves the characterisation of relationships between different actors both horizontally and vertically. The vertical dimension of multi-level governance recognises that national governments cannot effectively implement national climate adaptation and mitigation strategies without working closely with sub-regional and local governments as the agents of change. These should align to the overall global frameworks that govern climate change at international level including the United Nations Framework Convention on Climate Change (UNFCCC). On the other hand, local authorities will also have limited effectiveness if they operate without the assistance and guidance of central governments. Central governments provide the legal and institutional frameworks within which local governments and local authorities nest. Horizontal governance ensures that institutions that operate at the same level must have similar objectives as defined by the governing framework even though their working spaces and strategic thrusts may be different. That integration would bring synchrony in the way institutions at regional, national, and local level operate.

Zimbabwe also has a myriad of legal frameworks for climate governance. Legislation imposes limits on the way in which the present generation enjoy the natural resources. Ross (2010) points out that legislation directed at the implementation of sustainable development could potentially address many of the current shortcomings by increasing the priority, support and protection afforded sustainable development across government(s) as a long-term policy objective. Legislation also have a significant symbolic and educational impact in making people understand what is at stake. The legal framework for sustainable development in Zimbabwe is in line with the international treaties and protocols, which include the UNFCCC, the Kyoto Protocol, the Montreal Protocol, and the Paris Agreement. At the national level, there is the Constitution, which define the basic rights of citizens and governing architecture of nation-states and other major organisational and membership-based bodies, are supreme and provide the parameters that laws must conform to and comply. Importantly, governance is not merely contingent on institutions such as policies, laws, and constitutions. These institutions establish the formal 'rules of the game' in terms of the definition and exercise of rights over natural resources (Chirisa, 2013).

In line with the Constitution, Zimbabwe has a number of laws that aim to reduce water pollution and ensure that development is sustainable. The laws are there to manage natural resources in terms of their exploitation and conservation. These laws can be enabling or restrictive but their basic aim is to ensure in face of scarcity of natural resources and their economic importance, that natural resources are properly managed to ensure that they contribute to development process of the state and its populace. One of the laws is the Environmental Management Act, which gives the Environment Management Agency (EMA) the mandate is to supervise over natural resources usage and conservation. Nevertheless, the laws, on their own, are not effective as there is the need for enforcement of the laws through policy instruments. According to Salaman (2002), a policy instrument refers the means of government intervention in society in order to accomplish goals or to solve problems. The behavioural assumption underlying a policy instrument is that it attempts to get people do things that they might not otherwise have done. In short, policy instruments are tools at the disposal of government for public action or the mechanisms for dealing with public problems (problems like environmental degradation). Governments have several types of policy instruments at their disposal. Traditionally, three main types of policy instruments are distinguishable in the policy instrument theories and these are regulation instruments (sticks), economic policy instruments (carrots) and information instruments (sermons). Regulations are laws that should be complied with, failure of which sanctions or penalties follow (Stone, 2001). The economic instruments (carrots) are the rewards that help to reinforce desirable behaviour. In Zimbabwe, there are economic instruments for promoting the use of renewable energy sources, for example allowing free importation of solar equipment. According to Endres (2011), to stimulate the diffusion of new and less polluting technologies, governments frequently use these financial incentives. The informative instruments (sermons) are also knowledge tools. They mainly aim at providing information to change or encourage certain behaviour. Sermons or soft policy instruments rely on cooperation and reduce the need for coercive action (Bax, 2011). Sermons can be in the form of advertisements, leaflets, and posters, for examples advertisements encouraging people to prevent water pollution.

The motivation for having a multi-level and multi-stakeholder framework climate governance in Zimbabwe is to enhance success because the concept of sustainable development, which practitioners seek to attain, is elusive. Mensah (2019) points out that notwithstanding its pervasiveness and popularity, murmurs of disenchantment about the concept of sustainable development are rife as people continue to ask questions about its meaning, the feasible pathways, and the implications for development theory and practice. According to Dernbach and Mintz (2011), sustainable development provides a framework for humans to live and prosper in harmony with nature rather than living, as they have done for centuries, at nature's expense. Nonetheless, sustainability lacks an adequate or supportive legal foundation, in spite of the many environmental and natural resources laws that exist. PPCPs, because of their community-focused and multi-stakeholder approach, could help to enhance the formulation and implementation of sound governance practices to attain sustainable development.

THE RELEVANCE OF PARTICIPATIVE PROCESSES
IN CLIMATE SELF-GOVERNANCE IN ZIMBABWE

In the agriculture, there are complex risks, which require the active inter-action of multiple actors. More so, almost all the interviewees in this study pointed out that there is recognition of multi-stakeholder partnerships as vehicles for addressing complex problems and facilitating development, and this pushes for the adoption of PPCPs in the agricultural sector. One of the agricultural experts interviewed argued that the challenges in the Zimbabwean climate and agricultural sectors require active interaction by the tripartite actors, that is, the government, private sector, and community.

Moreover, the COVID-19 pandemic wreaked havoc throughout the world, and response action came from multiple stakeholders. Therefore, collaborative action is necessary in all sectors, including in agriculture, and PPCPs are a feasible financing option. The main theory guiding this re-search, the GNT, emphasises the resolution of complex problems through a multi-stakeholder approach. Therefore, sound climate change gover-nance in Zimbabwe could be through PPCPs. Literature agrees with this finding, and Klijn and Koppenjan (2012) argue that the resolutions of com-plex problems that are encountered in the contemporary involve multiple actors. Osborne (2010) also pointed out that operating in isolation, gov-ernments, businesses, and the community are often not able to tackle the complex societal challenges as they lack the resources or problem-solving capacities to do so. The complexity of the contemporary challenges and interdependencies between actors result in intensive interactions between actors (Dickinson, 2016).

All the participants in this study shared the same sentiment that multi-stakeholder collaborations were essential for sound climate governance. One of the participants from Chimanimani Rural District Council (RDC) argued that collaborations enhance access to finance, they enhance access to mod-ern technology and equipment and they enhance transparency because of involvement of multiple stakeholders. Moreover, multi-stakeholder collabo-rations allow for private companies to participate in national development, and enhances development of good water infrastructure The findings of this study agrees with literature as Post et al (2021) argue that collaborations are promoted as the most logical solution to a variety of service delivery and de-velopment problems, and are often presented as 'technical', politically neu-tral solutions. There is application of the principle of collective responsibility and the idea that governments alone cannot facilitate development.

According to the United Nations (2022), the challenges confronting the contemporary world are complex and unprecedented. A multi-stakeholder approach could help in the addressing the complex challenges (Mensah, 2019). There are a number of preconditions for sound management of water resources, including transparency and accountability through participatory mechanisms appropriate to local realities, needs and wishes. Some of the conditions for the success of multi-stakeholder collaborations include a clear and comprehensive legal framework (Hearne, 2009). In a similar vein, Post et al. (2021) argue that there is need for a country to create a policy framework for multi-stakeholder activity and regulation. Moreover, on the onset, there must be a clear definition of the partners' roles based on evaluation of, character, behaviour, education, background and skills. The diverse actors must be guided by well-defined objectives, clear division of roles and responsibilities, risk allocation, and other transaction elements (which asset changes hands under what provisions), to be agreed upon between the partners in advance. In addition, there is need for capacity building of all the actors so that they perform their roles.

ZIMBABWEAN COMMUNITIES AS LEVELLING MECHANISMS FOR ENSURING EQUITY IN CLIMATE GOVERNANCE

This study showed that eliciting the help and input of local communities is beneficial. As the main consumers or the end users, the local community may have in their possession local knowledge or information that would enable the partnerships to place its skills and finances into appropriate developments. In addition, the emergence of PPCPs is in line with prescriptions of the NPG paradigm, which emphasises the principle of collective responsibility and the idea that governments alone cannot facilitate development. The multi-stakeholder partnerships aim to enhance information exchange, increase possibilities for negotiation and advocacy and inclusive decision-making across sectors, all of which can enhance good governance. It is worthy to note that development is about people and there is need to ensure participation of all stakeholders in any intervention.

In a case study of Tsholotsho District, Bowora (2017) found that whilst there were considerable variations in the extent of community collective action, there are some striking examples of self-help initiatives, with locals making and fitting their own spare parts or collecting money to buy new ones. Categories of community collective action can also be in the form of regulatory tasks of the water points using the water supply system. Although the rules and regulations tend to be informal and unwritten, they are well known and enforceable within the community. The regulations on the use of water resources also include the establishment of conflict management strategies. In one village, for example, the Water Committee decided on the rules, which if any user was to breach, he or she appears before a village court (Bowora, 2017).

The need for increased community participation in climate governance in Zimbabwean communities has become a central element of policy. The involvement of villagers in sustainable development is an important part of collective community action and a determining factor of their subsequent capacity for maintenance. The collective action begins in the Ward, with the initial contacting of the community through the Ward Councillor, Ward Development Committees (WADCO), Village Development Committees (VIDCOs) and the traditional leaders. A Ward mobilisation meeting discusses the needs, priorities and capabilities of the ward in relation to the development of its water supply and other natural resources.

These duties are set out in a standard list of roles and constitutional provisions for the functioning of the Water Committee. The committee is theoretically subservient to the VIDCO from which it derives its authority. The Water Committee holds regular meetings and submits reports on its activities to the VIDCO. It is the responsibility of the secretary to keep records of these meetings and activities of any maintenance requirements relating to the borehole. After two years, there are elections to select a new committee. The system of community collective action is, therefore, in accord with the borehole sinking and development programme, making it difficult to isolate aspects purely relating to maintenance. Bowora (2017) observe that, to manage domestic water supplies in rural Zimbabwe, communities must also be willing to bear the cost of engaging independent service providers as regards the repairs of the water points. Poor rural people face a series of interconnected natural resource management challenges. They are in the front line of climate change impacts; the ecosystems and biodiversity on which they rely are increasingly degraded; their access to suitable agricultural land is declining in both quantity and quality, their forest resources are increasingly restricted, and degraded. The poor rural people are directly and indirectly dependent on natural resources for their livelihoods, relying on a suite of key natural assets from ecosystem and biodiversity goods and services to provide food, fuel and fibre. However, the continued forest and other natural resource degradation undermine the poor rural people's access to forest resources.

Moreover, the Zimbabwean government has integrated legislative and institutional management with participatory management approach in managing natural resources. Such approaches include co-management. The wisdom of devising resource management strategies shared by government and local communities is gaining recognition as ecosystems and communal property are increasingly threatened. These strategies, commonly known as joint management or 'co-management' schemes, are being utilised in such places as Chimanimani and Shamva Districts, where local resource management systems are still very important. Loosely defined (there is no generally accepted definition) co-management means the sharing of power and responsibility for resource management between the government and local resource users. Each partner plays an important role with government contributing administrative assistance and/or scientific expertise and enabling legislation while the local resource users provide knowledge of traditional management systems and practices developed from years of experience in the local environment. In some cases, local resource users are delegated legislative authority and may even share jurisdiction over resources with government.

Community Based Natural Resources Management (CBNRM) is also an option implemented in Zimbabwe. The government uses the CBNRM which is much or less similar to co-management. The evidence of use of CBNRM is government's Communal Areas Management Programme for Indigenous Resources (CAMPFIRE) programme, which covers the whole country and its design is in a way that encourages or enhances capacity of local people to benefit fully from natural resources in their areas like in wildlife. CBNRM is an approach that gives communities full or partial control over decisions regarding natural resources, such as water, forests, pastures, communal lands, protected areas, and fisheries. The extent of CBNRM control can range from community consultations to joint management to full decision-making and benefit collection responsibility, using tools such as joint management plans, community management plans, stakeholder consultations and workshops, and communal land tenure rights. Community-based institutions are key to any CBNRM project and selecting and building the capacity of local institutions is critical. The selection process must ensure transparency and accountability and minimise conflict. Together with decentralisation reforms which the government has carried out which has resulted in the creation of institutions like Rural District Councils. These institutions among others have a mandate to integrate CBNRM as it ensures stakeholder participation, increases sustainability, and provides a forum for conflict resolution. Such a community-based approach often leads to equitable and sustainable natural resource management, because there is proximity to resources, equity, capacity, biodiversity, cost-effectiveness, and development philosophy.

The advantage of the participatory management systems is that it brings project ownership to the people involved and the people stand more likely to benefit socially or economically. For instance, the Shamva District community has managed to accrue benefits as it has managed to benefit from some of the eucalyptus plantations, which the government is emphasising, that they be set at every school. The policy states that the community should be in support and should co-manage it with school authorities. From these small-scale plantations, they can benefit by buying gum poles at artificially low prices, which they can use for roofing and other imperative uses. In addition, some of the respondents highlighted that they benefited from the eucalyptus situated at schools.

Thus, the government use of management system of 'sustained yield approach' has proven to be effective in maintaining forests in the district. Participatory management resulted in assuming of some form of community ownership of water sources. The community co-manages small water bodies.

The other options are watershed restoration and maintenance, are potential sources of substantial financing to support rural communities' management of their natural assets, and to provide benefits to downstream water users or other communities. Nevertheless, while it may be simple enough to identify those who provide environmental services and the beneficiaries of those services, creating contractual relationships between them has proved thorny. Moreover, there might be a need for serous action on forestry to support and promote secure access to and sustainable management of forests, with a particular focus on incentives and participatory forest management. There is also a need for the introduction of an ecosystem approach, restoration and development of protected areas. Equally essential is the need for the development of value chains for sustainable and renewable natural products and development of certification schemes for sustainable forest management, strengthening of tenure rights to forest resources and governance systems of local communities, and further the investment in diversified agroforestry systems. Building of the capacity of local institutions to participate in and benefit from existing and emerging carbon and ecosystem markets is very important.

LESSONS LEARNED

Collaborative action is necessary to address the complex and unprecedented challenges, and PPCPs are a feasible option for sound climate governance. The multi-stakeholder collaborations are helpful in reducing information asymmetries, build transparency, and trust. Active involvement of as many partners as possible is important. There is a need to have devolved structures in the PPCPs arrangements, down to the district, ward, and village levels. The PPCPs could lead to inclusive solutions, mutual understanding, transparency, empowerment, accountability, and cross-sectional coordination. Innovations are necessary because the future of the world and sustainable development lies in 'policy innovation for transformative change', as espoused by the United Nations Research Institute for Social Development (UNRISD, 2016), the United Nations (2020), and other scholars like Pradeu et al. (2016), Chataway et al. (2017), as well as Mensah (2019).

The contemporary world faces a number of grand social and environmental challenges such as social inequality and climate change, which require innovative approaches. According to Abuiyada (2018), societies across the world are facing many complex and interwoven challenges, poverty, inequality, environmental degradation, demographic change, discrimination and violence that threaten efforts to enable people everywhere to live a peaceful, decent and dignified life on a healthy planet. Therefore, new policy prescriptions are necessary to ensure sustainable development, and Zimbabwean communities, like the rest of the world have adopted multistakeholder action as a way of addressing challenges like water shortage and environmental degradation. Informed by literature, the Governance Network Theory, and experiences in selected Zimbabwean Districts, this study proposes a PPCP framework for sound climate governance in Zimbabwe, and in ALAC, which is in Figure 8.3.

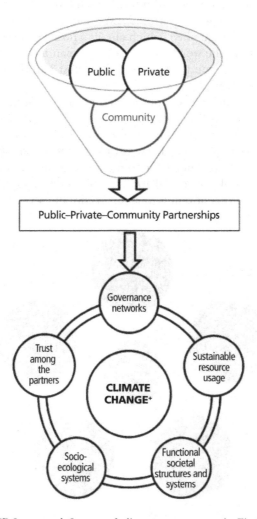

Figure 8.3   PPCP framework for sound climate governance in Zimbabwe.

According to Figure 8.3, PPCPs are feasible for sound climate governance in Zimbabwe. There is tripartite action between the governmental entities, private institutions, and the community. Governance networks between these three actors result in mutually beneficial outcomes and appropriate utilisation of the livelihood assets. Collaboration helps to ensure that there is optimal utilisation of the livelihood assets and allowing for the regeneration of the 'interest-bearing' resources like forests. Equally important is the need for functional societal structures and systems. The socio-ecological systems ought to allow for awareness building, education, and cultivation of desirable behaviour at the individual, interpersonal, community, organisational, national, and international levels. There is also a need to cultivate trust among the partners. Eventually, there would be climate change+ (climate change plus), which entails a positive modification of the climate. In addition, the word "trust" can be expanded to mean "truthfulness," "relationships," "understanding," "steadfast," and "tolerance," as illustrated in Figure 8.4.

According to Figure 8.4, on the preceding page, truthfulness is an element of trust, and, in line with the ancient writings in *The Apocrypha: I Esdras 4:41*, "...Great is Truth and mighty above all things." In this regard, all the stakeholders in the PPCPs ought to be true to one another, as lies are toxic. There is also a need for sound relationships among the actors in the PPCPs. It is also essential for the stakeholders to understand the concerns and positions of fellow partners in the PPCPs. Moreover, vibrant

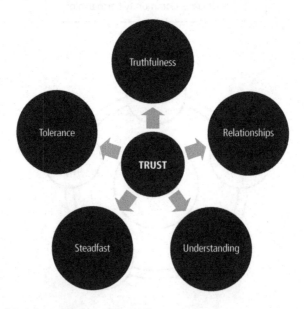

Figure 8.4 Meaning of the word "trust."

multi-stakeholder partnerships ensure strength. There is security in numbers (organised groups) and when risks like droughts come, the PPCPs would remain resolute. Finally, tolerance and respect of the views of the stakeholders is essential. It is essential for all the actors to be open and willing to adjust their positions. Being rigid is toxic.

IMPLICATIONS FOR POLICY AND THEORY

The Zimbabwean government may need an integrated legislative and institutional management with participatory management approach in managing natural resources. The wisdom of devising resource management strategies shared by government and local communities is gaining recognition as ecosystems and communal property are increasingly threatened. These strategies, commonly known as joint management or 'co-management' schemes, are being utilised in communities, where local resource management systems are still very important. Loosely defined (there is no generally accepted definition) co-management means the sharing of power and responsibility for resource management between the government and local resource users. Each partner plays an important role with government contributing administrative assistance and/or scientific expertise and enabling legislation while the local resource users provide knowledge of traditional management systems and practices developed from years of experience in the local environment. In some cases, local resource users are delegated legislative authority and may even share jurisdiction over resources with government.

There may also be a need to increase investment in research, education and extension. Training and education of stakeholder's helps people understand how to prevent and reduce adverse environmental effects associated with deforestation and forestry activities and take appropriate action when possible. Research substantiates it and helps to understand the problem, its cause and mitigation. In line with the New Public Governance, there is a need to ensure commitment towards citizen engagement in governance, in general, and climate governance, in particular. Much of the focus ought to be on society-centred, instead of state-centred policymaking.

CONCLUSION

This study showed that PPCPs are a feasible option for sound climate governance in Zimbabwe. A multi-level framework for climate governance is in place in Zimbabwe, but there is need to ensure that the local community becomes prominent in the quest to ensure positive modification of the climate (climate change+). Local communities are at the centre of resource exploitation and their actions distinguish success and failure. The elusive concept of sustainable development will be a really if there is commitment towards citizen engagement, as espoused by the New Public Governance Paradigm.

REFERENCES

Abuiyada, R. (2018). Traditional development theories have failed to address the needs of the majority of people at grassroots levels with reference to GAD. *International Journal of Business and Social Science, 9*(9), 115–119.

Adnyana, I. B., Anwar, N., Soemitro, R. A. A., & Utomo, C. I. (2015). Critical success factors of public–private–community partnerships in Bali tourism infrastructure development. *Journal of Sustainable Development, 8*(6), 208–215.

African Union Commission. (2015). *Agenda 2063: The Africa we want.*

AtKisson, A. (2015). *Multi-stakeholder partnerships in the post-2015 development era: Sharing knowledge and expertise to support the achievement of the Sustainable Development Goals: Background paper.* United Nations Department of Economic and Social Affairs.

Bax, C. (2011). *Policy instruments for managing European Union road safety targets: Carrots, sticks, or sermons?* Institute for Road Safety Research.

Bowora, J. (2017). *Collective planning and management of rural water infrastructure in developing countries: A case study of community boreholes in Chiredzi District, Zimbabwe.* University of Zimbabwe.

Chambers, R. (1995). *Poverty and livelihoods: Whose reality counts?* Discussion Paper 347, Institute of Development Studies.

Chataway, J., Daniels, C., Kanger, L., Ramirez, M., Schot, J., & Steinmueller. E. (2017). *Developing and enacting transformative innovation policy: A comparative study.* Paper prepared for 8th International Sustainability Transitions Conference, 18–21 June 2107, Gothenburg, Sweden.

Chinyanga, B. (2017), *Institutional arrangements for climate change governance in SADC countries: The case of South Africa and Zimbabwe* (Masters dissertation). University of Zimbabwe.

Chirisa, I. (2013). Politics in environmental stewardship in Zimbabwe: Reflections on Ruwa and Epworth. *African Journal of History and Culture, 5*(1), 13–26.

Dernbach, J. C., & Mintz, J. A. (2011). Environmental laws and sustainability: An introduction. *Sustainability, 3*(1), 531–540.

Dickinson, H. (2016). *From new public management to new public governance: The implications for a "new public service."* In H. Dickinson (Ed.), *The three sector solution: Delivering public policy in collaboration with not-for-profits and business* (pp. 41–48). The Australian National University Press.

Food and Agriculture Organisation. (2017). *Supporting responsible investments in agriculture and food systems: Overview of the FAO Umbrella Programme.* Food and Agriculture Organisation.

Gatzweiter, F. W., & Van Brain, J. (2016). *Innovation for marginalised smallholder farmers and development: An overview and implications for policy and research.* In E. W. Gatzweiter & J. Van Brain (Eds.), *Technological and institutional innovations for marginalised smallholders in agricultural development* (pp. 1–21). Food and Agriculture Organisation.

Government of Zimbabwe. (1982). *Parks and Wildlife Act [Chapter 20:14].* Government Printer.

Government of Zimbabwe. (1985). *Hazardous Substances and Articles Act [Chapter 15:05].* Government Printer.

Government of Zimbabwe. (1987). *Communal Land Forest Produce Act [Chapter 19:04].* Government Printer.

Government of Zimbabwe. (1990). *Forest Act [Chapter 19:05].* Government Printer.

Government of Zimbabwe. (1998). *Water Act [Chapter 20:24].* Government Printer.

Government of Zimbabwe. (2001). *Atmospheric Pollution Prevention Act [Chapter 20:03].* Government Printer.

Government of Zimbabwe. (2003). *Environmental Management Act [Chapter 20:27].* Government Printer.

Government of Zimbabwe. (2003). *Zimbabwe National Water Authority Act [Chapter 20:25].* Government Printer.

Government of Zimbabwe. (2013). *Constitution of Zimbabwe, Amendment Number 20 (Act 2013).* Government Printer.

Government of Zimbabwe. (2020). *National development strategy 1: January 2021–December 2025: Towards a prosperous and empowered upper middle income society by 2030.* Ministry of Finance and Economic Development.

Guizar, I., Gonzalez-Vega, C., & Miranda, M. (2015). *Uneven influence of credit and savings deposits on the dynamics of technology decisions and poverty traps.* The Ohio State University.

Hodge, G. A., & Greve, C. (2007). Public–private partnerships: An international performance review. *Public Administration Review, 67*(3), 545–558.

Kanniainen, J. P. (2017). *New public governance and new public management in the funding and contractual steering of the Finnish higher education system.* University of Tampere.

Kashwan, P., Biermann, F., Gupta, A., & Okereke, C. (2020). Planetary justice: Prioritizing the poor in earth system governance. *Earth System Governance, 6*(2020) 100075.

Klijn, E. H., & Koppenjan, J. (2012). Governance network theory: Past, present and future. *Policy and Politics, 40*(4), 187–206.

Klinsky, S. (2018). An initial scoping of transitional justice for global climate governance. *Climate Policy, 18*(6), 752–765.

Koppenjan, J. (2015, June 27). *New public governance: A framework.* International Summer School on Smart Networks and Sustainable Partnerships.

Lankford, B. A., Makin, I., Matthews, N., Noble, A., McCornick, P. G., & Shah, T. (2016). A compact to revitalise large-scale irrigation systems using a leadership–partnership–ownership theory of change. *Water Alternatives, 9*(1), 1–32.

Maung, T. M. (2013). *Facilitating public–private–community partnerships for effective governance of forest lands.* Ministry of Environmental Conservation and Forestry.

McQuaid, R. W. (2010). *Theory of organisational partnerships: Partnerships advantages, disadvantages, and success factors.* In S. P. Osborne (Ed.), *The new public governance: Emerging perspectives on the theory and practice of public governance* (pp. 127–148). Routledge.

Mensah, J. (2019). Sustainable development: Meaning, history, principles, pillars, and implications for human action: Literature review. *Cogent Social Sciences, 5*(1), 1–21.

Mugambiwa, S. S., & Rukema, J. R. (2019). Rethinking indigenous climate governance through climate change and variability discourse by a Zimbabwean rural community. *International Journal of Climate Change Strategies and Management, 11*(5), 730–743.

Musarandega, H. (2019). *Understanding climate change and vulnerability: Unpacking adaptation strategies for smallholder farmers in Chimanimani District, Zimbabwe* (Doctor of Philosophy Thesis). University of Fort Hare.

Mutambisi, T., Chanza, N., Matamanda, A. R., Ncube, R., & Chirisa, I. (2021). *Climate change adaptation in Southern Africa: Universalistic science or indigenous knowledge or hybrid.* In W. L. Filho, N. Oguge, D. Ayal, L. Adeleke, & I. da Silva (Eds.), *African handbook of climate change adaptation* (1751–1766). https://doi.org/10.1007/978-3-030-42091-8_8-1

Neely, C., Bourne, M., Chesterman, S., Kouplevatskaya-Buttoud, I., Bojic, D., & Vallée, D. (2017). *Implementing 2030 agenda for food and agriculture: Accelerating impact through cross-sectorial coordination at the country level.* Food and Agriculture Organisation and and World Agroforestry Centre.

Obama, B. (2015). *Remarks by President Obama at the first session of COP 21.* Retrieved April 23, 2023 from https://obamawhitehouse.archives.gov/thepress-office/2015/11/30/remarks-president-obama-first-session-cop21

Opolski, K., Modezelewski, P., and Kocia, A. (2013). New approach to network in public administration. *Journal of Applied Business and Economics, 14*(5), 99–109.

Osborne, S. P. (2010). *The new public governance: A suitable case for treatment.* In S. P. Osborne (Ed.), *The new public governance: Emerging perspectives on the theory and practice of public governance* (pp. 1–16). Routledge.

Osborne, S. P., Radnor, Z., & Nasi, G. (2012). A new theory for public service management? Toward a public service dominant approach. *American Review of Public Administration, 43*(2), 135–158.

Perez, M. C. (2015). *Public–private–community partnerships for renewable energy cooperatives.* Wageningen University.

Pestoff, V., & Brandsen, T. (2010). *Public governance and the third sector: Opportunities for co-production and innovation.* In S. P., Osborne (Ed.), *The new public governance: Emerging perspectives on the theory and practice of public governance* (pp. 223–236). Routledge.

Post, L., Schmitz, A., Issa, T., & Oehmke, J. (2021). Enabling the environment for private sector investment: Impact on food security and poverty. *Journal of Agricultural and Food Industrial Organisation, 19*(1), 25–37.

Pradeu, T., Laplane, L., Prevot, K., Hoquet, T., Reynaud, V., Fusco, G., Minelli, A., Orgogozo, V., & Vervoort, M. (2016). *Defining development: Current topics in developmental biology.* Elsevier.

Ross, A. (2010). It is time to get serious: Why legislation is needed to make sustainable development a reality in the United Kingdom. *Sustainability, 2*(1), 1101–1127.

Salaman, L. M. (2002). *The tools of government: A guide to the governance.* Oxford University.

Sathiyah, V. (2013). *History, identity, representation: Public–private–community partnerships and the Batlokoa community.* University of KwaZulu Natal.

Sharma, A. (2017). Precaution and post-caution in the Paris Agreement: Adaptation, loss and damage and finance. *Climate Policy, 17*(1), 33–47.

Stone, D. (2001). *Policy paradox: The art of political decision making.* Norton and Company.

Stranadko, N. (2022). Global climate governance: Rising trend of translateral cooperation. *International Environment Agreements.* Springer.

The Apocrypha: *I Esdras 4:41* (no date) https://www.kingjamesbibleonline.org/1-Esdras-4-41.

Torfing, J., & Triantafillou, P. (2013). *What's in a name? Grasping new public governance as a political–administrative system.* Paper presented at the ECPR General Conference in Bordeaux, September 4–7 2013.

Towards Sustainable Use of Resources Organisation. (2020). *Building resilience to natural disasters in populated African mountain ecosystems: The case of tropical cyclone Idai in Chimanimani, Zimbabwe. A report on environmental impact, and climate resilience building strategies.* TSURO Trust.

United Nations. (2023). *Progress towards the sustainable development goals: Towards a rescue plan for people and planet: Report of the Secretary-General (Special edition).* General Assembly Economic and Social Council.

United Nations. (2022). *The sustainable development goals report 2022.* United Nations Publications Office.

United Nations Research Institute for Social Development. (2016). *Policy innovations for transformative change: Implementing the 2030 Agenda for Sustainable Development.* United Nations Research Institute for Social Development.

Vangen, S., & Huxham, C. (2010). *Introducing the theory of collaborative advantage.* In S. P. Osborne (Ed.), *The new public governance: Emerging perspectives on the theory and practice of public governance* (pp. 163–184). Routledge.

Whaley, L., & Cleaver, F. (2017). Can functionality save the community management model of rural water supply. *Resources and Rural Development, 1*(1), pp. 58–66.

World Bank. (2016). *World development indicators 2016: Featuring the sustainable development goals.* World Bank.

World Commission on Environment and Development. (1987). *Our common future: From one earth to one world.* United Nations Documents.

CHAPTER 9

ENVIRONMENTAL VULNERABILITY AND DISASTER PREVENTION

A Study of Local Governments in Mexico

Manlio F. Castillo
Centro de Investigación y Docencia Económicas

Heidi Jane M. Smith
Universidad Iberoamericana

ABSTRACT

Disaster management requires institutions to help alleviate the effects of climate change on individual citizens. The chapter explores the institutional factors to explain how local Mexican governments can develop better (or worse) instruments to prevent and manage disasters. Using a national survey done in 2020 to over 2,000 municipal governments, the chapter performs an Optimal Scaling Regression to explore the factors that explain why local Mexican governments develop better or worse instruments to manage Disasters Triggered by Natural Hazards. When considering these instruments, we

Climate Governance in International and Comparative Perspective, pages 185–206
Copyright © 2024 by Information Age Publishing
www.infoagepub.com
185

evaluate which institutional factors are important and group them into three areas: exposure and public infrastructure, fiscal autonomy and management capacity, and social participation. The findings show that governments were more likely to have disaster and urban development plans if they also had these institutional factors. Thus, describing how local governments are crucial to help prevent and manage disasters in Latin America.

The countries of Latin America (Latam) constantly suffer the devastating effects of different natural phenomena (earthquakes, hurricanes, tsunamis, landslides, etc.). Many studies show that catastrophes can become even worse based on the actions are associated more with more or less governments' intervention, omissions, or errors than the actual natural phenomena. Despite the evidence and the recurrence of catastrophic events, the Latam region's local governments have made limited and differentiated progress in strengthening their emergency management policies and mechanisms associated with environmental vulnerability. Moreover, disaster management systems in the region are predominantly reactive (not preventive), centralized, and hierarchical, and many of them are headed by the armed forces. This kind of organization excludes citizen and community participation in disaster prevention, response, and resilience tasks.

Additional to the limitations of a hierarchical and centralized structure, Latam countries also lack stable financial mechanisms that provide sufficient funds for disaster prevention and response. Where governments, especially in emerging and developing economies, do not have adequate resources to pay for public policies, social and health, and general citizen welfare, less funding is appropriated to disasters prevention or relief.

Disaster management and recovery, as well as climate change policies, are expensive and implicitly promote inequality and not necessarily growth (Hof et al., 2010; Lomborg, 2020). Although carbon markets are used, and funding is given to developing countries and emerging economies, like Mexico, by international financial institutions to invest in mitigation projects, such funding does not necessarily include adaptation or disaster prevention projects. Even less funding has been intended to understand the costs for an economy to transition into clean energy. For which policies to promote these changes, we still do not know how to deal with on the whole, but rather have created regulations on specific sectors to mitigate effects or promote the purchase of fuel-efficient stoves, cars, and alternative energy sources such as the use of wind, solar, and nuclear energy, which can benefit the environment more than the use of fossil fuels.

At least in Mexico, cities or municipalities with more financing, planning, and development are more capable of addressing climate change problems (Castillo et al., 2021; Ramírez & Smith, 2016). However, financial risk management tends to reinforce the centralization of disaster management since it can offer the possibility of transferring funding risks from the local to the federal government. This funding acts as both a preventive (*ex-ante*) and reaction (*ex-post*) mechanism when Disasters Triggered by Natural Hazards (DTNH) occur. For example, in Mexico, the Catastrophic Bond contracted by the Federal Government (IBRD/FONDEN 2017) was successfully activated for the earthquake that occurred on September 7th, 2017, in Mexico, and the resources were delivered to the two epicenters in Oaxaca and Chiapas. The bond was activated again on September 19th, 2017, when a new earthquake left hundreds dead and others stranded for months (Albarrán, 2017) in central Mexico.[1]

Latin American countries, particularly Mexico, are prone to generating DTNH (De la Fuente, 2010) due to the geographical location and geological characteristics of the region but also to the capacities and intervention tools developed (or not) by governments at all levels to prevent, address, and promote recovery of an area affected by DTNH. In such context, the chapter explores the factors that explain why local Mexican governments develop better or worse instruments to manage DTNH. It should be noted that the profound social and resource inequities prevailing in Latam countries make it necessary to study the region in its specific context since the models, problems, and experiences of disaster management in developed countries are not fully representative of what happens in the subcontinent.

Up to now, the literature has documented the exposure to natural phenomena (i.e., geographic and climatic conditions), the availability of financial resources, the investment in social infrastructure, and the management capacities of governments as the central variables to explain the effectiveness of government tools to manage DTNH. However, little is known about how community and social participation can encourage demand and improve government actions in the DTNH policy arena. Especially in Mexico and Latam, citizens' demands and involvement in disaster prevention and management have been scarcely studied even though this variable could decisively help to understand better why interventions of governments tend to fail before each new natural event, accumulating each year enormous material and human losses (Freeman et al., 2002; Pielke et al., 2003).

As other disaster literature argues, public demand for natural hazard interventions is a central element for governments to take action (Drews & van den Bergh, 2016). Collective effectiveness is crucial from prevention to the resilience stage for disaster management. Therefore, the involvement of the local populations in the tasks of planning and implementing emergency and disaster management instruments is vital to ensure their effectiveness and appropriate collaborative management (Cvetkovic & Grbic, 2021; Doğulu, 2018; Motta & Rohrman, 2021; Teka & Vogt, 2010). Based on this literature, the chapter argues and tests the hypothesis that opening spaces for social participation in elaborating local planning instruments also contribute to more and better tools for managing DTNH.

The chapter is outlined as the following. First, a brief overview of the management of DTNH in Mexico and the role of local governments in this process is provided. The next section presents a literature review to identify the factors associated with a better performance of local governments in the prevention and care of DTNH. Such factors include natural exposure, investment in public infrastructure, fiscal strength, management capacity, and social participation. Finally, the chapter presents the study's methodology, analyzes the findings, and offers some conclusions and recommendations.

DISASTER MANAGEMENT IN MEXICO: AN OVERVIEW

A significant paradigm shift in disaster management can be found in recent years. The focus on the damage that a disaster can cause has evolved into a focus on risk management, i.e., on finding the causes of the damage. The Sendai Framework for Disaster Risk Reduction 2015–2030 was developed as an instrument promoted for governments to focus on risk management, not disaster management. The Sendai Framework is closely linked to the UN 2030 Agenda, the Paris Agreements, and the New Urban Agenda (INCYTU, 2019).

The Sendai Framework aims to reduce four areas related to DTNH: the number of deaths, the affected population, the economic loss (GDP), and the damage to critical infrastructure and interruption of essential services (United Nations, 2015). Based on the Sendai Framework, four types of disaster risk management can be identified (INCYTU, 2019): (a) prospective management, where measures are adopted to avoid the creation of new risks; (b) corrective management, where improvement of the existing disaster risk reduction is promoted to reduce the current risks identified; (c) reactive management, where the action is taken to deal with disasters once they occur; and finally, (d) compensatory management, which seeks the creation and adoption of financial instruments in the event of possible disasters, insurance, funds, bonds, etc.

According to this classification, the nature of disaster risk management in Mexico is reactive. Before the current federal administration, the country had made some progress in adopting a fund that provided some certainty in managing DTNH. The Natural Disaster Fund (Fonden, in Spanish) was created in 1999 thanks to a World Bank loan. Before the creation of Fonden, the federal, state, and municipal governments had to face disasters by redirecting public spending initially intended for other purposes (México Evalúa, 2021). The main problem with this system was that, frequently, the resources were unavailable at the optimal time (World Bank, 2012). The creation of the Fonden made it possible to guarantee the availability of financial resources, although it was not a corruption-free program.

Fonden was created in 1996 as a budgetary instrument administered by the Ministry of Finance and Public Credit (SHCP), forming part of Branch 23, to ensure that resources were available to finance reconstruction after a disaster, including the Restoration of damaged public infrastructure, low-income housing, forests, and natural areas. In 1999, its first Operating Rules were created, becoming a financial instrument coordinated by the Ministry of the Interior (SEGOB). During the year 2000, the figure of emergency declarations was created, as well as the Sectoral Damage Assessment Committees with the participation of the different levels of government to carry out joint assessments and determine the amounts necessary to respond to the disaster.

On the other hand, the transfers of federal resources for DTNH have a significant political component in that these variables have considerable weight when explaining the government's response to emergencies due to natural phenomena. Politicization distorts the application of financial resources according to the intensity and occurrence of the phenomena and also due to an inefficient implementation of recovery measures after a catastrophic event.

Despite its importance for disaster management in Mexico, the Fonden was criticized for alleged irregularities in its operation and allocation of resources. In 2020, the Mexican government announced the disappearance of the Fonden as part of restructuring natural disaster response mechanisms. With the elimination of the Fonden, Mexico has returned to the past and has become more vulnerable to natural disasters. The elimination of the Fonden represents the loss of an important tool for disaster relief and has raised concerns among civil society and subnational governments.

To replace the Fonden, the current federal administration of Andrés Manuel López Obrador (2018–2024) created the Program for the Attention of Emergencies due to Natural Hazards (PAENH). Since it is a program linked to the federal budget, the PAENH management tends to be slow and highly bureaucratic. The management of these disaster programs is a clear example of the reactive nature of the Mexican government's response to DTNH. For instance, between 2015 and 2019, only 1% of public spending was allocated to disaster prevention, while the remaining 99% was devoted to reactive disaster response (Guadarrama & Suárez, 2019). The authors also note that risk reduction is not implemented in the design of public policies or urban development, even though a formal directive exists.

To access the PAENH resources, states and municipalities must submit a Disaster Declaration request to the National Coordination for Civil Protection (NCCP). The NCCP determines whether or not to approve the Disaster Declaration and the amount of federal aid to be provided. In the request, states and municipalities must demonstrate that they do not have sufficient financial or operational capacity to deal with the disaster. Additionally, they must prove the existence of state insurance and declare the non-duplication of expenses when federal funds are authorized.

The NCCP is the administrative unit in charge of PAENH. One of the primary responsibilities of the NCCP is to buy inputs such as food, water for human consumption, shelter, medical care instruments, and medicines. The NCCP is also responsible for maintaining the local government's finances while the affected municipality regains its financial and operational capacity. The disaster declaration lasts seven days and can be renewed at the request of the municipality or state. Once the disaster has been declared, the NCCP buys the necessary supplies and distributes them in collaboration with local governments, the Armed Forces, or the National Guard. This task aims to provide disaster victims with the minimum necessities for survival. However, the budget availability is subject to change, according to the federal government's decisions.

The Failures of PAENH

The PAENH appears to be a program with relatively clear rules. However, its design and implementation suffer from numerous flaws. The five main weaknesses of the program are described in the following paragraphs and illustrated with some significant examples often replicated in many other cases.

Cooperation Failures

Although the PAENH guidelines order cooperation between federal, state, and municipal governments, collaboration is difficult to achieve. Instead, a high degree of disorder and improvisation is often observed. For example, the municipality of Tequila, in Jalisco state, was declared a disaster area in July 2022, according to the program guidelines (see DOF, 2021). However, by December of the same year, the federal government had not delivered the financial resources that should have been allocated to reconstruct houses and public infrastructure. It was also reported that the local government issued money for the rehabilitation of roads even though the state government should have covered it (Figueiras, 2022).

Focus on Procedures, not People

In September 2022, some municipalities of Guerrero state were affected by tropical storm "Lester." Six people died due to "Lester," which also caused road collapses, river overflows, flooding, and collapsed bridges (Covarrubias, 2022). However, the state government could not access PAENH funds because the number of millimeters of rainfall required to qualify as a natural disaster under the program's guidelines was not exceeded

(Covarrubias, 2022). Therefore, the program is not designed to deal with a disaster based on the damage caused to the population but rather prioritizes the fulfillment of bureaucratic requirements. As a result, assistance to the affected population is hampered, and damage may increase.

Uncertain Access to Financial Resources

Concerning the insurance and bonds that each state must take out to cover risks and disasters, a problem similar to the previous one arises. Despite severe damage to communities, and even when the financial capacity of local governments is exceeded, insurance and bonds do not guarantee access to financial resources if all the metrics agreed upon at the time of contracting are not met. This type of insurance was initially conceived as a complement to Fonden but is now used as a substitute due to the uncertainty created by PAENH (México Evalúa, 2021).

The Politicization of Decisions and Lack of Transparency

As in the case of Fonden, access to PAENH resources is granted based on uncertain, non-transparent criteria and with a worrying political component that distorts the government's response to disasters. The distortion introduced by political variables in allocating disaster relief resources is not exclusive to Mexico. Extensive literature documents that it also occurs in developed countries. For instance, Garrett and Sobel (2003) and Schmidtlein, Finch, and Cutter (2008) find that disaster declarations in the United States respond to a greater extent to the influence of political factors than to the intensity and occurrence of meteorological phenomena. Moreover, political factors reported in the academic literature that may influence the amount of public resources allocated to respond to DTNH include the partisan membership of the executive Husted & Nickerson, 2014; Sylves & Búzás, 2007), the possibility of the re-election of local rulers (Downton & Pielke, 2001; Garrett & Sobel, 2003; Kriner & Reeves, 2015; Reeves, 2011; Salkowe & Chakraborty, 2009; Sylves & Búzás, 2007), and the level of competition in the electoral contest (Reeves, 2011): the most contended competitions receive more declaratory statements by executives which therefore heighten the problem (Garrett & Sobel, 2003; Gasper, 2015; Kriner & Reeves, 2015; Sainz-Santamaria & Anderson, 2013; Stramp, 2013). The electoral base of a state or subnational entity (Gasper & Reeves, 2011) and electoral processes in the current year (Gasper, 2015) also influence the allocation of catastrophic funds. Due to the institutional fragility of their governments, political factors are critical in Mexico and Latam.

Failures in Equity, Targeting, and Victims' Identification

A key weakness of PAENH is the identification of victims within a disaster zone. It is possible that the program's resources are not intended for all people affected by a disaster, even if they live in a municipality that has been declared a disaster zone. For example, the municipality of Pichucalco, in Chiapas state, received a disaster declaration and access to PAEAN due to the effects of tropical storm "Karl" in December 2022 (INFOBAE, 2023). However, the residents of the towns of Platanar and La Crimea (in the same municipality) took over the mayor's offices in protest, accusing the authorities of not including them among the victims of "Karl." In other words, no mechanism guarantees the program's targeting and coverage of all disaster victims. Consequently, the program fails to serve the affected population equitably.

Problems and Contradictions of Civil Protection Management

In addition to the PAENH, the General Law for Civil Protection (GLCP) is the other essential instrument for disaster management in Mexico. Firstly, the GLCP mandates aligning civil protection policies with the National Development Plan (NDP) and the National Program for Civil Protection. But, unlike the last three NDPs, the current one does not reference integrated risk management or establish lines of action for protection against DTNH (Guadarrama & Suárez, 2019).

The GLCP provides the framework for federal, state, and local disaster response coordination and promotes a comprehensive approach to risk management, although these mandates are often not successfully implemented. Promoting social participation to build resilient communities and adapt to climate change is a priority of the GLCP. Furthermore, the GLCP requires states to standardize local civil protection regulations and have an administrative body for risk management. However, despite what the law prescribes, civil protection bodies in each state have different hierarchies and severe limitations to influence civil protection policies (Guadarrama & Suárez, 2019) or promote social involvement in disaster prevention and management.

Regarding funding, the law orders states to take out disaster insurances and other risk transfer instruments. Each state must create a civil protection fund with its own resources, although small federal contributions to these funds are allowed. The GLCP also provides for the implementation of financial instruments and budgetary programs for risk management, such as the PAENH.

Finally, the GLCP establishes the National System for Civil Protection (NSCP), composed of federal agencies, state and municipal civil protection agencies, civil society, community organizations, media representatives, and educational institutions. The coordinating body of the NSCP is the National Council for Civil Protection, headed by the President of Mexico. It reflects a high degree of centralization of risk management decision-making.

The Role of Local Governments

Although state laws dictate the implementation of a comprehensive risk management approach (including the integration of comprehensive risk management into territorial planning instruments, financial instruments for emergency and disaster services, the involvement of the population in some phases of risk management, and the professionalization of civil protection personnel), local governments in Mexico rarely develop adequate policy instruments to fulfill this mandate, making them almost entirely dependent on federal guidelines and assistance. As shown in Table 9.1, the percentage of municipalities with tools to prevent and manage emergencies is relatively low. Some develop only the basic programs, such as urban development plans or others. However, the lack of all the provisions is a significant problem, as they should all be interconnected for integrated management of risks and DTNH.

In practice, state and municipal authorities assume the role of implementers of the main actions defined by the federal government. Prevention and attention to DTNH are similar to what Castillo et al. (2021) describe for climate change policy: although regulations dictate that each level of government (federal, state, and local) has specific functions and must develop policies within its jurisdiction, actually it is a highly centralized system, with the federal government defining goals, processes, and funding, while most local governments ignore the problem. At best, some municipalities take disparate actions without defined objectives or appropriate information.

TABLE 9.1 Percentage of Mexican Municipalities With Plans for Disaster Prevention and Response

Plan or program	All municipalities	Municipalities with pop. ≥ 100,000 inhabitants
Urban development plan or program	30.7%	68.8%
Natural risk atlas	16.5%	49.6%
City development program	11.9%	40.2%
Ecological management program	4.8%	16.2%
Land regularization program	6.5%	20.1%
Ecological and land-use program	6.4%	19.7%
Program for the management of natural areas under municipal jurisdiction	3.8%	14.5%
Climate action plan or program	2.0%	7.3%

Source: Authors' calculation based on the National Census of Municipal Governments and Demarcations of Mexico City 2021.

Table 9.1 also shows that the municipalities that have made the most progress in developing DTNH management tools are the most urbanized, where most of the population is concentrated. Nevertheless, there are still significant differences in the development and quality of such tools, even among municipalities with high risks of DTNH or with enough financial resources to invest in measures to prevent, respond to, and build resilience to disasters. The following section explores the factors that may explain why some local governments have better tools for managing DTNH than others.

WHAT MAKES LOCAL GOVERNMENTS BETTER EQUIPPED TO DEAL WITH DISASTERS?

In the following paragraphs, we explain the main factors that encourage local governments to develop better tools for prevention and intervention in response to DTNH. Particular emphasis is placed on the importance of collaboration and social participation in creating effective disaster management tools. A hypothesis is proposed regarding each of the factors identified.

Exposure and Public Infrastructure

Risks and disasters are social constructions resulting from human interaction with the environment. They can by no means be considered as belonging to a singularity within the category of "natural" (INCYTU, 2019). The term "catastrophe" derives from the inability of societies to prevent or reduce risks and accidents (Oliver-Smith, 2009). Catastrophe risk is the probability that a natural or man-made phenomenon will cause damage and loss. One factor explaining this kind of risk is exposure. Exposure is the presence of people or assets in a state of vulnerability, suffering from a hazard, or located in a vulnerable area (INCYTU, 2019). On the other hand, vulnerability is the susceptibility of a population to the adverse effects of a threat or hazard through economic, social, institutional, environmental, or health variables (Alcántara-Ayala et al., 2019; INCYTU, 2019). A fundamental proposition is that governments of municipalities more exposed to potentially catastrophic natural phenomena (e.g., located on coasts, in seismic zones, or near volcanoes) have more incentives to develop tools to prevent and intervene in disasters since they may have a more significant number of vulnerable populations.

Furthermore, vulnerability is not just about geography or climate. Adequate infrastructure (e.g., monitoring systems, flood barriers, improvements to structures such as buildings and bridges, drainage and water management systems, etc.) can significantly reduce some of the risks that vulnerable populations face because of their location. In addition to increased risk, poor planning can deepen localities' impoverishment and economic stagnation (INCYTU, 2019). Conversely, investing more in social infrastructure may contribute to economic growth and reduce poverty and economic losses in the event of a disaster.

Therefore, the first two hypotheses are:

H1: *Local governments are incentivized to create DTNH management tools as municipalities' exposure to natural hazards increases.*

H2: *Increased DTNH preparedness, and thus better DTNH management tools, are associated with more investment in public infrastructure.*

Fiscal Autonomy and Management Capacity

Fiscal strength is one of the elements that make it possible to cover the impact of a DTNH and the financing of the recovery. For example, budget shortfalls are linked to the lack of financial reserves to cope with the consequences of catastrophes. To cover the expenses for recovery (INCYTU, 2019), bonds, loans, or special taxes are used in these cases.

The ability of a local government to allocate resources to DTNH prevention, care, and resilience increases with its fiscal capacity and autonomy. However, resources from local taxes are not the only ones available to governments. Fiscal strength also helps to get private insurance on better terms and access to other public or private financial instruments, such as bonds, specialized funds, and disaster recovery and insurance programs (OECD/ The World Bank, 2019). Especially in an environment where federal resources are highly bureaucratized (see Takeda & Helms, 2006) or provide inadequate incentives (see Donahue & Joyce, 2001; Schneider, 1992), greater freedom in the use of fiscal resources makes it more likely that resources will be invested in the development of appropriate disaster management tools. That is the basis of the following hypothesis:

H3: *Local governments with greater fiscal autonomy have better tools to manage DTNH.*

Nevertheless, if local governments do not have the material, human and organizational resources to manage DTNH, the availability of financial resources may be of little use. In addition to the existence of plans, programs, and financial resources, local governments' organizational capacity to use their tools and resources appropriately is crucial to preventing disasters or recovering affected areas. It is necessary to study the dynamics of this variable since not all governments effectively implement financial or management strategies to achieve these objectives (OECD/The World Bank, 2019), especially in Mexico and Latin America. Thus, a fourth hypothesis is:

H4: *Local governments with better management capacities have better tools to prevent, respond to, and recover from DTNH.*

Social Participation

One of the critical factors limiting the ability of local governments to mitigate or prevent the effects of DTNH is the lack of collaborative governance arrangements for this purpose. Collaborative governance for disaster management involves building trust among government, community, and social actors, sharing ideas, providing leadership, and collaborating among businesses, civil society organizations, individuals, academics, and different levels of government (Pasquini et al., 2021). Through coordination and collaboration, public agencies, social groups, and stakeholders may use their strengths to address and solve problems more efficiently (Lein et al., 2009).

The key to collaborative networks for emergency and disaster response is to maintain their flexibility and ability to adapt to changing conditions (Bryson et al., 2015), as well as the cognitive accuracy of the participants in the networks so that their perceptions of the relationships between actors are as realistic as possible (Choi & Brower, 2006). In some cases, explicit plans establishing participants' intentions, capabilities, payments, and retaliations can contribute to effective collaboration through formal agreements between authorities and community groups, stakeholders, or social organizations. By identifying potential tensions between participants in advance and finding ways to address and resolve disputes, these programs can help resolve conflicts before disasters occur (Pasquini et al., 2021).

The literature highlights community capital as a central factor in improving the prevention and mitigation of several natural hazards a social group face (Wolkow & Ferguson, 2001). Social and community participation complements the weaknesses or shortcomings of governmental organizations, allowing for a cohesive intervention of problems and improving social resilience (Lein et al., 2009).

Community members are at the heart of appropriate disaster management, as they directly experience the risk and impact of DTNH. Community participation is essential for the strengthening of political legitimacy and the expansion of mitigation and resilience capacities. Therefore, the study of community factors contributes to improving our understanding of DTNH management. Thus, the final hypothesis is as follows:

H5: *Local governments have better tools for managing DTNH when they allow social participation in their creation.*

METHODOLOGY

A cross-sectional database was constructed to test the hypotheses with information from 2,098 Mexican municipalities out of 2,467. Missing values preventing calculation were discarded. The data come mainly from the National Census of Municipal Governments and Demarcations of Mexico City 2021 (CMG), which includes information from 2020.[2] The marginalization data come from the National Population Council; the municipal population data from the Census of Population and Housing 2020; and the financial information from the State and Municipal Public Finance Statistics 2020, a survey produced by the National Institute of Statistics and Geography.

Since the econometric model includes categorical and continuous variables, the optimal scaling method was used to estimate the regression. Optimal Scaling Regression (OSR) allows the analysis of nonlinear relationships between different types of variables by transforming categorical data to maximize multiple correlations (Apon, 2020; Meulman et al., 1998, 2019). To use this estimation method, continuous variables were discretized by assigning ranks to cases based on a normal distribution. The variables used in the model are described in the following paragraphs.

Dependent variable: Management tools for disaster prevention and response. The CMG codes as 1 or 0 whether local governments have each of the following eight disaster prevention and response tools: (a) Urban development plan or program; (b) city development program; (c) ecological management program; (d) ecological and land-use program; (e) natural risk atlas; (f) program for the management of natural areas under municipal jurisdiction; (g) land regularization program; and (h) climate action plan or program. Since these tools are complementary, it is assumed that each Mexican municipality is better prepared to manage DTNH if it has all or most of them than if it has none or few of them. This variable is ordinal, and its values range from zero to eight since the worst case is that a municipality has none of the plans or programs, and the best case is that it has all of them.

Independent variables. Each independent variable is related to one of the possible factors that could explain the quality of municipal tools for DTNH management: exposition, fiscal autonomy, management capacity, public infrastructure, and social participation. A welfare index was added as a control variable (see Table 9.2).

Management capacity is measured by the ability of local governments to take action to address and prevent DTNH. The CMG tracks up to 39 actions

TABLE 9.2 Independent Variables		
Issue	Variable	Type
Exposition	% Population in irregular settlements	Scale
Fiscal autonomy	% Property tax collection (exclusive of municipal governments)	Scale
Management capacity	Level of local actions to address and prevent disasters	Ordinal
Public infrastructure	Public investment per capita	Scale
Social participation	Space for social participation	Ordinal
Welfare level	Welfare index	Scale

Source: The authors.

that local governments can take to (a) address and prevent irregular settlements, (b) prevent settlements in at-risk areas, and (c) improve and maintain urban centers. The census only reports whether each municipality carries out each of the 39 actions it monitors, coding each action as 1 or 0. The number of actions in the three previous topics was summed for each municipality to obtain a summary measure. This variable is ordinal, and its values may range from 0 to 39 since the worst case is that the municipality has no capacity to manage or prevent DTN; the best case is that it can carry out the 39 actions defined in the census. Note that no Mexican municipality can implement the 39 actions in the survey, as shown in Table 9.3.

TABLE 9.3 Descriptive Statistics						
Variable	N	Min	Max	Mean	S.D.	Variance
Management tools for disaster prevention and response	2,466	0	8	0.8	1.4	1.9
% Population in irregular settlements	2,337	0	101.5	0.5	3.8	14.4
% Property tax collection	2,450	0	105.0	44.2	33.6	1,128.4
Level of local actions to address and prevent disasters	2,466	0	31	5.8	6.5	42.7
Public investment per capita (2018 pesos)	2,239	51.94	42,843.50	1,919.84	2,047.01	4,190,258.19
Space for social participation	2,467	0	6	0.5	1.2	1.5
Welfare index	2,467	21.4	62.4	53.9	3.9	15.2
Valid N	2,098					

Source: The authors.

Social participation is measured by the spaces local governments open for citizens to participate in formulating instruments for the prevention and management of DTNH. The CMG records with 1 and 0 whether the community is involved in the discussion and planning related to the following topics and management tools: (a) Land use planning; (b) ecological planning; (c) urban development; (d) program for the management of natural areas; (e) settlements in risk areas; and (f) irregular settlements. This variable is ordinal, and its values may range from 0 to 6 since the worst case is that the municipality does not allow social participation in disaster management programs. The best case is that the local government opens space for social participation in the six topics.

FINDINGS AND DISCUSSION

Contingency tables were used to explore the relationship between the DTNH management tools introduced in the model. As expected, the evidence suggests that municipalities with an urban development plan or program are more likely to generate the rest of the instruments for preventing and attending to environmental risks and DTNH: the ecological management program, the ecological and land-use program, the natural risk atlas, the program for the management of natural areas under municipal jurisdiction, the land regularization program, and the climate action plan or program.

The urban development plans or programs are the minimum regulatory documents in which the Mexican municipalities outline the scenarios and objectives for local development. Therefore, all other planning instruments related to disaster prevention and management should refer to this document. However, despite its importance, not all municipalities in Mexico have an urban development plan.

More importantly, the evidence also indicates that municipalities reporting more social participation in urban development planning also report more social involvement in the following issues: land use, ecological plans, the management of protected natural areas, settlements in risk zones, and irregular settlements. This information suggests, first, that there is community interest in discussing and developing these tools with the government. With an appropriate approach by authorities, the social interest can be transformed into effective citizen participation in disaster prevention, control, and resilience strategies. Second, urban development plans can be a "gateway" to promote social participation in other local public actions related to environmental risk prevention and response.

The results of the econometric model estimation are presented in Table 9.4. First, the evidence shows that exposure to natural hazards is an incentive for local governments to develop management tools for disaster

TABLE 9.4 OSR: Estimation Results

Variable	Typified coefficients		
	Beta	Typical error estimation	Sig.
Population in irregular settlements	0.064	0.022	0.004***
Property tax collection	0.072	0.019	0.000***
Level of local actions to address and prevent disasters	0.232	0.032	0.000***
Public investment per capita	−0.076	0.021	0.000***
Space for social participation	0.435	0.025	0.000***
Welfare index	0.035	0.022	0.114
n 2,098			
R^2 0.439			
R^2 (corrected) 0.434			
F 85.461***			

Source: The authors.

prevention and response (H1). Despite this fact, it is important to remember that more vulnerability due to exposure does not automatically lead to the creation of such tools; financial and managerial resources are also crucial for this purpose.

Second, Table 9.4 shows that more investment in public infrastructure is negatively related to the existence of disaster management tools, which contradicts H2. A possible explanation for this result is that it reflects the fact that more resources are invested in the most marginalized municipalities, which also have poorer programs for managing DTNH. The issues of prevention and resilience to environmental risks are usually not a priority for the local governments of these municipalities, as they have more pressing needs, such as providing basic public services.

The variable of public investment covers not only the infrastructure directly related to risk and disaster prevention but also all types of infrastructure built in the municipality. It would be necessary to precisely identify the investments made in disaster infrastructure to analyze whether the effect is maintained and to explain how it relates to the management of DTNH.

Third, it cannot be denied that local governments with greater autonomy and fiscal strength tend to develop better DTNH management tools. As noted in H3, more significant own-source revenue may allow local governments to allocate resources to prevent, respond to, and build resilience to disasters. Such fiscal autonomy is particularly important in a context where the procedures for obtaining funding from the federal government have become more bureaucratic due to how PAENH operates, as it allows local governments some flexibility to respond quickly to emergencies. Whether the rigidity and problems of PAENH encourage the creation of state and local disaster funds should be further explored.

Nor can it be denied that local governments with better management capacities have better tools for DTNH prevention, response, and recovery (H4). The Mexican municipalities that perform best in developing their anti-disaster policies have been able to manage more concrete actions to address and prevent disasters.

Finally, it is undeniable that local governments have better tools for the management of DTNH when they allow social participation in their design and implementation (H5). In addition to the social interest in such tools, the appropriate channeling of this social demand can significantly improve local government policies, making social and community participation in the planning and implementation processes crucial.

CONCLUSIONS AND POLICY IMPLICATIONS

The experience of disaster management in the Global South differs from that of developed countries due to the geographical and economic context and the social inequalities that prevail in the region. Latam countries have not been able to develop adequate management and financial instruments to prevent and cope with DTNH. The response to these disasters remains reactive, militarized, highly hierarchical, and not flexible enough to adapt to the uncertainty generated by natural phenomena. In addition, Latam's emergency and disaster systems have not achieved virtuous interactions with vulnerable communities to ensure effective collective action in emergencies. The case of local governments in Mexico illustrates situations that may arise in different Latam countries.

The review presented in this chapter shows the urgency of promoting a paradigm shift in risk and disaster management in Mexico and Latam from a reactive to a prospective vision to reduce human and material losses due to DTNH. The framework for positioning this transformation is climate change and social development policies, as well as all national, regional, and local efforts in both areas.

According to the evidence, some factors contributing to explain that local Mexican governments have better planning tools aimed at managing environmental risk and DTNH are the level of natural exposure of municipalities to natural phenomena, a higher own revenue collection, more capacity for action by the local public administration, and more spaces for citizen participation in formulating disaster planning and management instruments. These results reinforce the need to overcome the centralized, non-transparent, and exclusive nature of the design and operation of the mechanisms for the prevention and management of DTNH. Public intervention in the face of disasters could be more effective and reduce human and economic losses by promoting more significant social and community participation. There is also an urgent need to establish a financing system (through an institutional context of the Mexican federalist system) to deal with disasters and their consequences. This system should diversify the sources of public and private resources to provide certainty to governments and vulnerable populations.

In addition, other measures are necessary to improve disaster prevention and management. First, it is essential to coordinate local, regional, and national organizations involved in risk and disaster management and to standardize their regulations to provide clear and consistent objectives for decisions and actions. It is also necessary that the regulations clearly define each government's operational and financial responsibilities.

Second, coordination with armed forces is necessary but insufficient. Military forces tend to focus on some care and recovery tasks but less on prevention. Moreover, in cases where rapid action or adaptation to a specific context is needed, the rigidity of military command and hierarchies can be counterproductive.

Third, investment in social infrastructure must be seen from a risk management perspective, not just a social policy perspective. Risk and disaster policy is a way of broadening the instruments for social development and poverty reduction in addition to climate change.

Finally, reviewing and redesigning existing programs from a comprehensive risk management perspective is crucial. The shortcomings identified in the Mexican PAENH point to some critical aspects to be addressed: a focus on affected or vulnerable populations, rapid and coordinated attention by authorities at all levels, timely access to sufficient resources, equitable treatment of victims of DTNH, and transparency in the criteria for allocating financial resources.

NOTES

1. On September 7, 2017, 102 people were killed, around 900 were injured and more than 2 million people were affected in southern Mexico by an 8.2 Mw earthquake epicentered in the state of Chiapas. A few days later, on September 19, another 7.1 magnitude earthquake struck the state of Puebla, causing 370 deaths, more than 7,000 injured people, and extensive material loses in central Mexico, including Mexico City.

2. The CMG 2021 "aims to generate statistical and geographic information on the management and performance of the institutions that constitute the public administration of each municipality and the territorial units of Mexico City, specifically in the functions of government, public safety, civil justice,

water, sanitation, urban solid waste and the environment, with the purpose of linking it to the government's work in the process of designing, implementing, monitoring and evaluating national public policies in all those subjects" (INEGI, 2021).

REFERENCES

Acuerdo que establece los lineamientos del Programa para la Atención de Emergencias por Amenazas Naturales, Diario Oficial de la Federación (2021). https://www.dof.gob.mx/nota_detalle.php?codigo=5626632&fecha=16/08/2021#gsc.tab=0

Albarrán, E. (2017, September 20). Bono catastrófico no se activará por sismo menor a 8: Hacienda [Catastrophic bonus will not be activated by earthquake less than 8 degrees: Ministry of Finance]. *El Economista.* https://www.eleconomista.com.mx/economia/Bono-catastrofico-no-se-activara-por-sismo-menor-a-8-Hacienda—20170920-0129.html

Alcántara-Ayala, I., Garza Salinas, M., López García, A., Magaña Rueda, V., Oropeza Orozco, O., Puente Aguilar, S., Rodríguez Velázquez, D., Lucatello, S., Ruiz Rivera, N., Tena Núñez, R. A., Urzúa Venegas, M., & Vázquez Rangel, G. (2019). Gestión Integral de Riesgo de Desastres en México: Reflexiones, retos y propuestas de transformación de la política pública desde la academia [Comprehensive disaster risk management in Mexico: Reflections, challenges, and proposals for transforming public policy from academia]. *Investigaciones Geográficas, 98,* 1–17. https://doi.org/10.14350/rig.59784

Apon, L. (2020). *Optimal scaling regression with factor-by-curve interactions* [Master thesis]. Leiden University.

Bryson, J. M., Crosby, B. C., & Stone, M. M. (2015). Designing and implementing cross-sector collaborations: Needed and challenging. *Public Administration Review, 75*(5), 647–663.

Castillo, M. F., Ramírez, E. E., & Smith, H. J. (2021). Políticas prodensificación y cambio climático: los desafíos de las ciudades mexicanas [Pro-densification policies and climate change: the challenges of Mexican cities]. *Sobre México Temas de Economía, 2*(3), 1–29. https://doi.org/10.48102/rsm.vi3.79

Choi, S. O., & Brower, R. S. (2006). When practice matters more than governments plans: A network analysis of local emergency management. *Administration & Society, 37*(6), 651–678.

Covarrubias, A. (2022, September 23). Guerrero no pedirá declaratoria de desastre [Guerrero will not request a disaster declaration]. *El Sol de Acapulco.* https://www.elsoldeacapulco.com.mx/local/estado/guerrero-no-pedira-declaratoria-de-desastre-8930816.html

Cvetkovic, V., & Grbic, L. (2021). Public perception of climate change and its impact on natural disasters. *Journal of the Geographical Institute Jovan Cvijic, SASA, 71*(1), 43–58. https://doi.org/10.2298/IJGI2101043C

De la Fuente, A. (2010). Natural disasters and poverty in Latin America: Welfare impacts and social protection solutions. *Well-Being and Social Policy, 6*(1), 1–15.

Doğulu, C. (2018). Assessing the social psychological aspects of vulnerability and resilience for natural hazards: Theoretical and methodological contributions. *Geophysical Research Abstracts, 21*, 1593.

Donahue, A. K., & Joyce, P. G. (2001). A framework for analyzing emergency management with an application to federal budgeting. *Public Administration Review, 61*(6), 728–740. https://doi.org/10.1111/0033-3352.00143

Downton, M. W., & Pielke, R. A. (2001). Discretion without accountability: Politics, flood damage, and climate. *Natural Hazards Review, 2*(4), 157–166. https://doi.org/10.1061/(ASCE)1527-6988(2001)2:4(157)

Drews, S., & van den Bergh, J. C. J. M. (2016). What explains public support for climate policies? A review of empirical and experimental studies. *Climate Policy, 16*(7), 855–876. https://doi.org/10.1080/14693062.2015.1058240

Figueiras, M. (2022, December 17). Termina el 2022 y recursos para atender contingencias por lluvias no llegaron [Resources to address contingencies due to rain did not arrive by the end of 2022]. *El Sol de Orizaba.*

Freeman, P. K., Martin, L. A., Linnerooth-Bayer, J., Warner, K., & Pflug, G. (2002). *Sistemas nacionales para la gestión integral del riesgo de desastres. Estrategias financieras para la reconstrucción en caso de desastres naturales* [National systems for comprehensive disaster risk management. Financial strategies for reconstruction in the event of natural disasters].

Garrett, T. A., & Sobel, R. S. (2003). The political economy of FEMA disaster payments. *Economic Inquiry, 41*(3), 496–509. https://doi.org/10.1093/ei/cbg023

Gasper, J. T. (2015). The politics of denying aid: An analysis of disaster declaration turndowns. *Journal of Public Management, 22*(7), 1–7.

Gasper, J. T., & Reeves, A. (2011). Make it rain? Retrospection and the attentive electorate in the context of natural disasters. *American Journal of Political Science, 55*(2), 340–355. https://doi.org/10.1111/j.1540-5907.2010.00503.x

Guadarrama, M., & Suárez, R. (2019). Más vale prevenir que lamentar. El diseño institucional de la Protección Civil en México ["Better safe than sorry." The institutional design of Civil Protection in Mexico]. In IMCO (Ed.), *Índice de Competitividad Urbana 2020* (pp. 50–65). IMCO. https://imco.org.mx/pub_indices/wp-content/uploads/2020/11/MA%cc%81S-VALE-PREVENIR-QUE-LAMENTAR.-EL-DISEN%cc%83O-INSTITUCIONAL-DE-LA-PROTECCIO%cc%81N-CIVIL-EN-ME%cc%81XICO.pdf

Hof, A. F., den Elzen, M. G. J., & van Vuuren, D. P. (2010). Including adaptation costs and climate change damages in evaluating post-2012 burden-sharing regimes. *Mitigation and Adaptation Strategies for Global Change, 15*(1), 19–40. https://doi.org/10.1007/s11027-009-9201-x

Husted, T., & Nickerson, D. (2014). Political economy of presidential disaster declarations and federal disaster assistance. *Public Finance Review, 42*(1), 35–57. https://doi.org/10.1177/1091142113496131

INCYTU. (2019). *Gestión de riesgo de desastres* [Disaster risk management]. https://foroconsultivo.org.mx/INCyTU/documentos/Completa/INCYTU_19-033.pdf

INEGI. (2021). *Subsistema de Información de Gobierno, Seguridad Pública e Impartición de Justicia* [Subsystem for Information on Government, Public Safety, and Administration of Justice]. Censo Nacional de Gobiernos Municipales y Demarcaciones Territoriales de La Ciudad de México 2021. https://www. inegi.org.mx/programas/cngmd/2021/

INFOBAE. (2023, January 5). Pobladores de Juárez, Chiapas, incendiaron el Ayuntamiento por falta de pagos del programa Bienestar [Residents of Juárez, Chiapas, burned down the City Hall due to lack of payments from the "Bienestar" program]. *INFOBAE.* https://www.infobae.com/america/mexico/2023/01/05/pobladores-de-juarez-chiapas-incendiaron-el-ayuntamiento-por-falta-de-pagos-del-programa-bienestar/

Kriner, D. L., & Reeves, A. (2015). Presidential particularism and divide-the-dollar politics. *American Political Science Review, 109*(1), 155–171. https://doi.org/10.1017/S0003055414000598

Lein, L., Angel, R., Bell, H., & Beausoleil, J. (2009). The state and civil society response to disaster: The challenge of coordination. *Organization & Environment, 22*(4), 448–457. https://doi.org/10.1177/1086026609347190

Lomborg, B. (2020). Welfare in the 21st century: Increasing development, reducing inequality, the impact of climate change, and the cost of climate policies. *Technological Forecasting and Social Change, 156,* 119981. https://doi.org/10.1016/j.techfore.2020.119981

Meulman, J. J., Hubert, L. J., & Heiser, W. J. (1998). The data theory scaling system. In A. Rizzi, M. Vichi, & H. Bock (Eds.), *Studies in classification, data analysis, and knowledge organization* (pp. 489–496). Springer. https://doi.org/10.1007/978-3-642-72253-0_66

Meulman, J. J., van der Kooij, A. J., & Duisters, K. L. W. (2019). ROS regression: Integrating regularization with optimal scaling regression. *Statistical Science, 34*(3), 361–390. https://doi.org/10.1214/19-STS697

México Evalúa. (2021). *Fondo contra desastres: una política más que viaja al pasado* [Disaster fund: Another policy traveling back in time]. https://www.mexicoevalua.org/fondo-contra-desastres-una-politica-mas-que-viaja-al-pasado/

Motta, M., & Rohrman, A. (2021). Quaking in their boots? Inaccurate perceptions of seismic hazard and public policy inaction. *Behavioural Public Policy, 5*(3), 301–317. https://doi.org/10.1017/bpp.2019.18

OECD/The World Bank. (2019). *Fiscal resilience to natural disasters.* OECD Publishing. https://doi.org/10.1787/27a4198a-en

Oliver-Smith, A. (2009). Anthropology and political economy of disasters. In E. C. Jones & A. D. Murphy (Eds.), *The political economy of hazards and disasters* (pp. 11–30). AltaMira Press.

Pasquini, L., Petrik, D., Nyamakura, B., Strachan, K., Spires, M., Shackleton, S., & Ziervogel, G. (2021). Effective collaborative climate change governance in urban areas. In F. J. Carrillo & C. Garner (Eds.), *City preparedness for the climate crisis* (pp. 209–223). Edward Elgar Publishing. https://doi.org/10.4337/9781800883666

Pielke, R. A., Rubiera, J., Landsea, C., Fernández, M. L., & Klein, R. (2003). Hurricane vulnerability in Latin America and the Caribbean: Normalized damage and loss potentials. *Natural Hazards Review, 4*(3), 101–114. https://doi.org/10.1061/(ASCE)1527-6988(2003)4:3(101)

Ramírez, E. E., & Smith, H. J. M. (2016). What encourages cities to become sustainable? Measuring the effectiveness of implementing local adaptation policies. *International Journal of Public Administration, 39*(10), 718–728.

Reeves, A. (2011). Political disaster: Unilateral powers, electoral incentives, and presidential disaster declarations. *The Journal of Politics, 73*(4), 1142–1151. https://doi.org/10.1017/S0022381611000843

Sainz-Santamaria, J., & Anderson, S. E. (2013). The electoral politics of disaster preparedness. *Risk, Hazards & Crisis in Public Policy, 4*(4), 234–249. https://doi.org/10.1002/rhc3.12044

Salkowe, R. S., & Chakraborty, J. (2009). Federal disaster relief in the U.S.: The role of political partisanship and preference in presidential disaster declarations and turndowns. *Journal of Homeland Security and Emergency Management, 6*(1). https://doi.org/10.2202/1547-7355.1562

Schmidtlein, M. C., Finch, C., & Cutter, S. L. (2008). Disaster declarations and major hazard occurrences in the United States. *The Professional Geographer, 60*(1), 1–14. https://doi.org/10.1080/00330120701715143

Schneider, S. K. (1992). Governmental response to disasters: The conflict between bureaucratic procedures and emergent norms. *Public Administration Review, 52*(2), 135. https://doi.org/10.2307/976467

Stramp, N. R. (2013). The contemporary presidency: Presidents profiting from disasters: Evidence of presidential distributive politics. *Presidential Studies Quarterly, 43*(4), 839–865. https://doi.org/10.1111/psq.12069

Sylves, R., & Búzás, Z. I. (2007). Presidential disaster declaration decisions, 1953–2003: What influences odds of approval? *State and Local Government Review, 39*(1), 3–15. https://doi.org/10.1177/0160323X0703900102

Takeda, M. B., & Helms, M. M. (2006). "Bureaucracy, meet catastrophe." *International Journal of Public Sector Management, 19*(4), 397–411. https://doi.org/10.1108/09513550610669211

Teka, O., & Vogt, J. (2010). Social perception of natural risks by local residents in developing countries—The example of the coastal area of Benin. *The Social Science Journal, 47*(1), 215–224. https://doi.org/10.1016/j.soscij.2009.07.005

United Nations. (2015). *Sendai framework for disaster risk reduction 2015–2030.* https://www.undrr.org/quick/11409

Wolkow, K. E., & Ferguson, H. B. (2001). Community factors in the development of resiliency: Considerations and future directions. *Community Mental Health Journal, 37*(6), 489–498. https://doi.org/10.1023/A:1017574028567

World Bank. (2012). *FONDEN Mexico's natural disaster fund—A review.* https://openknowledge.worldbank.org/entities/publication/8351219a-b806-559e-8d4d-62de1aa16374/full

CHAPTER 10

PARTICIPATORY CLIMATE CHANGE GOVERNANCE IN ENABLING LEADERSHIP FOR CLIMATE ACTION AND SUSTAINABLE DEVELOPMENT IN SOUTH AFRICA

S. A. Mthuli
University of KwaZulu-Natal–South Africa

ABSTRACT

In 2018, a then 15-year-old Swedish schoolgirl by the name of Greta Thunberg challenged the climate status quo, beginning a global youth movement for climate action. In South Africa, whilst the country's progressive Constitution gives recognition to the environment, it is beset with several intricate issues which are fiercely contested by Civil Society Organizations (CSOs). Thus, this chapter attempts to explore what strategy can be used to mobilise South Africa's CSOs and youth to collectively lead climate change governance for action and sustainable development. Climate change understanding–which informs action–is limited at local government level. Where it does exist, it is new, and implementation is slow due to many factors including policy mis-alignment and minimal prioritization because of a lack of leadership and participation resulting in limited climate change governance. This is worsened by the fact that an understanding of climate change is not well grounded at grassroot level (Ward level) in South Africa's municipalities. This chapter's contribution lies in the argument that the lowest existing local participatory legislated structure is a suitable space to foster climate change understanding and debate which can inform action and leadership for climate action and sustainable development.

The focus on climate change came to take on a geophysical, rather than a bioecological form in global governance because it emerged from a dynamic, interactive process between states and scientists. In the 1950s, state agencies steered and accelerated the development of the geophysical sciences, which set the discursive frame within which climate politics now plays a role (Allan, 2019). By the end of the 1980s, the threat of climate change had entered the policy arena and the basic scientific conclusions about the causes and dimensions of the potential human impact on the climate were sufficient to bring pressure to bear and to encourage action to be taken at an international level (Soltau, 2009). This was located primarily in the developed world because as of the 1950s many countries in the developing world were either recently independent or still fighting for independence from their colonial masters. This was the case in most of sub-Saharan Africa were many countries were still in the tight grip of colonial rule which only slightly loosened in the mid-1960s. Post independent sub-Saharan Africa was than marred by neo-colonialization and civil conflict, *coupe de tats*, leadership traps and poor public sector performance (Mthuli, Singh & Reddy, 2022; Mthuli Singh & Reddy, 2023). Thus, the natural environment never took centre stage because of the political endeavours at the time. By the 1990s, scientists and experts in the Western world responded to the States' requests to study carbon sinks by expanding the climate concept to include a focus on greenhouse gases and land-use practices (Allan, 2017). Climate change not only affects livelihoods (with the hardest hit being the most destitute populations), but also undermines public administration interventions in both the developed, less developed and developing world. Climate change is 'the wicked problem of problems', a super-wicked problem with Vordermayer-Riemer (2020) asserting that:

> The impacts of climate change are unlikely to be felt first by the nations that have primarily contributed to it, ... there is a considerable likelihood that climate change impacts will be distributed in an unequal fashion at the expense of low-income states. The consequences of climate change may not only affect the very existence of state territory but also impact in various ways on soils,

water resources, agricultural production and sources of food, and hence aggravate the harsh living conditions of people. (p. 277)

Public administration practitioners', scholars and researchers have long been accused of ignoring critical areas, such as public education (Raffel, 2007) and perhaps the same can be said of climate change and climate governance even though it affects policy implementation at all ends, both directly and indirectly. This study's method of enquiry was qualitative in nature and sampled secondary data on the subject matter from Google Scholar and the Taylor & Francis search engine online databases, providing a quantitative content analysis to reach conclusions (Mthuli et al., 2022; Mthuli et al., 2023) on a strategy to mobilise South Africa's CSOs and youth to collectively lead climate change governance for action and sustainable development in the country.

THE ENVIRONMENT, PUBLIC ADMINISTRATION AND SOUTH AFRICA

The Public Administration and management discipline has been slow to address climate change and governance and Pollitt (2015, p. 181) is of the view that "academic public management research appears to have been slow to address these issues. Yet potentially there are several strong points of contact between climate change issues and current public management research themes" possibly relating to public sector performance and efficiency (see Mthuli et al., 2023). In recent times, Public Administration research and public administration practise have changed and advanced immensely (see McDonald III, Hall, O'Flynn & van Thiel, 2022), the latter no more so than in the last half a century, starting at a slow pace with its consequences being severely felt today and undermining public administration interventions in both the developed and developing worlds. This trajectory began with the conceptualization of terms such as people-centred development which was later referred to as sustainable human development (see Peezy, 1992; Ul Haq, 1995; Rogers, Jalal & Boyd, 2012). Presently, whilst there is a stronger emphasis on climate change governance directed at the human and environmental development agenda, it nevertheless has strong political underpinnings. Therefore, while the environment is not a new political issue or political *per se*, "history is indeed political" and there are "bound to be various opposing views. Many of the views are often used to serve nationalist, mythological ideologies and political agendas" (Manaka, 2022, p. 1). Scholars such as McDonald III et al. (2022), exploring emerging windows of opportunity in new research for the field of Public Administration, point to a shift in public sector values away from efficiency and effectiveness to equity in areas such as comparative administration, climate change and social equity. According to Knieling and Leal-Filho (2013), climate change globally and locally is a concern and climate change governance is an emerging area that is related to the behaviour of private actors, businesses, civil organizations and non-governmental organizations, as well as the state and public administration systems. This is the case in both the developed and developing world, such as in South Africa which is a reasonably young democracy having transitioned in 1994. Post 1994, the environment and local participatory governance processes were recognised as critical to the development of the country and in redressing the legacy of Apartheid as evidenced by its legislative and regulatory frameworks. Despite the several post-apartheid (democratic) policy interventions that the South African government introduced to redress and correct the wrongs of the past, evidence of factors such as inequality, poverty and unemployment are still visible (Biyela, Nzimakwe, Mthuli & Khambule, 2018), often referred to as the 'Triple challenge'. It therefore can be argued that climate change is another factor undermining gains and efforts directed at addressing slow socio-economic development, high levels of unemployment, poverty and growing inequality between the rich and poor not only in South Africa but in the rest of the developing world as well. Evidence of the country's commitment to climate change governance includes its Climate Change Bill introduced to Parliament in February 2023 (see Government of the Republic of South Africa, 2022) and the proposed setting up of a National Climate Centre in an existing institution (because are a number of existing institutions that play a role in climate services in the country and already perform some of the functions associated with the proposed centre). Such functions relate to the collation of climate information, data, products and applications, and to the facilitation of climate research and development in the country, according to the national Department of Forestry, Fisheries and the Environment (DFFE, 2019).

In Sub-Saharan Africa, it is the poorest and most vulnerable citizens that stand to suffer the most from a lack of governmental action and/or weak coordination with regards to climate action and governance. Questions of climate change governance deal with both mitigation and adaptation whist at the same time trying to devise effective ways of managing the consequences of these measures across the different sectors (Knieling & Leal-Filho, 2013), as well as across government layers at national, provincial and local levels. Furthermore, local government is government closest to the people where climate change education ought to enable sustainable development and debate ought to be facilitated through existing participatory systems. Scholars such as Susskind and Kim (2022) are of the view that governments at the local level, such as in cities around the world, because of failed efforts to build municipal capacity to formulate and implement climate adaptation plans "will not be able to cope with climate change impacts until they enhance their capacity to adapt" (p. 593). The impact will be felt more severely in the developing world, especially in small rural towns and in poor communities such as those prevalent in South Africa. A country ravaged by the Triple challenge (Mthuli, Singh & Reddy, *In-press*) has the least capacity to cope, hence a new approach is needed in these areas that is driven by local government, engaging and empowering key stakeholders and the youth into participatory governance processes and climate change adaption and mitigation into integrated development planning, were all stakeholders collectively acknowledge climate change and its intricacies. This is vital in climate change governance and efforts in South Africa. Thereby making participatory climate change governance, like e-governance (see Khanyile, Nzimakwe & Mthuli, 2021) another intervention to further deepen both its liberal democracy and local governability with the recognition that in pursuit of development, the environment and the state are imminence.

SHIFTS IN DEVELOPMENT, THE ENVIRONMENT AND THE STATE: A THEORETICAL AND POLITICAL PERSPECTIVE

The need and drive for human progress inter alia development can be traced to ancient times. In more recent times, post-World War Two saw the re-emergence of development as well as the emergence of development studies, when according to Williams (2014) the Western developed world's attention turned to widespread poverty in the nations referred to as that of the Third world. These were colonies or those newly independent states, at which the Cold War lent legitimacy to the calls for development through state involvement, this national development through state-managed economic change (Williams, 2014). This made the peoples development and the state's role an intertwined political issue. However, this changed fast because of the shift from an "embedded liberalism" to the "neo-liberalism" order which called for a move away from state driven national development to more private driven initiatives related to the public downsizing or offloading of state assets by privatizing, this changing the role and function of the state (see Williams, 2014; Hughes, 2018; Khunoethe et al., 2021; Mthuli et al., 2022). This creates a self-regulating market and the most controversial in public sector management for performance under the New Public Management paradigm (see Khunoethe, Reddy & Mthuli, 2021; Mthuli et al., 2022) driven by prominent Western political leaders. On the other hand, the emergence of the environment as a political issue is also not new. The people-centred development paradigm according to Ul Haq (1995) was a forerunner of the current concept of sustainable human development as a marriage between the concepts of environment and development. Sustainable development which was once the concern mainly of environmental specialists (Rogers, Jalal & Boyd, 2012) after the 1980s witnessed governments and development agents around the world change the way they thought of the environment and development with the belief that the two were no longer mutually exclusive and that a healthy environment was pivotal to sustainable development (Peezy, 1992). Since then, achieving the goal of sustainable development (SD) and sustainability has proven to be a problematic endeavour because of definitions and ambiguity (Gardener 2014). At supranational bureaucracy agency level, the United Nations Commission on Sustainable Development (CSD) was created in 1992 under the auspices of the Economic and Social Council as a direct result of the United Nations Conference on Environment and Development. This was on the backfoot of the recognition by top economists and planners in the 1990s who began to recognize that economic development which erodes natural capital is often unsuccessful because this degrades resources upon which future growth is dependent (Peezy 1992). The term 'sustainable' in sustainable development, while been used for over 30s years, has witnessed little practical implementation until recently, especially with regard to the fiercely debated notion of 'sustainable' economic policies. SD has many challenges tied to global environmental issues with sustainable development indicators for example pointing to the role of international financial institutions in global warming and related problems. The multidisciplinary literature on SD reveals a lack of a comprehensive theoretical framework for understanding SD and its complexities (Jabareen, 2008). Nonetheless, today the need for SD is at its highest with the world in a state of climate crisis which conflicts with economic development thus requiring a compromise relating to net-zero/net-free products and services. While what inspired early development theory may have been poverty alienation and human well-being located in economic and political factors. As well as market exchange based on the experiences of Western industrialised society (see Peezy, 1992; Ul Haq, 1995; Rogers, Jalal & Boyd, 2012; Williams, 2014; Allan, 2017; Manaka, 2022; Smith & Jacques, 2022). In the future this could be said also of public administration because of the recent developments and debates, both scientific and political, from the carbon-markets on the environment and environmental administration, process therein.

The Politics of the Environment and Public Administration

Environmentalism is informed by green thinking, which is an ideology in its own right and it has been viewed this way for a long time (see Heywood,

2017). It has also long been divided because the politics of the environment have always had competing interests and values resulting in two approaches, i.e., the reformist and the radical. The difference between the radical and reformist positions is that, on the one hand, the reformist position is anthropocentric, in that it is human centred and based on the notion that protecting the environment is primarily to the benefit of humans. This position is thus an attempt to depoliticise the environment suggesting that:

> ...environmental protection can be effectively incorporated within the political and economic structures of modern industrial society... [also that] ...economic growth and environmental protection are not necessarily incompatible objectives. Economic development must be sustainable...

While on the other hand, the reformists positing that:

> ...fundamental economic, social and political change–nothing less, is than the creation of a new kind of society with different institutions and values. (Garner, 2019, pp. 10–11)

These approaches have been challenged by what Garner (2019) calls the "'Promethean' or 'cornucopian' approach which denies the existence of acute environmental problems" (p. 12). The 'Promethean' or 'cornucopian' discourses have been a dominant mode of thinking for most of the last century (Garner, 2019). However, with climate change and global warming clearly on the increase and thus becoming an intractable political problem, Brunnée and Toope (2010) ask "how does one get states and political leaders to prioritize the issue, nationally and internationally?" (p. 126). Since the turn of this century the focus on climate change adaptation, rather than mitigation, has become more prominent (McNamara & Buggy, 2017). At the same time, Public Administration theory and practice has hardly faced up to the problem of environmental management (see Lieber, 1970). In Europe, a report was released in 2019 that was the basis for the development of the EU EMAS (Eco-Management and Audit Scheme) titled 'Best environmental management practices for public administration sector' outlining the best practices that could be adopted by all types of public administration, including those at a local level. Overall, the report can be used by all public administration entities as a source of information to identify relevant actions to take to improve their environmental performance (Canfora, Antonopoulos, Dri, Gaudillat & Schönberger, 2019).

Recent history has shown that the dominant social paradigm (DSP) as a cluster of values, beliefs and ideals are connected to environmental problems and policies, the view of society, individual responsibility (guided by our attitudes towards the environment) and government and regulatory measures (see Smith & Jacques, 2022). However, this varies between the developed and developing world. Nevertheless, further to that the DSP points out that perceptions regarding a society or individual's responsibility—in this case, to environmental problems—informs government regulatory measures. Considering that the DSP's definition may vary and its influence may encompass "the acceptance of laissez-faire capitalism, individualism, growth, and progress and a faith in some science and technology" (Smith & Jacques, 2022, p. 12). In Africa, the African Youth Charter's Article 13 speaks to Education and Skills Development and states the need for "the development of respect for the environment and natural resources", while Article 19 speaks to Sustainable Development and Protection of the Environment and also states that the Unions parties ought to ensure the use of sustainable methods to improve young people's lives while vesting in youths' interest by protecting the natural environment based on the notion that they are the inheritors of the environment. Further to that, also to:

a) Encourage the media, youth organisations, in partnership with national and international organisations, to produce, exchange and disseminate information on environmental preservation and best practices to protect the environment; b) Train youth in the use of technologies that protect and conserve the environment; c) Support youth organisations in instituting programmes that encourage environmental preservation such as waste reduction, recycling and tree planting programmes; d) Facilitate youth participation in the design, implementation and evaluation of environmental policies including the conservation of African natural resources at local, national, regional and international levels ....

South Africa has heeded this call legislatively and regulatorily but the processes and implementation thereof is complex and presents many challenges that remain under researched.

South Africa's Legislative Framework on the Environment

To understand South Africa's socio-economic context of climate governance, one ought to reflect first on the historical context of the country as it transitioned from Apartheid to democracy in 1994—which marked a pivotal turning point of the state formed in 1910. The country remains faced with many socio-economic problems borne not only of Apartheid but also colonialization. Therefore, there was a portentous need for the democratic government, led by Nelson Mandela and others, to play in readdressing problems of the past. In doing so, the democratic constitution, the Constitution of the Republic of South Africa 1996a, reigns supreme and recognizes that the cornerstone of the country's democracy is its Bill of Rights which is Chapter Two of the Constitution. Section 24 of the Bill of Rights speaks to the environment, in that:

Everyone has the right—

(a) to an environment that is not harmful to their health or wellbeing; and

(b) to have the environment protected, for the benefit of present and future generations, through reasonable legislative and other measures that—

(i) prevent pollution and ecological degradation;

(ii) promote conservation; and

(iii) secure ecologically sustainable development and use of natural resources while promoting justifiable economic and social development. (Constitution of the Republic of South Africa, 1996a, p. 9)

South Africa to date is in the process of having a Climate Change Act, seen as urgent and one of the most critical legislations of the day especially due to the outdated 2015 climate action plan to cut emissions. Currently at the Bill stage, the Climate Change Bill is under public hearing at the National Assembly (see Parliamentary Monitoring Group [PMG] 2022; Bega, 2022). The Act once promulgated ought to enable the country to make great strides in climate change adaptation, as well as climate mitigation, informed by its participatory system. Thus, the environment and human rights researchers and scholars alike in South Africa and outside, have written extensively on the environment in the country and its impact on its people, business and development, and vice versa. According to Bond, Dada and Erion (2009), South Africa is a signatory of the UNFCCC (the United Nations Framework Convention on Climate Change) and Montreal Protocol and the country via the National Environmental Management Air Quality Act has enacted legally binding air pollution regulations. Nevertheless, while climate change was seen as a serious problem around the world, most South Africans were largely ignorant of its effects during the mid-2000s (Bond et al., 2009) especially with energy efficiency low and millions of rural households having to burn wood for their fuel and energy needs which was more of a concern as compared to the deforestation and health problems it often caused.

Further to that, Bond et al. (2009), referencing the GlobeScan 2006 measure, asserted that the government, media, and business had kept ordinary South Africans in the dark with regard to the environment and related policy choice implications, "with civil society making uneven efforts to address the deficit" (p. 11). According to the national Department of Environmental Affairs [DEA] (2014), South Africa also slipped in terms of rankings (according to indicators related to its ecological footprint and the environmental sustainability index), resulting in pressures on the environmental system and weakness in the system to deal with those pressures. Nevertheless, a just collective climate governance is possible in South Africa with its continued policy commitment to the environment. However, it needs to be acknowledged that it would also remain threatened by the Triple challenge of unemployment, poverty and inequality. Today, while society remains polarized on the issue of climate change, there is an increasing but slow consensus thereof due to the frequent occurrence of natural disasters. These disasters are becoming ever more severe and impact not only natural but human systems as well, thus undermining local administrations, which are government closest to the people and which are most often under resourced.

LOCAL GOVERNMENT AND THE ENVIRONMENT

Local government pertains to local administration and is vital in all aspects of life, considering also that it is government closest to the people. In the EU for example, although for all types of public administration the focus has been on local authorities, local administrations (of which there are a high number in the EU) are seen as having the highest potential for replicability of best environmental management practices at local level. These best practices relate to all types of public administrations and include but are not limited to, "making office buildings more environmentally sustainable, minimising the impacts of meetings and events organised, promoting sustainable commuting and business travel, or adopting green public procurement" (Canfora, 2019, p. 2). Furthermore, the EU's Best Environmental Management Practice for the Public Administration Sector report in relation to local government according to Canfora (2019):

> ...identifies best practices that encompass their policy/regulatory/planning role as well as their role in providing key services to residents in the fields of sustainable energy and climate change, mobility, local ambient air quality, land use, water supply and municipal waste-water treatment, noise pollution, green urban areas, green public procurement and environmental education and dissemination of information to citizen and businesses. Alongside best environmental management practices, the report also identifies suitable sector specific environmental performance indicators that can be used to measure and track performance in the areas addressed by each best practice, and, when possible, benchmarks of excellence, corresponding to the level of performance achieved by frontrunners. (p. 2)

Such an approach is yet to gain attention in local authorities in South Africa and the rest of sub-Saharan Africa. In South Africa however scholars such as Merle-Sowman & Brown (2006) have pointed to that nonetheless:

Local government has become the intended focal point for addressing the socio-economic needs of local communities and sustainable service delivery, with the principal tool for achieving these developmental objectives, the Integrated Development Plan (IDP). (p. 695)

The IDP in the country is a local tool to guide development objectives at local authority level and Ruwanza & Shackleton (2016) point out that the IDP ought to also be balanced against environmental demands with a sustainable development paradigm. Due to this, the natural environment, its intricacies and relation to human well-being and development are enshrined in the cornerstone to the country's democracy, i.e., its Bill of Rights as found in the Republic's Constitution, one of the most revered in the world.

Local Government and the Environment in South Africa

In South Africa, considerable information, knowledge and guidance on the environment can be found in various environmental policy documents. These documents facilitate a better understanding of the environment, and concepts, issues and standards related to implementation for responsible environmental management in the country, including at local government level (DFFE, n.d). Also, further to that:

Environmental management has seen significant conceptual development... and has become progressively diversified at policy level in an attempt to adequately deal with emerging fields of practice such as climate change, sustainable design, integrated environmental management, and strategic environmental planning. (p. 52)

In South Africa, public service provision at a local government level entails addressing the basic needs of communities, such as the provision of clean water and sanitation, electricity and basic infrastructure as is necessary for socio-economic development. However, this is guided by the Integrated Development Plan (IDP) and the financial and the administrative capacity of a municipality (Constitution of the Republic of South Africa, 1996b). However, on many occasions, provincial or national government has had to intervene (as prescribed by the Constitution) to assist local government in meeting their Constitutional obligations (Khunoethe et al., 2021; Khumalo, Mthuli & Singh, 2019; Mkhize, Nzimakwe & Mthuli, 2022).

The objectives of the IDP are recognised as "strategic objectives" to be achieved by a municipality according to the Local Government: Municipal Systems Act 2000. It consists of a "Process Plan" which highlights the milestones and activities that need to be achieved by a municipality (Khunoethe et al., 2021) thus making the IDP, as prescribed by the Municipal Systems Act, a tool to realize its goals. However, "under the 2001 Regulation which requires all Spatial Development Frameworks (SDFs) to undergo a strategic environmental assessment (SEA), a glaring omission is sustainability as an objective. Consequently, sustainable development is not a stipulated aim of an IDP or SDF" (DFFE n.d, p. 52) while scholars such as Wüst (2022) express that the IDP is a forward planning mechanism and " is the 'integrated' and 'strategic' plan prepared for the term of office of a newly elected council. By contrast, the SDF is generally regarded as the longer-term plan focusing more on spatial planning" (p. 54). The SDF is hence aimed at guiding the overall spatial distribution of current and desirable municipal land use so as to give effect to the vision, goals and objectives of its IDP by identifying what development ought to take place and how much of it is needed or wanted. The SDF can thus be seen as a component of the IDP. Furthermore, there are several other plans linked to the IDP & SDF relating to the environment such as the Environmental Management Framework, the Air Quality Management Plan, and an Integrated Waste Management Plan (see Wüst, 2022).

Gardener (2014) reviewing existing IDP Assessment Frameworks (AFs), observed that IDP AFs used by the national and a particular provincial government to assess IDPs from a sustainability perspective, found that they do not sufficiently measure sustainability issues in district, metropolitan and cross-border regional plans. He also noted that the AFs lacked depth from a sustainability perspective. Wüst (2022) however argues that the practical reality is that the IDP & SDF are mostly used as two separate tools because of the preparation processes and timelines but nonetheless are perpetual characteristics of the country's local government and planning landscape. Although local government is mandated to deal with issues of local scale relating to municipal roads, water supply and sanitation, storm water management, refuse and solid waste disposal and environmental issues such as air and water pollution, they also need to execute projects in accordance with their IDPs and Special Development Frameworks (SDFs) (DFFE, n.d).

There are at the same time in the country "various representative structures that exist, set up to facilitate the communication within government structures, between bureaucracies and civil society and these structures are aimed at bi-directional information flow, in order to have a general developmental practice, support environmental management", and vice versa (DFFE, n.d, p. 47), including at grass-roots community and local government level. However, many environmental policy documents in South Africa end up having never making their way to a larger audience or practitioners outside of the institutions that created them (see DFFE, n.d) despite this being in a country where access to information, especially at local government, is seen as critical considering that it is perceived as government closet to the people (Khumalo et al., 2019). Local government is thus critical to public service delivery even though in South Africa it often experiences the most challenges in both public service delivery provision and in sharing and integrating related information to enable an effective participatory governance system (Khanyile et al., 2021). In the environmental management field, environmental policies and guidelines usually only get utilised by "individuals within the sector, rather than those people or institutions outside the field that have control over the 'actioning' of the recommendations and controls" (DFFE, n.d, p. 52). For example, the DFFE (n.d) asserts that:

> . . . it is an open question whether the recent South African Risk and Vulnerability Atlas (SARVA) drafted by the Department of Science and Technology (DST), will be seen and used by rural development practitioners who, understandably, view climate change as something that the 'environmentalists' need to deal with, not knowing that it directly affects the feasibility of farming and biodiversity management. (p. 52)

People outside the field include rural people, businesses and civil society organisations (CSOs) in communities. The mission of most CSOs in South Africa are in some way or the other tied to basic human Rights and public service delivery expectations because they are better positioned to provide knowledge of basic Rights and the practical implications on the ground thereof (Auditor General South Africa [AGSA], 2023).

Communities within the participatory governance framework ought to also have a say in their governance, whether or not it includes the general natural environment or their immediate natural environment. However, more critical to this is communities understanding and the flow of information to them relating to the environment. Evidence has been found that points to the role of public officials in gatekeeping information as critical in enabling an ideal participatory governance environment in a rural setting (see Khanyile et al., 2021).

Consequently, further to the issues of the environment at local government level in South Africa, sustainable development is not required content of an IDP or SDF and the Municipal Systems Act requires the integrated formulation of IDPs and SDFs but "linkages between the two tends to be tenuous in practice" (DFFE, n.d, p. 52), with Harrison (2006) concluding that local government in post-apartheid South Africa is less effective than was anticipated in 1996 because planning instruments such as the IDP have "shown their limitations in assisting local governments to be more effective" (DFFE, n.d, p. 52). Todes (2004) thus concludes that although the focus of IDPs should be on integration and a multi-sectoral approach to development is a positive aspect of the IDP approach, more emphasis needs to be given to environmental issues, as well as a planning approaches that take into account the context, (i.e., social, economic and political dynamics) of the local authority (Todes, 2004). The advent of South Africa's democracy brought with it new systems of governance and also transformed planning and decision-making processes. At the same time, the principles of sustainability, integration, participation, social and environmental justice have long been placed on the country's political agenda, but prioritization has remained questionable (see Sowman & Brown, 2006) and the role of CSOs also warrants further debate (see Bond et al., 2009).

CIVIL SOCIETY ORGANIZATIONS AND THE ENVIRONMENT IN SOUTH AFRICA

In democracies, the "constant back-and-forth between government, civil society or formally organized institutions such as CSOs or individuals" is commonplace (DFFE, n.d, p. 47). This is the very nature of democracy and its many other intricacies. This back-and-forth takes place in a democratic environment, where issues and views, whether important or not, are seen as vital and topical and are thus played out through the varied participatory processes at the various government levels. However, at an individual, household or community level this is not always the case as these voices are not always easily 'heard'. In such instances, civil society has a responsibility to articulate these interests or concerns into interest groups which then represent the collective voice of the people. Thus, government's role is to aggregate a variety of interests from society and to then decide which of these interests should take priority (DFFE, n.d, p. 47).

CSOs in the developed and developing world have taken a stand on the issue of climate change and fiercely debated related issues including the misleading 'greenwash'- a carbon market that permits trade in pollution rights and NGO co-optation as well as failed politicking (see Bond et al., 2009). Interest groups remain important influencers of the decision-making process and together with the people, legitimatize government decisions in a participatory system. In South Africa, CSOs and interest groups have long been critical in climate change issues in the country and according to Bond et al. (2009) the first critics of climate trading were not those that had the most to lose from exposing varied scam reduction projects, but rather activist groups in the early 2000s. Bond et al. (2009) states that in the country, prior to the early 2000s, "through networks like the Climate Action Network (CAN) and the South African Climate Action Network (SACAN)" whom since as early as "1997 accepted carbon trading as a necessary evil" (p. 23). The carbon market and trading were seen as an emerging from environmental injustices and the role played by civil society activism and advocacy to mitigate such injustices in the country. Furthermore, evidence has suggested that those with political power usually or sometime have conflicting interests such as in business dealings that further ensures the "failure of the environmental justice critique to penetrate the realm of policy" (Bond et al., 2009, p. 17).

A good quality environment has not been placed above material consumption as evidenced by the CDM's in South Africa and the contradiction to Section 24 of the Bill of Rights. While South Africa was not included in the Kyoto Protocol Annex one list of countries, Bond et al. (2009) expressed that corporations, officials and uncritical NGOs were "promoting the Kyoto Protocol's CDM as a way to continue South Africa's hedonistic output of greenhouse gasses, while earning profits in the process" (p. 13). Also in the country, because mainstreaming environmental issues in IDPs is still low with low budget allocations and a lack of inclusion of vision and mission statements that speak to environmental issues in municipalities, for the "mainstreaming of environmental issues to be effective in IDPs", there ought to be both proactive, multi-faceted bottom-up as well as top-down approaches (Ruwanza & Shackleton, 2016, p. 28). In South Africa, local governments are central and located opportunely for implementing local adaption to the impacts of climate change although little research exists to understand the conditions under which a municipality is able to initiate the process of mainstreaming climate adaption, one of many factors being the need for environmental champions within the political leadership (Pasquini, Ziervogel, Cowling & Shearing, 2015). The local youth, just like CSOs in municipalities, can also lead and be those champions in South Africa as observed in the developed world.

THE YOUTH AND THE ENVIRONMENT IN SOUTH AFRICA

The South African national DEA (2014) reiterated that the management of the environmental issues in the country tends to be left to environmental departments and professionals but there are capacity limitations in the country within the environmental sector. In addition, many role players do not thoroughly understand nor appreciate the significance of the environment. However, the tools and mechanisms do exist to ensure that environmental management, if facilitated well, along with ample public participation can aid in changing attitudes, values and behaviours, as well as associated adaptions in order for role players to understand their responsivities and rights, while working towards achieving sustainability (DEA, 2014). Further to that, the DEA (2014) points to environmental awareness campaigns and capacity building programmes as pivotal in effectively communicating information about key environmental issues and how these issues are impacted by development and the concept of sustainable development (DEA, 2014).

Pasquini et al. (2015) observed various factors at local government that enable action to mainstream climate change adaption based on two case studies in the Western Cape province of South Africa. Some of these factors include environmental champions within the political leadership, resources, and political stability (Pasquini et al., 2015). Hence the role of political leadership can never be under stated, as is further evidenced by the current state of water scarcity in the country. The Blue Drop & Green Drop Reports have shown a lack of commitment from the political leadership of municipalities despite directives from Department of Water and Sanitation with regard to pollution and quality of water throughout the country (see BusinessTech, 2023; SANews [South African News], 2023; South African Government, 2023). South Africa's current political local government landscape has transformed immensely post the 2016 Local Government Elections (LGLs), resulting in a high political competitiveness leading to coalitions to govern and political instability and fiercely contested LGEs. Hence, the current landscape presents an opportunity for political environmental champions, although much attention is away from the environment and focuses immensely on political control which an environmental champion can leverage as their cause for control. This political instability regarding the environment, according to Pasquini et al. (2015) remains under-researched. Therefore, this brings us to the question of what strategy can be used to mobilise youth and CSOs in South Africa to collectively lead climate change governance for sustainable development? One that is thus argued for is grass root, participatory and collective climate change governance which would also strengthen South Africa's troubled liberal democracy.

PARTICIPATORY CLIMATE CHANGE LOCAL GOVERNANCE: AN INFORMED COLLECTIVE EFFORT PERSPECTIVE FROM THE GROUND IN SOUTH AFRICA

Climate change has been happening for some time with its undeniable consequences felt across the world and will continue to be felt for a long time to come, hence the need for mitigation and adaptation strategies for the present and for the future. Climate change, like public sector performance and management in South Africa, is a wicked issue (see Mthuli et al., 2023), however the ultimate wicked problem is a public policy and administration issue. It, according to Pollitt (2016) "...will directly affect a vast range of government functions, from building regulations to flood defences; from agricultural policy to public health; from border controls to emergency services, and from energy policy through transport policy to the insurance industry and international diplomacy" (p. 78). Climate change is complex, multifaceted, eludes easy categorization, is planetary in scope and inter-generational in its implications. It involves everything—thus, addressing it calls upon everyone across the world. It is also then a "classic collective action problem. It can only be solved if all states, or at least the major greenhouse gas emitters, cooperate" as the phenomenon of human-induced climate change and its dangerous implications are now beyond doubt, which was not always the case (Brunnée & Toope, 2010, p. 126).

The approach and systems in place to deal with climate change need immediate attention and in South Africa the DEA (2014) expressed that, "the system should further facilitate effective climate change mitigation and adaptation...and the realization of sustainable human communities, recognizing that while in the country there are various strategies focusing specifically on climate change, such as the climate change response plan" (pp. 28–29), there remains no accepted set of indicators to measure the success of actions towards sustainable development and sustainability in the country.

Climate change is here to stay for a long time, unless drastic measures are taken to address it related to mitigation and adaptation in South Africa. This starts by ensuring participatory climate change governance at local government is led and driven by youth and CSOs that organise for local climate policies and whom are representative of all citizens in the country regardless of race, religion, educational and economic standing because these can threaten participatory climate change governance, making climate change and sustainable development to be seen as a 'privileged' subject. Societal social-economic imbalances are a contextual factor in the country that ought to seen as a threat to its democratic functioning and future endeavours. Thus, climate change integrated into participatory governance processes at local government, driven by youth and CSOs can be a breeding ground for productive ideas and debates on the matter, whilst also fostering social capital critical for environmental education and learning on the matter. This is supported by Jorgenson et al. (2019) who are of the view that children and the youth being involved in climate change action and education points explicitly to collective action and multi-actor networks. Further to that, Krasny et al. (2015) suggest that:

> ...social capital presents a framework for how environmental education (EE) programs can bring youth and adults together to create the conditions that enable collective action, as a complement to ongoing work in EE focusing on individual behaviors. (p. 1)

South Africa's third democratic parliament (from 2004–2009) was characterised by a strong move towards community participation, which requires local government to consult all sectors of its community on services being offered or about to be rendered as supported by the White Paper on Transforming Public Service Delivery (1997), wherein the first of the eight *Batho Pele* principles involves 'consultation' as being a priority. The principle of consultation refers to the idea that the views of communities are to be sourced on the nature of services being provided, as well as on the quality of such services (see Booysen, 2009; Khanyile et al., 2021), as well as on the environment. Thus, participatory climate change governance would add value to Integrated Environmental Management if well-grounded at Ward Committee level.

In the country there are legislations and regularity frameworks that guide community participation and systems/methods in local government, and these exist to deepen democracy, ensure governability and continuously improve service delivery in the county (Khaynile et al., 2021). However, for this to happen, the environment also needs to be considered. Legislative and regulatory interventions are threatened if the natural environment is not understood and considered, hence there is a need to reconcile different interests and values. These are further undermined by the Triple challenges in South Africa as was argued earlier. Participatory democratic local governance ought to inform participatory climate change and for it to work, there ought also to be equity, equality, and a political and economic balance in and between communities. Communication also must be a two-way channel to balance power with development and to ensure substantial progress. These can be seen as some of the by-products of participatory democracy as entrenched in the IDPs to ensure positive progress in South Africa's local government.

Nevertheless, the imbalance that exists, that of socio-economic development in South Africa's populace and existing competing interests (some borne of this economic imbalance) will undermine participatory climate change governance interventions. The existing values and scarcity of resources on the side of government and the state require political leaders that have an understanding of the critical role of the state, its existence and statecraft which are vital in addressing public sector issues (see Mthuli et al., 2023), including in this case climate change and governance. Hence, redefining the political character of the issue of climate change is critical in South Africa. In this regard, Garners (2019) reminds us that it is imperative to note that,

> ... the impact of climate change has differed, and will continue to differ, from state to state, and from community to community, and the costs of dealing with it are going to be similarly diverse [...] the effect of climate change impact upon people, groups, classes, nations and regions vary differently. (p. 7)

Furthermore, he contends that climate change is a political issue because government and the state are central in dealing with the implications of climate change (Garner, 2019). However, recent evidence and the development of the EU EMAS shows that it goes beyond climate change being simply a public administration issue, but a policy implementation activity informed by government policy choices as well. The South African government's role, policy choices and evident solutions to addressing climate change, such as the CDM's can be seen as a way to articulate Garners point. Competing claims, discourses and environmental thought have divided the approaches to environmentalism between the radicals and the reformists as explained earlier. However, Garner (2019, p12) is of the view that, "it is possible to hold position in both camps." Barry (1994;1999) cited in Garner (2019, p. 11) is of the view that this has remained unhelpful in reaching what could be seen a just sustainable development. Climate change is indeed a wicked problem facing public administration today and as Alford & Head (2017) assert, there need not be so much attention on solving this wicked problem but rather in making more progress towards better managing the problem.

Science and technology in South Africa to better understand climate change can enable participatory climate governance. Crafting solutions or possible solutions begins by acknowledging that there is a need for widespread youth public learning. This can be facilitated by integrating climate change knowledge in already existing local governance structures such as Ward Committees which are chaired by Councillors (see Khanyile et al., 2021). The DEA (2014) has pointed out that young people need to be explicitly involved in environmental issues through integrating knowledge of environmental rights and even the National Environmental Management Act [NEMA] 1998 (Act No. 107 of 1998) principles into national educational programmes, as well as awareness and capacity needs of role players such as NGOs involved in development (DEA, 2014). However, Eilam (2022) points out that climate change is scarcely addressed in school curricula globally and school graduates are mostly uneducated about it. This can be seen as undermining any current and future mechanisms to manage, monitor and evaluate climate change, especially since the involvement of the youth is critical in ensuring sustainable futures for themselves and future generations. Further to that, the lack of youth understanding of climate change and participation will undermine any facilitation of youth participation in the design, implementation, and evaluation of environmental policies at local and national level, as noted in the African Youth Charter's Article 13 and 19. Hence grass-roots climate change governance should be more strongly cultivated through science and technology and facilitated though education, understanding and awareness at a local level in South Africa. This will contribute to ensuring that not only the governance system but also those at Ward level, including the youth and CSOs, will collaborate to adapt to and/or mitigate the effects of climate change and contribute to more resilient futures. As Pasquini et al (2015) assert that access to a knowledge base and social networks have been observed to have a positive effect on environmental mainstreaming. Furthermore, a reformist stance would enable a unified policy approach to climate change governance. However, in the South African context, this remains threatened by various socio-economic development issues tied to the country's triple challenges of inequality, unemployment, and poverty.

Climate change is thus unfortunately a political issue although some or many may think otherwise, because to overcome it and its related issues requires organized power such as that of the youth and others in society to realise not only sustainable development that is just but that is based on existing participatory climate governance systems in the developing world. In current times, this is also threatened by misinformation and Smith & Jacques (2022) believe that this may pose a serious threat to future environmental policy formulation. South Africa's local government, as a microcosm of the country's Triple challenges, is no exception to such issues pertaining to information, its availability, its trustworthiness and the flow thereof. Participatory climate change local governance, can thus inform behaviour changes through choice, ensure integrated learning, facilitate behavioural change debate, informed decision making and planning at grass root level, thereby contributing to policy and regulatory alignment in South Africa. South Africa's continued policy commitment to the environment can be realised by ensuring participatory local governance processes that would contribute to collective climate governance and eventually to more democratic processes in the facilitation of climate self-governance. This could perhaps create a strong 'back to nature' movement amongst South Africa's youth, similar to the German experience and the green ideology of the 1960s and will assist in entrenching the idea of sustainable development in the country as originally advanced in the 1987 Brundtland Report and Rio 'Earth Summit' in 1992 (see Heywood, 2017).

CONCLUSION

Africa, because of its projected youth population growth in the near further, should seize the opportunity and drive climate change policy. In South Africa, a strategy that can be used to mobilise youth and CSOs in South Africa to collectively lead climate change governance for sustainable development is one that is at grass root level, participatory and collective. This would also aid in strengthening South Africa's troubled liberal democracy, it is argued. Integrating a focus on climate change mitigation and adaption in an already existing participatory governance system—enabled by climate change education—would be key in influencing the dominant social paradigm and related values, beliefs and ideals connected to environmental problems and policies. It will also spur understanding and awareness regarding individual's climate change risk and/or mitigation perceptions amongst those who have influence or will have an influence on policy making in the country–which are those at grass-roots level, and can influence policy choices based on location, rather than race, gender, educational and/or economic standing. This could also contribute to developing an accepted set of indicators to measure the success of actions towards sustainable development and sustainability in the country, as well as enabling the co-production of knowledge at grass root level and thus aid in overcoming the failure of the environmental justice critique to penetrate the realm of policy. An area for future research could be to empirically understand the effect of political instability on the environment as well as to examine the role and potential of political environmental champions who can leverage their cause for control in coalitions.

REFERENCES

Alford, J., & Head, B. W. (2017). Wicked and less wicked problems: A typology and a contingency framework. *Policy and Society, 36*(3), 397–413. https://doi.org/10.1080/14494035.2017.1361634

Allan, B. (2017). Producing the climate: States, scientists, and the constitution of global governance objects. *International Organization, 71*(1), 131–162. https://doi.org/10.1017/S0020818316000321

Auditor General of South Africa. (2023). *Audit information/civil society organisations.* Retrieved June 9, 2023 from https://www.agsa.co.za/AuditInformation/CivilSocietyOrganisations.aspx

Bega, S. (2022, May 19) Climate change bill: One of the most important draft laws to cross the desks of SA's lawmakers. *Mail & Guardian.* Retrieved June 9, 2023 from https://mg.co.za/environment/2022-05-19-climate-change-bill- one-of-the-most-important-draft-laws-to-cross-the-desks-of-sas-lawmakers

Biyela, A. C., Nzimakwe, T. I., Mthuli, S. A., & Khambule, I. (2018). Assessing the role of intergovernmental relations in strategic planning for economic development at local government level: A case study of Umkhanyakude District Municipality. *Journal of Gender, Information and Development in Africa, 7*(2), 221–239.

Bond, P., Dada , R., & Erion, G. (2009). Introduction. In P. Bond, R. Dada, & G. Erion (Eds), *Climate change, carbon trading and civil society: Negative returns on South African investments* (pp. 1–32). University of KwaZulu-Natal Press.

Booysen, S. (2009). Public participation in a democratic South Africa: From popular mobilisation to structured co-optation and protest. *Politeia, 28*(1), 1–27.

Brunnée, J., & Toope, S. (2010). *Legitimacy and legality in international law: an interactional account.* Cambridge University Press. https://doi.org/10.1017/CBO9780511781261.006

BusinessTech. (2023). *Municipalities face criminal charges over water quality in South Africa.* Retrieved June 20, 2023 from https://businesstech.co.za/news/government/694289/municipalities-face-criminal-charges-over-water-quality-in-south-africa

Canfora, P., Antonopoulos, I. S., Dri, M., Gaudillat, P., & Schönberger, H. (2019). *Best environmental management practice for the public administration sector.* JRC Science for Policy Report, EUR 29705 EN, Publications Office of the European Union, Luxembourg, 2019, ISBN 978-92-76-01442-3, https://doi.org/10.2760/952965, JRC116121. Retrieved from: https://data.europa.eu/doi/10.2760/952965 (Accessed: 12 May 2023).

Constitution of the Republic of South Africa. (1996a). *Chapter two: Bill of rights.* Retrieved June 2, 2023 from https://www.justice.gov.za/legislation/constitution/saconstitution-web- eng.pdf

Constitution of the Republic of South Africa. (1996b). *Chapter seven: Local government.* Retrieved June 2, 2023 from https://www.justice.gov.za/legislation/constitution/saconstitution- web-eng.pdf

Department of Environmental Affairs. (2014). *The environmental impact assessment and management strategy for South Africa–2015.* Retrieved June 2, 2023 from https://www.dffe.gov.za/sites/default/files/docs/eiams_environmentalimpact _manage mentstra tegy.pdf

Department of Forestry, Fisheries & the Environment. (2019). National climate change adaptation strategy. Retrieved June 9, 2023 from https://www.dffe.gov.za/sites/default/files/docs/nationalclimatechange_adaptationstrat egy_ue 10november2019.pdf

Department of Forestry, Fisheries & the Environment. (n.d). Governance (Chapter 4). Retrieved May 7, 2023 from https://www.dffe.gov.za/sites/default/files/reports/environmentoutlook_chapter4.pdf

Eilam, E. (2022). Climate change education: The problem with walking away from disciplines. *Studies in Science Education, 58*(2) 231–264, https://doi.org/10.10 80/03057267.2021.2011589

Gardener, R. D. (2014). *Sustainable regional development: Developing a sustainability assessment framework for district and metropolitan integrated development plans* (Masters dissertation, Stellenbosch: Stellenbosch University). Retrieved June 24, 2023 from https://scholar.sun.ac.za/items/05461303-9741-4fc4-8228 -7b643f370a0b

Garner, R. (2019). *Environmental political through: Interests, values and inclusion.* Red Globe Press, Springer Nature Limited.

Government of the Republic of South Africa. (2022). Climate change bill. Retrieved June 9, 2023 from https://www.gov.za/sites/default/files/gcis_document/202203/b9-2022.pdf

Heywood, A. (2017). *Political ideologies: An introduction* (6th ed.). Red Globe Press.

Jorgenson, S. N., Stephens, J. C., & White, B. (2019). Environmental education in transition: A critical review of recent research on climate change and energy education. *The Journal of Environmental Education, 50*(3), 160–171. https://doi.org/10.1080/00958964.2019.1604478

Khanyile, M., Nzimakwe, T. I., & Mthuli, S. A. (2021). Challenges of innovations in participatory governance in uMshwathi local municipality in Kwazulu-Natal. *Journal of Public Administration, 56*(4.1), 972–985.

Krasny, M. E., Kalbacker, L., Stedman, R. C., & Russ, A. (2015). Measuring social capital among youth: Applications in environmental education. *Environmental Education Research, 21*(1), 1–23. https://doi.org/10.1080/13504622.2013.843647

Khumalo, S. M., Mthuli, S. A., & Singh, N. (2019). Economic development through the local informal economy in sustaining livelihoods: The case of the rural coastal town of Mtubatuba. *Journal of Public Administration, 54*(4–1), 772–789.

Khunoethe, H., Reddy, P. S., & Mthuli, S. A. (2021). Performance management and the integrated development plan of the Msunduzi municipality in South Africa. *NISPAcee Journal of Public Administration and Policy, 14*(2), 161–187.

Knieling, J., & Leal-Filho, W. (2013). Climate change governance: The challenge for politics and public administration, enterprises and civil society. In J. Knieling & W. Leal Filho (Eds.), *Climate Change Governance.* Climate Change Management. Springer. https://doi.org/10.1007/978-3-642-29831-8_1

Lieber, H. (1970). Public administration and environmental quality. *Public Administration Review, 30*(3), 277–286. https://doi.org/10.2307/974044

McDonald, B. D. III, Hall, J. L., O'Flynn, J., & van Thiel, S. (2022). The future of public administration research: An editor's perspective. *Public Administration, 100*(1), pp. 59–71. https://doi.org/10. 1111/padm.12829

McNamara, K. E., & Buggy, L. (2017). Community-based climate change adaptation: A review of academic literature. *Local Environment, 22*(4), 443–460, https://doi.org/10.1080/13549839.2016.1216954

Mkhize, L. N., Nzimakwe, T. I., & Mthuli, S. A. (2021). Bureaucrats' views on performance management in a KwaZulu-Natal Provincial government department. *Administratio Publica, 21*(1), 83–101.

Mthuli, S. A., Singh, N., & Reddy, P. S. (2022). Political leadership for improved performance in South Africa: A shift from public management to new public governance. In C. Jones, P. Pillay, P. S. Reddy, & S. Zondi (Eds.), *Lessons from political leadership in Africa: Towards inspirational and transformational leaders* (pp. 15–35). Cambridge Scholars Publishing. https://books.google.co.za/books?id=y_5fEAAAQBAJ&lpg=PA15&ots=teHU-if6—&dq=mthuli%20sa&lr&pg=PA15#v=onepage&q=mthuli%20sa&f=false

Mthuli, S. A., Singh, N., & Reddy, P. S. (2023). Political leadership and Ubuntu for public sector performance in South Africa. In R. Elkington, F. W. Ngunjiri, G. J. Burgess, X. Majola, E. Schwella, & N. de Klerk (Eds.), *African leadership: Powerful paradigms for the 21st century* (pp. 27–43). Emerald Publishing Limited. https://doi.org/10.1108/978-1-80117-045-120231004

Sowman, M., &. Brown, A. L (2006). Mainstreaming environmental sustainability into South Africa's integrated development planning process. *Journal of Environmental Planning and Management, 49*(5), 695–712. https://doi.org/10.1080/09640560600849988

Parliamentary Monitoring Group. (2022). *Climate change bill: Public hearings.* Retrieved June 9, 2023, from https://pmg.org.za/committee-meeting/35596/

Pasquini, L., Ziervogel, G., Cowling, R. M., & Shearing, C. (2015). What enables local governments to mainstream climate change adaptation? Lessons learned from two municipal case studies in the Western Cape, South Africa. *Climate and Development, 7*(1), 60–70. https://doi.org/10.1080/17565529.2014.886994

Pollitt, C. (2015). Wickedness will not wait: Climate change and public management research. *Public Money & Management, 35*(3), 181–186. https://doi.org/10.10 80/09540962.2015.1027490

Pollitt, C. (2016). Debate: Climate change—The ultimate wicked issue. *Public Money & Management, 36*(2), 78–80.

Raffel, J. A. (2007). Why has public administration ignored public education, and does it matter? *Public Administration Review, 67*, 135–151. https://doi.org/10.1111/j.1540- 6210.2006.00703.x

Ruwanza, S., & Shackleton, C. M. (2016). Incorporation of environmental issues in South Africa's municipal integrated development plans. *International Journal of Sustainable Development & World Ecology, 23*(1), 28–39. https://doi.org/10.10 80/13504509.2015.1062161

Smith, Z. A. (2022). *The environmental policy paradox.* Taylor & Francis.

Soltau, F. (2009). *Fairness in international climate change law and policy.* Cambridge University Press. https://doi.org/10.1017/CBO9780511635403.003

South African Government. (2023). *Minister Senzo Mchunu on release of Blue Drop Watch Report, No Drop Watch Report, and Green Drop Watch Report.* Retrieved June 20, 2023, from https://www.gov.za/speeches/minister-senzo-mchunu -release-blue-drop-watch-report-no-drop-watch-report-and-green-drop

South African News. (2023). *Report shows decline in drinking water quality.* Retrieved June 20, 2023, from https://www.sanews.gov.za/south-africa/report -shows- decline- drinking-water-quality

Susskind, L., & Kim A. (2022). Building local capacity to adapt to climate change, *Climate Policy, 22*(5), 593–606. https://doi.org/10.1080/14693062.2021.187 4860

Wüst, F. (2022). The South African IDP and SDF contextualised in relation to global conceptions of forward planning—A review. *Town and Regional Planning, 80*(1), 54–65.

Vordermayer-Riemer, M. (2020). *Non-regression in international environmental law: Human rights doctrine and the promises of comparative international law.* Intersentia. https://doi.org/10.1017/9781839701221.008

CHAPTER 11

OCEAN HEALTH PART I

The Concept, Governance, and South African Context

Thean Potgieter
University of the Free State

ABSTRACT

The economic and welfare growth the world experienced during the previous century might have been unprecedented, but it was unequal, resulted in climate change and was to the long-term detriment of the environment. Human existence depends on ocean health as the oceans absorb carbon dioxide, regulate the global climate and recycle nutrients. Oceans are crucial for global trade, provide vital sources of food and hydrocarbons, while coastlines are exceptional habitable and leisure spaces.

Oceans-related economic activities contribute substantially to the GDP of South Africa and are crucial to socio-economic growth, development and transformation. But it is threatened by climate change, unsustainable ocean management, poor governance and maritime security challenges. Utilising ocean resources must therefore be environmentally sustainable, well governed and aimed at the preservation fragile marine ecosystems—the so-called blue economy approach.

Climate Governance in International and Comparative Perspective, pages 231–249
Copyright © 2024 by Information Age Publishing
www.infoagepub.com

This chapter is the first of two on ocean health and places emphasis on methodological aspects, a review of relevant sources, the importance of the oceans to the planet, climate change, and blue economic governance. Although Part II is more concerned with the anthropogenic influences, the two parts form a whole.

Destruction lurks beneath the beauty of the ocean surface. It might not be obvious to most observers as 'going to the sea' is synonymous with vacation, recuperation and a 'good time'—yet the health of our oceans, the harbinger of life, is deteriorating. The economic and welfare growth the world experienced during the previous century might have been unprecedented, but it was unequal and came at a real cost to the environment. This is to our long-term detriment and why concerns about oceans' health need precedence are glaringly noticeable in the global and African maritime domain.

As human beings our existence is inextricably linked to the ocean and coastal ecosystems. The sea covers more than 70% of earth's surface, it absorbs carbon dioxide, regulate global climate and temperatures, and recycle nutrients amongst others. Oceans are crucial to the global economy as they carry more than 80% of global trade, their health is vital for a food secure future, they provide roughly a third of the global supply of hydrocarbons, and coastal regions provide exceptional habitable and leisure spaces. However, evidence is mounting about the adverse ecological impact of climate change on oceans, coasts, communities, and those dependent on the sea for their livelihood.

The story of sub-Saharan Africa is also a maritime one. The sea made interaction with the rest of the world possible and as such contributed to forging the complex socio-economic character of the continent. The sea is important to the well-being and development of the continent, it contributes to livelihoods, protein, and the cultural experiences of African societies. But maritime economic activities must be environmentally sustainable and its associated developments equitable—essentially parallel to the 'green economy' notion of development being socially inclusive and ecologically balanced.

Governments, both internationally and on the African continent, increasingly emphasize the contribution that ocean resources could make to economic growth and development. But many challenges inhibit the opportunities of communities and societies to fully benefit from the immense wealth and capacity of our oceans. Although the economic potential of South Africa's extensive resource rich maritime domain is substantial, this wealth and geography is also a scourge because it is difficult to govern and provides ample opportunities for illicit activities.

Due to the vulnerability of our oceans and multiple threats (ranging from climate change to the illicit exploitation of ocean resources) global cooperation provides the best chances of preserving healthy oceans and balancing economic activities with sustainability and equity. Nations therefore concluded global pacts, linking environmental care to ending poverty. This is inherent to the 2030 Agenda for Sustainable Development of the United Nations (UN), which includes a specific emphasis on conservation and the sustainable use of the oceans. Yet due to insufficient environmental care and the unscrupulous plundering of marine resources progress is not encouraging.

The focus of this chapter is on ocean health and South African blue economic governance within the international and regional context. Inherent to this is sustaining healthy oceans and thriving coastal communities, which are threatened by climate change and human interference. This chapter is the first of two on ocean health. In Part I emphasis is placed on methodological aspects, a review of relevant available sources, the importance of the oceans to the planet, climate change, environmental threats, the economic potential and governance of the maritime domain. Part II is essentially about the anthropogenic influences on ocean health with specific reference South Africa. But the two parts form a whole—the one could not be appreciated without the other. Comprehending the impact of climate change and anthropogenic actions on the oceans and communities, necessitates an appreciation of the role of oceans in the well-being of the planet as well as the who, why and what of climate change. This I endeavoured to allude to throughout.

EXPLAINING THE RESEARCH APPROACH

The methodology selected had to be the most suitable for contributing towards an analysis of ocean health in the South African context. But the dilemma is, how could sense be made from the complex and changing physical phenomena in the ocean environment, the impact of humans and society, as well as our efforts to properly govern the greater natural processes in a context of human voraciousness? To that effect this analysis of reality is an endeavour to contribute to our ability to make sense through utilising knowledge from various spheres and specific empirical instances of theoretical phenomena (Wessels & Potgieter, 2021, pp. 9–20). A diversity of relevant thematic foci thus had to be relied upon, including from areas such as oceanography, biology, geology, climatology, history, public administration, sociology, geography, economics, politics, and security studies.

The scholarly approach of this chapter and its results could be interpreted as interdisciplinary. Conceptually the study design and methodology are not limited to any one field, as it links and integrates theoretical frameworks from various disciplines and the analysis depends on perspectives and skills specific of these disciplines (Tobi & Kampen, 2018, p. 1210). It draws appropriate ideas and concepts from across related and unrelated disciplines (or separate branches of expertise) and applies diverse theoretical lenses to understanding complex problems beyond the limitations of common boundaries with a view of creating greater inclusive insight together with actionable knowledge (Appleby, 2019).

Interdisciplinarity is necessitated by the relevance of both natural sciences and humanities for a study aimed at the understanding the interrelationships between the physical and living marine environment, the anthropogenic impact on the oceans, the blue economy as well as the governing of such diversity. The complementary and integrated knowledge from different disciplines is vital for answering complex questions, greater understanding and producing new insights (Wernli & Darbellay, 2016, pp. 4–13).

REVIEWING SELECTED SOURCES

Interdisciplinary research on ocean health is challenging due to the differences in scientific cultures when analysing data and reporting research results (specifically between the natural and social sciences). The systematic review of relevant literature for these two chapters confirmed the general impression that finding a symbiosis between diverse paradigms or epistemologies, or perhaps even the tension between positivism and constructivism, poses unique challenges.

The extensive scope of the topic at hand required an exploration of literature relevant to the interconnections between disciplines. Implicit to such an interdisciplinary approach, is the need to study diverse sources. Many of sources are relevant to the content of both chapters (Part I and Part II) on ocean health. Although it is difficult to highlight only some sources from the vast body of literature representing different scientific paradigms, a few would nonetheless be acknowledged.

Preference fell on scholarly sources, and despite the dichotomy of different scientific approaches, scholarly quality was usually obvious to identify. An important measure was to place emphasis on accredited publications and sources from academic publishing houses. Various studies consulted on the physical oceans' environment, climate change and extreme weather were based on considerable empirical research and data analysis. An invaluable source is *The Oceans: A Deep History* by Eelco Rohling (2017). It provided significant insights into the 4.4-billion-year natural history of the oceans as well as their crucial contribution towards sustaining life on the planet as we know it.

Cooley and Schoeman (2022a, 2022b) were the coordinating lead authors of a large research team that produced two pivotal pieces on oceans and coastal ecosystems and their services. It formed part of a substantial interdisciplinary study by an intergovernmental panel on climate change and its impact. The sections pertaining to the oceans resulted from extensively assessments, including new laboratory studies, field observations and process studies, model simulations, as well as Indigenous and local knowledge. It highlighted risks and provided evidence on the impact of climate change and anthropogenic activities on ocean and coastal systems as well as on human communities and activities. The authors also examined potential solutions to these incredible threats to our existence, ranging from ocean governance issues to ecological and human adaptation.

Many sources by international organisations (such as the UN and the World Bank) and reputable non-governmental organisations (NGO) such as Oxfam and the World Wildlife Foundation (WWF) were utilised, given contributions were scientifically sound, by renowned specialists or from reputable institutions. An Oxfam report (Cole, 2015) highlights the unprecedented global environmental changes and social inequality akin to our current situation. It emphasises that South Africa's future depends on its ability to address environmental challenges and social deprivation, together with creating safe and humane spaces for people to live in.

The discussions on the use and abuse of living resources and pollution depended on data from South African government sources, international organisations, NGOs and news outlets. Data on harvesting living resources are, amongst others, available from the UN Food and Agriculture Organisation (FAO), the World Bank, Statistics South Africa (StatsSA) and the South African government department responsible for fisheries and the environment. FAO provides public access to vast statistical databases that have been maintained for more than six decades. Its flagship publication is the biennial State of World Fisheries and Aquaculture report. It provides detailed comparative data on global aquatic resources, the trade in seafood products as well as illegal, unreported, and unregulated (IUU) fishing (FAO, 2020; FAO, 2022). However, when analysing statistics, a major challenge is the discrepancies between different datasets—specifically on catches, IUU fishing and poaching. Even official sources differ, as data provided by Statistics South Africa (StatsSA) is different from other government data (StatsSA, 2020; StatsSA, 2021; DEFF, 2020). This is a major shortcoming and addressing it is crucial for effectively governing living resources.

In assessing IUU fishing and poaching FAO data was of relevance, but more detail on the South African case was required. The report by De Greef and Haysom (2022) on the illicit flow of abalone from South Africa provides such depth. It elucidates the disconcerting extend of an illegal trade which is at its highest level ever, has caused the collapse of the resource, and has profoundly influenced corruption, the erosion of state institutions and criminalisation. News items and various websites (including newspapers, popular science magazines, and NGOs) assisted with piecing together such stories. But due to the challenges associated with information on the

Internet, it was always necessary to closely scrutinise sources. The test for reliability depended on the credentials of the author, the originator or the sponsor of the information, the currency of the information at its publication date and the purpose of a specific website.

Utilising such diverge sources contributed towards greater insight and understanding of natural phenomena as well as the analysis of the risks and opportunities relevant to ocean health. It linked these great global patterns with the South African context.

OCEAN HEALTH, CLIMATE CHANGE AND COMMUNITIES

Scientists estimate that the earth was formed about 4.6 billion years ago. The earliest indications of single-celled creatures are about 4.28 billion years old, whereas multicellular life took longer and developed about a billion years ago. Mammals appeared roughly 66 million years ago, modern humans only about 200,000 years ago, while our settled lifestyle is around 11,000 years old—a very short period in the history of our planet (Rohling, 2017; Heinberg, 2021).

Although evidence suggests that the oceans formed about 4 billion years ago, their shapes continued to evolve due to the movement of continents, coastal erosion, sedimentation as well as the impact of humans during the last few millennia. However, the oceans are crucial for sustaining life as we know it since the small fraction of the sun's energy that reaches earth across vast tracks of space is sufficient to prevent most of the ocean water on the planet from freezing (Rohling, 2017; Heinberg, 2021).

The average depth of our oceans is about 3,700 m (oceans are the deepest at trenches associated with subduction zones), whereas the continental shelves (at the edges where continents continue under water) are less than 200 m deep. Some continental shelves are very small, but in some cases (as between Australia and New Guinea or the islands and peninsulas of Southeast Asia) shelves could be massive. There are various, but inconclusive, theories on the origin of water and the atmosphere of the planet. Salt developed from dissolved minerals due to chemical weathering, while the high levels of carbon dioxide (CO_2) in the early atmosphere caused acid rain and rock corrosion (Rohling, 2017). The salinity of our oceans and spatial gradients effects the movement of the atmosphere and oceans. Circulation is key to the movement of heat, the ingredients of life (such as oxygen) and causes atmospheric-surface pressure systems. Through the powerful surface water flows around the world virtually the whole surface ocean is connected. This whole ecosystem could be severely disrupted by rising temperatures and large freshwater flows due to the melting of ice caps.

In the 21st century our planet and its peoples must buffer and manage the serious legacy challenges that resulted from insufficient environmental care, ranging from the depletion of resources, overharvesting of species, excessive waste, and climate change, to habitat and biodiversity loss. Specifically damning is our dependency on fossil fuels (since the Industrial Revolution commenced in the eighteenth century), the use of toxic chemicals, and atomic weapons test, amongst others. Although substantial scientific developments certainly contributed to the quality of human life, it increased greenhouse gas emissions and caused higher temperatures, altered the chemical configuration of the atmosphere and oceans, destroyed ecosystems, contributed to mass extinction of species, the wide presence of plastics and higher radioactivity. If these trends continue to accelerate, it will "constitute a turning point for all life on Earth" (Heinberg, 2021).

The greenhouse effect is natural to earth as greenhouse gases, such as water vapour (H_2O), carbon dioxide (CO_2), methane (CH_4) and nitrous oxide (N_2O), are present in the atmosphere. Greenhouse gases allow solar radiation to reach earth and absorb infrared radiation emitted by Earth, which heats earth's surface. It insulates the planet and makes it liveable—without it, earth would be on average 33°C colder. But burning fossil fuels (coal, oil and gas) release large amounts of carbon dioxide (CO_2) into the air and contribute to trapping more heat in the atmosphere, which causes global warning. As fossil fuels represent about 75% of global greenhouse gas emissions and 90% of all CO_2 emissions, it is the largest contributor to climate change (UN, n.d.).

Scientists warn that global warming will increase due to greenhouse gas emissions and concentrations of CO_2 in the atmosphere. Predictions are that global surface temperature rise would be up to 4°C, while international leaders target a 1.5°C average increase. Greater temperature rise result in climate change, and as more weather disasters is its effect, the socio-economic cost would still become substantially higher. The past three decades saw more "billion-dollar weather related disasters" than ever before, with fifty in 2020 alone (Le Roux, 2022). As South Africa is one of largest contributors of greenhouse gas emissions (currently 14th in the world), its persistent reliance on fossil fuels and extractive industries does not sufficiently address the urgent requirements for reducing emissions.

Climate researchers are convinced that global warming adversely affected ocean health. As the oceans warm it aggravates eutrophication (nutrient enrichment with high plant and algae growth) and deoxygenation (or hypoxia—low levels of oxygen), which together with ocean acidification (reduced carbonates) have adversely affected the physical and chemical composition of oceans (Cooley & Schoeman, 2022a, pp. 381–386). It altered the cyclic and seasonal characteristics of marine habitats, causing a biological shift in marine ecosystems (Huggett et al., 2022, p. 81). The result is 'dead zones' with reduced habitats, mass mortalities and disruption of fish populations in coastal marine eco-systems and on the verges of continental shelves (Cooley & Schoeman, 2022a, pp. 395–397, 410–415). As a result of an increase in ocean temperatures, global ice loss and sea-water expansion, the global mean sea level rose around 0.16 m between 1902 and 2015. But the quicker recent rate (3.6 cm per year between 2006 and 2015) is the highest on record and of significant concern (Scholes & Engelbrecht, 2021, p. 3).

Climate change exacerbates the effects of non-climate anthropogenic drivers such as the degradation of the marine environment, pollution, overfishing, eutrophication, and invasive foreign species. Small and microscopic oceanic life (referred to as zooplankton) is crucial for marine ecosystems and carbon cycling, and since they quickly reflect thermal variations, they are also reliable indicators of climate related changes (Huggett et al., 2022, p. 81). This has specifically been evident in the climate-driven decline in global zooplankton biomass. Such changes can have profound implications as it can result in declining food quality for fish and biomass loss of pelagic fish (Heneghan et al., 2023). As the human exploitation of oceans are growing, within the context of climate change it is crucial to study key ocean indicators to assess changes in the oceans, inform policy and improve ocean governance.

Climate change has adversely affected South Africa's ecosystems, economy, development opportunities and livelihoods. The increase in average temperatures since 1999 is double the global rate, which is simultaneous with changing rainfall seasons, regular droughts, stronger winds, and flooding. This impacts negatively on biodiversity, marine stocks, water security, food production and job security (Cole, 2015). In addition, extreme temperatures place more strain on health systems while also affecting service delivery, infrastructure, and emergency services.

In terms of environmental sustainability and social justice South Africa is not 'healthy'. As South Africa is one of the most unequal societies in the world, the living conditions of many people is dire, and the imperative for government to prioritise addressing poverty is obvious. Due to urbanisation, population growth and industrial developments, the country face considerable challenges associated with environmental care, climate change, land productivity, sea and air quality and freshwater provisioning (Cole, 2015).

South Africa (as other coastal countries) is experiencing more intense rainfall and more extreme weather at shorter intervals which increases the risk of climate related disasters and flooding in urban and coastal areas. Together with sea-level rise it causes saltwater intrusion and as areas become uninhabitable, whole communities require relocation (Le Roux, 2022). Heat waves are more frequent with protracted drought and desertification in some areas together with increases in average near surface temperatures. Agriculture faces major risks due to drastically warmer and drier futures

(Scholes & Engelbrecht, 2022, pp. 1–4). As a result, ecosystems are changing due to overused and degrading land, which in combination with oceans stresses due to climate change, poor oceans management, and biodiversity losses, threatens food security, jobs and cause financial losses.

Since 2010 South Africa has more regularly experienced extreme weather with a disrupted socio-economic impact. Well-known examples include the Knysna fire in 2017, the drought that ravaged the Western Cape and Eastern Cape (up to 2018 and 2023 respectively), the 2019 floods in Gauteng, the cyclones, storms, severe rainfall and floods in KwaZulu-Natal in 2022 and 2023, as well as the Western Cape storms and floods in 2023.

For two days (April 11–12, 2022) penetrating rains equal to the annual rainfall, fell in the provinces of KwaZulu-Natal and the Eastern Cape. It caused floods and landslides, resulted in the deaths of at least 450 people, destroyed more than 12,000 houses and forced about 40,000 people to flee their homes (Tandon, 2022). The flood caused much damage as the region was still recovering for two large storms and three tropical cyclones that hit southeast Africa in a period of six weeks earlier in 2022 (Burke, 2022). The combined effect of poor spatial development (with marginalised groups living in areas more vulnerability to flooding due to structural inequality), poor infrastructure maintenance by authorities and insufficient integrated disaster management have exacerbated the impact. The associated socio-economic losses are significant and is increasing, which probably makes the April 2022 KwaZulu-Natal floods South Africa's most costly weather-related disaster ever (Durban Chamber of Commerce and Industry, 2022).

Coastal flooding in the Western Cape has become more frequent with higher loss of life and economic cost. This is due to a combination of natural events (large low pressures and cold fronts with big storms and very high rainfall) and human factors such as vegetative overgrowth in waterways, land pollution, clogged drainage systems, overpopulation, and non-adherence to environmental laws through the settling on vulnerable ecosystems (Dube et al., 2022, S453, S462).

During June 2023 coastal regions in the Western Cape, Eastern Cape and KwaZulu Natal again received severe rain. Low-laying areas in the Western Cape were flooded, while considerable storm damage occurred. Local authorities in coastal areas often provide insufficient early warning and disaster information to prepare local communities. Studies have shown that a more proactive and responsive sharing of information is required to improve governance processes as well as community resilience and adaptability in climate related disasters (Busayo & Kalumba, 2021). An enhanced focus on coastal flooding is necessary as the "knowledge of the actual extent of climate change risk to coastal areas remains a challenge [and] ... addressing flood resilience in that context is problematic" (Dube et al., 2022, p. 454).

The impact of the floods on communities and the environment was exacerbated by governance failures of local authorities. Poor wastewater management and infrastructure maintenance often pollute coastal waters which is to the detriment of marine ecosystems, negatively affects economic activities such tourism and seafood harvesting, while it holds considerable risks to human health. Due to insufficiently maintained infrastructure and poor spatial planning, natural disasters such as the floods in KwaZulu-Natal caused untold harm to people, massive destruction to the environment and resulted in harmful off-flow of debris and pollutants into the oceans. A major challenge is the lack of consequence management as executive managers, councils and officials are not held to account for not governing as they were supposed to.

As the oceans supports human well-being through regulating planetary cycles and ecosystems, the extensive living resources of the oceans and various maritime related occupations are vulnerable to the impacts of over-harvesting, pollution and climate change. Climate related changes (such as algal blooms and overgrowth, and marine pathogen) together with pollutants, toxins and harmful run-offs into the ocean poses risks to seafood safety, exposes humans to harmful pollutants and cause economic losses. The geographic spread of marine-borne pathogens and toxins are exacerbated by climate change as they influence marine food cycles, while posing risks to human health and ecosystems (Cooley & Schoeman, 2022a, pp. 382, 457–462).

Fishing and aquaculture are specifically at risk as recent harmful algal blooms linked to climate change and marine pathogen have resulted in economic losses to fisheries and aquaculture in Asia, North America and South America. Since 2016 it caused estimated losses of U.S. $800 million to salmon farms in Chile, the closure of the Dungeness crab and razor clam fisheries in the United States, harmed fishing, and disrupted communities in the Pacific United States (Cooley & Schoeman, 2022a, p. 469).

The so-called 'red tides' common to the Benguela upwelling region of the Cape West Coast, emanate from a dense accumulation of microscopic algae, which could be poisonous to humans as the algae containing toxins. The low oxygen levels of such accumulations cause marine mortalities and intermittent strandings of rock lobster (locally known as crayfish or *kreef*). In 1997 and 2000 hundreds of tons of crayfish 'walked-out' in Elands Bay and in February 2023 a red tide caused the walkout of five tons of crayfish in the greater St Helena Bay area. DFFE activated a West Coast Rock Lobster Contingency Plan for government agencies in all spheres and communities to cooperate in rescuing and rehabilitating life crayfish and coordinate clean-up operations (DFFE, 2023).

Climate change will also reduce the abundance and quality of other natural marine products such as preservatives, dietary supplements, sponges, cosmetic products, pearls, choral and jewellery. Harmful algal blooms resulted in a decline in tourism and coastal properties in the United States, UK and France, while higher ocean temperatures and nutrient enrichment have caused an increase in floating sargassum (brown macroalgae or seaweed) in the central Atlantic Ocean. It disrupted tourism on beaches in the Caribbean and Mexico and cleanup costs ran into millions of dollars. Although such changes often induce tourists to move to different locations, the communities depending on tourism lose livelihoods and it contributes to inequality (Cooley & Schoeman, 2022a, pp. 467–469).

Greater governance focus on both nature conservation and addressing the impacts of climate change is required. The wetlands, seagrasses, salt lagoon and marshy areas of coastal ecosystems store much carbon removed from the atmosphere, while they also protect coastlines from erosion and the impact of storms (Independent Philanthropy Association of South Africa, 2022, p. 44). Strategically located marine protected areas in ecological sensitive areas can contribute much towards coastal protection and mitigating the impact of climate change, ensuring biodiversity, and building carbon-storing or 'blue carbon' ecosystems.

MARITIME GOVERNANCE AND THE BLUE ECONOMY CONTEXT

Although utilising ocean and coastal spaces for economic growth and development is an enduring leitmotif in human history, balancing economic activities with protecting vulnerable ecosystems is more recent and has become an issue of growing importance. The resource wealth of the oceans provides the "natural capital" necessary for wealth creation, but unscrupulous business interest and greed increasingly threatens the environment. The global necessity to combine economic pursuits with promoting healthy oceans highlights the important distinction between an ocean or maritime economy and the blue economy—with the latter implying the necessity for economic growth and development to be balanced with sustaining ocean health and the environment (Potgieter, 2018, p. 51). However, sustainable blue economic growth and development require good governance, environmental care, and maritime security.

Governance in the maritime domain could be seen as the "the legislative, institutional and implementation mechanisms aimed at regulating activities in the maritime space and coordinating the involvement of state and non-state actors" (Potgieter, 2021, chapter 7). These could be mandatory provisions made in national laws and regulations, or voluntary such as codes of conduct, agreements, and regional and international frameworks.

The emphasis on blue economic development within the South African and regional context ran concurrent with a growing international focus. But many risks pertain to the global maritime environment, including climate change, natural disasters, marine pollution, IUU fishing and insecurity. Good international, regional and national governance will contribute towards mitigating these risks and balancing the inherent tension that exist between three crucial aspects namely socially equitable economic development, environmental sustainability and security. Environmental degradation can lead to poverty and the marginalisation of coastal communities, stimulate illicit activities, insecurity and can cause radicalisation (Potgieter, 2018, p. 51).

The primary international governance framework is the UN Convention on the Law of the Sea (UNCLOS) accepted in 1982. It makes provision for the sustainable development of the world's oceans, national jurisdiction over ocean spaces, access to the seas, environmental care, utilisation of both the living and non-living resources of the oceans, research as well as how to address differences stemming from it (UN, 1982). The International Maritime Organization (IMO) is the specialized UN agency responsible for safety and security of shipping, prevention of marine and atmospheric pollution by ships, waste dumping and pollution at sea (IMO, 2023).

In general states have consensus on the importance of international agreements on maintaining good order at sea, maritime safety and security and managing transboundary marine resources for sustaining marine species, specifically in the context of climate induced distributional patterns. UNCLOS is supported by a plethora of international and regional agreements and international structures such as the UN Framework Convention on Climate Change (UNFCCC), the UN Convention on Biological Diversity (CBD), the Convention on Wetlands (Ramsar) as well as various Regional Seas Conventions and Action Plans (Cooley & Schoeman, 2022b, 3SM 27, 40; FAO, 2023).

The UN Sustainable Development Goals (SDGs) is a call to end poverty through sustainable and inclusive economic development and to protect the environment. The blue economy ideal is inherent to Goal 14 on the sustainable use of marine resources and the conservation of oceans, which is closely coupled with Goal 13 on fighting climate change and limiting emissions (UNECA, 2016, p. 9). After the UN Assembly declared 2021–2030 the decade of Ocean Science for Sustainable Development ('the Ocean Decade') a High-Level Panel for a Sustainable Ocean Economy was created by 16 states (including Ghana and Namibia from Africa) to link ocean health and wealth through advancing environmentally sustainable models for the development of the ocean economy (Loureiro et al., 2022, pp. 11–12).

In March 2023 a landmark UN treaty, the High Seas Treaty aimed at protecting the world's oceans, was agreed upon. It provides for global rules to protect biodiversity in international waters, the fair and equitable sharing of resources, and the creation of protected marine areas beyond national jurisdiction—implying that about 30% of the oceans could be protected from damaging human activities (Kim & Treisman, 2023). The fact that there was sufficient political will to agree upon the various frameworks indicated above is of note and they provide important targets to strive to. But implementation occurs with mixed measures of success as the examples of the pledges at the 2017 Paris Agreement to reduce the greenhouse gas emissions indicates. The pledge made at the 2011 CBD to improve the management of marine resources through sustainable ocean harvests, protect 10% of global ocean with MPAs, and ease anthropogenic pressure by 2020 were not met as only 7.74% of marine areas are protected (Cooley & Schoeman, 2022b, 3SM 40).

Although South Africa and other countries in the region are signatories to UNCLOS, to really give meaning to it require regional cooperation, consistent action, the domestication of international legal prescripts and maritime capacity. This specifically pertain to governing the marine environment and its living and non-living resources, marine infrastructure, maritime security, and coordination within the national, regional and international spheres. The idea of blue economic development and enhanced maritime security received a regional boost with the adoption of the 2050 Africa Integrated Maritime Strategy (referred to as AIMS) by the African Union. AIMS places emphasis on the importance of blue economic growth and maritime security to "foster increased wealth-creation from Africa's oceans and seas by developing a sustainable thriving blue economy in a secure and environmentally manner", while improving the well-being of citizens and "reducing marine environmental risks as well as ecological biodiversity deficiencies" (African Union, 2012).

Regional cooperation remains important. The peaceful, productive and environmentally sustainable use of the Indian Ocean and its resources was reiterated by members of the Indian Ocean Rim Association (IORA). It formed part of the of Perth Principles, a Declaration IORA member states (including South Africa) accepted on 1 November 2013, and formed the basis for the 2015 Declaration on "Enhancing Blue Economy Cooperation for Sustainable Development in the Indian Ocean Region." It emphasised that states must protect their maritime environments to ensure sustainable development, as well as the importance of more blue economic investments in fisheries and aquaculture, renewable ocean energy, seaport development, shipping as well as offshore hydrocarbons and seabed minerals. Seventeen Indian Ocean states (including eight African states) singed this Declaration at the September 2015 IORA meeting in Mauritius (IORA 2015).

During the past two decades South Africa consistently contributed to regional maritime agreements, specifically on economic cooperation, fighting IUU fishing, stimulating ocean health, as well as regional security cooperation (on issues of piracy, smuggling, borders, and terrorism). Due to insufficient political will and limited resources various actors have and are still benefitting immensely (often unscrupulously) from Africa's rich maritime resources (Masie and Bond, 2018, p. 319). Blue economic development and inter-regional maritime governance therefore require greater and consistent South-South, and specifically inter-Africa, cooperation.

South Africa has the longest coastline in Africa (close to 3,000 km), and inclusive of the Prince Edward Islands the country's Exclusive Economic Zone (EEZ) is about 1,553,000 square km, which could increase to about four million square km if the continental shelf extension claims are recognised (South Africa, 2012). Together with South Africa's rich ocean resources and its extensive maritime interests, coastal goods and services (in various sectors) contribute about 35% to the country's gross domestic product (GDP). Although this is crucial to socio-economic growth and transformation, prioritising maritime issues from a development and governance perspective is a recent development in South Africa. The focus on maritime sector growth and development was highlighted in the National Development Plan 2030, and permeated into an ambitious ocean economy initiative launched in 2014 as part of *Operation Phakisa* (Department of Planning, Monitoring and Evaluation, n.d.; Operation Phakisa, 2014). Its focus areas include maritime transport, manufacturing, offshore oil and gas, small harbour development, coastal and marine tourism, aquaculture, protection services and governance. In 2010 the ocean economy provided about 316,000 employment opportunities and contributed roughly R54 billion to GDP, which Phakisa planners thought could by 2033 rise to more than a million jobs and a R177 billion contribution to GDP (Potgieter, 2018, pp. 54–55).

Despite global economic conditions, poorer than expected South African economic performance and the COVID-19 pandemic, amongst others, Operation Phakisa did result in some new investments and contributed towards streamlining maritime governance processes. Through a cooperative effort by DFFE and the Department of Science & Innovation a National Oceans and Coastal Information Management System (OCIMS) platform came into being to provide ocean and coastal data services. Small harbour development projects emanating from Operation Phakisa supports small-scale fishing, coastal vessel activities and more inclusive maritime developments (Loureiro et al., 2022, pp. 11–12). But in the main improved trade and shipbuilding, infrastructure development and oil and gas initiatives remain a combination of opportunities, growth, false starts, and frustrations. South Africa certainly has the wherewithal for performing considerably better in terms of its blue economic growth and development. But this will depend on investors' confidence, good governance, maritime security, and environmental sustainability.

LESSONS OF RELEVANCE FOR THE WAY FORWARD

An interdisciplinary research approach is relevant for a study aimed at making scholarly sense of the intricate and changing physical phenomena and human influences pertinent to the ocean environment. As increasingly complex challenges pertaining to ocean health must be well managed, gaining inclusive insight, actionable knowledge and devising appropriate interventions require wisdom that is beyond the limitations of one sphere of academic endeavour.

In the context of climate change, decisionmakers need to be more serious about governing ecosystems and ocean resources. Important considerations include how changes in temperature, weather and the chemistry of oceans alter the geographic movement of species and what, in combination with anthropogenic influences are the socio-economic consequences to communities. This must be combined with scientific evidence, political will, budgets, public education, and a community approach to limit vulnerabilities to changes in oceans and ecosystems. If ecosystems are well managed it will bolster their ability to adapt and limit the impact of future climate change.

The detrimental impact that the floods in coastal provinces (specifically in KwaZulu-Natal in 2022) had on communities and the environment was exacerbated by the governance failures of local authorities. Specifically poor wastewater management, infrastructure maintenance, and errors in spatial development caused considerable damage, polluted coastal waters and marine ecosystems, negatively affected economic activities, and posed considerable risks to human health. Yet there were no consequences for executive managers, councils, and officials that failed to govern as they were supposed to.

CONCLUSION

Climate change combined with stresses induced by human activity has resulted in changes to fragile marine ecosystems across the world. As ocean surfaces around the world is virtually connected by powerful surface water flows, higher temperatures and large freshwater flows from melting of ice caps could cause severe disruptions to marine biodiversity, and with it the character, nature and livelihoods of societies and coastal communities.

The various global initiatives and agreements that places ocean health centrally in the climate debate and endeavours to conserve biodiversity contribute towards developing coherent and common frameworks for environmentally sustainable blue economic activities. However, it did not yet result in the truly transformative changes necessary.

To address the impacts of climate change, a greater focus on nature conservation, specifically coastal ecosystems and marine biodiversity is required. Wetlands, salt lagoon, and coastal ecosystems have a crucial role to play in storing carbon as well as protecting coastlines from erosion and storms. Greater governance and policy focus is needed to create and sustain strategically located marine protected areas in ecological sensitive areas. This will not only contribute towards coastal protection, but mitigate the impact of climate change, safeguard biodiversity, and build carbon-storing or 'blue carbon' ecosystems.

South Africa faces considerable challenges due to a combination of climate change, high levels of inequality and poverty, urbanisation, population growth, and industrial developments. Climate factors such as rising sea levels, more severe storms combined with lacking oceans governance threatens quality of life, the environment, food security, health, biodiversity, and marine ecosystems. The recent floods (in 2022 and 2023) in three of South Africa's coastal provinces bears testimony to the devasting impact of climate change when combined with poorly maintained infrastructure, inadequate governance and incorrect spatial development.

As Africa's rich ocean resources are of much economic value to the continent, contributes to food security and provides many livelihoods, more emphasis on the blue economic growth and development are necessary. But for economic benefits derived from the oceans to be environmentally sustainable, it must be in the context of international and regional cooperation, good maritime governance, environmental care, and sound business approaches to chart a collective sustainable development course for the region.

This chapter placed the spotlight on climate change, ocean health and South African blue economic governance within the international and regional context. But as ocean health is severely impacted upon by anthropogenic factors, the next chapter (Part II) will endeavour to explore how humans influenced, and gravely affected, living resources and ocean health with specific reference to the South African case.

REFERENCES

African Union. (2012). *2050 Africa's integrated maritime strategy (2050 AIMS)*. African Union. https://au.int/sites/default/files/documents/30929-doc-2050_aim_strategy_eng_0.pdf

Appleby, M. (2019). *What are the benefits of interdisciplinary study?* OpenLearn, The Open University. https://www.open.edu/openlearn/education-development/what-are-the-benefits-interdisciplinary-study

Burke, J. (2022, April 24). After the relentless rain, South Africa sounds the alarm on the climate crisis. *The Guardian.* https://www.theguardian.com/world/2022/apr/24/south-africa-floods-rain-climate-crisis-extreme-weather

Busayo, E. T., & Kalumba, A. M. (2021). Recommendations for linking climate change adaptation and disaster risk reduction in urban coastal zones: Lessons from East London, South Africa. *Ocean & Coastal Management, 203*, 105454. https://doi.org/10.1016/j.ocecoaman.2020.105454

Cole, M. (2015). *Is South Africa operating in a safe and just space? Using the doughnut model to explore environmental sustainability and social justice.* Oxfam Research Report. https://policy-practice.oxfam.org/resources/is-south-africa-operating-in-a-safe-and-just-space-using-the-doughnut-model-to-555842/

Cooley, S. R., & Schoeman, D. S. (Coordinating lead authors). (2022a). Oceans and coastal ecosystems and their services. In H.-O. Pörtner, D. C. Roberts, M. Tignor, E. S. Poloczanska, K. Mintenbeck, A. Alegría, M. Craig, S. Langsdorf, S. Löschke, V. Möller, A. Okem, & B. Rama (Eds.), *Climate change 2022: Impacts, adaptation and vulnerability. Contribution of Working Group II to the Sixth Assessment Report of the Intergovernmental Panel on Climate Change* (pp. 379–550). Cambridge University Press. https://https://doi.org/10.1017/9781009325844.005

Cooley, S. R., & Schoeman, D. S. (Coordinating lead authors). (2022b). Oceans and coastal ecosystems and their services. Supplementary material. In H.-O. Pörtner, D. C. Roberts, M. Tignor, E. S. Poloczanska, K. Mintenbeck, A. Alegría, M. Craig, S. Langsdorf, S. Löschke, V. Möller, A. Okem, & B. Rama (Eds.), *Climate change 2022: Impacts, adaptation and vulnerability. Contribution of Working Group II to the Sixth Assessment Report of the Intergovernmental Panel on Climate Change* (3SM-1–3SM-68). Cambridge University Press. https://www.ipcc.ch/report/ar6/wg2/

De Greef, K., & Haysom, S. (2022). *Disrupting abalone harms. Illicit flows of H. midae from South Africa to East Asia.* Global Initiative Against Transnational Organized Crime. https://globalinitiative.net/analysis/abalone-south-africa-east-asia/

Department Environment, Forestry and Fisheries, South Africa. (2020). *Status of the South African marine fishery resources 2020.* https://www.dffe.gov.za/sites/default/files/docs/publications/statusofsouthafrican_marinefisheryresources2020.pdf

Department of Forestry, Fisheries and the Environment, South Africa. (2023, February 8). *West coast rock lobster contingency plan activated following marine species walkouts.* DFFE Media Release. https://www.dffe.gov.za/mediarelease/creecy_westcoastrocklobster_redtides

Department of Planning, Monitoring and Evaluation. (n.d.). *Operation Phakisa.* http://www.operationphakisa.gov.za/Pages/Home.aspx

Dube, K., Nhamo, G., & Chikodzi, D. (2022). Flooding trends and their impacts on coastal communities of Western Cape Province, South Africa. *GeoJournal 87*(Suppl 4), 453–468. https://doi.org/10.1007/s10708-021-10460-z

Durban Chamber of Commerce and Industry. (2022, July 14). *The human cost of climate change. Flood risks and management in urban landscapes* [Paper presentation]. Academy of Science of South Africa (ASSAf) Presidential Roundtables. https://research.assaf.org.za/handle/20.500.11911/162

Food and Agriculture Organisation. (2020). *The state of world fisheries and aqua-culture 2020. Sustainability in action.* https://doi.org/10.4060/ca9229en

Food and Agriculture Organisation. (2022). *The state of world fisheries and aqua-culture 2022. Towards blue transformation.* https://doi.org/10.4060/cc0461en

Food and Agriculture Organisation. (2023). *FAO treaties database.* https://www.fao.org/treaties/results/details/en/c/TRE-000003/

Heinberg, R. (2021). *Power. Limits and prospect of human survival.* New Society Publishers. https://www.perlego.com/book/2195420/power-limits-and-prospects-for-human-survival-pdf

Heneghan, R. F., Everett, J. D., Blanchard, J. L., Sykes, P., & Richardson, A. J. (2023). Climate-driven zooplankton shifts cause large-scale declines in food quality for fish. *Nature Climate Change, 13,* 470–477. https://doi.org/10.1038/s41558-023-01630-7

Huggett, J. A., Groeneveld, J. C., Singh, S. P., Willows-Munro, S., Govender, A., Cedras, R., & Deyzel, S. H. P. (2022). Metabarcoding of zooplankton to de-rive indicators of pelagic ecosystem status. *South African Journal of Science, 118*(11/12), 81–84. https://doi.org/10.17159/sajs.2022/12977

Independent Philanthropy Association of South Africa. (2022, February). *The climate crisis—A toolkit and resource pack for funders in South Africa.* https://ipa-sa.org.za/wp-content/uploads/2022/02/IPASA-TOOLKIT-Version-24-Feb-2022.pdf

Indian Ocean Rim Association. (2015, September 3). *Mauritius declaration on blue economy.* https://www.iora.int/media/8216/iora-mauritius-declaration-on-blue-economy.pdf

International Maritime Organization. (2023). *Introduction to IMO.* https://www.imo.org/en/About/Pages/Default.aspx

Kim, J., & Treisman, R. (2023, March 7). What to know about the new U.N. high seas treaty—And the next steps for the accord. *NRP News.* https://www.npr.org/2023/03/07/1161196476/un-high-seas-treaty-international-waters

Le Roux, A. (2022, July 14). *SADC's climate risks and trends. African futures and innova-tion* [Paper presentation]. ASSAf Presidential Roundtables. https://research.assaf.org.za/handle/20.500.11911/162

Loureiro, T. G., Du Plessis, N., & Findlay, K. (2022). Into the blue—The blue economy model in Operation Phakisa "Unlocking the Ocean Economy" programme. *South African Journal of Science, 118*(11/12), 11–14. https://doi.org/10.17159/sajs.2022/14664

Masie, D., & Bond, P. (2018). Eco-capitalist crises in the "blue economy": Operation Phakisa's small, slow failures. In V. Satgar (Ed.), *The climate crisis: South African and global democratic eco-socialist alternatives* (pp. 314–337). Wits University.

Operation Phakisa. (2014). *Unlocking the economic potential of South Africa's oceans: Marine protection services and governance executive summary.* South African gov-ernment. http://tinyurl.com/y8vpfkpj

Potgieter, T. (2018). Oceans economy, blue economy, and security: notes on the South African potential and developments. *Journal of the Indian Ocean Region, 14*(1), 49–70. https://doi.org/10.1080/19480881.2018.1410962

Potgieter, T. (2021). South Africa: The blue economy experience. In D. Sparks (Ed.), *The blue economy in Sub-Saharan Africa,* Taylor and Francis. https://

www.perlego.com/book/2567169/the-blue-economy-in-subsaharan-africa
-working-for-a-sustainable-future-pdf

Rohling, E. J. (2017). *The Oceans. A deep history.* Princeton University Press. https://
www.perlego.com/book/740076/the-oceans-a-deep-history-pdf

Scholes, R., & Engelbrecht, F. (2021). *Climate impacts in Southern Africa during the 21st
century.* Centre for Environmental Rights. https://cer.org.za/wp-content/
uploads/2021/09/Climate-impacts-in-South-Africa_Final_September_2021
.FINAL_.pdf

Statistics South Africa. (2020). *Census of ocean (marine) fisheries and related services
industry, 2018. Financial and production statistics.* http://www.statssa.gov.za/?
page_id=1856&PPN=13-00-01&SCH=7917

Statistics South Africa. (2021). *Ocean (marine) fisheries and related services industry,
2019.* http://www.statssa.gov.za/publications/Report-13-00-00/Report-13-00
-002019.pdf.

Tobi, H., & Kampen, J. K. (2018). Research design: The methodology for interdisciplinary research framework. *Quality & Quantity, 52,* 1209–1225. https://doi.
org/10.1007/s11135-017-0513-8

United Nations. (1982). *United Nations convention on the law of the sea (UNCLOS).* http://
www.un.org/depts/los/convention_agreements/texts/unclos/part6.htm

United Nations. (n.d.). *Causes and effects of climate change.* UN Climate Action.
https://www.un.org/en/climatechange#

United Nations Economic Commission for Africa. (2016). *Africa's blue economy: A
policy handbook.* https://www.uneca.org/sites/default/files/PublicationFiles/
blueeco-policy-handbook_en.pdf

Wepener, V., & Degger, N. (2019). South Africa. In C. Sheppard (Ed.), *World seas:
An environmental evaluation* (pp. 101–119). Academic Press. https://doi.org/
10.1016/B978-0-08-100853-9.00006-3

Wessels, J., & Potgieter, T. (2021). Case studies as an approach to challenges in
public administration. In J. Wessels, T. Potgieter, & T. Naidoo (Eds.), *Public
administration challenges: Cases from Africa.* Juta.

Wernli, D., & Darbellay, F. (2016). *Interdisciplinarity and the 21st century research-
intensive university.* LERU Technical Report. https://doi.org/10.13140/RG
.2.2.21578.16321

World Bank. (2023). *Fisheries. World Bank indicator South Africa.* World Bank Group.
https://data.worldbank.org/indicator/ER.FSH.PROD.MT?locations=ZA

CHAPTER 12

OCEAN HEALTH PART II

The South African Policy Experience and Lessons Learned in Sustaining Ocean Life

Thean Potgieter
University of the Free State

ABSTRACT

The anthropogenic effect on the oceans is vast and worsening. Our ocean ecosystems are vulnerable to multiple serious threats, ranging from climate change to overharvesting and pollution. Despite the emphasis governments place on the crucial contribution of living ocean resources to food security and economic growth and development, many challenges inhibit communities and societies to benefit appropriately from this ocean wealth.

This situation is of much relevance to South Africa due to the abundance of living ocean resources and the dependency of many communities on its harvest. But this inherent wealth is difficult to govern, while illicit activities and overharvesting are driving marine species to extinction. Preserving the oceans require global and regional cooperation, good governance, maritime

Climate Governance in International and Comparative Perspective, pages 251–276
Copyright © 2024 by Information Age Publishing
www.infoagepub.com

security, and sustainable economic activities in line with collective international agreements—yet progress is wanting.

This chapter represents an endeavour to analyse, determine and understand the adverse effect of anthropogenic influences on our living oceans with an emphasis on the use and overharvesting of living ocean resources and the threat of pollution. The South African policy experience and lessons learned will be highlighted throughout.

Closer inspection reveals that our ocean ecosystems are vulnerable to multiple serious threats, ranging from climate change to overharvesting and pollution. Despite the emphasis governments place on the crucial contribution of living ocean resources to food security and economic growth and development, many challenges inhibit the opportunities of communities and societies to benefit appropriately from this ocean wealth.

The anthropogenic effect on the oceans is vast and worsening. We have caused climate change, contributed to the depletion of marine resources, polluted the environment, and have destroyed livelihoods. As our efforts to preserve ocean health, ensure equity and have environmentally sustainable economic activities require global cooperation, nations made collective agreements aimed at ending poverty and saving the environment. This is expressed in the Sustainable Development Goals (SDG) of the United Nations (UN) which includes a focus on the environmentally sustainable use of the oceans—but the anticipated progress is wanting.

This situation is of much relevance to South Africa due to the economic value of its maritime domain, as well as the dependency of many communities on the ocean and its harvest. But this inherent wealth is difficult to govern and faces many risks. Illicit activities and overharvesting are driving marine species to extinction while causing serious damage to biodiversity.

The two chapters on ocean health is an effort to make sense of the varied natural and anthropogenic factors impacting on, and threatening, ocean health. In Part I the focus is on ocean health and governance. Part II represents a more directed endeavour to analyse, determine and understand the adverse effect of anthropogenic influences on our living oceans with an emphasis on the use and overharvesting of living ocean resources and the threat of pollution. The South African policy experience and lessons learned will be highlighted throughout.

The research approach utilised in these chapters (see the discussion in Part I) required an exploration of literature relevant to the interconnections between disciplines. Inherent to such an interdisciplinary approach is the need to study diverse sources, which contributed towards building an understanding of the link between natural phenomena and the risks and opportunities relevant to ocean health, as well as our endeavours to govern it.

FISHING AND AQUACULTURE

Societies globally depend on both the protein and economic benefits associated with harvesting living marine resources. Oceans and inland waters provide more than 3.3 billion people with at least 20% of their protein, and livelihoods to about 60 million people (Cooley & Schoeman, 2022, p. 469). In Africa more than 200 million people depend on the oceans for vital nutrition, fish provides about 22% of the protein, and an income to more than 10 million people (African Union, 2012, p. 8). As a result, changes in ocean health, the climate and the availability of fish would influence billions. The most important objective of fishery management is therefore to limit the overexploitation of fish stocks and retain biodiversity which provides the best chances of buffering climate change.

Biologically sustainable global fish stocks have declined from 90% in 1974 to 65.8% in 2017 resulting in a growing reliance on aquaculture (UN Food and Agriculture Organization [FAO], 2020a). World Bank data indicates that the total global fish production was 216,872,258 million tons (metric) in 2021, of which 91,177,191 tons were capture fish production (down from about 97 million tons in 2018). Global catches have remained around 90 million tons from 1994 onwards (compared to 58,3 million tons in 1980). But global aquaculture production has consistently increased from about 43 million tons in 2000, to 78 million tons in 2010, 104 million tons in 2015, 120 million tons in 2019 and 126 million tons in 2021 (World Bank, 2023).

Due to South Africa's long coastline and the meeting of the Agulhas and Benguela Currents, the country boasts unique marine ecosystems. The warm Agulhas Current emanates from the confluence of the Mozambique and East Madagascar Currents and flows south along the South African East Coast. The cold Benguela Current flows north along West Coast and comes from the upwelling in the deep Atlantic Ocean which in terms of biomass makes it one of the most productive global ecosystems. As South African coastal waters are rich in marine biodiversity commercial, subsistence and recreational fisheries can catch more than 630 different marine species.

Allowable catches in South Africa (including industrial to small scale and recreational fishing) are managed through permits issued by a government department—formally the Department of Environment, Forestry and Fisheries (DEFF) and since April 2021 the Department of Forestry, Fisheries and the Environment (DFFE). UN data indicates that South African catches have declined: in 2000 the commercial catch was 674,117 tons, compared to about 612,200 tons in 2016 (FAO, 2018; FAO, 2022a, p. 54). In 2020 total fisheries production (including aquaculture) were 612 456 tons, with capture fish production being 602,703 tons (see Figure 12.1).

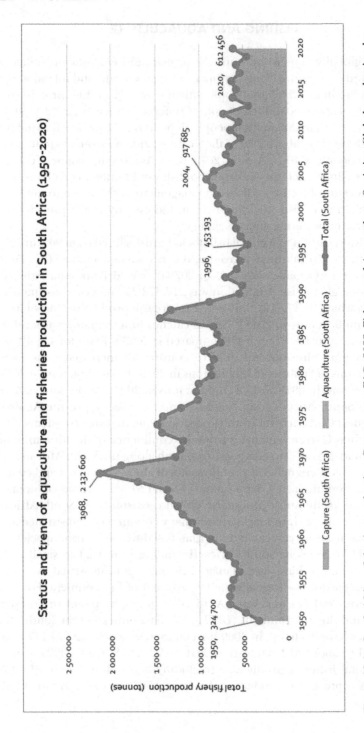

Figure 12.1 South African fisheries production. Data source: FAO, 2022. Fishery and Aquaculture Statistics. Global production by production source 1950–2020 (FishStatJ; https://www.fao.org/fishery/en/statistics/software/fishstatj)

Capture fish production is projected to decline further to 522,000 tons in 2030 (FAO, 2022b, p. 213).

Although data from various sources differ, the trend remain. The World Bank reported that total fisheries production was 501,855 tons in 2021, compared to 458,340 tons in 2019, 724,174 tons in 2012, and 917,685 tons in 2004. In 2021 the capture fish production was about 491,330 tons, compared to 622,090 tons in 2016, 719,247 tons in 2012, 911,731 tons in 2004 (World Bank, 2023). This is much lower than the highest catch on record: 2.1 million tons in 1968. Aquaculture production rose from 2,819 tons in 2000, to 5,895 tons in 2005, 5,927 tons in 2012, 6,730 tons in 2015, 9,224 tons in 2019, 9,753 tons in 2020, and 10,525 tons in 2021. Though not comparable to global averages, production is expected to increase to between 12,000 and 15,000 tons by 2030 (FAO, 2022b, p. 213; FAO, 2022a, pp. 61–62; World Bank, 2023).

As about half of South Africa's annual catches are consumed locally, the country is a net exporter of fish. Exports were valued at U.S. $598 million in 2017 (Wepener & Degger, 2019, p. 110; FAO, 2018). In 2019 South Africa's total income in ocean (marine) fisheries and related services industry was South African Rand R17,6 billion, an 8,1% increase compared to the reported 2017 income of R15 billion. Although 68.5% was concentrated in the sector's top ten companies according to Statistics South Africa (Stats-SA), the concentration ratio fell from 80% in 2014 (StatsSA, 2020). The available employment statistics for the sector is confusing. StatsSA indicates that 16,744 persons were employed in the ocean fishing industry in 2019, of which 12,738 (76,1%) were in permanent employment or proprietors (StatsSA, 2021, p. 5). But data still on the DFFE website (and older sources) indicate that up to 30,000 people could be directly and 80,000 indirectly employed in the fishing industry (Wepener & Degger, 2019, p. 110; FAO, 2018). World Wildlife Foundation (WWF) data indicated that 29,233 subsistence fishers from 147 communities were mostly harvesting line fish, mussels, rock lobster (crayfish) and abalone (WWF, 2018).

Regarding the status of fish stocks, the latest assessment by government indicates an improvement with 61% of stocks "not to be of concern" and 39% "of concern" in 2020 estimations, compared to 2012 estimations of 46% of the stocks "not to be of concern", 49% in 2014 and 52% in 2016. The over-exploited species decreased by 4% between 2016 and 2018, which could be ascribed to better governance efforts by the Department, specifically increased monitoring and assessing of fish stocks (DEFF, 2020, p. 1; WWF, 2018).

Worry aspects are that the number of stocks "in an optimal state" rose to 21 in 2020 (from 15 in 2012). Also, well-known fish stocks that are "depleted" include small pelagic fishes such as the sardines, white stumpnose, geelbek, while the "heavily depleted" category include abalone, crayfish, black musselcrackers, southern bluefin tuna, yellowfin tuna, bigeye tuna, soupfin shark, great hammerhead shark, and harders (DEFF, 2020, pp, 1–9, 69–76, 82–111; FAO, 2022b, p. 53).

Climate change is also negatively impacting on fishing as warming resulted in the migration of fish stocks towards higher latitudes. The West Coast rock lobster (crayfish) migrated south and east into abalone fishing areas, whereas migration patterns have also been detected amongst geelbek and various shark species. As large pelagic species migrate across the EEZ of various countries, these resources are regulated by regional and international Fisheries Management Organizations (DEFF, 2020, pp. 8, 38, 71).

The complex migration of various species poses challenges to the management of fisheries, has influenced the diversity of harvests, and caused economic losses to commercial, small-scale and subsistence fishers. Small scale and subsistence fisheries are most affected by climate change, as large-scale fishing operations limit their exposure through reliance on technology, diversity in harvests and greater mobility in following migrating species. Their operations, however, require changing regulations and governance (Cooley & Schoeman, 2022, pp. 456–458, 469).

IUU fishing is a global threat to fish stocks, it degrades ecosystems, result in major biodiversity losses, and the efforts of states to manage their living resources in a sustainable way. As it undermines state authority legitimate industrial and small-scale fisheries are at an unfair disadvantage, while the food security and socio-economic well-being of coastal communities are at risk. Members of such communities could themselves venture into IUU activities, whereas vulnerable people (migrant workers, women, and children) are more exposed to modern slavery, bondage, forced labour and other abuses often associated with IUU fishing and the wider criminal networks that drive it (Couper et al., 2015, Chapter 6; FAO, 2022b, p. 143).

The extent of IUU fishing is appalling as it accounts for about a fifth of global catches (11 to 26 million tons), is worth between U.S. $10 to 23.5 billion per year, while overall annual financial losses resulting from it is probably $50 billion or more. It contributes to more than 90% of global fisheries stocks being fully exploited, overexploited, or depleted (FAO, 2020). Following timber and mining, IUU fishing globally ranks as the third most lucrative natural resource crime (Daniels et al., 2022, pp. 6, 53). Developing countries with lacking maritime security capacity and pervasive corruption, are most affected by IUU fishing and are losing billions (in USD) annually.

Foreign fishing fleets (amongst others from the EU, China, South Korea, Japan, and Cuba) operate in great numbers around West Africa, mostly as part of bilateral agreements on the payment of fees to coastal states. But significant IUU fishing is occurring as one in four fish consumed globally could be from IUU fishing in Africa. About 48.9% of the vessels known to be involved in IUU fishing were found around Africa and the continent's economic losses linked to illicit financial flows could be up to U.S. $11.49

billion (Couper et al., 2015, Chapter 4). West Africa is the epicentre as 37% of the total catches in the region could be IUU fishing and losses from illicit financial flows might be U.S. $9.4 billion. Tax revenue losses is estimated at 20% of this amount, while IUU fishing threatens the livelihood and food security of the whole region (FAO, 2020; Daniels et al., 2022, pp. 6, 18–20). The free trade port of Las Palmas (on Gran Canaria) is important to shipping and the local economy, but it handles about 400,000 tons of fish per year and is reportedly also used for the transhipment of most IUU catches from West African waters (Couper et al., 2015, Chapter 7).

Southern and East Africa lose about South African Rand 12.2 billion annually due to IUU fishing, which is a severe threat to sustainable fish stocks and ocean health in the region. Due to insufficient resources for proper policing of the territorial waters and EEZs of countries in the region, various not-for-profit and international conservations organisations are engaged. *Stop Illegal Fishing* and *Sea Shephard* have cooperated with Tanzania's government to increase monitoring and achieved some successes. In 2018 they identified a Chinese flagged fishing vessel (*Tai Hong No 1*) with many shark fins onboard and the *Buah Naga No 1* (Malaysian flagged) violating Tanzania's regulations. As part of a cooperative monitoring operation between South Africa, Tanzania, Kenya, and Mozambique in 2009, the South African environmental protection vessel *Sarah Baartman* apprehended the *Tawariq 1* fishing illegally in Tanzanian waters with 260 tons of frozen fish (mostly tuna and shark fins) on board (Mwaijande, 2021).

IUU fishing is mostly associated with distant water fishing fleets. Due to dwindling catches or regulations, many fishing fleets have moved operations away from their own shores to areas with less surveillance and control. About 90% of global distant water fishing are conducted by five countries (China, Taiwan, Japan, Spain and South Korea)—operations that would probably not be profitable if not for government subsidies (Daniels et al., 2022, p. 8). It is estimated that a third of the vessels engaged in IUU fishing are Chinese flagged, while about 8.76% have flags of convenience. Nearly a quarter of the vessels involved are from the ten top IUU fishing companies, of which eight are from China (led by the Nasdaq-listed Pingtan Marine Enterprise), and one each from Colombia and Spain (Daniels et al., 2022, pp. 15, 25).

The WWF reported that between 2015 and 2021 IUU fishing of tuna and shrimp (including prawn) in the Southwest Indian Ocean (off the coasts of Kenya, Madagascar, Mozambique, South Africa, and Tanzania) resulted in income losses of around U.S. $142.8 million per year as roughly 36% of catches (including 48.7% of the tuna and 26.4% of the shrimp) were probably IUU fishing (WWF, 2023). It is crucial that the market destinations of these products and key stakeholders the region act, and that the EU and China do more to ensure that imports are not the result of IUU fishing.

Although recent studies (FAO, 2022b, p. 179) showed that fishing stocks threatened by overfishing were systematically being rebuild in several global cases (including some South African stocks), the scourge of IUU fishing remains. Specifically at risk are states with limited capacity to protect their marine sources and enforce the law within their maritime zones. In conjunction with environmental care, fighting IUU fishing require adherence to international law, global agreements, appropriate regional arrangements, political will, good governance, and national law enforcement capacity.

Proper governance firstly depends on the frameworks for sustainable fisheries provided by international law. It starts with UNCLOS, an international maritime law framework that provides for all activities, including utilising living resources and conservation (UN, 1982). This was followed by the 1995 Code of Conduct for Responsible Fisheries and the international guidelines emanating from it. Albeit a voluntary instrument, it is highly used and provides for sustainability and responsive management of marine resources to protect ecosystem and biodiversity. It also resulted in two legally binding agreements on the responsibilities of flag states on the high seas and the *Port State Measures to Prevent, Deter and Eliminate IUU Fishing*, referred to as the Port State Measures Agreement (FAO, 2016). The Port State Measures Agreement was approved at a FAO conference in 2009 and came into force in 2016. By 2023 it was ratified by 76 parties including South Africa, but not by China, India, Tanzania, and others (FAO, 2023). As countries can prevent the proceeds from IUU fishing being offloaded at a port of entry, this agreement fights IUU fishing and place some restrictions on supply chains. To further enhance its implementation FAO is developing guidelines for monitoring transhipment and the movement of catches to prevent IUU catches from becoming part of supply chains (FAO, 2022b, pp. 128–129). These policing measures are also supported by a World Trade Organisation Trade Facilitation Agreement (entering into force in 2017) to expedite customs clearance and movement of legal goods across borders (FAO, 2022b, p. 102).

Transparency in supply chain is crucial for limiting the proceeds of IUU fishing from reaching key consumers (such as Europe, USA, Japan, and China). Despite advanced IUU regulations as much as 15% of fish entering the EU and USA markets could be illegal and, in the USA, it annually amounts to about U.S. $ 2 billion (Daniels et al., 2022, p. 8). Fighting IUU fishing requires certifying catches to indicate to buyers and consumers that they originate from a legal source. But more needs to be done to streamline processes, standardise certification and promote compliance.

Crucial to implementing the Port State Measures Agreement is the adherence by port states to national, regional and international fishery regulations (including permissions by regional fisheries management organizations) in allowing foreign fishing vessels to offload potential IUU catches.

But information about products going through many ports is insufficient, due to uneven and often fragmented control by national and regional actors. In a study on the frequency of port visits by fishing vessels, ports were identified, and ranked, based on their IUU fishing risk (Hosch et al., 2019). Positional data obtained from the Automatic Identification System (AIS) transmitted by fishing vessels were used and the authors maintained that poor governance increases the likelihood of visits by vessels engaged in IUU fishing. Various African ports often visited by such vessels include Tema, Abidjan, Walvis Bay, Dakar and Nouadhibou on the West Coast of the continent, Port Louis, and Port Victoria in the Indian Ocean coast as well as Cape Town. Although ports in Asia posed the highest risks, ports in Africa is 'mid-range' and high risks countries identified are Gabon, Senegal and Sao Tome and Principe (Hosch et al., 2019, pp. 20–24, 38).

South Africa is obliged to deter IUU based on global and regional agreements it is party to, as well as in terms of its national regulatory frameworks. According to the WWF, as less than 0.5% of marine ecosystems are protected in South Africa, IUU fishing and poaching is driving many South African marine species to extinction, while it causes socio-economic stress, considerable biodiversity damage and costs the country billions (Bhana, 2020, p. 105). IUU fishing vessels would catch in prohibited waters (often with illegal gear), land it illegally in a port (not necessary their flag country) or transfer it to factory ships at sea. Infringements in South African waters often occur as the large sea area and incidences of corruption, poor policing resources and inaction pose considerable challenges to law enforcement and identifying and apprehending suspicious vessels.

Despite IUU fishing ravaging South Africa's living marine resources, there are many examples of vessels flagged by various states being apprehended. In October 2013 the patrol vessel *Victoria Mxenge* apprehended seven ships for IUU fishing with tuna, swordfish, dolphin, and sharks on board. These vessels, and three more ships belonging to the same unknown owners (probably a syndicate from Taiwan) were placed under arrest in Cape Town. Their 75 Indonesian crewmembers were confined on board in atrocious conditions, without money, food, fresh water, or proper sanitary facilities. They indicated they were not paid for five years, and various charities kept them alive for months. The cargoes or the ships could not be sold to support the crews (because illegal catches might not be marketed). After two of the ships 'escaped' from Cape Town, Interpol was requested to issue a 'purple notice' to recapture the vessels, but as the owners altered their names and registration this did not occur. This case illustrates the unscrupulousness of IUU fishing syndicates who "regard seafarers as merely factors of production to be disposed of at will" (Couper et al., 2015, Chapter 10).

In 2016 there were two incidents involving Chinese vessels. Three Chinese vessels were found in the South African EEZ with illegal fishing gear, no valid fishing licences and 600 tons of chokka (Cape Hope squid) on board. Two vessels attempted to escape, but all three were eventually impounded in East London. Their fine (about R700,000 per vessel) could not be seen as an effective deterrent when compared against the potential value of fully laden trawlers (Potgieter, 2021, Chapter 7). After nine identical trawlers were pictured fishing in marine protected areas with their AIS turned off and carrying no flags, insufficient official action caused public outrage. Only one vessel (the *Lu Huang Yuan Yu 186*) was eventually apprehended by the *Victoria Mxenge*, and a chase by naval vessels also failed to catch the other eight vessels before leaving the South African EEZ. As the *Lu Huang Yuan Yu 186* did not have the required fishing permits or licences she was only charged with entering South Africa's maritime zones and violating lawful commands by fishery control (Bhana, 2020, pp. 84–90). This incident again indicates that much IUU fishing occurs within South Africa's large EEZ and that it is difficult to apprehend suspect vessels.

Technological innovations provide new tools for fighting IUU. The free *Skylight* satellite monitoring and artificial intelligence platform uses imagery (from the *Sentinel 1* satellite of the European Space Agency), advance software analysis and machine learning for maritime monitoring to detect "surface suspicious activity in real-time" (Godfrey, 2022). It collaborates with lower-income coastal states in fighting IUU fishing by providing fisheries intelligence and capacity building. In West Africa *Skylight* assisted with monitoring maritime protection areas and identified suspicious vessels. In August 2021 the fishing vessels *Torng Tay No. 1* requested permission to enter the Port of Durban, but local authorities were notified through *Skylight* that the vessel loitered at sea, indicating a possible transhipment of fish to another vessel (transhipments contributes to IUU catches landing on the market). The *Torng Tay No. 1* was fined R50,000 by South African authorities for underreporting on catches—insignificant considering the problem— but a repetition of the offence would result in a fine of R500,000 (Godfrey, 2022). As part of a pilot project that commenced at the end of 2022 South Africa is placing hydrophones to combat IUU fishing at strategic locations in marine protected areas. Law enforcement will be notified of possible illegal activities if suspicious noises are detected and identified (Africa Defence Forum, 2022).

The vulnerabilities, inequality and injustices associated with fisheries governance in South Africa was highlighted by the COVID-19 pandemic. The implementation of pandemic measures placed restrictions on fishing activities and mobility, caused a loss of markets and made the sale of fish products very difficult. As the international export market for high value species such as lobster 'crashed', many small-scale fishers lost their seasonal income (Sowman et al., 2021). Aquaculture, specifically shellfish production of mussels and oysters (with catering as its major market) was exceptionally hard hit by movement restrictions, and the dramatic decline in the tourism and hospitality industry. This resulted in aquaculture operations shutting down as the cost of doing business simply became too high. The precarious situation was worsened as such (often small scale) enterprises as well as their employees received limited social protection and emergency relief from government. The crisis placed emphasis on urgent requirements for greater transformation and better governance in the sector to address the socio-economic challenges and injustices poor coastal communities face.

THE PERFECT STORM: ANNIHILATION OF ABALONE AND CRAYFISH IN THE WILD

The illegal trade in wildlife products, including exotic woods and plants, endangered species, ivory, and rhinoceros' horn is part of an extensive global crime industry, while wildlife products are often smuggled together with illegal narcotics and other contraband (Haenlein & Smith, 2017). Various high value marine products from part of such illicit trading networks. Abalone, sea cucumber, shark fin and fish maw are delicacies in Cantonese cuisine, has been associated with royalty for thousands of years and eating it indicates status.

An abalone species (*Haliotis midae* or *perlemoen* to locals) found only in South African waters, has been massively poached for over three decades. Illicit abalone is harvested by divers (often from small boats or from the coast), dried and then smuggled. As much as 43% of it might be traded through a few non-abalone-producing countries in sub-Saharan Africa, before being shipped mostly to Hong Kong for entry into markets in Asia (Okes et al., 2018, pp. 1, 18). It is estimated that a third of the dried abalone entering Hong Kong could be abalone poached in South Africa. Legal produce undergoes rigorous health checks, they are cleaned, properly dried, canned or shipped alive. As poached abalone is dried (often under dubious conditions) and are not subject to the same health scrutiny, they could pose health threats (Givetash, 2023; Wagner, 2021, p. 23). The illicit trade is very harmful to South Africa as the species has been driven to extinction in the wild, while it forms part of a lucrative criminal economy that is associated with violence, turf wars, corruption, and the erosion of state institutions (De Greef & Haysom, 2022, pp. 1, 10).

The opening-up of the Chinese economy and its growing middle class stimulated demand for abalone. It coincided with the end of apartheid and flourishing international trade after decades of sanctions. This facilitated the illegal abalone trade and by the 1990s adaptive organised crime and Chinese triad gangs engaged in drug markets had also established an illegal abalone trade along the Overberg coast. As law enforcement agencies in South Africa underwent "periods of rapid (often unsuccessful) reform to enable them to meet the challenges facing the new democracy" such infringements were not sufficiently addressed at the time (De Greef & Haysom, 2022, p. 17).

After the price of abalone spiked in 2005, criminal business involvement in this lucrative market from Hong Kong, Taiwan and mainland China increased even more (Haenlein & Smith, 2017). As local gangs around Cape Town were recruited, abalone poaching and trafficking became intertwined with other forms of international organised crime. According to De Greef and Haysom (2022, pp. 18–19) violent rivalry erupted between criminal groups from Taiwan and Hong Kong as abalone was also traded for methaqualone (used for manufacturing mandrax) and methamphetamine (or *tik*). It stimulated drug abuse, gang violence, shootouts, arson, while there were even rumours of killed rivals being fed to great white sharks. Illicit abalone became imbedded into criminal networks also smuggling illegal commodities such as rhino horn across Southern Africa and Asia (Okes et al., 2018, pp. 28–35; Grobler, 2019).

Some of the abalone is moved from Hong Kong through Vietnam to China. As it is recorded as imports from Hong Kong there is insufficient proof that it is acknowledged as abalone from South Africa. This could be similar to the trafficking of shark fin and sea cucumber. In addition, the constant legal flow of abalone from South Africa (specifically from growing aquaculture) combined with corruption provides cover for smuggling as documents are often falsified to conceal illicit shipments (De Greef & Haysom, 2022, pp. 14, 36).

Within a context of high unemployment and poverty, abalone poaching offers an important source of income to many in coastal communities. As they have the requisite skills to act as divers, crews, or do maintenance, remuneration are better than in legal employment. With poached abalone fetching high prices, each member of a boat crew could earn more than R200,000 (or U.S. $14,000) in one operation. Such revenues could then be used to buy drugs and weapons to increase the area and power of criminal groups. Higher demand, specifically due to COVID-19, pushed the abalone price to more than U.S. $63 (R1,000) per kg in 2020 and boosted poaching. As a result, members from poor coastal communities form the crucial network that enables organised crime to poach, process and export of abalone (Givetash, 2023; De Greef & Haysom, 2022, p. 19). As this has done immeasurable ecological and socio-economic damage and law enforcement is unable to contain it extreme solutions have even suggested—such as allowing the illegal harvesting of abalone to continue beyond natural supply levels so that it is no longer viable to criminals (Pinnock, 2022).

In the late 1980s law enforcement in South Africa became aware of criminal gangs from Hong Kong and Taiwan being involved in illegal shark fin and abalone trade. During the 1990s members of the South African Police Service (SAPS) attended training in Hong Kong and Macau on Chinese triads operations. However, despite South African investigators providing sufficient evidence and wishing to cooperate with officials in Hong Kong, no South African suspects have been conclusive linked to criminal networks in Hong Kong. Actions were hindered as environmental offences were until recently not part of Hong Kong laws against organized crime (De Greef & Haysom 2022, pp. 35–36).

A detailed report published in January 2023 in the *South China Post* indicates that the fight against illegal abalone is drawing attention in the Far East (Givetash, 2023). It highlighted that illegal abalone (known as 'white gold' in Asian markets) is very lucrative to criminal gangs and could be worth between U.S. $60 and $120 million per annum. Due to the high value of abalone, legal farms in the Western Cape require improved security as they are also targeted by poachers.

Crayfish (West Coast rock lobster or *kreef*) has also been ruthlessly targeted by poachers. Its habitat is rocky coastal areas between Walvis Bay in Namibia and East London in South Africa. Crayfish is commercially harvested in depths of up to 100 m. Nearshore harvesting is from small boats or from the shore, with nets or diving by recreational and small-scale fishers. Sadly, this once abundant delicatessen and priced global export commodity were overharvested and poached to the point of extinction. To sustain the source the DFFE have reduced allowable catches from more than 3,000 tons in 2005–2006 to 700 tons in 2021–2022. It is only 0.4% of South African catches in mass, but 9.2% in value. The legal industry is worth more than R500 million per year and provide 4,200 jobs, but representatives of coastal communities are pleading for reducing catches to prevent the source from collapsing completely as it would have disastrous consequences for communities.

Small-scale poaching is mostly opportunistic, for local consumption, or aimed at earning cash. But of considerable concern is the large-scale poaching for organised transnational crime, which is made worse by DEFF's poor monitoring and corruption amongst officials (DEFF, 2020, p. 107; Cochrane, 2022). The so-called Bengis case (involving Arnold Bengis, David Bengis and Jeffrey Noll) is of much relevance. Up to 2001 they made huge profits from illegally harvesting and trading large quantities of crayfish and Patagonian toothfish through Hout Bay Fishing in South Africa and companies in the United States. South African fishery inspectors were bribed while fishery reports and export documents were falsified to hide the overharvesting. Moreover, previously disadvantaged South Africans (without valid USA working permits) worked for low wages at their fish processing facility in Portland, Maine (United States Attorney's Office, 2013). According to environmental experts the crayfish stock was in "free fall" and its "terminal decline was only halted when...bringing the illicit activities to an abrupt halt" (Dutot, 2021, p. 3).

South African prosecution of Hout Bay Fishing commenced in 2002. They paid fines, lost their licences, confiscation orders (of around U.S. $7 million) were affected, and the business closed (Bhana, 2020, pp. 82–84). Evidence suggested that their profits were more than U.S. $60 million, but there were considerable forensic difficulties in tracing it in the USA and in foreign trusts. Successful prosecution nonetheless occurred in the USA, which resulted in Arnold Bengis receiving a 57-month prison sentence and the amounts recovered through confiscation orders amounted to over U.S. $20 million (Dutot, 2021, pp. 4–6). The Bengis case is an important example for global law enforcement and high value asset recovery from illicit fishing activities.

Claims are often made that poachers have infiltrated law enforcement in the Western Cape, that officials are bribed, or have personal (or family) relationships with poachers. Given the extend of criminal activities, catching poachers is not a priority while antipoaching operations are difficult due to the long and rocky coastline, many hidden coves, and the number of available officials. South Africa does not have a coastguard and the South African Navy (with warships and submarines) is not responsible for coastal protecting within 100 m of the shore. Although the Police and DFFE is responsible, both lack specialist law enforcement capacity and equipment, while the Police have wide crime prevention responsibilities.

Despite continued SAPS actions, it seems that law enforcement has lost the fight against abalone and crayfish poaching. Annual abalone exports rose with about 8% between 2009 and 2016 (to 5,500 tons) of which probably a half is poached. As the illegal trade has become an inherent part of the economy of some coastal communities, stopping it would be difficult (Givetash, 2023). In 2019 the SAPS seized illegal abalone worth more than U.S. $160 million, while two Chinese nationals were deported as part of their sentencing deal—various Chinese nationals were apprehended and sentence during earlier cases (Oakes et al., 2018, p. 28). In July 2022 the Minister of Police, Bheki Cele, indicated that in a five-year period the Police have confiscated 441,847 abalone, 15,089 crayfish tails and 8,301 whole crayfish poached only from the delicate Overstrand marine ecosystem (Bryant, 2022).

Although only prominent cases reach South African newspapers, reports published between January and June 2023 are indicative of major abalone and crayfish poaching activities continuing. In January law enforcement confiscated more than two tons of illegal abalone (worth R4.9 million) in Bloemfontein, about 2.5 tons (worth R4.5 million) in Cape Town, 4,867 abalones in Korsten and 1,620 abalones in Gqeberha (Montsho, 2023; Mahamba, 2023; Francke, 2023a; Naidoo, 2023). In February Cape Town law enforcement seized illegal abalone and thousands of crayfish probably worth R9 million in separate incidents. In March crayfish tails (worth about

R1.1 million) were confiscated in Vredenburg and thousands of poached abalones in the Eastern Cape. Abalones seized in Western and Eastern Cape during April and May were worth about R6.3 million, and about R5.2 million during June (Staff Reporter, 2023; Duba, 2023; Francke, 2023b; Nene, 2023; SAPS, 2023). Such successes are the preverbal 'drop in the bucket' and even though more than 35 persons were arrested, others would quickly take their places as the risks are low and the rewards high.

Abalone and crayfish stocks in the wild are under severe threat. International trade data indicated that the illegal abalone harvesting increased with 47% between 2017 and 2018. Abalone stock is "heavily depleted" and even the marine protected area in Betty's Bay provides no sanctuary as the mean density of abalone is about 1% of its 1990s level. The combination of allow-able catches (96 tons per annum), poaching and ecological changes have all but disseminated the abalone stock in the wild (DEFF, 2020, pp. 1, 8–9).

The declining crayfish stock is attributed to constant poaching, mass strandings due to 'red tide' (see the previous chapter), as well as the migration of the species into the area east of Cape Hangklip (DEFF, 2020, pp. 110–111). Estimates made in 2020 indicate that crayfish stocks were only 1.8% of pre-fished levels and as it could disappear altogether, considerable concern remain about extensive poaching (DEFF, 2020, p. 2). The decline of the crayfish stock was devastating to West Coast coastal communities and caused economic hardships. Solutions are simple, but difficult to achieve: better policing, and well-managed small-scale fisheries.

POLLUTION AND ENVIRONMENTAL CARE

The physical, chemical, and biological effect of anthropogenic activities on the environment have put marine ecosystems under immense pressure. These include overloading oceans with coastal run-off containing terrestrial nutrients (eutrophication) as well as chemical, biological and physical pollutants, toxins, and pathogens that also disturbs natural light (Cooley & Schoeman, 2022, p. 386). In addition, poor coastal developments and the hardening of shorelines have caused habitat destruction.

In 1974 the oceanographer Willard Bascom warned that human waste disposal into the oceans will harm marine life (Petrik & Ojemaye, 2022). This has indeed occurred as population growth and development around the world and in sub-Saharan Africa has resulted in both greater wastewater generation and pollution, aggravated by challenges associated with poor wastewater management (Onu et al., 2023). Extensive studies on the nature of anthropogenic pollution provide evidence of its threat to coastlines, water quality, marine life (from mangroves and coral reefs to animals, food chains and plankton) and fisheries. It substantially reduces the self-purification mechanism of marine ecosystems and pose serious risks to economic, social, and cultural conditions (Mousavi et al., 2023). Global ocean pollution includes plastics, pesticides, disinfectants, medical waste, antibiotics, and chemicals, which harms marine life and humans. Some elements in dumped products causes challenges ranging from feminisation to reproductive impairments, antibiotic resistance and endocrine disruption in marine life and humans (Petrik & Ojemaye, 2022). Although artificial ammonia-based fertilizers substantially improve crop yields, fertilizer run-off could act as a nutrient to algae and create "dead zones" around river mouths. If the dense algae blooms sink and decompose, it reduces the oxygen available to fish and other marine life (Heinberg, 2021).

Plastic packaging provides common and affordable solutions to all industries, and in the food industry it probably prevents many thousands of tons of food going to waste every year. But plastics are a main source of global ocean pollution. Great floating lumps of plastics are to be found in the Pacific Ocean and it is estimated if current trends continue, plastics in the oceans will outweigh the fish remaining fish by 2050 (Heinberg, 2021). The main challenge with plastic pollution is not large noticeable (and easily removable) pieces but the leaching of small pieces of harmful organic chemicals that could, amongst others, cause cancer, contribute to various illnesses and lead to fertility problems in both humans and animals. Plastic pollution has probably affected all oceanic species—from getting entangled in it, to swallowing it.

Nuclear power is efficient in generating electricity, but when things go awry it could result in terrible radioactive pollution. The melted-down nuclear reactors at Fukushima, in Japan, showed that the immediate vicinity could be radioactive for millennia. Besides, there would probably be no other solution than to dump millions of litres of radioactive water stored in tanks at Fukushima into the ocean (Heinberg, 2021). In 2021 when the South African power utility Eskom announced plans to extend the operational span of Koeberg (a nuclear power station on the coast north of Cape Town) over two decades at a cost of R20 billion, some civil society organisations called for protests. But protests against Koeberg are not new. Although the National Key Points Act stipulates that nobody is allowed within one nautical mile of Koeberg, six Greenpeace protesters entered the small harbour seaward of it with boats and scaled the buildings in August 2002 (Ekron, 2002). As the sea was the easiest way to gain access to the site, protecting the ocean approaches to Koeberg is a maritime security issue.

South Africa has been in the grips of an electricity crisis for years. Additional power could be generated by docking floating power stations (Karpowership) at Richards Bay, Saldanha Bay and the port of Ngqura. In 2020 during the COVID-19 emergency, Karpowership attempted to bypass the complex processes for environmental permits under the National Environmental

Management Act. Exemptions is possible in extreme cases where fast action could prevent an even bigger disaster—purported to be emergency power for medical care during the pandemic. However, as the impression was that Karpowership planned to use this exemption to provide electricity for several years, it was not granted due to concerns about the impact on birds, sea life and the local fishing industry, amongst others (Comrie 2023).

The saga did not end there. In January 2023 Karpowership was selected to provide 1,220MW of electricity to Eskom as part of the Risk Mitigation Independent Power Producer Procurement Programme. Despite the South African President, Cyril Ramaphosa, and other ministers emphasising Karpowership's contribution to alleviating the worsening electricity crisis, multiple civil society organisations continued to oppose it. Karpowership still had no environmental permits by early June 2023. The company indicated they would continue with efforts to gain permission to operate and amend the environmental impact assessment to provide for generic environmental management of substations and transmission infrastructure (Engel 2023; Carnie 2023). Resistance essentially focussed on environmental protection, locking the country (as with Ghana and Lebanon) into long-term contracts that are not cost-effective, and limiting the country's choice of energy sources. A R200 billion investment that would "sail away" after two decades was not seen as a solution.

The growing pollution of South Africa's marine environment is very concerning for ecosystems and the safety of harvested seafood. The presence of sewage in the oceans is a major concern as it is often not 'dumped' but result from insufficient wastewater management processes. Tests on seawater and marine life around the Cape Peninsula, for example, has shown faecal contamination as well as the presence of harmful medical compounds. A study conducted in the United States found that up to 104 prescribed pharmaceuticals were present in edible fish, whereas a study in False Bay pointed to the presence of numerous pharmaceuticals pesticides, industrial chemicals, and personal care products in seawater and in edible marine species (such as snoek). As such compounds do not degrade quickly, they could do serious harm to biodiversity over time (Petrik & Ojemaye, 2022).

The seawater quality along the coast of KwaZulu-Natal, specifically the rivers around eThekwini (Durban) and the city's beaches has been concerning for a long time. From January 2022 onwards water quality tests showed critical levels of Escherichia coli (E. coli) which is a health hazard to humans and the marine environment. It resulted in many of the city's beaches being closed in mid-2022, and again just before the holiday season at the end of the year. Major risks to human health include "cholera, hepatitis and other waterborne diseases transmitted via exposure to sewage bacteria and pathogens in affected rivers as well as the ocean" (Du Plessis, 2022). Over decades decreasing water quality was evident in increased algal blooms that caused large-scale environmental destruction (such as massive fish deaths) and the closure of beaches. This contamination emanated from sewage, litter, and toxic substances (generated by industries, human settlements, and farms) carried from the land to the sea and is closely associated with the inability of the local government to deliver quality services to its residents and businesses. The large-scale flooding that occurred in KwaZulu-Natal in April 2022 exacerbated an already dire situation as it damaged infrastructure, wastewater treatment plants and water supply, while vandalism also caused pumps to breakdown (Du Plessis, 2022). This pervasive problem of sewage pollution by the eThekwini municipality served as an example of how poor water governance can pose hazards to human health, lead to substantial environmental degradation, and cause economic losses due to a decline in tourism.

The garbage, emissions, sewage, and ballast water of ships contribute much to pollution. Ballast water dumping and the pumping of bilges is an important issue because untreated water could introduce invasive marine species and contain fossil fuels or various chemicals. The damage that could be done to habitats and ecosystems was illustrated along the rocky West Coast shores as the introduction of the Mediterranean mussel displaced the native mussel species and altered the coastal habitat (Robertson, 2015; Rantsoabe, 2014, pp. 8–10). As about two million tons of ballast water is dumped off the South African coast annually, regulations are in line with international best practice and stipulates that "ballast water to be exchanged deep sea prior to entering territorial waters" (South African Maritime Safety Authority [SAMSA], 2016). Compliance and monitoring though, remain an issue of global concern.

Oil pollution is a great danger to much of South Africa's pristine coastline and rich biodiversity, specifically as vessels often come to grief in the adverse sea conditions. Oil destroys the thermal insulation of seabirds and marine mammals, its toxic polycyclic aromatic hydrocarbons could cause smothering, while the chemical toxicity in the ocean may well result in habitat loss for organisms and species critical to the ecology. South Africa's largest ever oil spill occurred 122 km north-west of Cape Town in August 1983. The *Castillo de Bellver* (with roughly 250,000 tons of oil onboard) burned, exploded, and broke in two. As much of the oil burned or moved seaward, the most visible impact was about 1,500 oiled gannets from a nearby island. Yet much concern remains as the bow section of the ship sank while still containing up to 100,000 tons of oil. The West Coast with its sensitive bird colonies experienced two more important oil spills: In 1994 the bulk carrier *Apollo Sea* sank in a storm and 2,000 tons of bunker oil on board impacted on about 10,000 penguins. When the bulk carrier *Treasure* sank in 2000, the 1,300 tons of bunker oil on board contaminated more than 20,000 penguins, of which 2,000 died and about 19,000 were relocated (Rantsoabe, 2014, pp. 30–31).

On May 23, 2022 a spillage occurred during a transfer of oil between two tankers off the port of Ngqura (also known as Coega). SAMSA, the DFFE and Transnet National Ports Authority (TNPA) immediately commenced with joint cleanup operations. Five oil recovery vessels recovered the oil and by the afternoon of May 24 the heavy oil was removed and only patches of light oil sheen could be seen in Algoa Bay. None of the oil reached the beaches, or the bird islands in Algoa Bay. To contain the spillage the two vessels were kept side-to-side immediately after the incident and were only separated on May 25 (SAMSA, 2022).

Errors in coastal planning and execution such as the creation of hard coastal structures and dredging could do much damage. Dredging is common in coastal waters, specifically in the construction of ports and for ensuring safety of navigation, but it could result in habitat loss, hydraulic entrainment, release contaminants, sedimentation, loss of oxygen and underwater noise (Wenger et al., 2017, pp. 967–969).

The larger Saldanha Bay (including the Langebaan Lagoon catchment area) is ecologically sensitive and of considerable conservation interests, but the construction of a port in the Bay during the 1970s resulted in significant physical and hydrodynamic changes, while also causing ecological damage to marine and coastal habitats (Saldanha Bay Water Quality Forum Trust, 2023). The hard barriers erected for the harbour (a 3.1 km long causeway, a 900 m long jetty and a breakwater of 1.7 km) as well as extensive dredging (relocating up to 30 million m3 of soil) disrupted the natural water flow, which caused important changes in the bathymetry and sediment characteristics of the Bay. The average depth increasing with between 0.39 m and 3.203 m, sedimentation loss totalled about 49 million m3, while erosion and siltation processes occurred on some Langebaan beaches (Henrico & Bezuidenhout 2020, pp. 236–242). To increase natural sand deposits and augment the beaches, the local authority, Saldanha Bay Municipality, constructed a groyne embankment (Du Toit et al., 2022, pp. 325, 338). The Saldanha Bay case clearly indicates the impact harbour construction could have on the natural characteristics of coastal areas and beach formation. At the time insufficient care was taken with the environmental impact of the initial construction and dredging. Coastal governance has improved since, but challenges remain in an environment where commercial port activities, aquaculture, tourism and a sensitive ecology must be balanced.

CONCLUDING REMARKS AND ENDURING LESSONS

These chapters on ocean health endeavoured to analyse the South African experience within the global context of climate change combined with the effect of humans. It is argued that an interdisciplinary approach is relevant for making sense of such complex challenges and the relationships between multifaceted phenomena.

Climate change is a threat to coastal habitats and fragile marine ecosystems across the world. Together with the stresses induced by human activity, it will continue to adversely affect marine biodiversity, and with it the character, nature, livelihoods and prosperity of societies and coastal communities. As sufficient scientific evidence exists to justify the global concerns about ocean health, decisionmakers need to have the political will to put global agreements into action as it will bolster the capacity of communities and the oceans to adapt.

The SDGs call for an end to poverty through sustainable and inclusive economic development and environmental protection. Linking these concepts are pertinent, as poor socio-economic conditions could provide impetus for environmental crimes. IUU fishing and the relentless abalone and crayfish poaching in South Africa is disruptive to marine lifecycles and does untold damage to biodiversity. But as members from poor coastal communities are conducting much of the poaching operations, they have become the crucial enablers making it possible for transnational organised criminal networks to flourish. Such links are probably also applicable to various other environmental crimes.

The fact that some of South Africa's marine species have been driven to the point of extinction is both shocking and extremely sad. The political economy facilitating the illegal abalone and crayfish trade is symptomatic of poor socio-economic conditions as an antecedent for involvement in organized crime, but it is further facilitated by government corruption, human voracity, well-organised crime syndicates as well as professional financial and logistical service providers. As in many other spheres of human activity in South Africa, a culture of corruption and poor governance are at the root of such problems, and it is creating the space for it to fester.

The profound harms associated with the illegal abalone and crayfish trade, is well document and despite decades of anti-poaching efforts, illegal harvesting seem to be at its highest-ever levels as a myriad of South African and transnational groups are competing for profits. In addition, countries in the region (to whom abalone is not indigenous) export it. Existing controls have failed to balance socio-economic and environmental requirements, while the realistic prospect of establishing control seems limited. Special efforts are therefore required to safeguarding the resource and partner with communities in the process. In essence a radical shift in state responses is needed.

Comprehensive maritime security is crucial, but it receives insufficient attention in discourses such as this. Given the challenges and complexities associated with the region, as well as the large number of state and other role-players involved, protection of the oceans and sustainable economic development requires improved state capacity to enforce national and international regulatory frameworks. States have the same responsibilities in their territorial waters as on land, and the sovereign rights to explore, exploit and conserve resources in their EEZ. These responsibilities are distributed to various agencies, but international and regional cooperation is crucial for addressing the destructive effect of maritime governance failures.

As substantial ill begotten financial gains are associated with illicit activities in the maritime domain, the threat it poses is poignantly illustrated by the destructive impact of IUU fishing on global ecosystems, communities, and food security. Regional authorities need to be mindful of migrating marine species when establishing allowable catches and support each other with all means possible. These include fighting IUU fishing through sharing information, surveillance and maritime security capability, building capacity, while also implementing the Port State Measures Agreement with vigour (to limit the offloading of illicit catches in another country). Broad strategic approaches, proper international, regional, and local multi-level governance and good maritime security ultimately provide some of the essential ingredients for sustaining healthy oceans and conserving marine biodiversity.

The increasing pollution of South Africa's marine environment due to human activities is concerning for marine ecosystems and the safety of harvested seafood. Key sources of pollution are coastal run-off containing terrestrial nutrients (eutrophication) as well as chemical, biological and physical pollutants, toxins, and pathogens. In addition, pollution associated with shipping (such as oil spills and dumping of ballast water) as well as coastal developments have caused habitat destruction and remain constant sources of concern. As with other concerns relating to ocean health, good governance is crucial—specifically pertaining to wastewater management, navigation safety, monitoring shipping activities, enforcing regulations, and smart coastal developments. The danger of oil spills remains. Guarding against oil spills require maintaining a clean-up capacity through good symbiosis between the public and private sector role players.

If ocean health is not sustained the potential impact of higher temperatures, pollution, and an altered chemical configuration of the atmosphere and oceans will, amongst others, destroy ecosystems and cause the mass extinction of species, which could spell "a turning point for all life on Earth" (Heinberg, 2021).

REFERENCES

Africa Defence Forum. (2022, August 30). South African academy wields technology against illegal fishing. *ADF News.* https://adf-magazine.com/2022/08/south-african-academy-wields-technology-against-illegal-fishing/

African Union. (2012). *2050 Africa's integrated maritime strategy (2050 AIMS)*. African Union. https://au.int/sites/default/files/documents/30929-doc-2050_aim_strategy_eng_0.pdf

Bhana, S. (2020). *A critical analysis of the legal framework to deter illegal, unreported and unregulated fishing in South Africa's maritime zones* [Master's Thesis, University of KwaZulu-Natal]. Durban. https://researchspace.ukzn.ac.za/handle/10413/19804

Bryant, D. (2022, July 11). Poaching of abalone and crayfish out of control in Overstrand. Statement by DA Shadow Minister of Environment, Forestry and Fisheries. *Polity Media Statement*. https://www.polity.org.za/article/poaching-of-abalone-and-crayfish-out-of-control-in-overstrand-2022-07-11

Carnie, T. (2023, June 1). Saldanha harbour powerships plan "fatally flawed," says Environmental Affairs. *Daily Maverick*. https://www.dailymaverick.co.za/article/2023-06-01-saldanha-harbour-powerships-plan-fatally-flawed-says-environmental-affairs/

Cochrane, K. (2022, May 4). Abalone and rock lobster stocks are under severe threat—Here's how to preserve them. *Daily Maverick*. https://www.dailymaverick.co.za/article/2022-05-04-abalone-and-rock-lobster-stocks-are-under-severe-threat-heres-how-to-preserve-them/

Cooley, S. R., & Schoeman, D. S. (2022). Oceans and coastal ecosystems and their services. In H.-O. Pörtner, D. C. Roberts, M. Tignor, E. S. Poloczanska, K. Mintenbeck, A. Alegría, M. Craig, S. Langsdorf, S. Löschke, V. Möller, A. Okem, & B. Rama (Eds.), *Climate Change 2022: Impacts, adaptation and vulnerability. Contribution of Working Group II to the Sixth Assessment Report of the Intergovernmental Panel on Climate Change* (pp. 379–550). Cambridge University Press. https://doi.org/10.1017/9781009325844.005

Comrie, S. (2023, January 27). No charges for environmental crimes—NPA won't prosecute Karpowership case. *Daily Maverick*. https://www.dailymaverick.co.za/article/2023-01-27-no-charges-for-environmental-crimes-npa-wont-prosecute-karpowership-case/

Couper, A., Smith, H., & Ciceri, B. (2015). *Fishers and plunderers: Theft, slavery and violence at sea*. Pluto Press. https://www.perlego.com/book/664466/fishers-and-plunderers-theft-slavery-and-violence-at-sea-pdf

Daniels, A., Kohonen, M., Gutman, N., & Thiam, M. (2022). *Fishy networks: Uncovering the companies and individuals behind illegal fishing globally*. Financial Transparency Coalition. https://financialtransparency.org/reports/fishy-networks-uncovering-companies-individuals-behind-illegal-fishing-globally/

De Greef, K., & Haysom, S. (2022). *Disrupting abalone harms. Illicit flows of* H. midae *from South Africa to East Asia*. Global Initiative Against Transnational Organized Crime. https://globalinitiative.net/analysis/abalone-south-africa-east-asia/

Department Environment, Forestry and Fisheries, South Africa. (2020). *Status of the South African marine fishery resources 2020*. https://www.dffe.gov.za/sites/default/files/docs/publications/statusofsouthafrican_marinefisheryresources2020.pdf

Duba, S. (2023, May 5). Suspects face court for illegal possession of abalone worth R72,000. *IOL News*. https://www.iol.co.za/news/crime-and-courts/suspects

-face-court-for-illegal-possession-of-abalone-worth-r72-000-88cc8182-9d53
-4cdd-a18d-4132bd7414e8

Du Plessis, A. (2022, December 11). Durban coastline: Sewage polluted beaches pose threat to holiday makers and the environment. *The Conversation.* https:// theconversation.com/durban-coastline-sewage-polluted-beaches-pose-threat -to-holiday-makers-and-the-environment-196244

Du Toit, L., Henrico, I., Bezuidenhout, J., & Mtshawu, B. (2022). Analysing the changes in the bathymetry of Saldanha Bay between the years 1977 and 2021. *South African Journal of Geomatics, 11*(2), 325–339. http://dx.doi.org/10.4314/ sajg.v11i2.11

Dutot, C. (2021). *Hout Bay and the illegal lobster trade: a case study in recovering illicit proceeds of IUU fishing and wildlife trafficking.* Green Corruption Case Study, Basel Institute on Governance. https://baselgovernance.org/sites/default/ files/2022-08/220929_case-study-06.pdf

Ekron, Z. (2002, August 26). Easy access to Koeberg. *News24.* https://www.news24 .com/news24/easy-access-to-koeberg-20020826

Engel, K. (2023, June 1). Karpowership appeals rejection of Saldanha application extension, not backing down. *Cape Argus.* https://www.iol.co.za/capeargus/ news/karpowership-appeals-rejection-of-saldanha-application-extension-not -backing-down-dcae401f-e372-4ecc-8d55-61a6950e9342

Food and Agriculture Organisation of the United Nations. (2016). *Agreement on Port State Measures (PSMA). Parties to the PSMA.* http://www.fao.org/port -state-measures/background/parties-psma/en/

Food and Agriculture Organisation of the United Nations. (2018). *Fishery and aquaculture country profiles: The Republic of South Africa.* http://www.fao.org/ fishery/facp/ZAF/en

Food and Agriculture Organisation of the United Nations. (2020). *The state of world fisheries and aquaculture 2020. Sustainability in action.* https://doi.org/10.4060/ ca9229en

Food and Agriculture Organisation of the United Nations. (2022a). *Aquaculture growth potential in South Africa.* FAO Factsheet. https://www.fao.org/3/ cc3102en/cc3102en.pdf

Food and Agriculture Organisation of the United Nations. (2022b). *The state of world fisheries and aquaculture 2022. Towards blue transformation.* https://doi .org/10.4060/cc0461en

Food and Agriculture Organisation of the United Nations. (2023). *FAO treaties database.* https://www.fao.org/treaties/results/details/en/c/TRE-000003/

Francke, R-L. (2023a, January 18). Foul smell leads Eastern Cape police to poached abalone worth R700k. *IOL News.* https://www.iol.co.za/news/south-africa/ eastern-cape/foul-smell-leads-eastern-cape-police-to-poached-abalone-worth -r700k-b9d9fcce-ed21-408d-aa33-7ffd034b3c19

Francke, R-L. (2023b, April 25). Six bust in Cape Town for R2.8m worth of abalone. *IOL News.* https://www.iol.co.za/news/crime-and-courts/look-six-bust -in-cape-town-for-r28m-worth-of-abalone-2d15b671-4f80-43a2-a1ed-2a2c-4ca34d10

Givetash, L. (2023, January 28). How South Africa's illegal abalone trade enriches gangs while threatening safety. *South China Morning Post.* https://www.scmp

.com/magazines/post-magazine/long-reads/article/3208034/white-gold
-how-south-africas-illegal-abalone-trade-enriches-gangs-while-threatening
-safety-asian

Godfrey, M. (2022, July 15). Skylight's Ted Schmitt: Technology can turn tide of
war against IUU. *Seafoodsource News*. https://www.seafoodsource.com/news/
environment-sustainability/skylights-ted-schmitt-technology-can-turn-tide-of
-war-against-iuu

Grobler, J. (2019). Exposing the abalone-rhino poaching links. *Oxpeckers Environmental
Journalism*. https://oxpeckers.org/2019/03/abalone-rhino-poaching-links/

Haenlein, C., & Smith, M. L. R. (2017). *Poaching, wildlife trafficking and security in Africa*. Taylor and Francis. https://www.perlego.com/book/1578034/poaching
-wildlife-trafficking-and-security-in-africa-myths-and-realities-pdf

Heinberg, R. (2021). *Power. Limits and prospect of human survival*. New Society Publishers.
https://www.perlego.com/book/2195420/power-limits-and-prospects-for
-human-survival-pdf

Henrico, I., & Bezuidenhout, J. (2020). Determining the change in the bathymetry of Saldanha Bay due to the harbour construction in the seventies. *South
African Journal of Geomatics, 9*(2), 236–249. http://dx.doi.org/10.4314/sajg
.v9i1.16

Hosch, G., Soule, B., Schofield, M., Thomas, T., Kilgour, C., & Huntington, T.
(2019). Any port in a storm: Vessel activity and the risk of IUU-caught fish
passing through the world's most important fishing ports. *Journal of Ocean and
Coastal Economics, 6*(1), Article 1. https://doi.org/10.15351/2373-8456.1097

Mahamba, C. (2023, January 4). Four suspects arrested for running illegal abalone processing facility. *IOL News*. https://www.iol.co.za/the-star/news/four
-suspects-arrested-for-running-illegal-abalone-processing-facility-a7b61f8b
-e2f3-4de3-90a0-c717cf97c652

Montsho, M. (2023, January 23). Police seize R5m worth of abalone in clandestine Bloemfontein laboratory complete with hidden escape exit. *DFA News*.
https://www.dfa.co.za/south-african-news/police-seize-r5-million-worth
-of-abalone-at-clandestine-lab-in-bloem-e1bca612-fed9-4d45-8669-80f5586b
9da8/

Mousavi, S. H., Kavianpour, M. R., & Alcaraz, J. L. G. (2023). The impacts of dumping sites on the marine environment: a system dynamics approach. *Applied
Water Science, 13*(109). https://doi.org/10.1007/s13201-023-01910-9

Mwaijande, F. (2021). Managing the blue economy. A case study of Tanzania. In
D. Sparks (Ed.), *The blue economy in sub-Saharan Africa*, Taylor and Francis.
https://www.perlego.com/book/2567169/the-blue-economy-in-subsaharan
-africa-working-for-a-sustainable-future-pdf

Naidoo, D. (2023, January 13). Gqeberha man arrested for possession of 1,620 units of
protected abalone. *IOL News*. https://www.iol.co.za/news/environment/look
-gqeberha-man-arrested-for-possession-of-1-620-units-of-protected-abalone
-f0c75279-bd46-41bc-8431-3856d3fb6522

Nene, N. (2023, February 7). Multi-million Rand abalone bust in Cape Town. *Eyewit-
ness News*. https://ewn.co.za/2023/02/07/multi-million-rand-abalone-bust
-in-cape-town

Okes, N., Bürgener, M., Moneron, S., & Rademeyer, J. (2018). *Empty shells. An assessment of abalone poaching and trade from southern Africa.* TRAFFIC International. https://www.traffic.org/site/assets/files/11065/empty_shells.pdf

Onu, M.A., Ayeleru, O.O., Oboirien, B., & Olubambi, P.A. (2023). Challenges of wastewater generation and management in sub-Saharan Africa: A review. *Environmental Challenges, 11.* https://doi.org/10.1016/j.envc.2023.100686

Petrik, L., & Ojemaye, C.Y. (2022, May 29). Marine life in a South African bay is full of chemical pollutants. *The Conversation.* https://theconversation.com/marine-life-in-a-south-african-bay-is-full-of-chemical-pollutants-182791

Pinnock, D. (2022, February 24). Let abalone go extinct—this might be the radical solution South Africa needs. *Daily Maverick.* https://www.dailymaverick.co.za/article/2022-02-24-let-abalone-go-extinct-this-might-be-the-radical-solution-south-africa-needs-new-report-suggests/

Potgieter, T. (2021). South Africa: The blue economy experience. In D. Sparks (Ed.), *The blue economy in Sub-Saharan Africa,* Taylor and Francis. https://www.perlego.com/book/2567169/the-blue-economy-in-subsaharan-africa-working-for-a-sustainable-future-pdf

Rantsoabe, S. (2014). *Review of South Africa's marine pollution prevention measures, particularly those regarding vessel-source oil pollution.* [Master of Science Dissertation, World Maritime University]. Malmo. https://commons.wmu.se/cgi/viewcontent.cgi?article=1475&context=all_dissertations

Robertson, T. (2015). Marine invasions in South Africa: patterns and trends. *Quest, 11*(2), 44–45. http://academic.sun.ac.za/cib/quest/articles/P44-45.Marine Invasions.pdf

Saldanha Bay Water Quality Forum Trust. (2023). *The challenge of addressing cumulative impacts.* https://sbwqft.org.za/

Sowman, M., Sunde, J., Pereira, T., Snow, B., Mbatha, P., & and James, A. (2021). Unmasking governance failures: The impact of COVID-19 on small-scale fishing communities in South Africa. *Marine Policy, 133,* 104713. https://doi.org/10.1016/j.marpol.2021.104713

South African Maritime Safety Authority. (2016, February 9). *Marine notice no. 10 of 2016.* https://www.samsa.org.za/Pages/Search-MarineNotice.aspx

South African Maritime Safety Authority. (2022, May 27). *Oil spill incident vessels separated as mop up continues in Algoa Bay. The 10th Province.* https://blog.samsa.org.za/2022/05/27/oil-spill-incident-vessels-separated-as-mop-up-continues-in-algoa-bay/

South African Police Service. (2023, March 3). *Media statement. Office of the Provincial Commissioner Eastern Cape.* https://www.saps.gov.za/newsroom/msspeech detail.php?nid=45060

Staff Reporter. (May 2023, May 3). Abalone poacher nabbed in Makhanda. *Grocott's Mail.* https://grocotts.ru.ac.za/2023/06/06/abalone-poacher-nabbed-in-makhanda/

Statistics South Africa. (2020). *Census of ocean (marine) fisheries and related services industry, 2018. Financial and production statistics.* http://www.statssa.gov.za/?page_id=1856&PPN=13-00-01&SCH=7917

Statistics South Africa. (2021). *Ocean (marine) fisheries and related services industry, 2019.* http://www.statssa.gov.za/publications/Report-13-00-00/Report-13-00-00 2019.pdf

United Nations. (1982). *United Nations Convention on the Law of the Sea.* http://www .un.org/depts/los/convention_agreements/texts/unclos/part6.htm

United States Attorney's Office. (2013, June 14). *Officers of fishing and seafood corporations ordered to pay nearly $22.5 million to South Africa for illegally harvesting rock lobster and smuggling it into the United States.* Press Release, Department of Justice. https://www.justice.gov/usao-sdny/pr/officers-fishing-and-seafood -corporations-ordered-pay-nearly-225-million-south-africa

Wagner, N. (2021). Tragedy of the abalone commons. *AgriProbe, 18*(2), 22–25. https://doi.org/10.10520/ejc-agriprob-v18-n2-a11

Wepener, V., & Degger, N. (2019). South Africa. In C. Sheppard (Ed.), *World seas: An environmental evaluation* (pp. 101–119). Academic Press. https://doi.org/10.1016/ B978-0-08-100853-9.00006-3

World Bank. (2023). *Fisheries. World Bank indicator South Africa.* World Bank Group. https://data.worldbank.org/indicator/ER.FSH.PROD.MT?locations=ZA

World Wildlife Foundation. (2018). *2018 Oceans scorecard.* WWF. https://wwfafrica. awsassets.panda.org/downloads/wwf_sa_ocean_scorecard_2018.pdf

World Wildlife Foundation. (2023, May 4). *US$142.8 million potentially lost each year to illicit fishing in the South West Indian Ocean.* https://www.wwf.eu/?10270441/ US1428-million-potentially-lost-each-year-to-illicit-fishing-in-the-South-West -Indian-Ocean

Wenger, A. S., Harvey, E., Wilson, S., Rawson, C., Newman, S. J., Clarke, D., Saunders, B. J., Browne, N., Travers, M. J., Mcilwain, J. L., Erftemeijer, P. L. A., Hobbs, J-P. A., Mclean, D., Depczynski, M., & Evans, R. D. (2017). A critical analysis of the direct effects of dredging on fish. *Fish and Fisheries, 18*, 967–985. https://doi.org/10.1111/faf.12218

CHAPTER 13

CLIMATE CHANGE GOVERNANCE AND INSTITUTIONAL STRUCTURES IN EGYPT PRE- AND POST-COP27

Laila El Baradei
American University at Cairo, Egypt

Shaimaa Sabbah
American University at Cairo, Egypt

In September 2022 when we started writing this book chapter, a lot of commotion was taking place about Climate Change in Egypt because of the forthcoming COP27 Conference scheduled to take place in the city of Sharm El Sheikh in November 2022. Government, private sector, academics, civil society and media are all talking Climate Change. The Conference took place as scheduled and was reportedly a huge success. Reasons for the acclaimed COP27 success was mainly attributed to the last-minute

Climate Governance in International and Comparative Perspective, pages 277–318
Copyright © 2024 by Information Age Publishing
www.infoagepub.com

agreement reached, for the establishment of a 'Loss and Damage Fund', to compensate countries suffering from major climate change related crises (COP27.eg, 2022, Nov. 20).

Climate change is a global challenge that is expected to have negative impacts on many countries, including Egypt. Amongst the predicted negative impacts of climate change on Egypt as listed in the international climate report of 2018 are: an increasing level of drought affecting the productivity of agricultural lands, and rising levels of the Mediterranean Sea waters that may lead to the drowning of large areas of the northern coast of Egypt under sea water (Egypt State Information Service, 2022). The low delta lands, overlooking the Mediterranean Sea, are at risk of being flooded as a result of expected sea level rises. The gravity of the impact has been reported as an estimated reduction in cultivated land by 0.99 million acres by 2030, and a loss in at least 30% of the food production capacity in the Delta region by 2030 (Egypt First Updated INDC, 2022, June 8). Egypt was characterized as "highly vulnerable to climate change impacts," and ranked 107/181 countries by the 2019 ND-GAIN Index[1] that assesses the level of vulnerability of different countries to climate change and ranks them accordingly (World Bank, 2021).

Although concern with climate change has been on the Government of Egypt's agenda for several decades now, yet the hosting of the COP27 Conference in Sharm El Sheikh has gotten many parties to escalate their efforts in trying to respond to the eminent challenges. COP27 represented a great window of opportunity for moving forward with efforts to mitigate and abate climate change impacts in Egypt. A main challenge is to make sure the governance and institutional structure dealing with climate change action, that got solidified as a result of the COP27 incentive, remains in place, and continues to be similarly effective post the Conference.

The purpose of the current book chapter is to capitalize on the rising attention to Climate Change by all Egyptian government bodies in preparation for Egypt's hosting COP27 in November 2022, investigate the role of the different government, private and civil society actors preparing for the Conference, focus on the governance and institutional setup in Egypt dealing with climate change issues, and assess to what extent this rising level of concern with climate change is likely to continue into the future.

The main research question is: What are the main features of the governance structure and the institutional setup guiding efforts towards the mitigation and adaptation of climate change in Egypt pre and post COP27? And to what extent have efforts escalated in preparation for COP27, and what are the probabilities for sustaining these efforts?

It is important to note her that in our analysis of climate change governance, we examine the role of government as a main player, but we also try to explore the roles of the various stakeholder groups in the climate change governance ecosystem in Egypt. The chapter is divided into eight main sections covering the introduction; methodology; COP27 in Sharm El Sheikh and its importance; a review of climate change governance in Egypt pre-COP27, including ratified treaties, laws, and organizational structures; examples of climate change policies and projects implemented in Egypt pre-COP27; then a review of climate change governance COP27 related changes, covering Egypt's National Climate Change Strategy (NCCS) and the Intended Nationally Determined Contributions Report (INDC); an examination of some of Egypt's commitments and announced relevant projects in preparation for COP27, and a final section where we try to assess the quality of climate change governance in Egypt.

Throughout the chapter, the focus is on the role of COP27 in spurring and solidifying climate change governance efforts. The question remains whether climate change will continue to be on top of the Government of Egypt policy agenda and whether sufficient financing will be made available to make this happen.

METHODOLOGY

The methodology is a documentation analysis reviewing Egypt's climate change governance documents, literature, projects and initiatives pre and post COP27. The Documentation analysis explored information about major Climate Change related governmental projects announced in the media and the official governmental websites, before COP27, and those were juxtaposed with those announced after COP27.

Additionally, the researchers attempt a two-pronged assessment of the quality of climate change governance in Egypt. The first level of assessment is through a stakeholder analysis and an examination of the different roles played by each of the stakeholder groups, as mentioned in the published literature and media reports. The stakeholder groups investigated covered international organizations, such as the World Bank, local civil society organizations, higher education institutes, the private sector, nongovernmental organizations, consultancy firms, and media organizations.

The second level of assessment was inspired by existing frameworks already developed by different researchers and entities to assess the quality of climate governance in different countries around the globe. These frameworks were developed either by international civil society organizations or by universities. They include the Climate Action Tracker, the Oxford Policy Management Framework for Assessing Climate Change Governance, and the C40 Working Group for Assessing Good Climate Governance. Details about the three climate governance assessment models are found in the study Annex 1.

After reviewing the three different tools used for assessing the quality of climate governance, the authors decided to focus in their assessment on seven specific dimensions, perceived as most relevant to the Egyptian context, namely:

- political commitment and its sustainability,
- the policy and institutional framework,
- the stakeholder engagement,
- finance and investment,
- awareness and capacity building,
- cross departmental arrangements,
- and monitoring and evaluation capabilities.

Figure 13.1 depicts the basis for the assessment methodology used, and how the three climate governance assessment tools developed by different international civil society organizations and academic institutions were

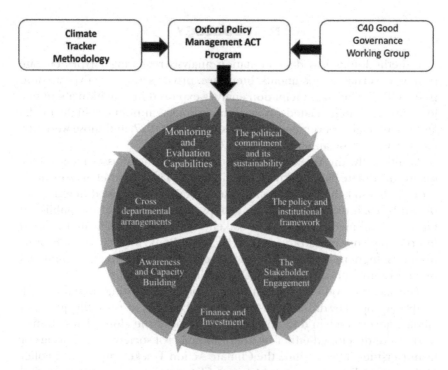

Figure 13.1   Methodology used for assessing the quality of climate governance in Egypt. *Source:* Inspired from the three climate governance assessment tools developed by Climate Tracker, Oxford Policy Management ACT Program, and C40 Good Governance Working Group.

used to determine the dimensions implemented in the field work of the current study.

Next, the researchers conducted in-depth interviews with 5 key experts, and stakeholder groups representatives about the seven key determining dimensions of climate change governance as operationalized by the researchers. Each interview continued for 1–1.5 hours. The interviewees covered a current parliamentary member who had submitted a draft law on climate change in 2022, a sustainability manager at a higher education institute, a senior government manager from the health sector responsible for accrediting green hospitals, a free-lance consultant who had worked on sustainability issues, and a private sector manager in a Green Building Sustainability firm. The interview guide used for the study is found in the study Annex 2. The researchers also participated in a USAID workshop on Climate Change Governance in Egypt (USAID, 2023, May 29) and capitalized on the opinions of 5 additional experts in attendance, representing top level leaders at EEAA and at the Egyptian National Planning Institute. The experts' opinions were quoted anonymously and analyzed, to further apply the seven assessment dimensions.

LITERATURE REVIEW

Because this chapter is situated within a book about climate change governance in the Global South, it is expected that there will be ample coverage of the published literature dealing with the concepts and theories of climate change governance. What is provided here is a brief synopsis of what major issues are covered and discussed. The literature reviewed is organized based on the scope of analysis, starting with the global level, transnational level, national, and then municipal level. Also encompassed in the review are issues related to citizens engagement, institutional theories, effect of organizational culture and importance of effective coordination.

Multiple studies focus on global climate governance, the developments and changes it went through (Kahler & Avant (2017), its potential impact on starting conflicts and instabilities (Uexkull & Buhaug (2021), and the importance of adopting a bottom-up approach where citizens and NGOs are more involved (Leal-Arcas, 2018). On a transnational level, how networks develop for the purpose of adapting to climate change is discussed, with pointers towards the need for support to be provided by governments to facilitate their work (Giest & Howlett, 2013). Existing transnational regulatory networks for climate change are mapped and loopholes identified (Quirico, 2012). Comparative approaches are utilized to identify how different countries come up with institutional innovations at the national level to deal with climate change challenges. Bauer et al. (2012), in comparing between ten OECD countries, point out to how there is a need for both vertical and horizontal integration of climate change adaptation policies, and how soft steering and coordination mechanisms are of prime importance.

Focusing on the national level, Matthews (2011) points to how the government will still play a central role in climate change governance, but that there is a need for effective intra-governmental coordination to achieve the desired policy objectives. The importance of inter-sectoral coordination is asserted by Al-Zu'bi (2016) after analyzing climate change governance in Jordan. The idea of collaboration between government and citizens is also emphasized as an important prerequisite for effective adaptation to climate change and for reducing risks on the most vulnerable groups in society (Brink & Wamsler, 2018). Within the scope of analyzing institutional frameworks related to climate change, a Common Pool Resource Theory (CPRT) is used by scholars for the purpose of figuring out the soundness and effectiveness of institutions used to manage shared resources, such as water resources. Issues covered within the CPRT include how rules are implemented, and to what extent they lead to equitable sharing of burdens and benefits (Schlager & Heikkila, 2011). Meanwhile, Patterson et al. (2019) discuss how adaptation to climate change requires changes to existing institutions and attempt to explain how and why this happens. Inderberg (2011) explains how organizational culture affects the ability to adapt to climate change.

Other scholars, such as Granberg & Elander (2007) zoom in on the local level and analyze how municipalities in Sweden deal with climate change through different types of initiatives, including networking between different municipalities within Sweden and also with other municipalities in other countries. Similarly, Schreurs (2010) analyze the role of urban and regional governments in East Asia in responding to climate change and how they develop their local climate change action plans, and how they join networks at various levels.

Because climate change governance is an issue of global concern, the literature is rife with analysis that tackles its various aspects, using different approaches and lenses. However, due to the high level of its complexity, there is still room for learning and improving on what gets to be implemented.

THE COP27 IN SHARM EL SHEIKH CITY

Zooming in on the situation in Egypt and COP27, it is important to note that very high expectations were awaiting COP27 held in the city of Sharm El Sheikh in Egypt from the 6–18 of November 2022. The central theme of the Conference was: "Together for Implementation" (De Melo et al., 2023). Egypt's Minister of International Cooperation affirmed that at COP27 the intention was to move from the pledges made earlier at COP26 into tangible projects with secured climate finance especially for developing countries (The Guardian, 2022, May 25).

The African continent was looking up to Egypt as a representative leading forth the discussions of global climate change, with Africa now taking a more proactive role by having Egypt preside over COP27. A lot of high hopes were raised in Egypt's ability to influence the international climate change agenda on behalf of African nations, where despite their negligible contributions to GHG emissions, they are perceived as most vulnerable, and the harm is compounded by the general prevalent weak governance capabilities and poverty conditions (Mantlana & Jegede, 2022; Busby et al., 2014).

In setting the scene for COP27, the city of Sharm El Sheikh was all polished and tidied up in preparation, with the objective of presenting it as a 'green city'. In that regard, the city intensified its reliance on natural gas and eco-friendly electrical means, awarded green stars to nearly 120 of its 160 hotels, established an integrated system for solid waste management in the city, and established a solar cell plant on the roof top of Sharm El Sheikh's airport (Samir, 2022, Aug 25). The solar energy operated airport was reported as only one step towards its transformation into being totally environmentally friendly (Egypt Today, 2022, July 9).

Post COP27, the Egyptian Minister of Environment reported that there were nearly 51 thousand participants, 120 heads of states, 168% more events than the earlier three COPs and that it was the largest COP after COP 15 (Fouad, 2023, March 12). Meanwhile, the Minister of Planning and Economic Development in Egypt, was reported to have met with her Emirati counterpart, since COP28 will be held in the UAE, and offered to share with her the Egyptian experience in organizing COP27 and the lessons learnt (MoPED, 2022, Nov. 16). Thus, the government side of the story shows that they are proud of what was accomplished at COP27.

Government reports about COP27 saluted the achievements realized on the national, regional and international levels. At the national level, the government secured funding for its climate change projects and signed agreements with a value amounting to $83 billion. At the African level, it was an opportunity for emphasizing Egypt's political leadership position among its African peers. And at the global level everyone was proud of the newly established Loss and Damage Fund (SIS, 2022, Nov., 23).

CLIMATE CHANGE GOVERNANCE IN EGYPT PRE-COP27

Ratified International Treaties and Conventions:

Egypt ratified the United Nations Framework Convention on Climate Change (UNFCCC) in 1994 and then ratified the related Paris Agreement in 2017 and since 2017 Egypt is committed to preparing Biennial Update Reports (BURs) and submitting these reports to the Conference of Parties (COP) of the UNFCCC when they meet (Egypt BUR, 2018). The first BUR report was submitted to the UNFCCC in 2018, but earlier than that Egypt had submitted the Initial National Communication on Climate Change in 1999 (Egypt BUR, 2018). Thus, the official interest and concern with Climate Change in Egypt has always been there at least since the 1990s, although the issue was not necessarily on the top of the government's policy agenda as it is now.

Environmental Laws and Government Organizational Structures

The regular governance structure for managing environmental affairs in Egypt is headed by the Egyptian Environmental Affairs Agency (EEAA), which reports to the Minister of State for Environmental Affairs. EEAA was established by virtue of Law 4 for 1994 to replace the earlier version of an environmental agency present since 1982. It was mandated with formulating public policies and plans to protect and develop the environment through coordination with relevant bodies, and with implementing some pilot environmental projects (Law 4 for 1994, article 5). Thus, it was very clear in the law that the main role of EEAA was to coordinate with other line ministries and relevant bodies, and to have a limited executive role. In order for EEAA to achieve its objectives, a long list of responsibilities was articulated in Law 4/1994 covering, among inter alia, its responsibilities in preparing draft environmental laws, developing the national environmental protection plan, setting environmental standards and guidelines, following up on implementation of international and regional projects related to the environment, and issuing an annual state of the environment report. The law covers three main sections dealing with the protection of land, air and water environments from pollution, but does not clearly discuss climate change. Law 4/1994 was later amended partially by virtue of Law 9/2009, and notably in one of the amended articles, article 47 repeated, a clear mention is made of 'ozone depleting substances' and how it is prohibited to either trade, use, import or possess (Law 9 for 2009, article 47 repeated). A second amendment to Law 4/1994 was through Law 105 for 2015 where the main change involved securing more resources for the Environmental Protection Fund situated at EEAA (Law 105/2015).

The position of a Minister of State for Environmental Affairs was created in 1997, by virtue of Presidential Decree no. 275/1997, when one morning the director of the EEAA arrived to office to find a newly appointed minister seated in his very office. The decree 275/1997 mentioned by name the first Minister of State for Environmental Affairs for Egypt, Nadia Makram Ebeid, and enumerated five main responsibilities related to the position, covering: representing EEAA, developing and reinforcing work systems at EEAA, coordinating with other ministries for the implementation of the Egyptian Environmental Protection Law, finalizing the organizational structure for EEAA and revamping the responsibilities of the Chief Executive Officer of EEAA unless otherwise delegated (Presidential Decree 275/1997).

EEAA is the implementing arm for the Ministry of the Environment. In 1996, a Climate Change unit was established within EEAA and then in 2009 upgraded to a Central Department (CCCD). The Climate change Central Department at EEAA is the technical secretariat to the NCCC and the focal point for the UNFCCC. It is also responsible for coordinating climate change work with the relevant government ministries and organizations. It also works on collaborating with the private sector, civil society, academia, international organizations, media and other stakeholders for all climate change measures that require following a participatory approach (Egypt First Updated INDC, 2022, June 8).

A National Climate Change committee was formed in 1997 then in 2015 upgraded to a National Climate Change Council (NCCC) (Egypt BUR, 2018; Prime Ministerial Decree 1912 for 2015). The NCCC is headed by the Minister of the Environment and has the Director of the Egyptian Environmental Affairs Agency (EEAA) as the Vice President. Members in the council include representatives from fifteen ministries, plus representatives from national security agency, the Central Agency for Organization and Administration (CAOA), the General Federation of NGOs, the Information and Decision Support Center (IDSC)—a think tank affiliated to the Cabinet of Ministers–plus three local experts to be selected by the Minister of the Environment. The NCCC is mandated with formulating and updating the national climate change plan, steering climate change activities in Egypt and integrating climate change in all national development plans, proposing and following up on budgetary allocations needed for climate change mitigation and abatement projects within each ministry concerned with climate change, approving projects presented for funding to the Green Climate Fund, and approving the Intended Nationally Determined Contributions related to Climate Change. The NCCC is required to meet at least once bi-monthly, and its members deserve an honorarium equivalent to EGP500 for each session attended to be disbursed from the NCCC's budget (Prime Ministerial Decree 1912 for 2015).

In 2019, the NCCC organizational structure was revised by virtue of another Prime Ministerial Decree (1129 for 2019), whereby the most important change introduced was having it the Prime Minister preside over the council and the Minister of the Environment as the Secretariat. According to the new structure, the NCCC is organized into a Supreme Committee, an Executive Bureau and technical working groups. The Supreme Committee has representation of nine ministries, which are: The Ministry of Foreign

Affairs, Ministry of Investment and International Cooperation, Ministry of Water Resources and Irrigation, Ministry of Planning and Economic Development, Ministry of Finance, Ministry of Environment, Ministry of Agriculture and Land Reclamation, Ministry of Defense and the secretary of the Ministerial Group for Services (Prime Ministerial Decree 1129 for 2019 article 3). The Executive Bureau is headed by the Director of the Egyptian Environmental Affairs Agency (EEAA), has as a secretariat the focal person for the UNCCC and eleven government representatives, including many of those key ministries represented in the Council plus a representative from the National Security Agency. The Executive Bureau supervises the work of the technical groups and acts as the liaison between these groups and the Supreme Committee; while the technical working groups prepare the needed technical reports and raise them to the Executive Bureau (Egypt NCCS, 2022; Prime Ministerial Decree 1129 for 2019). The Council is required to meet at least once every year, and not every two months as was stated in the earlier decree related to its establishment. It is also mandated with reporting regularly to the Cabinet of Ministers regarding progress achieved.

Theoretically, a National Measuring, Reporting and Verification System (MRV) should report to the NCCC and have affiliated MRV units in relevant ministries responsible for reporting on climate change mitigation and adaptation actions. Not all ministries have MRV units established. At the Ministry of Petroleum, for example, the unit is called the "Climate Change and Energy Efficiency Unit", while other ministries have sustainable development units that report on the achievement of the SDGs as a whole, and do not have climate change focused units (Egypt NCCS, 2022). The NCCS 2022 proposes an overall M&E framework, with templates and suggested indicators to be used by the different ministries, but the system has not been put to the test yet. Check Figure 13.2 for the Institutional Structure of Government led Climate Governance in Egypt.

EXAMPLES OF CLIMATE CHANGE POLICIES AND PROJECTS IMPLEMENTED IN EGYPT PRE-COP27

According to Egypt's First Updated Nationally Determined Contributions Report (2022, June 8), a list of climate policies and projects were implemented Pre-COP27, since 2015 to the date of publishing the report in 2022. These policies and projects included:

Energy Sector

- *Energy Policy Reforms* including a substantial drop in energy subsidies from 6% of GDP in FY 2012/2013 to 0.3% of GDP in FY 2019/2020;

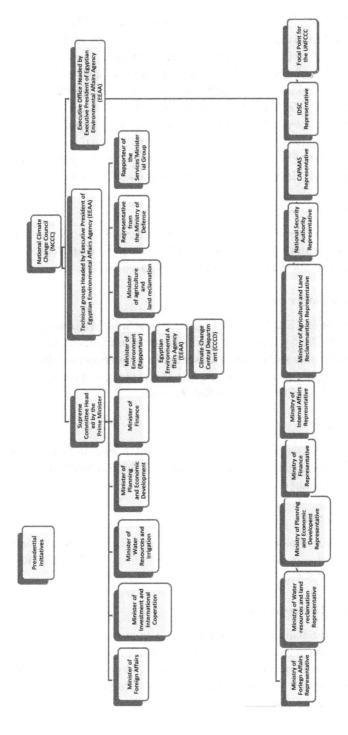

Figure 13.2 Institutional structure of government led climate governance in Egypt. *Source:* Developed by Authors based on NCCS and NCCC mandates.

- *Renewable Energy investments* increased significantly. The total installed wind and solar power plants increased by 340% from FY 2015/2016 to FY 2019/2020. Among the renewable energy projects implemented were the *Benban* Solar Park in Aswan (1465 MW), *Assuit* hydropower plant (32MW), *Kom Ombo* Solar Plant (26 MW) and the *Gabal El-Zeit* Wind Power Plant (480 MW);
- *Energy Efficiency on Demand Side*. Despite a growing population, the total consumption of electricity has reportedly dropped in FY2019/2020 compared to FY 2018/2019, but no exact figures were given for the extent of the decrease in consumption in the NDI report.

Transportation Sector

- The goal of *Low Carbon Transport* was addressed through the extension of the underground metro and the implementation of a stage 4 for its operation through an 11.5 km additional line.

Solid Waste Management Sector

- In the *Solid Waste Management sector,* a new Solid Waste Regulation Law 202/2020 was issued in addition to the implementation of four new pilot solid waste infrastructure projects in four different governorates, two in Upper Egypt, *Assuit, Qena,* and two in the Delta region, that of *Gharbiya and Kafr El Sheikh.*

Finance Sector

- The first *Sovereign Green Bonds* were issued in 2020 by Egypt's Ministry of Finance for the value of $750 million and listed in London Stock Exchange.

Agricultural Sector

- A number of national projects aiming at enhancing adaptation to climate change were implemented, including: The Building Resilient Food Security Systems in Southern Egypt (2013–2018).

Coastal Zone Management Sector

- One of the national implemented projects to adapt to climate change was the Enhancing Climate Change Adaptation in the North Coast and the Nile Delta Regions in Egypt Project (2018-2024) (Egypt's Updated Nationally Determined Contributions Report, 2022, June 8).

Healthcare Sector:

- *Certification of Green Healthcare Facilities in Egypt:* The Egyptian General Authority for Healthcare Accreditation and Regulation (GAHAR) for the first time in 2022 announced new certification standards for green health facilities. The announcement was made pre-COP27 in tandem with the governmental objective to promote for sustainable and green buildings (Nassar, 2022, May 5).
- Development of Green Healthcare Facilities Standards (2022) for certifying Green Hospitals with Excellency certificate after getting accreditation to be regulated under the Universal Health Coverage in Egypt. These standards focus mainly on different actions to be carried out by these healthcare facilities to ensure saving more energy, optimum medical waste management, and decreasing CO_2 emissions.

CLIMATE CHANGE GOVERNANCE
COP27-RELATED CHANGES

Egypt's National Climate Change Strategy 2050

Ever since the decision was made to host COP27 in Egypt, a lot of enthusiastic and positive steps have been taken by the government and by various parties to show serious commitment towards the cause. On top of those steps was the launch of Egypt's National Climate Change Strategy 2050 (NCCS) in May 2022 (Egypt NCCS, 2022) after being adopted and approved by the Prime Minister (Prime Ministerial Decree 1860 for 2022). This is considered the first comprehensive national climate change strategy that tries to provide the roadmap to be followed by Egypt in meeting the challenges of climate change, and calls on cooperation between the government, the private sector and civil society in doing so. Five main goals have been listed as part of the Egypt NCCS. The first two goals were stated as follows: "Goal 1: Achieving sustainable economic growth and low emission development in various sectors, and Goal 2: Enhancing adaptive capacity and resilience to climate change and alleviating the associated negative impacts" ((Egypt NCCS, 2022, p. 19). These two goals were identified as the most important national goals. Goal 3 calls for enhancing Climate Change Action Governance. Goal 4 deals with enhancing climate financing and finally Goal 5 deals with "enhancing scientific research, technology transfer, knowledge management and awareness to combat climate change" (Egypt NCCS, 2022, p. 35).

Egypt Intended Nationally Determined Contributions

Following Egypt's ratification of the UNFCCC in 1994, and the adoption of the Paris Agreement in 2015, Egypt followed these steps with the issuance of a document clarifying its Intended Nationally Determined Contributions (INDC). This first version of the INDC was criticized for being too general, lacking in quantification, and presenting the commitments as conditional on the receipt of international funding, estimated in the first INDC as $73 billion (Abdallah, 2020).

A few months before COP27, in June 2022, the Egyptian government issued an updated version of this earlier document, covering the period from 2015–2030. The Updated NDC is reported by government as being in line with Egypt's Sustainable Development Strategy: Vision 2030, NCCS, the National Strategy for Adaptation to Climate Change, and other sector strategies and plans (Egypt First Updated NDC, 2022, June 8). Notably now the updated document is referred to as the Nationally Determined Contributions, no longer just intended. The main changes introduced to the updated NDC in 2022 include a revision of the earlier identified mitigation and adaptation measures of the first INDC; documenting the adaptation measures implemented from 2015–2020, and adding more mitigation and adaptation measures categorized by sector (Egypt First Updated NDC, 2022, June 8; Ahramonline, 2022, July 21).

Egypt's updated INDC 2022 lists a large number of measures that would lead to better mitigation and adaptation to climate change impact. It also points out to the fact that Egypt alone does not have the financial means to carry out all what is required to implement the planned interventions. An estimated sum of $246 billion was presented as the total cost of both Mitigation and Adaptation projects in need of finance from the international community up to the year 2030; $196 billion for mitigation interventions and $50 billion for adaptation interventions. Additionally, a strong call out was made to the donor community to help as well with capacity building and technology transfer to enable the implementation of Egypt's 'ambitious' NDC 2022 (Egypt First Updated INDC, 2022, June 8). Post-COP27, the Egyptian Ministry of Environment made it clear that there are three main prerequisites for dealing with climate change in Egypt and meeting our commitments: climate finance, technology transfer and capacity building (Fouad, 2023, March 12).

Although the updated version of the NDC was perceived as an improvement over the earlier version which did not include any emission reduction targets at all, yet it was also criticized due to a number of aspects, namely: not focusing sufficiently on mitigation, not covering comprehensively all sectors where mitigation efforts will be directed, lacking clear implementation plans, and only partially fulfilling transparency requirements (Ellahamy, 2022).

Interestingly, a Second Updated Nationally Determined Contributions Report was issued in June 2023 with the same figure requested to finance implementation of the listed Adaptation and Mitigation projects, as in the 2022 report, amounting to USD 246 billion (Egypt Second Updated INDC, 2023, June 26).

EXAMPLE OF EGYPT'S COMMITMENTS AND ANNOUNCED PROJECTS IN PREPARATION FOR COP27

A long list of Climate Change initiatives, programs and initiatives were started in many government agencies right before COP27. Examples of these initiatives in the various sectors are listed below:

Agriculture

- *The Presidential Initiative of Planting 100 Million Trees* nationwide was announced pre-COP27. The plan is to plant trees along highways and to establish more parks and forests in 9900 locations covering a total estimated area of 6.6K feddans. The objective behind the Initiative is to improve air quality, reduce greenhouse gases and achieve economic outcomes; the last objective to be achieved partially through using waste water to plan wood forests (Nader, 2022, Aug. 8).
- The *'Future of Egypt Project for Sustainable Agriculture'* targeting the reclamation of 1,050K feddans: The project is ongoing and is situated along the new highway of Rod El Farag-El Dabaa. Irrigation depends on ground water and treated wastewater. Already 350K feddans have been reclaimed (Presidency.eg, 2022, May).

Investment

- *Egypt's Sovereign Fund for Investment and Development was announced as investing in green projects,* such as the Hydrogen and Green Ammonia Production Project (MoPED, 2022, May 19).
- Egypt announces the issuance of *Green Bonds* worth $750 for the first time in Africa and the Middle East ((MoPED, 2022, May 19).
- Launching of the *"National Initiative for Green Smart Projects in Governorates"* by five different government entities: Ministry of Planning and Economic Development, Ministry of Environment, Ministry of Information and Communication Technology, Ministry of International Cooperation, and the National Council for Women. The initiative aimed at encouraging the implementation of 18 green projects nationwide through a competition, where the best projects that meet the selection criteria get access to funding and get to

announced during COP27 (MCIT, 2022, Aug. 30). "The National Initiative for Green Smart Projects in Governorates" (NIGSP) was implemented, a list of green smart projects in all 27 governorates was created and the best six projects were announced during COP27 (MCIT, 2022, Dec., 6).

Health Sector

- Certification of Shefaa Al Orman Hospital to be the first hospital in Egypt to get the Green Excellence certification (Shefaa Al Orman Hospital and Green Marketing—Training Bakers, 2022).

Sustainable Cities

- *Sharm El Sheikh City as a Eco-Friendly City*: The Minister of Environment announced that the plan is to transform the city of Sharm El Sheikh into being an eco-friendly city that has buses operating on either natural gas or electricity, and where tourism is green and parks are expanded (Egypt Today, 2022, Aug 23).
- *New Mansoura Smart City*: In December 2022, the first phase of the New Mansoura Smart City was inaugurated by the President of the Republic. It is located along the Mediterranean coast, extending for over 15 Km. and targets the accommodation of nearly 1.5 million citizens. It is planned as one of many other smart cities nationwide and was announced as being in conformance with Egypt's NCCS as it preserves energy and encourages digitalization (Ahram online, 2022, Dec. 1).

ASSESSING THE QUALITY OF CLIMATE CHANGE GOVERNANCE IN EGYPT

Climate Governance Stakeholders Groups Analysis for Non-Governmental Players

In exploring the climate change governance structure for Egypt, it is important to note that the exploration is not only restricted to government, but also covers the role of the private sector, civil society, media, international organizations and other stakeholders. Table 13.1 lists some examples of climate change stakeholder groups in Egypt, briefly examines the roles they play, and suggests possible fortification/enhancement measures:

Using the Climate Change Governance Assessment Model

Climate Action Tracker Assessment Model was the only model that was applied to the Egyptian context. The overall rating for Egypt's efforts, according to the CAT Tracker, up till August 2022, were deemed "highly

TABLE 13.1		
Stakeholder Group	Role	Suggested Enhancement Measure/s
Egypt's UN Climate Change High Level Champion Mahmoud Mohie El Din	Dr. Mahmoud Mohie El Din, Former Minster of Investment in Egypt, an Executive Director at the International Monetary Fund, and the UN Special Envoy on Financing the 2030 Agenda, was designated as the COP27 High Level Climate Champion for Egypt. He spoke profusely in the media advocating for climate action at large, giving recommendations on the way forward, and occasionally hailing progress achieved by Egypt in that regard.	Continued championing of the case
Youth and Future Generations:	Young people were given space and were represented at COP27. The 10th of November 2022 was marked as Youth Day during COP27. During that day young people led workshops and activities that emphasize the role of youth in climate change action (Royi, 2022, Nov. 23).	• More meaningful and sustained participation needed at different levels. • Need for including climate change issues, problems, mitigation, and adaptation in educational curricula to raise youth awareness about climate change.
Higher Education Institutes:	Universities in Egypt had a role to play pre and during COP27. The American University in Cairo (AUC), a nonprofit, had a booth at the Green Zone in COP27 and used it to present climate change research prepared by its faculty and students. Climate Change related research was presented from different schools around campus. Similarly, a number of public universities were in attendance, such as Cairo University, Ain Shams University and a number of private universities, such as MSA.	• Continued involvement with Climate Change needed, more research, more awareness, better dissemination of research findings. • Need for more partnerships between higher education institutes and industry to work on implementation of projects related to climate change mitigation and adaptation that capitalize on evidence-based research.

(continued)

TABLE 13.1 (continued)		
Stakeholder Group	**Role**	**Suggested Enhancement Measure/s**
The World Bank	On November 8, 2022, the World Bank launched its Country Climate and Development Report (CCDR) for Egypt dwelling on the impact of climate change in Egypt, plus presenting recommendations for a low carbon growth model (World Bank, 2022, CCDR). The report was appreciated by government officials, including both the Minister of the Environment and the Minister of International Cooperation (The World Bank, 2022, Nov 8). The World Bank boasts that it is the largest "single source of financing for climate action in developing countries–$31.7 billion in FY22" (The World Bank, 2022, Nov. 3).	• More climate change finance funding needed to be directed to Egypt to help with implementation of its Nationally Determined Contributions and both adaptation and mitigation projects. • WB can work as a mediator between different Egyptian entities working on mitigating climate change actions and other financial institutions to facilitate funding. • WB can help fund capacity building programs for public sector employees regarding climate change issues and governance in Egypt.
Environmental Advocacy Non-Governmental Organizations:	There are nearly 200 NGOs in Egypt that focus on environmental protection at large (Abdel Monem & Lewis, 2020), but the exact number of those that focus on climate change is not clearly determined. The NGO sector in Egypt is strongly regulated, and possibly hampered, by the state (Herrold & Atia, 2016). Even Environmental NGOs require government permission before publishing any critical reports about the environment (Klein, 2022, Oct. 18).	• More active role needed, and more impactful projects and programs related to climate change. • NGOs can help in raising climate change awareness of public sector employees.
Media	The Media organizations in Egypt were very active in reporting on Climate Change issues and in covering COP27, but things quietened down nearly totally few months after COP27. This is especially the case because other priorities took precedence, including the difficult economic situation, the devaluation of the Egyptian pound and the rising inflation rates.	Sustained coverage of Climate Change issues and contribution to creating a higher level of awareness about climate change.

(continued)

TABLE 13.1 (continued)		
Stakeholder Group	Role	Suggested Enhancement Measure/s
Consultancy Firms	A number of private environmental consultancy firms are operating in Egypt. Their numbers increased in Egypt specially after the passing of the Egyptian Environmental Protection Law and the requirement for Environmental Impact Assessment studies (EIAs) to be presented to EEAA before the start of any project.	More evidence based and higher quality research and consultancy work.
The Private Sector Investors	There are many investment opportunities related to climate change, such as investments related to new and renewable energy. Many private sector companies are into establishing solar energy plants. The Egypt Renewable Energy Law now allows private sector solar plants to sell the surplus electricity produced to government (Abdel Monem & Lewis, 2020).	• More investments in green projects by the private sector • More public private partnerships to support climate change mitigation and adaptation projects.
Lay Citizens	Lay citizens in Egypt may not be fully aware of the importance of Climate Change and what it entails. They were exposed to the hike in media coverage around COP27, but other than that there are other priorities that concern them.	More environmentally active citizens needed.
Members of Parliament:	In November 2022, during the month when COP27 was being held, a new climate change draft law was presented to the lower house of parliament to enforce measures that would lead to net zero Green House Gas emissions by 2050. The law was presented by MP Amira Saber a member of the so-called Coordination Committee of Parties Youth Leaders and Politicians (Samir, 2022, Nov 24). The main drive for presenting the draft law was the lack of any Egyptian legislation that discusses climate change, but the bill has not been passed as yet.	Further pushing for Climate Change legislation.

(continued)

TABLE 13.1 (continued)		
Stakeholder Group	Role	Suggested Enhancement Measure/s
Political Activists & Human Rights Organizations	A number of vocal political activists used COP27 to communicate with the international audience—either attending in person or online- their criticism towards the Government of Egypt's performance regarding protection of political rights. International media picked up the message and articles were published on the use of COP27 to Greenwash reality (Klein, 2022, Oct. 18).	Government needs to provide more space for environmental activism to gain more international credibility.
Local Administration	Governorates and Local Administration units have an important role to play. In 2019, Giza governorate through donor funding and through using consultants, prepared a Giza Climate Change Strategy. However, it was criticized for being too general and for lacking clear implementation mechanisms (Eissa & Khalil, 2022). In preparation for COP27 many governorates and ministries were given clear instructions to develop their own Climate Change Strategies, but it was more of a repackaging of existing projects.	More decentralization and empowerment of the local administration is needed to enable them to play a clear role in adapting to climate change.

insufficient" (CAT, 2022, Aug 8). The reason for this negative rating being that the 2030 targets set for reducing emissions are minimal, and contrary to what the NDC document speculates, and contrary to what Egypt committed to in the Paris Agreement, there will be a rise in emissions by 2030, not a reduction (CAT, 2022, Aug 8).

Findings from the Climate Action Tracker Climate Governance Assessment point to the following: Political commitment is rated as 'poor'; the institutional framework is rated as 'neutral'; the policy process is rated as 'poor'; and stakeholder engagement is rated as 'poor' (CAT, 2022, March 28).

CAT Assessment of the Egyptian climate governance in Egypt called for a number of improvement mechanisms. To ensuring better political commitment, they called on political leaders to consider climate change more of a priority, empower EEAA and give it more authority and give the ministries who have climate change relevance more leadership roles. To have a more effective institutional framework, there is a need to build the capacity of both the MoE and EEAA, ensure better coordination between the central and local administration units involved with climate change measures. In order to have a more effective policy process, there is a need to have climate focused laws, ensure better implementation of the existing laws, submit a more quantified NDC with clearer targets, have a more transparent climate change data collection system in place across ministries. For more effective stakeholder engagement, improve existing consultation mechanisms, build more awareness, and allow civil society to advocate freely for climate action (CAT, 2022, March, 28).

These assessments were validated through the in-depth interviews conducted by the researchers with experts and representatives of stakeholder groups.

Opinions of Interviewees/Informed Experts

The following section presents some of the field experts' opinions about the seven dimensions of quality climate change governance in Egypt.

Political Commitment and its Sustainability

The opinion of several representatives of stakeholder groups was that 'climate change' is likely to remain on the government's agenda at least until the organization of COP28. The rationale was that, in addition to the high-level political will, this was a commitment expected from the COP27 hosting country.

> There was political will coming from the top of the GOE down to most of the institutions to mobilize their efforts towards climate change issues and hence projects. (Sustainability Manager at a Higher Education Institute, Interview, 2023, May 28)

> Although there are other pressing priorities, yet the probability that Climate Change remains on the Government of Egypt's agenda is high. COP28 will be held in the U.A.E. so there is regional pressure on Egypt to stay involved. (Independent Consultant, interview, 2023, May 27)

> Egypt will stay involved, and climate change will stay on its list of priorities, at least until the next COP28 is finalized in the U.A.E. This is what happened in the case of Scotland who hosted COP26 and stayed involved until Egypt finalized COP27. (Member of Egyptian Parliament, Interview, 2023, May 27)

Nevertheless, the level of commitment and enthusiasm is not expected to stay at the same high level as during COP27. For one thing there are more pressing economic issues being faced and occupying the GOE, and for another thing, COP27 represented the peak of everyone's attention, and it is expected that following that peak, things must go back to normal. Also, not all stakeholder groups were similarly interested in climate change following COP27.

> I think we are serious as a government about climate change but because of the difficult economic situation it is taking a back seat. The peak of attention to the issue of climate change was around the organization time for COP27. This attention now partially waned at the local level and is not expected that the attention will be as intensive as it was in 2022. (Member of Egyptian Parliament, Interview, 2023, May 27)

> Many stakeholder groups did not maintain the same level of action and attention to climate change issues once COP27 ended. (Independent Consultant, interview, 2023, May 27)

> Interest in climate change like a lot of other issues go through a peak state, then steady state, then downhill state. We are still at the steady state. (Green Health Care Facilities Project Director, Interview, 2023, April 30)

The Policy and Institutional Framework

In discussing the NCCS, there were several criticisms voiced by the experts, covering the excessive role played by government entities, the somewhat ambiguity of the plan and its missing out on statistical evidence. According to the National Planning Institute expert, there is over concentration in the NCCS on the governmental efforts. He pointed out that:

> Amongst the other entities that should be recognized and have more of a role are the private sector, the civil society, the Federation of Egyptian Industries, and the informal economy. The issue is not asking a person his/her opinion, but rather participation requires defining responsibilities, legislation, and accountability mechanisms. (Government Official, 2023, May 29, USAID Workshop)

This point of view notes that governance is not just about delineating the role of the different government organizations responsible but figuring out how the private sector and civil society can have space and mechanisms in place to play an effective role.

> I think the NCCS2050 is somewhat vague and lacks scientific evidence. (Independent Consultant, interview, 2023, May 27)

As for the existing National Climate Change Council (NCCC), one expert from the private sector working on sustainable buildings did not even know it existed and had never heard of the NCCS. Another expert, who is an MP, criticized its lack of empowerment.

I have not heard about the NCCC or the NCCS. (Engineer and Consultant in Green Buildings, 2023, May 15)

The National Climate Change Council is not sufficiently empowered. It just developed a strategy, but the responsibilities are not clear, and no set target dates. (Member of Egyptian Parliament, Interview, 2023, May 27)

As for the existing legal structure affecting climate change and whether there is a need for a dedicated law, a couple of experts stated, that more important than a new law, is guaranteeing effective enforcement.

We need a Climate Change law, but the important thing is implementation, having the capacity to implement and monitor. (Engineer and Consultant in Green Buildings, 2023, May 15)

Some laws are not enforced and some laws, such as the law for solid waste management, are difficult to understand. We need more awareness raising about the existing laws. (Green Health Care Facilities Project Director, Interview, 2023, April 30)

Moreover, the MP who proposed the new climate change law to parliament talked about how challenging and timely the process for the approval of the proposed law is likely to be. We cannot compare ourselves to the developed world, the number of MPs interested in climate change issues is limited, and the legislative process itself is quite time consuming and lengthy.

Our shift to the green economy is expected to be slower than the case in developed European countries for example. (Member of Egyptian Parliament, Interview, 2023, May 27)

The number of MPs who are interested in climate change issues can be counted on the one hand, which is a huge challenge. (Member of Egyptian Parliament, Interview, 2023, May 27)

The draft law has been transferred to various committees within parliament to discuss and to various ministries for consultation. It is a long process. We have a great legal structure for environmental issues but 'climate change' is not mentioned everywhere. The world is moving quickly, and we are moving very slowly. I think it is not considered a priority by the legislature. However, a positive thing is that the draft law on Carbon Markets is currently being discussed in the parliamentary committees. (Member of Egyptian Parliament, Interview, 2023, May 27)

The Stakeholder Engagement

According to the top leadership from the Egyptian Environmental Affairs Agency (EEAA), the Ministry of the Environment participated with all relevant ministries, academia, and civil society in preparing the NCCS.

It was developed in a participatory manner. (EEAA official, 2023, May 29,
USAID Workshop)

This is of course the point of view of the government. However, contrary to the official opinion by government, according to a well-informed
independent consultant, amongst the problems of the existing institutional
structure is the lack of a clear participation mechanism to secure societal
buy-in (Independent Consultant, 2023, May 29, USAID Workshop). Thus,
even though the original NCCS may have involved a participatory approach, there is no existing mechanism to ensure continued solicitation of
different stakeholders' opinions.

Some other experts were doubtful about the participatory process that
the government referred to as being part of the development of the NCCS.
Either they said that they were surprised to learn that there is a NCCS, or
they participated but were concerned about the extent their opinions will
be heeded to.

We woke up in the morning to realize there is a NCCS. They forgot to ask the
citizens or engage them. (Independent Consultant, interview, 2023, May 27)

GAHAR was not consulted during the preparation of the NCCS. (Green
Health Care Facilities Project Director, Interview, 2023, April 30)

Some stakeholders were asked their opinion about the strategy, but the issue is what happened with their answers. (Engineer and Consultant in Green
Buildings, 2023, May 15)

The government was perceived as having the dominant role in climate
change governance in Egypt. Most of the mega climate change projects are
government led.

The government is leading the shift to green economy. The mega projects
[related to climate change] the "Green Hydrogen Project" and "Benban" are
all government initiated. (Member of Egyptian Parliament, Interview, 2023,
May 27)

As for the role of the government entities, experts pointed out that
there were a few ministries dominating the scene, especially in organizing for COP27, mainly, the Ministry of Foreign Affairs and the Ministry of
Environment.

Amongst the leading ministries in organizing COP27 were the Ministry of the
Environment, the Ministry of Foreign Affairs, the Ministry of International
Cooperation, and the Ministry of Planning. (Green Health Care Facilities
Project Director, GAHAR Interview, 2023, April 30)

Not all government ministries are similarly interested and engaged in climate change. Those ministries that were involved early in COP27 were engaged such as the Ministry of Planning, Ministry of International Cooperation and Ministry of Social Solidarity. Other ministries may have had other priorities. (Independent Consultant, interview, 2023, May 27)

And in discussing the role of the private sector in climate change governance, it was noted that its role is still relatively limited.

The role of the private sector in that regard [climate change] is still limited. (Member of Egyptian Parliament, Interview, 2023, May 27)

Accordingly, several experts talked about the importance of providing economic incentives for that sector to be appropriately mobilized and involved. Incentives referred to included tax incentives, customs reductions, and easing of bureaucratic processes.

The private sector needs incentives to move to the Green Economy, at least tax incentives. We cannot impose [high] taxes on solar energy plates and then expect the private sector to move to using renewable energy. We must be consistent. (Member of Egyptian Parliament, Interview, 2023, May 27)

We need [as the private sector] from government to facilitate the importation of PV panels and exempt them from customs... The private sector needs to make profits. (Engineer and Consultant in Green Buildings, 2023, May 15)

The private sector will always prioritize economic considerations over environmental and climate change issues... only recently the government started offering "the golden license" to encourage private sector to start projects related to new and renewable energy... this is a kind of incentive to the private sector. (Independent Consultant, interview, 2023, May 27)

Positive incentives were deemed more attractive and more administratively feasible than negative measures. The MP pointed out how when imposing financial penalties on polluters for example, corruption may ensue when private sector entities try to avoid paying, and how challenges in collection arise as well when the polluters themselves are governmental entities.

In the law [Climate Change Law] I proposed, I suggested imposing higher penalties on industries who are intensive carbon emitters. I also suggested some positive incentives. Positive incentives are usually more effective penalties. The negative penalties are difficult to administer partly because of potential corruption and partly because many of the polluting industries belong to government. (Member of Egyptian Parliament, Interview, 2023, May 27)

Private sector entities can play different roles in climate change governance. There are consulting firms, there are multinationals supporting CSR initiatives related to climate change, and there are private banks financing climate change related projects.

> Our role [as private sector] when we work with government is to advise about the sustainability of buildings, the three pillars of sustainability: economic, environmental and social pillars.... The LEAD building system certification, guided by the World Green Building Council, requires us to guide government how to have certified green building. Nothing much changed about our role after COP27. (Engineer and Consultant in Green Buildings, 2023, May 15)

> GAHAR wants to transform all hospitals to become green and sustainable. The Green Healthcare standards were developed pre-COP27.We cooperated with the private sector. AstraZeneca and we had an MOU with GAHAR to support the training and the media campaigns to raise awareness. (Green Health Care Facilities Project Director, Interview, 2023, April 30)

> During COP27 AstraZeneca in cooperation with the Ministry of Health provided a mobile hospital to ensure the safety of the conference attendees. (Green Health Care Facilities Project Director, Interview, 2023, April 30)

> AstraZeneca assisted the Sharm El Sheikh International hospital in receiving International Green Accreditation through fixing infrastructure. They could have helped other hospitals in Sharm El Sheik, or in Southern Sinai, but they did not. Interest has waned. We are in the steady state. (Green Health Care Facilities Project Director, Interview, 2023, April 30)

> The banks have an important role to play. They pressure the real estate developers when they get financed by banks and must prove they implement green projects. (Engineer and Consultant in Green Buildings, 2023, May 15)

Higher education institutes were perceived to have an important role to play in creating awareness about climate change, doing research.

> Higher education institutes played an active role during COP27. Many organized competitions for students related to climate change, encouraged research projects, and organized workshops and symposia to enhance awareness. (Independent Consultant interview, 2023, May 27)

> Higher education institutes should offer obligatory courses about sustainability. Higher education institutes have started paying attention to climate change and doing research. Everything starts with research and then comes awareness and implementation. (Engineer and Consultant in Green Buildings, 2023, May 15)

The role of the media is very important as well in creating awareness, but its role can be improved if enough attention is provided to climate change issues, and if the concept is simplified to the public in ways they can understand and related to.

> The public media in preparation for COP27 dealt with stereotypical issues but did not discuss citizens' concerns, for example with the cutting of trees. The media will encourage people to ride bicycles in the morning, but it will not comment on the readiness of the roads. There are limits to what the public media can discuss. (Independent Consultant interview, 2023, May 27)

> The role of national media in preparation for COP27 was not very effective. They did not cover COP27 as intensively as they covered the Gouna Film Festival for example. (Green Health Care Facilities Project Director, Interview, 2023, April 30)

> People seem to think that the problem with climate change is just about rising temperatures. The role of media should be to highlight concrete examples for the impact of climate change such as rising sea level in Alexandria and how people may be asked to relocate. Give people doable and affordable behavior change examples...baby steps. (Engineer and Consultant in Green Buildings, 2023, May 15)

Civil society organizations were also perceived to have an important role. They do help create awareness but there is a need for better coordination of their activities.

> Nonprofit organizations have a very important role to play to raise awareness. (Engineer and Consultant in Green Buildings, 2023, May 15)

> Civil society organizations in line with COP27 were all working on climate change related issues, but there was no coordination, nor a clear strategy. They produced research but there was a lot of duplication in the issues tackled, and other issues were not discussed. They need to work on producing better quality research and implement more meaningful projects. Some just re-packaged existing projects under the umbrella of climate change. (Independent Consultant interview, 2023, May 27)

Finance and Investment

There is over dependence on foreign aid and consultants in the existing climate change institutional structure. To have a functional and sustainable climate change institutional structure, it is obvious there is a need for national sources of funding. According to EEAA Climate Change official:

> Dependence on foreign aid and consultants does not secure sustainability. (EEAA Official, 2023, May 29, USAID Workshop)

As for the probability of Egypt receiving the called for international financing to be able to implement the needed adaptation and mitigation measures outlined in its NCCS, the government representative responsible for "greening hospitals," and the private sector consultant working on 'Sustainable Buildings' were optimistic believing that because of the positive achievements on the ground, there is a good chance.

> I think that the concern with Greening will remain as a priority on the GOE agenda. The concern with the environment started before COP27. There is a 75% chance that Egypt will be able to get the needed financing because we have realized a lot of positive achievements. The culture of Egyptians has changed after COP27. Government officials have the willingness to realize changes and apply the green standards to all organizations. At GAHAR we have a plan to cover all Egypt with green accreditation by 2030. However, not all hospitals and healthcare facilities are at the same level of readiness. (Green Health Care Facilities Project Director, Interview, 2023, April 30)

> There is a high probability that Egypt will receive the needed funding. Maybe not all the requested funds, but a large chunk of it... Egypt is a country with lots of potential for implementing climate change projects. Egypt already has implemented a lot of climate change projects, like those renewable energy projects. It has lots of PV [photovoltaic] panels. (Engineer and Consultant in Green Buildings, 2023, May 15)

Others were more cautious in their speculations about financing the climate change agenda pointing out to the complexity of international financing at large, and the difficult economic and financial situation on a global level.

> International finance of climate change is complicated. (Member of Egyptian Parliament, Interview, 2023, May 27)

> Not sure whether Egypt will get the financing required for implementing the Climate Change Intended Nationally Determined Contributions, considering what the world is going through and the difficult economic situation for many nations. There may be no obligation for them to assist Egypt. (Independent Consultant interview, 2023, May 27)

> There is no guarantee that this exact figure will be received [Finance requested in the NCCS]. I believe that if the global community witnessed progress in our national sustainable oriented projects... climate finance will be positively affected. (Sustainability Manager at a Higher Education Institute, Interview, 2023, May 28)

The respondent from the private sector believed that unless we get the needed international financing, not all the planned for climate change projects will be implemented.

> I think that concern with climate change will remain on top of the GOE's agenda. If attention increases, it will be because of the financing that has been

made available and incentives that they may get from the World Bank. If no finance is available, the [climate change] projects will not be implemented. (Engineer and Consultant in Green Buildings, 2023, May 15)

Awareness and Capacity Building

When discussing awareness about climate change there are different constituents to consider. There were concerns voiced regarding the limited level of awareness of citizens at large and their lack of conviction about the importance of climate change issues especially when facing economic hardships.

> The laypeople may be somewhat sarcastic about the idea of GOE organizing COP27 and considering it a priority while they can witness the same government cutting out trees in many parts of Cairo. If people are to prioritize environmental versus economic priorities, they will go for the economic priorities because of the current situation. (Independent Consultant, interview, 2023, May 27)

> People's awareness about climate change and sustainability is still limited, especially of citizens in areas away from Cairo and Alexandria. COP27 helped stir the still waters, but we are still not there. (Member of Egyptian Parliament, Interview, 2023, May 27)

> People have no awareness about climate change. Logically since we are not an industrial nation, like China for example, there is limited awareness. The illogical part is that climate change can have a lot of indirect negative impacts on Egypt, and therefore there needs to be more awareness. It is the role of the government mostly to raise awareness among citizens. (Engineer and Consultant in Green Buildings, Interview, 2023, May 15)

> After COP27 citizens' awareness about climate change improved. Before COP27 the level of awareness was much less. Now there are clear guidelines by the President for all ministries to go green. It is a long journey, and we must complete it. (Green Health Care Facilities Project Director, Interview, 2023, April 30)

Interestingly, one of the interviewees working in a managerial position in a government organization responsible for Green Healthcare Facilities standards, was unaware of the NCCS 2050 and NDC reports. Another aspect related to awareness and capacity building of government employees. A serious problem that was pointed out by the Director of the Climate Change Unit at EEAA, was the difficulty of attracting and retaining qualified calibers in government.

> The four main ministries involved with the organization of COP27 [Ministry of Environment, Ministry of International Cooperation, Ministry of Foreign Affairs and Ministry of Planning] are highly aware about climate change is-

sues, but not the other ministries (Member of Egyptian Parliament, Interview, 2023, May 27)

In the M&E units, to attract calibers, we need to revise compensation. (EEAA Official, 2023, May 29, USAID Workshop)

This is a problem faced throughout the Egyptian government bureaucracy. Relatively low pay, compared to the private sector, makes it very difficult to attract and retain highly qualified calibers. The field of climate change is dynamic and continuously developing, so there is a need for the calibers who work in that area to be, not only qualified and competent about the field, but also capable of continuously learning and developing their capacities.

As for whether there are consistent capacity building and training efforts exerted by different stakeholder groups to elevate awareness and competencies related to climate change, experts mentioned a few concrete examples for workshops delivered in various government entities.

At the Ministry of Investment through the project Fekretak Sherketak [where the Ministry trains young people to build their own start-ups] some awareness sessions were given related to the circular economy concept. No Climate Change awareness training was given to the regular staff. (Independent Consultant interview, 2023, May 27)

At GAHAR we organized a training course on Green Health Standards to the organization staff. In the next phase, the training will be provided to hospitals—quality units—to raise awareness levels. We also implemented a social media campaign to raise awareness. We have an initiative 50*50 targeting the green accreditation of fifty hospitals in Egypt in 50 months. (Green Health Care Facilities Project Director, Interview, 2023, April 30)

A lot of training is being provided to government employees who will be moved to the New Administrative Capital, but the training does not cover "climate change" issues. (Green Health Care Facilities Project Director, Interview, 2023, April 30)

Only leaders in government have awareness about climate change, but this is not the case among employees. Employees are not concerned and do not care. They consider it an additional bureaucratic task or burden, most employees. I do not want to generalize. There may be awareness workshops that I do not know about. (Engineer and Consultant in Green Buildings, Interview, 2023, May 15)

The Office of Sustainability sends out a monthly e-newsletter to raise awareness and educate community members on how to be more positively active toward a sustainable present and future. (Sustainability Manager at a Higher Education Institute, Interview, 2023, May 28)

Cross Departmental Arrangements and Coordination

Knowledge sharing is key to effective coordination. This was perceived as missing when discussing climate change governance.

> The Knowledge infrastructure is fragmented. (EEAA Official, 2023, May 29, USAID Workshop

Similarly, coordination between different government entities involved with climate change governance and between government entities and other stakeholder groups is not as needs be.

> There is a problem with coordination between government organizations. (Independent Consultant, interview, 2023, May 27)

> Coordination is the missing magic factor. The NCCS is very broad and vague. (Member of Egyptian Parliament, Interview, 2023, May 27)

> In preparing for COP27, GAHAR dealt with separate entities. There wasn't a clear coordinating entity that we dealt with. (Green Health Care Facilities Project Director, Interview, 2023, April 30)

> Unfortunately, we work on separate islands. GAHAR and the Ministry of Health do not coordinate their work and sometimes duplication occurs. (Green Health Care Facilities Project Director, Interview, 2023, April 30)

> While the GOE has been involved in a set of initiatives associated with climate change adaptation and mitigation, a set of institutional challenges still exist...the availability and accuracy of data, alongside the lack of financial resources and cross sectoral coordination represent a major dilemma. (Sustainability Manager at a Higher Education Institute, Interview, 2023, May 28)

> There is a need for all parties to better coordinate their work. Some private sector companies, like Hassan Allam Construction Company, have appointed a Coordinator official. (Engineer and Consultant in Green Buildings, 2023, May 15)

Monitoring and Evaluation Capabilities

Although the NCCS includes a clarification of the M&E structure to be followed, yet in reality not all government organizations have these units operational.

> The M&E units in the ministries are not activated. The CAOA [Central Agency for Organization and Administration] needs to train the calibers in these units. (Former Director of the National Planning Institute, 2023, May 29, USAID Workshop)

This is a very important point that should be heeded. Again, plans are worthless, if they do not get to be translated into actions, and if those actions do not get to be monitored and evaluated.

We do not have an M&E unit at the Ministry of Investment focusing on Climate Change. (Independent Consultant interview, 2023, May 27)

GAHAR has a role in monitoring the performance of the Green accredited hospitals and healthcare facilities. If an unannounced evaluation mission finds lack of adherence, the accreditation may be cancelled (Green Health Care Facilities Project Director, Interview, 2023, April 30)

We do not issue an annual or semi-annual progress report on climate change file. If we have better and more accurate data, we will be able to come up with better policies. (Member of Egyptian Parliament, Interview, 2023, May 27)

[Our institution] produces the Carbon Footprint Report and the Sustainability Report as the two main MRV tools. (Sustainability Manager at a Higher Education Institute, Interview, 2023, May 28)

We have to do an Environmental Impact Assessment (EIA) by law before we start a project, but during the operation of a project there is nothing about monitoring climate change related procedures, guidelines, rules or regulations. (Engineer and Consultant in Green Buildings, 2023, May 15)

There is a problem with M&E in Egypt. We [the private sector] get a certificate from government that the electricity we bought from them is 'renewable', however we cannot ascertain that this is 100% true. It is also difficult to track where the received international funds go. There is also a lack of data and statistics coming out of government that would enable us to assess what is happening as regards climate change. (Engineer and Consultant in Green Buildings, 2023, May 15)

CONCLUSION

Climate Change has been on the GOE agenda for a long time. Starting 2015, there were a lot of serious climate action policies and projects being implemented on the ground in a number of diverse sectors, including the energy sector, transportation, solid waste management and integrated coastal zone management. Additionally, a few months before COP27 was organized in 2022, the GOE issued the National Climate Change Strategy 2050 and the Updated Nationally Determined Contributions Report.

The government's institutional structure for climate governance is clearly defined and intricate and headed by the National Climate Change Council (NCCC) that reports directly to the Cabinet of Ministers. It has representation from all relevant ministries and has a well described Monitoring and Evaluation System. However, the M&E system is not yet fully functional.

Egypt worked hard to prepare for the hosting of COP27. In parallel to its announced organization in Egypt, climate change issues took central stage on the government's agenda. In 2022, more climate action projects were announced and started in many sectors, such as in the investment sector, the agricultural sector, the health sector and the sustainable cities sector.

Climate Governance however is not the sole responsibility of the state. Many actors are involved and have an important role to play. The GOE is waiting for more international donors to finance its planned adaptation and mitigation projects and programs listed in its Updated Nationally Determined Contributions Report 2022. Universities and research centers need to continue playing an active role in creating awareness about the problem of climate change and producing more research and studies. Environmental NGOs need to continue their advocacy role and the government needs to give them more space to voice their concerns. There are lots of opportunities for private and international investors in the climate action domain. Consulting firms' services are in demand to advise both government and business about the needed sustainable actions. Members of the Egyptian parliament still need to decide on the Climate Change draft law they reviewed and hopefully approve it, if there is a real need for it. The media did intensive coverage of COP27 and of climate change issues, but attention seriously waned after the Conference ended. Coverage is not expected to be of the same level and intensity as during COP27, but they have to realize that dealing with climate change issues is not a one-off thing, and therefore they have a role in educating the public and monitoring the performance of different stakeholder groups.

More needs to be done. In 2023, with the stringent budgetary constraints facing the government, the currency devaluation and the rising inflation rates, it may be unlikely that climate change will continue to hold the same level of attention it managed to attract during COP27.

NOTE

1. More about the ND-GAIN Index can be found here: http://gain.nd.edu/our-work/country-index/

APPENDIX A
Details about Three Different Climate Governance Assessment Models

The Climate Action Tracker (CAT), an independent scientific project assessing and comparing between different countries' performance regarding climate change and the Paris Agreement commitments, has a methodology for assessing the quality of Climate Governance that looks into four main dimensions:

- political commitment,
- institutional frameworks in place,
- the policy formulation, implementation and review processes and
- the stakeholder engagement in policy development (CAT, 2021).

Each of the four dimensions in the CAT Assessment Tool, is measured by a number of relevant indicators. Following is the link to the full-fledged CAT Climate Governance Assessment Framework and its indicators: https://climateactiontracker.org/documents/865/2021-08_CAT_Climate Governance_MethodologyNote.pdf

When applied to Egypt, CAT used a methodology that only assessed the climate governance in Egypt pre COP27. It depended on developing a framework of critical elements in an iterative process. Categories and their corresponding criteria were developed depending on the literature and the CAT consortium's experience; therefore, it depends only on observations, it did not carry out interviews or take different stakeholders and experts' opinions.

The Oxford Policy Management Methodology: Another methodology for assessing the climate action governance was developed by Oxford Policy Management under the ACT program. It covers seven dimensions under three main themes: Foundations for action on climate change; Stakeholders for action on climate change; and mainstreaming of climate change. Each them is cascaded into a number of dimensions.

- Theme 1: The foundations for action on climate change covers two dimensions: one for evidence base and one for the policy framework;
- Theme 2: The Stakeholders for action on climate change covers three dimensions: awareness and understanding, political commitment and participation and influence;
- Theme 3: The Mainstreaming of climate action covers two dimensions: a dimension for institutional capacity and one for finance and investment (Oxford Policy Management, 2020).

This tool depends on reviewing documents and literature, plus selecting key informants and getting their inputs for different structured questions to evaluate the mentioned dimensions according to a scoring system for indicators for each dimension. This tool has not been applied for Egypt.

Following is the link to the full-fledged Oxford Policy Management Climate Governance Assessment tool: https://www.opml.co.uk/files/Publications/8617-action-on-climate-today-act/context-assessment-methodology.pdf?noredirect=1

The C40's Good Climate Governance Tool: This tool has not been applied for Egypt as well. C40 is a network of mayors from around the world who have come together to confront climate change. Their guide/tool lists ten factors that characterize good climate governance. The ten factors in brief are as follows: institutional arrangements, legal frameworks, mainstreamed climate policy, cross-departmental arrangements and action, vertical integration, budgetary mainstreaming, external governance, monitoring and transparent reporting, communication and engagement, and innovative solutions to capacity and resource challenges. These ten factors are then explained further in case studies from around the world to help other cities derive lessons and figure out ways of emulating (C40 Cities Climate Action Planning (2020).

Following is the link to more information about the C40s Good Climate Governance Tool: https://www.c40knowledgehub.org/s/article/Good-Climate-Governance-in-Practice?language=en_US

APPENDIX B
Interview Instrument:
Climate Governance and Institutional Structures in Egypt Pre- and Post-COP27 Interview Qs

Political Commitment

National Commitment:
1. What is the likelihood that climate change will remain on top of the GOE's agenda as it was in 2022 with the organization of the COP27 in Sharm?
2. If you think that it will not remain as a top item on the agenda, what may be some of the reasons behind that?

Global Commitment:
3. To what extent there is likelihood that Egypt will receive the climate finance called for which is estimated as $246 billion from now until 2030?

Stakeholder Engagement

1. What was the role of your organization in COP27? What actions were taken pre and post COP27 regarding this role?
2. To what extent lay citizens understand climate change?
3. Are ministries and gov. bodies fully aware of climate change?
4. What role does the private sector play in climate change? And how can it be improved?
5. What role do higher education institutions and research centers play in climate change? And how can it be improved?
6. What role should the media play related to climate change governance in Egypt?
7. What role do civil society organizations play and should play with regards to climate change in Egypt?
8. In your opinion, who are other stakeholders taking the lead in climate change governance pre and post COP27?
9. What are examples of the roles played by leading stakeholders?
10. Are there any potential areas of improvement in the roles played by the different stakeholder groups?
11. Do you think that the different stakeholder groups were sufficiently consulted in drafting the National Climate Change Strategy 2050?

Institutional Framework

12. According to your organization, what were the climate governance actions taken pre COP27? To what extent, these actions have been continued, or are likely to continue as planned?
13. What are examples of programs/projects implemented and/or launched pre COP27? How far they have they reached post COP27?
14. To what extent do you think these projects/policies were effective in climate change mitigation and adaptation interventions pre and post COP27? How?
15. Does your organization have a budget allocated for dealing with climate change issues?
16. Is the budget allocated focusing on mitigation or adaptation projects?
17. Did the allocated budget remain unchanged, pre and post COP27?
18. Are there hurdles to coordination between different actors in government?
19. What coordination regulations do you have at your organizations for managing coordination between your organization and other organizations working on climate change mitigation and adaptation?
20. Does your organization have partnerships with other organizations in different sectors for mitigation and adaptation processes?

21. What are examples of such partnerships carried out pre and post COP27?
22. Has your organization worked on capacity building for their staff regarding the climate change issues? How?

Policy Processes

1. To what extent do you think the National Climate Change Council NCCC is sufficiently empowered?
2. Do we need a new law that focuses on Climate Change? Do you think the draft Climate Change law presented to parliament in 2022 will be passed/approved anytime soon?
3. We need evidence for better policies. To what extent the NCCS2050 and the NDC were informed by evidence and data?
4. To what extent do you think we are ready for data collection and effective reporting on Climate Change?
5. Does your organization have a Monitoring, Reporting, and Verification (MRV) system for climate change? What is the product/output of this MRV? And does these were the same pre and post COP27?

Do you have any other comments you would like to make related to climate change governance in Egypt?

REFERENCES

Abdallah, L. (2020). Egypt's nationally determined contributions to Paris agreement: review and recommendations. *International Journal of Industry and Sustainable Development, 1*(1), 49–59.

Abdel Monem, M. A. S., Lewis, P. (2020). Governance and institutional structure of climate change in Egypt. In E. S. Ewis Omran & A. Negm (Eds.), *Climate change impacts on agriculture and food security in Egypt* (pp. 45–57). Springer Water. https://doi.org/10.1007/978-3-030-41629-4_3

Ahramonline. (2022, Dec. 1). President Sisi inaugurates first phase of New Mansoura smart city. *Ahram Online.* https://english.ahram.org.eg/News Content/1/1235/480856/Egypt/Urban—Transport/President-Sisi-inaugurates -first-phase-of-New-Mans.aspx

Ahramonline. (2022, July 21). Egypt submits updated climate commitments as per Paris agreement. *Ahram Online.* https://english.ahram.org.eg/NewsContent/ 1/2/471870/Egypt/Society/Egypt-submits-updated-climate-commitments -as-per-P.aspx

Al-Zu'bi, M. (2016). Jordan's climate change governance framework: From silos to an intersectoral approach. *Environment Systems & Decisions, 36*(3), 277–301.

Bauer, A., Feichtinger, J., & Steurer, R. (2012). The governance of climate change adaptation in 10 OECD countries: Challenges and approaches. *Journal of Environmental Policy & Planning, 14*(3), 279–304.

Brink, E., & Wamsler, C. (2018). Collaborative governance for climate change adaptation: Mapping citizen-municipality interactions. *Environmental Policy and Governance, 28*, 82–97.

Busby, J. W., Smith, T. G., & Krishnan, N. (2014). Climate security vulnerability in Africa Mapping 3.0. *Political Geography, 43*, 51–67.

Climate Action Tracker. (2022, Aug 8). *Egypt country summary.* https://climate actiontracker.org/countries/egypt/

Climate Action Tracker. (2022, March 28). *Climate governance in Egypt.* https:// climateactiontracker.org/publications/climate-governance-in-egypt/

Climate Action Tracker. (2021, Aug 16). *Climate governance series methodology.* https:// climateactiontracker.org/publications/climate-governance-methodology/

COP27.eg (2022, Nov. 20). *Historic climate deal sealed at COP27 as climate conference takes a leap to save lives and livelihoods.* Official Host Country Website. https:// cop27.eg/#/news/259/Historic%20Climate%20Deal%20Sealed%20a

de Melo, M. C., Fernandes, L. F. S., Pissarra, T. C. T., Valera, C. A., da Costa, A. M., Pacheco, F. A. L. (2023, May 15). The COP27 screened through the lens of water security. *Science of the Total Environment, 873.* https://doi.org/10.1016/j .scitotenv.2023.162303

Egypt BUR (2018). *Egypt's first biennial update report to the United Nations framework convention on climate change.* https://www.eeaa.gov.eg/portals/0/eeaaReports/N -CC/BUR%20Egypt%20EN.pdf

Egypt Environmental Protection Law Number 4 for 1994, *The Egyptian Gazette, 5,* 3 Feb. 1994. https://manshurat.org/node/13071

Egypt First Updated NDC. (2022, June 8). *Egypt's first updated nationally determined contributions.* https://unfccc.int/sites/default/files/NDC/2022-07/Egypt%20 Updated%20NDC.pdf

Egypt Second Updated NDC. (2023, June 26). *Egypt's second updated nationally determined contributions.* https://unfccc.int/sites/default/files/NDC/2023-06/ Egypts%20Updated%20First%20Nationally%20Determined%20Contribution %202030%20%28Second%20Update%29.pdf

Egypt Law Number 9 for 2009 Concerning the Amendment of Some Articles of Law 4 for 1994, *The Official Gazette,* Issue 9 repeated, 1 March, 2009. https:// manshurat.org/node/13092

Egypt Law 105/2015 Concerning the Amendment of Some Articles of Law 4 for 1994, *The Official Gazette,* Issue 42 Repeated, 19 October 2015. https://mans-hurat.org/node/13099

Egypt NCCS. (2022). Egypt National Climate Change Strategy 2050. Arab Republic of Egypt, Ministry of Environment. https://www.eeaa.gov.eg/portals/0/ eeaaReports/N-CC/EgyptNSCC-2050-Summary-En.pdf

Egypt Presidential Decree Number 275 for 1997 Concerning Identification of Responsibilities of the Minister of State for Environmental Affairs, *The Egyptian Gazette,* 7 August 1997. https://manshurat.org/node/13159

Egypt State Information Service. (2022, Sept. 4). *Egypt and climate change.* Retrieved September 15, 2022 from https://www.sis.gov.eg/Story/160255/Egypt-and -Climate-Change?lang=en-us

Egypt State Information Service. (2022, Nov. 23). *Egypt's multiple gains, contributions at COP27.* Retrieved June 17, 2023 from https://sis.gov.eg/Story/172928/ Egypt's-multiple-gains%2C-contributions-at-COP27?lang=en-us#:~:text= Wednesday%D8%8C%2023%20November%202022%20%2D%2007% 3A50%20PM&text=Egypt%20also%20succeeded%20at%20the,to%20reduce %20their%20harmful%20emissions

Egypt Today. (2022, July 9). *All you need to know about Sharm el-Sheikh Green Airport.* https:// infoweb-newsbank-com.libproxy.aucegypt.edu/apps/news/document -view?p=AWNB&t=pubname%3AEETC%21Egypt%2BToday%2B%2528Cairo %252C%2BEgypt%2529&sort=YMD_date%3AD&fld-base-0=alltext&max results=20&val-base-0=cop27%20and%20preparations%20in%20tourism%20 sector%20in%20egypt&docref=news/18B28FE678FA6530

Eissa, Y., & Heba Allah Essam E. Khalil (2022). Urban climate change governance withing centralized governments: A case study of Giza, Egypt. *Urban Forum, 33,* 197–221.

Ellahamy, H. A. (2022, Nov. 15). Egypt's nationally determined contributions: What are we missing out? *Alternative Policy Solutions.* https://aps.aucegypt .edu/en/articles/932/egypts-nationally-determined-contributions-what-are -we-missing-out

Fouad, Y. (2023, March 12). Opening speech by Minister Yasmine Fouad at the inauguration session of the Research and Creativity Convention (RCC) 2023. *The RCC 2023 Conference.* The American University in Cairo.

Giest, S., & Howlett, M. (2013). Comparative climate change governance: Lessons from European transnational municipal network management efforts. *Environmental Policy and Governance, 23*(6), 341–353.

Granberg, M., & Elander, I. (2007). Local governance and climate change reflections on the Swedish experience. *The International Journal of Justice and Sustainability, 12*(5), 537–548.

Herrold, C., & Atia, M. (2016). Competing rather than collaborating: Egyptian nongovernmental organizations in turbulence. *Nonprofit Policy Forum, 7*(3), 389–407.

How did Egypt benefit from COP27? webp (891×1280), 2022. Retrieved April 30, 2023, from https://idsc.gov.eg/Upload/InfoMedia/Attachment_A/702/%D9%83 %D9%8A%D9%81%20%D8%A7%D8%B3%D8%AA%D9%81%D8%A7%D8 %AF%D8%AA%20%D9%85%D8%B5%D8%B1%20%D9%85%D9%86%20 COP27.webp

How to measure governance: A new assessment tool. (2020, February 27). Oxford Policy Management. https://www.opml.co.uk/blog/how-to-measure-governance-a-new -assessment-tool

Inderberg, T. H. (2012). Governance for climate-change adaptive capacity in the Swedish electricity sector. *Public Management Review, 14*(7), 967–985.

Kahler, M., & Avant, D. (2017). *Innovations in global governance.* Working Paper. Council on Foreign Relations: International Institutions and Global Goverance

program. https://cdn.cfr.org/sites/default/files/report_pdf/Memo_Series_
Kahler_et_al_Global_Governance_OR_1.pdf

Klein, N. (2022, Oct. 18). Greenwashing a police state: The truth behind Egypt's COP27 masquerade. *The Guardian.* https://www.theguardian.com/environment/2022/oct/18/greenwashing-police-state-egypt-cop27-masquerade-naomi-klein-climate-crisis

Leal-Arcas, R. (2018). Re-thinking global climate change: A local bottom-up perspective. *The Journal of Diplomacy and International Relations, 20*(1), 4–12. https://ciaonet-org.libproxy.aucegypt.edu/record/59923

Mantlana, B., & Jegede, A. O. (2022). Understanding the multilateral negotiations on climate change ahead of COP27: Priorities for the African region. *South African Journal of International Affairs, 29*(3), 255–270.

Matthews, F. (2011). The capacity to co-ordinate—Whitehall, governance and the challenge of climate change. *Public Policy and Administration, 27*(2), 169–189.

Ministry of Communications and Information Technology. (2022, Dec., 26). *National Initiative for Green Smart Projects: Climate Solutions Made in Egypt.* Ministry of Communications and Information Technology, Media Center. https://mcit.gov.eg/en/Media_Center/Latest_News/News/66714

Ministry of Communications and Information Technology. (2022, Aug., 30). *National Initiative for Green Smart Projects in Governorates Kicks Off.* Ministry of Communications and Information Technology, Media Center. https://mcit.gov.eg/en/Media_Center/Latest_News/News/66426

Ministry of Planning and Economic Development. (2022, May 19). *PM Launches 2050 National Strategy for Climate Change.* Ministry of Planning and Economic Development, Media News. https://mped.gov.eg/singlenews?id=1184&type=previous&lang=en

Ministry of Planning and Economic Development. (2022, Nov. 22). *COP27 Highlights: Egypt's Minister of Planning Meets with the UAE Minister of Climate Change to Discuss Ways of Enhancing Cooperation.* Ministry of Planning and Economic Development, Media News. https://mped.gov.eg/singlenews?id=2573&type=previous&lang=en

Nader, N. (2022, Aug. 8). Egypt to plant 100 mln trees to combat climate change. *Ahramonline.* https://english.ahram.org.eg/NewsContent/1/2/472813/Egypt/Society/Egypt-to-plant—mln-trees-to-combat-climate-change.aspx

Nassar, M. (2022, May 31). GAHAR announces standards of 3-level certificate for green healthcare facilities in Egypt. *CSR Egypt News.* https://www.csregypt.com/en/gahar-announces-standards-of-3-level-certificate-for-green-health-care-facilities-in-egypt/

Patterson, J., de Voogt, D. L., & Sapianis, R. (2019). Beyond inputs and outputs: Process-oriented explanation of institutional change in climate adaptation governance. *Environmental Policy and Governance, 29*(5), 360–375.

Presidency.eg (2022, May). *The Inauguration of the Future of Egypt Project for Sustainable Agriculture.* The Arab Republic of Egypt Presidency website. Retrieved March 6, 2022 from https://www.presidency.eg/en/%D8%A7%D9%84%D9%85%D8%B4%D8%A7%D8%B1%D9%8A%D8%B9-%D8%A7%D9%84%D9%82%D9%88%D9%85%D9%8A%D8%A9/%D8%A7%D9%81%D8%AA%D8%A7%D8%AD-%D9%85%D8%B4%D8%B1%D9%88%D8%B9-%D9%85%D8

%B3%D8%AA%D9%82%D8%A8%D9%84-%D9%85%D8%B5%D8%B1-%D9
%84%D9%84%D8%B2%D8%B1%D8%A7%D8%B9%D8%A9-%D8%A7%D9
%84%D9%85%D8%B3%D8%AA%D8%AF%D8%A7%D9%85%D8%A9/

Prime Ministerial Decree 1912 for 2015. Establishment of the National Council for Climate Change, *The Official Gazette*, Issue 23, 6 August, 2015. https://manshurat
.org/node/14253

Prime Ministerial Decree 1129 for 2019. Responsibilities, Objectives and Structure of the National Council for Climate Change, *The Official Gazette*, Issue 18 (Repeated), 7 May 2019.

Prime Ministerial Decree 1860 for 2022. Adoption of the National Climate Change Strategy 2050, *The Official Gazette*, Issue 23, 9 June, 2022.

Quirico, O. (2012). Disentangling climate change governance: A legal perspective. *Review of European Community & International Environmental Law, 21*(2), 92–101.

Royi, Z. (2022, Nov 28). *COP27 creates new spaces for young people*. Climate and Development Knowledge Network. https://cdkn.org/story/cop27-creates-new
-spaces-young-people

Samir, S. (2022, Aug 25). How Egypt prepares to host COP27? *Egypt Today*. https://
www.egypttoday.com/Article/1/118624/How-Egypt-prepares-to-host-COP-27

Samir, S. (2022, Nov 24). New bill brought before Parliament to achieve net-zero emissions by 2025. *Egypt Today*. https://www.egypttoday.com/Article/1/120881/
New-bill-brought-before-Parliament-to-achieve-net-zero-emissions

Schlager, E., & Heikkila, T. (2011). "Left dry and dry? Climate change, common-pool resource theory, and the adaptability of western water compacts. *Pubic Administration Review, 71*(3), 461–470.

Schreurs, M. A. (2010). Multi-level governance and global climate change in East Asia. *Asian Economic Policy Review, 5*(1), 88–105.

Shefaa Al Orman Hospital and Green Marketing—Training Bakers. (2022, November 9). https://trainingbakers.com/2022/11/09/shefaa-al-orman-hospital
-and-green-marketing/

The Guardian. (2022, May 25). *Egypt sys climate finance must be top of agenda at Cop27 talks*. https://www.theguardian.com/environment/2022/may/25/
egypt-climate-finance-top-of-agenda-cop27-talks

Uexkull, N. von, & Buhaug, H. (2021). Is climate change driving global conflict? *Political Violence @ A Glance: Commentary and Analysis*. https://ciaonet-org
.libproxy.aucegypt.edu/record/67167

USAID. (2023, May 29). Climate change governance workshop. Organized by the *USAID Economic Governance Project*, Cairo: City Stars Inter-Continental Hotel. (Participant Observation).

World Bank. (2022, Nov 8). Egypt: Climate action can strengthen long-term growth. *World Bank Press Release*. https://www.worldbank.org/en/news/press
-release/2022/11/08/world-bank-climate-action-can-strengthen-egypt-s-long
-term-growth#:~:text=SHARM%20EL%2DSHEIKH%2C%20November%209
,with%20the%20Government%20of%20Egypt.

World Bank. (2022). *Egypt country climate and development report*. The World Bank: Middle East and North Africa, Nov. 8. https://documents1.worldbank.org/

curated/en/099510011012235419/pdf/P17729200725ff0170ba05031a8d4
ac26d7.pdf

World Bank. (2021). *Climate risk country profile: Egypt.* https://climateknowledge
portal.worldbank.org/sites/default/files/2021-04/15723-WB_Egypt%20
Country%20Profile-WEB-2_0.pdf

CHAPTER 14

INSTITUTIONAL RE-STRUCTURING OR INSTITUTIONS ANEW?

A Decision Process in the Republic of Suriname's National Climate Adaptation Planning

Kalim U. Shah
University of Delaware

ABSTRACT

The South American country of Suriname is concerned about its vulnerability to the impacts of climate change and has embarked on developing a national adaptation plan for the near to long term. This study was undertaken in support of that ongoing planning process. An institutional strengthening and capacity building assessment was completed through review of secondary information as well as a series of intensive national workshops and focus groups with national leaders and experts. The results demonstrate that a two-pronged restructuring approach required to implement national adaptation

Climate Governance in International and Comparative Perspective, pages 319–341
Copyright © 2024 by Information Age Publishing
www.infoagepub.com

319

activities. The first prong revolves around recommendations to strengthen existing government ministries and agencies that have existing mandates relevant to climate adaptation. The second prong revolves around the creation of a new national climate institute that is specifically mandated to administer the national adaptation plan, coordinate with the efforts of other allied government entities and other national stakeholders and take national lead on climate change efforts for Suriname. This study lays out many of the key findings that lead to this proposal and the critical threats and opportunities for making this restructuring successful. As a unique case study, it brings sunlight to the black box of how developing country governments grapple with institution building decisions. The principled options for restructuring are instructive to other countries embarking on such efforts to implement their climate change agendas. In sum it provides much needed insight to the challenge of modernizing public governance in developing countries and facing climate change risk through 'implementation ready' institutional arrangements.

Suriname's people, society, economy and environment are already affected by extreme weather and climate events and are under increasing risk from the impacts associated with climate change. Recent floods, for example, affected over 13,000 households in Suriname, particularly in Brokopondo and Sipaliwini districts, and "caused damage and loss valued at approximately SRD\$111 million across the housing, health, education, energy, transport, communications, agriculture, tourism, commerce and trade sectors" (Government of Suriname, 2013). Hurricanes, saltwater intrusion along the coast, droughts and crop loss, loss of biodiversity and flooding in the interior are all recognized impacts that Suriname must face due to climate change (National Climate Change Policy and Action Plan, 2015). Where there is high vulnerability and exposure to these types of climatic change, the risk of similar or more severe impacts in the future is high. Action is already being taken to address climate impacts, but more needs to be done.

Suriname's 2012–2016 National Development Plan, the 2013 Second National Communication to the United Nations Framework Convention on Climate Change (UNFCCC) and the 2012–2016 Environmental Policy Plan all recognize the significance of climate change impacts on Suriname and the opportunities for low carbon emission development. A National Adaptation Plan is the logical next step in enabling Suriname to build resilience to the impacts of a changing climate, providing a clear roadmap to respond to the challenges of a changing climate, seize opportunities for climate compatible development and attract climate finance.

According to the UNFCCC, countries develop National Adaptation Plans (NAPs) to guide their adaptation needs in the medium to long terms along with the necessary programs and strategies to meet these goals. The NAP is a process with the goals of reducing climate vulnerability through several means including adaptive capacity and resilience building, mainstreaming adaptation across relevant new and existing national policies and programs and making climate adaptation a central part of the national development planning process across sectors and governance levels. In developing the NAP, best practice suggests that Suriname work with the best available science, local knowledge and gender sensitivity. As with the broader concept of sustainability, adaptation should balance relevant social, economic and environmental priorities.

This study emerged as part of national adaptation planning process. Addressed here is the question of government's institutional capacity to oversee and implement the national adaptation plan successfully. Effective governance of national adaption requires an institutional structure with leadership and direction (Ampaire et al., 2017). Efficient implementation of the plan requires specialized expertise, coordination, and adequate resources. What sort of institutional and organizational structures will best be suited to national adaptation plan governance and implementation? Should existing structure be expanded and strengthened or are new institutions and institutional networks needed for progress?

COUNTRY BACKGROUND

Located in northeastern South America, the Republic of Suriname spans approximately 163,820 square kilometers and is bordered by French Guiana to the east, Guyana to the west, Brazil to the south, and the Atlantic Ocean to the north. The country's diverse landscape includes a fertile, low-lying coastal plain in the north, hilly terrains in the central region, and expansive tropical rainforests and savannas in the south. Ecologically, Suriname hosts diverse ecosystems, with over 80% of its territory covered by tropical rainforests. The remainder comprises coastal and marine ecosystems, freshwater environments, and savannas. Suriname is part of the Guiana Shield, one of Earth's oldest geological formations, and is home to an extensive array of biodiversity, including numerous endangered species.[1]

Suriname's climate is characterized as a tropical climate with a high humidity, which is generally controlled by the bi-annual passage of the Inter –Tropical Convergence Zone (ITCZ); once during the period December to February (known as the short-wet season), and the second time, during the months of May—mid August (long wet season). The periods in between are the short dry season (February to the end of April) and the long dry season (middle of August to the beginning of December). Suriname has three climate types, namely monsoon climate, tropical rainforest climate and a humid and dry climate (Government of Suriname, 2019). Another major condition determining the country's climate encompasses the surface conditions, such as the abundance of rivers and swamps and the presence of well-developed vegetation cover that produces large amounts of water vapor, which together with the local convection and orographic lifting along the hills and mountainous regions, also contribute to the relatively high precipitation in the country. In general, climatic conditions have remained almost the same throughout the year for decades, the variation of annual temperature is only 2–3°C. At the same time there is an insignificant change in rainfall as well, when excluding the extremely dry and wet years. This leads to the conclusion that the climate of Suriname is relatively stable. An example of an exceptionally wet year is 2006, when due to large amounts of rainfall significant areas along the upstream of rivers were inundated. However, it was also noted that such an event seems to re-occur every 25–75 years (Government of Suriname, 2019).

This small country is home to a population was estimated to be around 591,919 as of 2021,[2] primarily residing along the coastal regions and in the capital city, Paramaribo. Dutch serves as the official language, reflecting the diverse ethnic composition of the country. The country's economic landscape is dominated by sectors such as mining, agriculture, and forestry. The mining sector, primarily gold and bauxite, contributes significantly to the GDP and constitutes around 80% of the export earnings as of 2020, with gold accounting for about 70%. The agriculture sector, while smaller, is essential for employment and food security, with key products including rice, bananas, and palm kernels. Suriname was the ninth-largest global exporter of rice in 2019. The forestry sector, though accounting for only about 3% of the national GDP, plays a crucial role in maintaining Suriname's rich biodiversity and ecological balance (Ooft, 2016).

Considering climate change implications, Suriname faces major challenges, particularly in its vulnerable coastal region. Key climate indicators such as rising sea levels, increasing temperatures, and shifting precipitation patterns pose significant threats. Potential consequences include an increased frequency and severity of flooding events, which could impact agriculture, infrastructure, and residential areas. Additionally, climatic alterations may threaten Suriname's unique biodiversity and disrupt the forestry sector, which forms a key component of the country's economy (Government of Suriname, 2019).

LITERATURE REVIEW

The ongoing discourse regarding the most effective strategies for improving governance structures centers predominantly around two main strategies: the creation of new institutions and the reform of existing ones. The decision to adopt either approach hinges on numerous factors, including societal needs (Gani, 2011), cultural contexts (Kayalvizhi and Thenmozhi,

2018), historical trajectories (Khan, 2006), and resource availability (Shah and Rivera, 2007). I draw on two popular theoretical perspectives that may point to salient institutional options for developing countries like Suriname, facing the urgency of constructing an operational governance environment to enable effective climate action. Addressing climate change demands effective governance structures. In developing countries, where resource constraints and institutional inefficiencies are often more acute, the question of whether to create new governance institutions or reform existing ones becomes highly pertinent. To better understand these dynamics, this expanded review explores the theoretical underpinnings of these two strategies and assesses their potential implications for climate governance within developing nations.

From a new institutionalism theory perspective, institutions serve as the 'rules of the game,' helping to shape and direct the behavior of individuals and organizations within a society (Goldthau and Witte, 2010). New institutionalists suggest that instituting completely new governance structures can help foster an environment that promotes socio-political and economic advancement (Carrigan and Coglianese, 2011).

In many developing countries, particularly those only a few decades beyond colonial regimes, the formation of new institutions may be necessitated to move beyond those historical trajectories. Many of these countries bear the weight of a colonial past, where extant institutions reflect an archaic or incompatible governance system. This incongruity may often lead to institutional failure and necessitate the construction of new ones (Acemoglu et al., 2001). Hence, new institutionalism calls for establishing institutions tailored to a more decentralized, participatory, and inclusive form of governance (Shah, 2011). It also suggests that new governance structures should be decentralized, sharing power between local institutions and higher levels of governance, crafting dual leveled decision-making authority that leverages the local level which is better positioned to understand and respond to local needs, circumstances, and cultural contexts (Cook at al, 2017). Participatory and inclusive governance design also features prominently in the argument for new governance structures (Kelman, 2023). By promoting the establishment of institutions that prioritize participatory decision-making processes, NIT encourages wider stakeholder engagement, fostering deliberation, negotiation, and collaboration among diverse societal actors (Fung & Wright, 2001). This can lead to decision-making processes that better reflect the perspectives and interests of all stakeholders, promoting social justice and fairness. NIT underscores the role of institutions in distributing power and resources within a society. By advocating for the design of institutions that prioritize inclusivity, to give voice to traditionally marginalized groups, such as women, ethnic minorities, and the poor (Gaventa, 2006).

This inclusivity can foster social cohesion, equity, and the legitimacy of the governance process.

The creation of new institutions dedicated to climate change can catalyze more coherent and targeted policy responses. Developing nations including Suriname have adopted this approach, recognizing the multifaceted and cross-sectoral nature of climate change. Potentially, establishing dedicated climate change ministries or agencies can centralize their climate responses and ensure more streamlined decision-making processes (Dasgupta & Beard, 2007). However, the creation of new institutions can also present challenges related to inter-agency coordination, resource allocation, and potential overlaps in mandates (Peters, 2011).

In contrast, path dependency theory argues for the reform of existing institutions, suggesting that countries tend to remain consistent with their initial development path due to the inertia and increasing returns associated with existing institutional setups (Munck et al., 2014). Path dependency advocates argue that reforming and adapting current institutions is a more viable approach as these are firmly rooted in societal norms and command a degree of legitimacy and acceptance. In developing countries, where power structures are often deeply embedded, the reform approach could provide a realistic pathway towards institutional change. Incremental reform initiatives can be utilized to rectify power imbalances, combat corruption, and enhance the efficiency and accountability of the governance system, all while minimizing social disruption (Seekings, 2008).

Once a certain institutional path has been set, it tends to persist due to the self-reinforcing mechanisms of increasing returns, learning effects, and coordination effects (Pierson, 2000). These mechanisms often lead to resistance against drastic changes or the creation of entirely new institutions. Therefore, in developing countries where power structures and institutional arrangements are deeply embedded and characterized by longstanding traditions, social norms, or political arrangements, the task of creating entirely new institutions may face substantial resistance. Instead, reforming existing institutions by gradually introducing changes within their existing structures might provide a more viable and less disruptive pathway towards institutional transformation. By integrating new goals, such as sustainability or inclusivity, into the existing institutional framework, it's possible to steer these institutions towards better outcomes without entirely overturning established structures. This is not to say that reforming existing institutions is without challenges. As Mahoney and Thelen (2010) noted, there can still be resistance from those who benefit from the status quo, and the reform process might be slow and incremental. Nevertheless, given the often complex and deeply entrenched power structures in many developing countries, gradual reform of existing institutions may represent a more pragmatic approach to achieving meaningful and sustainable change.

Path dependency posits that once a certain institutional path has been set, it tends to persist due to the self-reinforcing mechanisms of increasing returns, learning effects, and coordination effects (Pierson, 2000). These mechanisms often lead to resistance against drastic changes or the creation of entirely new institutions. Therefore, in developing countries where power structures and institutional arrangements are deeply embedded and characterized by longstanding traditions, social norms, or political arrangements, the task of creating entirely new institutions may face substantial resistance and could potentially destabilize the political order. Instead, reforming existing institutions by gradually introducing changes within their existing structures might provide a more viable and less disruptive pathway towards institutional transformation. By integrating new goals, such as sustainability or inclusivity, into the existing institutional framework, it's possible to steer these institutions towards better outcomes without entirely overturning established structures.

Institutions often continue established paths due to increasing returns and inertia. Hence, embedding climate governance within existing institutional structures may prove more feasible and effective, considering their established legitimacy and societal acceptance (Shah, 2011). Developing countries could leverage this approach by reforming existing institutions like environmental, energy, or agricultural ministries to integrate climate change considerations into their operations. South Africa's Department of Environmental Affairs exemplifies how climate considerations have been integrated into an existing institutional structure (Seekings, 2008). Yet, the reform of existing institutions may also face obstacles related to resistance to change, bureaucratic inertia, and the struggle to shift long-established norms and practices (Mahoney & Thelen, 2010).

A third theoretical pathway, institutional hybridity, presents itself, with its advocates pointing to the limitations of the previous two perspectives (Zinecker, 2009). This approach involves an adaptive combination of existing institutions with newly introduced ones (Peters, 2011), offering a flexible, context-specific response to the multi-dimensional challenges faced by developing countries. It argues that this strategy enables institutional change while maintaining stability, especially in contexts where comprehensive reforms or new institution-building might face significant resistance (Pritchett et al., 2013). In the context of developing countries, institutional hybridity can be particularly beneficial due to several reasons including contextual flexibility, adapting existing institutions to local realities while also creating new ones to fill any governance gaps; resource optimization in resource-constrained settings; and balancing of continuity and change where political instabilities may be a potential risk. The result is a dynamic, context-specific model for climate governance (Peters, 2011). China's approach to climate governance illustrates the successful application of this hybrid approach. The government has reformed existing departments, such as its National Development and Reform Commission, while also creating new ones like the Ministry of Ecology and Environment (Lo, 2015). This hybrid model allows for a comprehensive response to climate change, utilizing the advantages of existing institutional knowledge and structures whilst facilitating innovative and targeted interventions.

APPROACH AND METHODOLOGY

Based on the above theoretical perspectives, a high-level panel of local governance experts agreed to consider five options when deciding on the institutional governance structure through which it's national climate change strategy should be deployed.

> *New institutional theory* supports Option A: The statutory formalization of a new national climate institute.
>
> *Path dependence theory* supports Option B: Develop a "clearinghouse arrangement" that will gradually evolve into a formal Institute; or Option C: Develop a "platform arrangement" that may or may not shift to a formal Institute.
>
> *Institutional hybridity* supports Option D: Utilize an existing ministries/agency where climate action functions can be added with later potential to branch into a new institute (Bureau of Statistics, Planning agency, national university etc.).

Over a period of eighteen months, three national consultation workshops were held in Suriname with the objective of receiving feedback from all major stakeholder groups on the design of the national adaptation plan. At each national workshop, representation was elicited from key technical and managerial representatives from twenty-eight ministries, agencies, commissions, and special purpose units of government that were identified as having some aspect of direct or indirect mandated responsibility for aspects related to climate change adaptation. In addition to government, there was invited representation from non-governmental and community organizations, business and industry and professional associations, multilateral agencies, and donor partners as well as local and foreign academia.

The first and second Workshops were dedicated to soliciting input from national stakeholders on (a) what would be the 'success factors' of a national climate governance organization?; and (b) what are the critical functions for such an organization? The third Workshop was dedicated to gathering the perspectives of national stakeholders on how 'success factors' and 'critical functions' would map on to each of the five organizational structure options available (see 2.0 above).

The first national workshop was developed after a comprehensive review of secondary information provided by several Surinamese sources including various government ministries, the local UNDP and IDB offices and local branches of international NGOs. These included over thirty documents dated from 2002 to present such as various national reports, sector specific reports, reports prepared in accord with multilateral environmental agreements to which Suriname is signatory, consultants' studies and technical articles. The workshop itself sought to solicit information related to seven themes for 'success with national adaptation' noted by the UNFCCC. These are: climate information; human and institutional capacities; long term vision and mandate; implementation ability; mainstreaming ability; participation and a monitoring and evaluation system (Dulal and Shah, 2014).

The second national workshop built on the findings and recommendation of the previous workshop. Relevant to this chapter, were the main suggestions for institutional re-structuring that emerged based on the thematic discussions and feedback as mentioned above. These two suggestions were: (a) national climate adaptation requires a centrally mandated entity to provide leadership, advice, national coordination and take responsibility for overall implementation and (b) national climate adaptation will only be successful if all the ministries and agencies currently contributing in their own ways to climate adaptation are strengthened in capacity and resources to lend their support to the new central entity. The second national workshop therefore explored these suggestions. This chapter primarily focuses on the first suggestion.

The third national workshop focused almost exclusively on considering what form a national climate change entity would take. Based on a review of different organizational forms utilized in other jurisdictions and with the functions of the would-be entity as delineated through the second workshop, this third workshop comes to some conclusions on what this national entity might look like and the form most likely to be able to deliver on the details of the national adaptation plan.

FINDINGS FROM WORKSHOPS: CHALLENGES AND PRIORITIES

Challenges

Climate information challenges include (a) accessing data and information from disparate ministries and agencies and (b) coordinating data and information being gathered by ministries, agencies and other organizations.

These issues as well as the institutional confusion and organizational tensions caused, continue having deleterious impacts on using climate data and information and planning for data and information gathering for climate adaptation purposes in the future. The Bureau of Statistics that already has a data collection mandate could, at least presently, provide the data management services for climate information. This could be coupled with the climate information and data collection being undertaken by ministries, where this would be coordinated by the cross-ministerial climate change committee. The Cabinet of the President that already coordinates national efforts that have to do with international environmental agreements and obligations can continue to play that role with respect to climate change.

Some key government ministries and agencies with a climate relevant mandate already exist and are active. These include the Ministry of Public Works, Transport and Communications, Ministry of Regional Development, Environment Coordination Unit in the Office of the President (NIMOS), Planning Office, National Coordination Center for Disaster Relief, Meteorological Services, Hydraulic Services and others. It is instructive however, that no one ministry or agency outside of the special purpose Climate Compatible Development Unit of the Cabinet of the President, has a distinct climate change mandate. The coordination of environment policy is a main responsibility of the Ministry of Labor, Technological Development and Environment. An anticipated Environmental Bill, once passed will however centrally position a special purpose agency of the Ministry, called NIMOS, with that responsibility.

Institutional roadblocks to mainstreaming include the lack of interdepartmental cooperation, clear mandates, and responsibilities. Other roadblocks are weak data gathering, national inventories and databases; and poor cross-sector communications among various ministries and agencies, with vital stakeholders and partners external to government who share climate adaptation responsibilities. Rationalize current institutional arrangements within and among key government ministries, units, and agencies to allow data access and coordination, activity collaboration, ease organizational tensions, uncertainties and tensions; and empower ministries to mainstream climate adaption in their operations while coordinating within a national system. The business and industry sector has not been well represented in climate adaptation planning or efforts.

There was advocacy for the passing of the Environmental Framework Bill which will empower NIMOS and other key government ministry stakeholders to move forward with many of the core elements of the NAP. These include mainstreaming climate change in individual ministry operations, coordinating, collaborating with the Climate Institute and most importantly, providing an approved basis upon which to rationalize institutional structures.

Stakeholders are very wary and cautious about wanting to avoid and now identify duplication of work of separate institutions. An overview is needed of the work that is already being done. There are deep concerns about what will be the legal framework for the Climate Institute and hence its standing, mandate, powers, authorities, and positioning within the broader national structure. There is a clear uncertainty of "who reports to whom on what" which perhaps sums up the current sentiments. Several non-governmental organizations made a strong and valid call to bring them closer into the planning fold as they possess data from their close and sometimes long-standing interactions with farmers, fishermen, villagers etc. on projects. They are also beginning to make use of 'citizen science' data collection techniques that could well feed into the Institute. From the experience of the IDB in identical situations where different institutions must come together it is important to identify how to bring them together. For the platform/Institute to be successful it becomes important for stakeholders to answer: 'What's in it for me?' as an underlying criterion for selecting different institution types.

Priority Functions

In the operational model, the first set of capabilities required will be to co-ordinate data inputs at the Institute that are generated by and collected from other key stakeholders. These key stakeholders are the ministries, agencies, units that are already mandated to collect the relevant data and information. The role of the Institute is less about data generation, but more about data acquisition. Several capabilities will be necessary here including data quality and standardization; data cleaning and processing; data storage and retrieval.

The statement above must come with a caveat given the example currently observed with the General Bureau of Statistics. The General Bureau of Statistics has by law the mandate to collect and produce data. They do not always receive the data that is required to produce for example environmental statistics. It would be convenient if there will be a collaboration between the ABS and Institute. But the point of concern here remains, that the Institute may not receive data from the various institutional partners, given the fact that even now and even with a firm mandate, the ABS has difficulties in receiving necessary data (the right data and quality data).

The second set of capabilities required, lies in the transformation role of the Institute. Here the data that is collected and acquired from stakeholder partners and supplemented by the activities of the Institute itself, are analyzed very specifically for climate change adaptation purpose. It will be important to emphasize that the data analyses to be performed by the

Institute will be unique to the institutional landscape in Suriname. Several capabilities will be necessary including climate scenario modelling, working with big data and large datasets, model downscaling, climate forecasting, computer simulation modelling.

The third set of capabilities that the Institute must build is in translating the technical outputs into useable products for its stakeholders and counterparts. This includes reports and briefs geared towards assisting policy makers in decision making processes. Here, the Institute will ensure that the technical models, scenarios, forecasts etc. generated are used to develop actionable and implementable recommendations and guidance for end users. These end users could be other government ministries, agencies and units that take this climate change information to its own stakeholders or end users could also include sector clients such as farmers associations, business, industry, non-governmental organizations, and international aid agencies.

ANALYSIS OF GOVERNANCE OPTIONS TO MEET CHALLENGES AND PRIORITY FUNCTIONS

Option A: The Statutory Formalization of a New National Climate Institute (an option supported by new institutional theory)

A statutory authority is typically set up by an Act of Parliament or other form of government declaration. This lays out the goals, objectives and often the life duration of the statutory agency as well as the governance system such as a board and the executive management. There will typically at minimum, be a signed memorandum of understanding between the Institute and key collaborators. One consideration that will arise may be under which line ministry, if any, the Institute would fall under.

Pros: The government may see it fit to delegate the Institute as a statutory authority for several reasons including: (a) The current constellation and arrangement of ministries and agencies that have formal and informal, direct and indirect, mandatory and voluntary responsibilities for aspects of climate change adaptation needs to be rationalized and harnessed into a more cohesive approach. That is, the Institute becomes an agent of efficiency in circumstances of limited resources. (b) A designated statutory Institute can provide legitimacy that climate change adaptation efforts are taken seriously and are not a political matter but one of national development significance. The Institute becomes further removed from political influence and possible interference and becomes an unbiased source of expertise to be tapped into by the various government ministries. (c)

Accountability—The jurisdiction of the Institute as statutory authority is set out by Act of Parliament. This, therefore, makes switching, sharing or evasion of responsibility more difficult. This becomes especially important in two main reasons. First, as the Institute engages in expert advisory services on climate change analysis of a highly technical nature that will be used to guide other ministries, agencies and other 'clients', there will be a reasonable expectation that the technical advice is sound. Second, as the Institute initiates its financial sustainability as well as accessing of funds, grants and revenue flows from national, regional, and international sources, strict measures of accounting as well as measuring return on investments become important. In the statutory structure these tasks may be more transparent than otherwise.

Cons: There are two main challenges or "cons" to consider with this roadmap option for the Institute: (a) The concept of the Institute in this highly formalized and institutionalized form with a firm mandate may be viewed with caution by other ministries and agencies that have staked a claim to climate change adaptation responsibilities through their own work plans. Some such ministries and agencies may be critical collaborators for the effective functioning of the Institute and any such apprehension will have to be surmounted. A typical solution to this would be to develop a shared governance model for the Institute where such critical ministry collaborators are invited to have governance positions to help chart the Institute forward (Shah et al., 2013). (b) As a statutory Institute the Institute will have more responsibility, if not full responsibility, for its financial sustainability. Should the Institute be provided with early funding through annual central government budgeting, then the Institute will quickly have to illustrate the national return on investment with project successes. Sustainable financing will have ot be the foremost priority of the newly Instituted Institute after the mandate and governance structures are cemented.

Option B: Develop a "Clearinghouse Arrangement" That Will Gradually Evolve Into a Formal Institute (an option supported by path dependency theory)

This can be considered as a 'soft launch' of the Institute concept rather than the full formal institutionalization. In this model, the Institute becomes the 'central node' with ministries, agencies and climate change related units acting as 'sub-nodes' in a networked system. There may or may not be memorandum of understanding between the Institute and sub-nodes; and more often such involvement is voluntary. The initial role of the clearinghouse is knowledge, data and information sharing (as opposed to new knowledge generation). Eventually as trust and legitimacy build through cooperation of collaborators, a decision may be made to shift the clearing-house model to a 'platform model'.

Pros: There are two main advantages of this clearinghouse approach. (a) By using the clearinghouse as an interim starting point, a willing and equitable coalition on likeminded and interested ministries, agencies, units and other stakeholders can be brought together less formally. This allows a certain amount of 'breathing room' for the coalition to solidify. It can also become clear, which partners are proactive and contributing and which are not; which partners are helpful toward the Institute objectives and which are antagonistic. (b) The clearinghouse mechanism does not require as much institutional, organizational, or financial investments as a fully formalized statutory Institute or a platform. However, it places Suriname in a position of having some primary apparatus to ramp up climate change co-ordination, analysis, and evaluation, should there be need and resources to do so.

Cons: There are two main challenges or disadvantages to this clearing house approach: these are: (a) The voluntary or less formalized nature of the clearinghouse mechanism becomes a 'double-edged sword', in that the Institute leadership cannot necessarily enforce the collaboration of partners (sub nodes). This becomes critical in terms of sharing of data and information collected, held by, and likely owned by these partners. (b) A clearinghouse fundamentally functions as a space and structured repository for data and information from disparate sources that once together, can prove useful to all the partners. The Institute leadership becomes the custodian and administrator in this model. In addition, the clearinghouse becomes a space to discuss, debate, share views, solicit assistance, advice, and collaboration among partners. This may not entirely meet the national gap that has been identified in terms of the Institute operational model (data co-ordination, data transformation, data translation). A clearinghouse does not typically engage in advanced level analyses. It would have to evolve towards a platform or Institute to do so.

Option C: Develop a "Platform Arrangement" That May or May not Shift to a Formal Institute (an option supported by path dependency theory)

There has been significant recent interest in the development partnership platforms and processes that involve multiple partners in efforts addressing climate change, given the cross sectoral and cross disciplinary nature of the challenge. In such voluntary partnership platforms, stakeholders have a shared understanding and pursue collective goals. They may also pursue individual goals under the auspices of the platform, should those be aligned with the platform objectives. If not aligned, partners may still pursue their own activities separate from the platform without repercussions. Typically, some level of responsibility should be shouldered and agreed to by partners. Participation is often also driven by the perceived benefits (to themselves and their mandates) that may emerge from the platform processes. Platforms are best suited for delivering equitable benefits for partners in the process on a long-term basis. Multi-stakeholder platforms create value to the partners who participate in the efforts. Such platforms are typically driven by partners who come together on a common vision of what must be done to solve climate change problems. Such platforms multiply efforts more than if partners pursue the same efforts individually. The main goal of a multi-stakeholder process is to see change in policy and implementation. The platform is therefore more proactive and knowledge generation biased than the clearinghouse but less formalized in its setting of goals and objectives, measures and mandates than the formalized Institute.

Pros: There are three main advantages of taking the platform mechanism approach. (a) Institute leadership can quickly identify specific issues that affect the climate change adaptation community and prioritize where to direct and focus the platform community's efforts and resources. (b) Carry out joint analysis and research which will better inform the policy formulation process and subsequent implementation. The Platform, unlike the clearinghouse, has enough structure and resources to promote some level of analytical work. Primarily, the analytical work is conducted based on the joint expertise and resources (e.g., shared databases) of the partners. This is different from the analytical reach of the formalized Institute which is far more than the sum of the expertise of partners. This is because the formalized Institute is actively mandated and positioned to develop its own advanced and unique analytical capabilities, perhaps to fill voids identified nationally. (c) The Platform allows easier, more efficient pooling of resources, talents, and other capabilities of a diverse range of stakeholders, thereby strengthening the capacity to effect policy and project implementation change. The sharing information on problems and solutions, promotes greater levels of understanding and trust between the various stakeholders. This also encourages debate and discourse leading to a united front on national policy issues and directions, which can lead to more effective national planning for climate change and minimized bureaucracy.

Cons: There are three main challenges or disadvantages of taking the platform mechanism approach. (a) The platform may be an end goal or an intermediary stage towards a full-fledged and formal Institute. If the latter is the intent, then there is a reasonably high risk, based on experiences in other country scenarios, that the evolution may not occur. This could be because of loss of momentum on the issues, non-realization of financing to move to the next step, fragmentation of partners because of policy or operational disagreements. Or, it could be that it is decided that all the goals of the Institute can actually be realized within the framework and structure of the platform. (b) The platform does not necessarily forward an agenda of advanced, unique analytical capabilities that is envisioned in the Institute concept. Again, analysis at platform level becomes what can be achieved based on the sum of the partners' expertise, not necessarily the development of unique and advantageous Institute expert capabilities. (c) The platform shifts partnership more toward a formalization and less towards voluntariness. With more formalization comes responsibilities of partners. Such responsibilities will need to be 'enforced' through the Platform governance structure. One downfall of platforms is often too closed and insular or too open and diffused leadership and management. This is unlike a formal Institute where governance is clearly mandated and institutionalized.

Option D: Utilize Existing Ministries/Agencies Where More Climate Action Functions can be Added With Later Potential to Branch Into a New Institute (an option supported by institutional hybridity theory)

The Surinamese government's starting point was a broad bias towards inaugurating a new National Climate Institute, without sufficiently considering what that would mean and if it would fit the expected functions. An option for moving toward this new governance mechanism while retaining and building from existing government ministries, agencies and commissions was presented as an option; essentially an institutional hybrid approach. This presents the possibility of working with the existing institutional structures and organizational relationships to achieve the objectives envisioned for an Institute.

Pros: (a) These could include the ability to co-opt existing resources and skilled personnel that exist at places such as the General Bureau of Statistics and the University to initiate the climate change adaptation strategy and build momentum towards larger and deeper efforts. (b) There is more support and less resistance from national leadership, to maintaining efforts within and among known institutions instead of creating a new one that could cause tensions around redistributions of mandates, power and resources (c) Some measure of avoidance of tangible and intangible start-up costs is possible. Tangible startup costs in terms of financial investments, equipment, hardware, software, talent acquisition etc. Intangible in terms of relationship building, partnership negotiations, some amount of acquired trust and legitimacy capital among stakeholders etc. This would mean that the Institute integrates into the Gonini platform and avoids many start up risks, as well as benefits from lessons already learned by the Gonini platform and the systems and structure put in place. Building on an existing agency could be more cost-effective as it already has a framework and operational structure in place. It might require fewer resources to scale up or reconfigure an existing entity than to start a new one from scratch, which often necessitates a significant investment in infrastructure, personnel, and technology. (d) Existing agencies have built up a wealth of experience and expertise over the years. This institutional knowledge can be invaluable when dealing with new but related responsibilities. (e) By utilizing an existing agency, it's possible to reduce the level of bureaucracy and the number of overlapping jurisdictions, which can often complicate government efficiency. Given the pre-existing infrastructure and framework, it might be faster to expand the functions of an existing agency rather than creating a new one. Time is often critical, especially when addressing urgent issues. (f) Existing agencies have established relationships with stakeholders, understand existing processes and policies, and are familiar with the legal and regulatory environment. This continuity can provide a stable platform for the new functions. It can be easier to build on the established public trust in a known entity rather than trying to establish legitimacy and credibility for a new agency.

Cons: (a) With the expected advancements in national climate change policy and the increasing climate change adaptation pilots and expected sophistication in projects, there is the need for expanded institutional capacity or institutional building. (b) Experience has suggested that in cases of devising national policies and strategies around new scientific and technical oriented phenomena (such as climate change), it is less effective to pile new expectations of performance on to existing institutions that are oftentimes already grappling with limitations and constraints in reaching their existing mandates (Daze et al., 2017). Additional mandates therefore stress the individual institutions involved and the relational network of institutions across the sector. (c) Another concern of note is that among the existing key institutions that could share in the responsibility of the national climate change mandate (in conjunction with or separate from the Climate Institute), none of them were set up to be implementation oriented. More particularly, even if they have acquired some implementation orientation, this does not extend to climate change adaptation. For example, the Bureau of Statistics in this discussion is a data collector and the University could be interpreted as a data analyzer. Neither is a core implementer. To cobble together the resources and skills of existing institutions to implement the vision, instead of a climate Institute, would be to handicap the effort from the start. (d) A cobble of existing institutions means it will be that much harder to hold that structure together and expect accountability for performance or achieving objectives. This would be especially so in a case where the climate change adaptation mandate is divided out to several existing institutions; and then the part mandates within each partner organization become subsumed in that organization's other programming (Azhoni et al., 2017). The mandate and accountability for it becomes co-mingled, less transparent, and less measurable. This could risk meeting national climate change adaptation objectives in an efficient way. (e) If an agency takes on too many responsibilities, it might dilute its core focus and expertise. The agency's main function could suffer as resources and attention are diverted to the new tasks. For example, the Forestry Division is focused on forest management for conservation and as an economic activity. There could be a conflict of interest or a reduction in its ability to adequately perform both tasks with a climate change functionality added on. On the extreme end, there could be potential conflict of interest if the new functions might require the agency to take decisions that could negatively impact their existing roles and responsibilities. (f) Existing government units in Suriname are already understaffed or operating at their full capacity, both in terms of personnel and resources. Adding more functions could overburden these units, leading to decreased efficiency and potentially lower quality outcomes. (g) Especially specialized government units may have established work cultures. If new functions do not align with the current culture, or if the organization is resistant to change, it can be challenging to implement new processes or expectations. This resistance can also lead to lower morale among employees, which can further hamper the effectiveness of the new functions. (h) The skills required for the new functions might not match the existing skill sets within the unit. This could lead to a need for extensive retraining, hiring new personnel, or even letting go of staff who can't adapt to the new tasks, all of which could have significant costs and time implications. While adding functions to an existing unit might initially seem to reduce bureaucracy, it can also add layers of complexity to the administrative processes. The new functions might require different procedures, reporting structures, or oversight mechanisms, leading to a more complicated organizational structure.

OUTCOME DECISION BY THE GOVERNMENT OF SURINAME

The government ultimately decided to approve Option D supporting the institutional hybridity theory and offering several compelling reasons for its selection.

By process of consultation, negotiation and agreement, government first reviewed Option A versus Option D. They agreed that Option won out for the following reasons. Firstly, utilizing existing ministries/agencies allows for the co-option of resources and skilled personnel that are already present in organizations such as the General Bureau of Statistics and the University. By initiating the climate change adaptation strategy within these established entities, the government can leverage their expertise and build momentum towards larger and deeper efforts. This approach avoids the need for creating a new institute from scratch and enables the government to make efficient use of existing capacities. Secondly, there is likely to be more support and less resistance from national leadership when efforts are maintained within known institutions rather than creating a new one. The establishment of a new institute could lead to tensions surrounding redistributions of mandates, power, and resources. By building on existing agencies, the government can work within established organizational relationships and minimize potential conflicts. Thirdly, utilizing an existing agency can save costs associated with tangible and intangible startup expenses. Starting a new institute requires significant investments in infrastructure, personnel, and technology. In contrast, building on an existing agency with a framework and operational structure already in place can be more cost-effective and reduce the need for additional resources. Furthermore, the institutional knowledge and experience accumulated by existing agencies are invaluable when dealing with new but related responsibilities, such as climate change adaptation. Fourthly, an existing agency provides continuity, established relationships with stakeholders, and familiarity with processes, policies, and the legal and regulatory environment. This stability and public trust in a known entity make it easier to establish legitimacy and credibility for the new functions. By expanding the functions of an existing agency, the government can avoid bureaucratic complications and overlapping jurisdictions that often impede government efficiency.

Despite potential challenges, Option D provides a more pragmatic and realistic approach for the government to manage its national climate governance than offered by Option A at this time. By leveraging existing capacities, relationships, and resources, the government can effectively address climate change adaptation while minimizing costs and complexities associated with creating a new institute. As the existing agencies gain experience and demonstrate success in climate action, there remains the potential to transition towards a dedicated National Climate Institute in the future, based on the knowledge and lessons learned from the hybrid approach.

Similarly, by process of consultation, negotiation and agreement, government then reviewed Option B and option C versus Option D. They agreed that Option D has distinct advantages over the latter two. Compared to Option B, which proposes the development of a clearinghouse arrangement that may evolve into a formal institute, Option D offers a more concrete and structured approach. While the clearinghouse model focuses primarily on knowledge sharing and data coordination, it lacks the ability to enforce collaboration among partners. In contrast, Option D allows for the utilization of existing ministries/agencies, which already have established mandates and operational structures. By leveraging these entities, the government can quickly mobilize resources and skilled personnel, initiating climate change adaptation strategies and building momentum more effectively. Option D also benefits from the support and leadership buy-in of national institutions, minimizing potential tensions surrounding redistributions of mandates, power, and resources. Additionally, Option D provides a more cost-effective alternative, as it avoids the need for significant financial investments, infrastructure development, and talent acquisition that would be required for a fully formalized statutory institute.

Compared to Option C, which suggests developing a platform arrangement with the potential to shift to a formal institute, Option D offers greater stability and accountability. The platform approach, while flexible and voluntary, carries the risk of losing momentum, experiencing funding limitations, or facing fragmentation among partners. Option D, on the other hand, allows for the consolidation of responsibilities within existing ministries/agencies, providing a more solid foundation for the implementation of climate action functions. By utilizing established institutions, the government can leverage their expertise, relationships with stakeholders, and familiarity with existing processes and policies. This continuity facilitates a stable platform for new functions, benefiting from the established public trust and credibility of these entities. Furthermore, Option D enables more effective coordination and streamlining of efforts by reducing bureaucracy and overlapping jurisdictions.

Overall therefore, the institutional hybridity approach of Option D provides a balance between leveraging existing capacities and expertise while gradually shaping a new institute that can address the evolving needs of climate change adaptation in Suriname.

CONCLUSIONS

This study provides unique insight to a governance conundrum that developing country governments often face—what formal structure and form should governance institutions take to fulfill their mandates? It is a particularly daunting question when the subject of governance is complex, multifaceted and far-reaching as climate change has been oft characterized. In Suriname's case, the institutional hybridity argument wins out over building of new institutions or path dependency. This study also fills a gap in the literature that is sparse on these investigations in developing countries, particularly small ones (Shah, 2022).

Now that the decision has been made by the Surinamese government, it will be interesting to follow how the plan is executed and to then evaluate if the institutional hybridity option works. A suggestion likely to gain traction is a major evaluation at the five-year juncture, done by a national climate steering committee. At the five-year juncture, the viability and need for a spun-off new national climate institute would again be revisited.

The decision to build new institutions or reform existing ones does not adhere to a universal formula. Instead, it requires a comprehensive understanding of the social, political, and historical complexities intrinsic to each case. Theoretical perspectives provide valuable guidance in this regard. As developing nations navigate their unique and complex challenges, a nuanced application of these theories, which takes local contexts into consideration, may pave the way for more effective and sustainable governance transformations. This case study offers two lessons they should take on board. One is that the complexity of the climate challenge does not have to translate into a similarly complex web of governance interrelations from the starting line. There can well be benefits to weaving new working relations within the existing governance fabrics to start contending with the issue, while still leaving opportunities to spin off new specialized or coordination-oriented institutions later as may be merited. Second, commitments to building of new institutions will generate well-worn problems for governments if such new institutions continue to be burdened by the systemic institutional weaknesses and challenges. Being dependent on existing specialist bureaus for climate data or project authorizations are oft cited examples. If the mechanics of interagency work are not resolved or at least brought into alignment and disciplined enough to service a new institution, the new institution is jeopardized. Governments must weigh if they are ready, willing, and able to do this.

ACKNOWLEDGEMENTS

The author draws on primary information he gathered from 2017–2019 as the National Adaptation Plan (NAP) Lead Consultant to the Government of the Republic of Suriname, sponsored by the United Nations Development Program under the Japan Caribbean Climate Change Program.

NOTES

1. About Suriname. https://www.discover-suriname.com/about-suriname
2. Population, total Suriname. https://data.worldbank.org/indicator/SP.POP .TOTL?locations=SR

REFERENCES

Acemoglu, D., Johnson, S., & Robinson, J. A. (2001). The colonial origins of comparative development: An empirical investigation. *American Economic Review, 91*(5), 1369–1401.

Ampaire, E. L., Jassogne, L., Providence, H., Acosta, M., Twyman, J., Winowiecki, L., & van Asten, P. (2017). Institutional challenges to climate change adaptation: A case study on policy action gaps in Uganda. *Environmental Science & Policy, 75*, 81–90.

Azhoni, A., Holman, I., & Jude, S. (2017). Adapting water management to climate change: Institutional involvement, inter-institutional networks and barriers in India. *Global Environmental Change, 44*, 144–157.

Carrigan, C., & Coglianese, C. (2011). The politics of regulation: From new institutionalism to new governance. *Annual Review of Political Science, 14*, 107–129.

Cook, N. J., Wright, G. D., & Andersson, K. P. (2017). Local politics of forest governance: Why NGO support can reduce local government responsiveness. *World Development, 92*, 203–214.

Dasgupta, S., & Beard, V. A. (2007). Community driven development, collective action and elite capture in Indonesia. *Development and Change, 38*(2), 229–249.

Dazé, A., Price-Kelly, H., & Rass, N., (2016). *Vertical integration in national adaptation plan (NAP) processes: A guidance note for linking national and sub-national adaptation processes.* International Institute for Sustainable Development.

Dulal, H. B., & Shah, K. U. (2014). "Climate-smart" social protection: Can it be achieved without a targeted household approach? *Environmental Development, 10*, 16–35.

Government of Suriname. (2013). *Second national communication on climate change.*

Government of Suriname (2019). *Suriname national adaptation plan.* https://www4 .unfccc.int/sites/NAPC/Documents/Parties/Suriname%20Final%20NAP_ apr%202020.pdf

Fung, A., & Wright, E. O. (2001). Deepening democracy: Innovations in empowered participatory governance. *Politics & Society, 29*(1), 5–41.

Gani, A. (2011). Governance and growth in developing countries. *Journal of Economic Issues, 45*(1), 19–40.

Gaventa, J. (2006). Finding the spaces for change: A power analysis. *IDS Bulletin, 37*(6), 23–33.

Goldthau, A., & Witte, J. M. (Eds.). (2010). *Global energy governance: The new rules of the game.* Brookings Institution Press.

Hall, P. A., & Taylor, R. C. (1996). Political science and the three new institutionalisms. *Political Studies, 44*(5), 936–957.

Kayalvizhi, P. N., & Thenmozhi, M. (2018). Does quality of innovation, culture and governance drive FDI?: Evidence from emerging markets. *Emerging Markets Review, 34*, 175–191.

Khan, M. H. (2006). Governance and development. *Eastern and Western Ideas for African Growth, 85.*

Lo, A. Y. (2015). Challenges to the development of a low carbon economy in China. *Energy Policy, 84*, 178–188.

Mahoney, J., & Thelen, K. (2010). A theory of gradual institutional change. *Explaining Institutional Change: Ambiguity, Agency, and Power*, 1–37.

Mimura, N., Pulwarty, R. S. Duc, D.M. Elshinnawy, I., Redsteer, M. H., Huang, H.Q., Nkem, J. N., & Sanchez Rodriguez, R. A. (2014). Adaptation planning and implementation. In *Climate Change 2014: Impacts, adaptation, and vulnerability. Part A: Global and sectoral aspects.* Contribution of Working Group II to the Fifth Assessment Report of the Intergovernmental New York, NY.

Munck af Rosenschöld, J., Rozema, J. G., & Frye-Levine, L. A. (2014). Institutional inertia and climate change: a review of the new institutionalist literature. *Wiley Interdisciplinary Reviews: Climate Change*, 5(5), 639–648.

Ooft, G. (2016). *Inflation and economic activity in Suriname* (No. 16/03). Centrale Bank van Suriname Working Paper Series.

Ostrom, E. (1990). *Governing the commons: The evolution of institutions for collective action.* Cambridge University Press.

Peters, B. G. (2011). *Institutional theory in political science: The new institutionalism.* Bloomsbury Publishing.

Pierson, P. (2000). Increasing returns, path dependence, and the study of politics. *American Political Science Review*, 94(2), 251–267.

Pritchett, L., Woolcock, M., & Andrews, M. (2013). Looking like a state: Techniques of persistent failure in state capability for implementation. *Journal of Development Studies*, 49(1), 1–18.

Seekings, J. (2008). The continuing salience of race: Discrimination and diversity in South Africa. *Journal of Contemporary African Studies*, 26(1), 1–25.

Shah, K. U. (2022). Preparing public health at the front lines: Effectiveness of training received by environmental health inspectors in the Caribbean. *International Review of Administrative Sciences*, 88(3), 826–842.

Shah, K. U., & Rivera, J. E. (2007). Export processing zones and corporate environmental performance in emerging economies: The case of the oil, gas, and chemical sectors of Trinidad and Tobago. *Policy Sciences*, 40, 265–285.

Shah, K. U. (2011). Corporate environmentalism in a small emerging economy: Stakeholder perceptions and the influence of firm characteristics. *Corporate Social Responsibility and Environmental Management*, 18(2), 80–90.

Warren, R. F., Wilby, R. L., Brown, K., Watkiss, P., Betts, R. A., Murphy, J. M., & Lowe, J. A. (2018). Advancing national climate change risk assessment to deliver national adaptation plans. *Philosophical Transactions of the Royal Society A*, 376(2121), 20170295.

Zinecker, H. (2009). Regime-hybridity in developing countries: Achievements and limitations of new research on transitions. *International Studies Review*, 11(2), 302–331.

PART IV

LOCAL GOVERNMENT PERSPECTIVES

CHAPTER 15

CLIMATE GOVERNANCE AND SEA LEVEL RISE

A Case Study of Alexandria City, Egypt

Laila El Baradei
American University at Cairo, Egypt

ABSTRACT

Policies may be formulated, legitimized, and disseminated but not necessarily implemented effectively. The Government of Egypt is aware of the Sea Level Rise problem and its negative impacts on the city of Alexandria. Strategies, laws, institutions, initiatives, projects, and measures have been put forward to tackle the problem. However, it still persists. The aim of the current article is to explore what climate governance looks like in the context of Sea Level Rise in a historic city, figure out what are the climate policy implementation challenges and how to overcome them. The study is exploratory in nature. The methodology used is three-pronged based on document analysis, a comparative analysis of other countries' experiences, and the implementation of an online survey with a non-random sample of educated Alexandria citizens. Findings point out a number of suggested prerequisites to overcome the policy implementation challenges, on top of which are: having genuine political

Climate Governance in International and Comparative Perspective, pages 345–377
Copyright © 2024 by Information Age Publishing
www.infoagepub.com
345

will, making available adequate financial resources, ensuring citizens' awareness and engagement, strict law enforcement, proactive long-term planning,
utilizing advanced evidence-based protection measures and learning from
other countries' experiences.

BACKGROUND

For several decades the Government of Egypt (GOE) has been discussing the
impact of climate change on the rising sea levels in the Mediterranean Sea
and the impact on the coastal shoreline of its second biggest city, Alexandria.
The aim of the current article is to try to assess the seriousness of the problem, review the successive policies and actions taken by the government over
time, compare the national policies with other international policies pursued
by governments trying to protect their shore cities, and figure out what works,
and what does not work; all the time with the aim of assessing the GOE policy
implementation challenges and trying to find a way out.

Alexandria is Egypt's second-largest city, following Cairo, the capital, with
a population estimated at nearly 5.5 million (CAPMAS, 2023) and a shoreline that extends for nearly seventy kilometers along the Mediterranean Sea.
Alexandria city is one of the 27 governorates—equivalent to municipalities—
in Egypt. It is a mostly urban governorate divided into nine districts (SIS,
2022). It is a major Summer resort for Egyptians and has Egypt's main shipping port responsible for nearly 80% of all exports and imports (MoPED,
2023). The city of Alexandria was founded by Alexander the Great in 332 BC
and includes a number of important historical landmarks, including the Bib-
liotheca Alexandrina, a modern library built to commemorate the ancient
library established by Alexander the Great in the old times (SIS, 2022). It
also has the *Montazah* Palace and gardens, built during the reign of Khedive
Abbas Helmy II more than a hundred years ago, and the *Qaytbay* Citadel,
built in 1477 by Sultan Ashraf Qaytbay (SIS, 2022). All three historical sites
mentioned are directly overlooking the Mediterranean Sea.

The main research problem is that despite the clear awareness of the negative impact of climate change on the eroding shores of the city of Alexandria and despite the policy statements, and implemented measures over the
years, the problem persists. Policies may be well formulated and articulated,
but when it comes to implementation on the ground, many challenges are
faced. Up till the present day, one can clearly witness the disappearing sandy
shores of Alexandria in many locations and the apparently futile impact of
the placed concrete boulders along the coastal line to protect the shores.
Alexandria has a lot of historic heritage sites that we cannot afford to lose.

Media sources, repeatedly come up with headlines screaming about the
threats posed to the city of Alexandria as a result of Climate Change. The

headlines try to raise the national consciousness about the importance of preserving and protecting the historic heritage of the city of Alexandria. Titles of news articles include announcements such as: "Experts reveal shocking scenarios for the sinking of Alexandria" (Egyptindependent.com, 2022, Nov. 15); "Egypt: Alexandria expected to sink by 2100" (Gbane, Africanews.com, 2022, Nov. 3); "Alexandria is in danger of drowning as a result of Climate Change" (France24.com, 2022, Nov. 2); "Boris Johnson warns that Alexandria will disappear" (Sadek, 2021, Nov. 17); "Climate Change: Will Alexandria, the City of the God be swallowed by the sea?" (Amara, BBC News, 2020, Oct. 2). The alarming thing is that these are all relatively new media articles, which means the problem is still observable and palpable.

Another point is that although there may be a plethora of studies delving into the technical measurements of the extent of the impact of climate change and sea level rise (SLR) on the shorelines of Alexandria, as will be evident in the literature review, much less has been written about the government of Egypt policies in that regard, and their level of effectiveness. The main research question is: What are the GOE adaptation policies to overcome the SLR problem in Alexandria, and why do these adaptation policies repeatedly fail?

METHODOLOGY

The methodology pursued involved a review of relevant policies and published literature, a comparative study of possible sea rise adaptation measures implemented in other countries, and an online survey directed to Alexandria citizens. The purpose of the online survey was to check citizens' level of awareness regarding the problem of SLR and its gravity, their knowledge about the government adaptation measures, the extent they perceive these measures as adequate, and what suggestions they have for overcoming many of the implementation challenges identified.

The survey was developed and shared with a non-random sample of Alexandria citizens. Respondents were reached out to through sharing the survey link by email and WhatsApp and by asking respondents to share the same link with other citizens in Alexandria, following the snowballing technique. The survey was purposively structured to be short and easy to answer to facilitate access. It was also made available in both English and Arabic. A total of 100 citizens anonymously responded to the survey during the month of April 2023, and then it was closed for analysis. They just had to ascertain that they were citizens of Alexandria and mention their current job/occupation. As shown in Figure 15. 1, the respondents pursued diverse occupations, whether as academics, managers, engineers, teachers, IT specialists, accountants, consultants, physicians, and others. What they

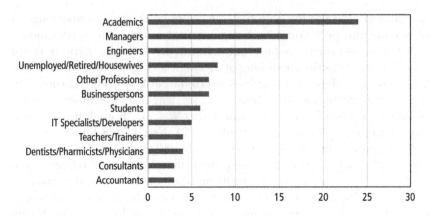

Figure 15.1 Occupation of survey respondents. *Source:* Based on results of online survey conducted by author.

all had in common, as evidenced by their professions, was they were highly educated.[1]

CONCEPTUAL FRAMEWORK

Figure 15.2 demonstrates the main factors that may contribute to overcoming the implementation challenges of the SLR Adaptation Policy. The factors were identified from the literature reviewed, and the field study was undertaken.

LITERATURE REVIEW

SLR is a global problem facing coastal cities worldwide. Accordingly, the phenomenon has been discussed at length in the literature, oftentimes focusing on specific coastal cities. Low-lying coastal zones, deltas, and small islands have been identified as most vulnerable to negative effects of SLR, including coastal erosion, more frequent storms, inundation of land, salinization of soils, ground and surface water, change of coastal ecosystem, and blocked drainage (Oppenheimer et al., 2019).

The Global mean sea level is reported as rising and accelerating mostly because of the melting glaciers and icebergs in the Arctic pole and in Antarctica. SLR is not uniform but varies from one region to another. The negative impacts of SLR are expected to substantially increase if there are no serious adaptation measures put in place (Oppenheimer et al., 2019). If nations do not commit to the Paris Agreement goal of reducing carbon

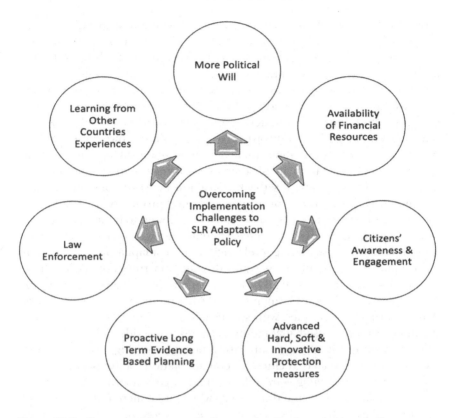

Figure 15.2 Conceptual framework: Overcoming implementation challenges of SLR adaptation policy. *Source:* Author's visualization based on the literature reviewed and results of fieldwork with Alexandria Citizens.

emissions, one of the estimations has it that by 2050 nearly 800 million people living in coastal cities could face a sea rise more than half a meter (C40 Cities, 2018).

Typologies for SLR Adaptation Measures

Coastal adaptation in response to SLR has been described as a 'wicked' and 'challenging' problem (Valente & Veloso-Gomes, 2020). The choice between adaptation alternatives is not purely a technical decision but is also political and subject to value judgments (Oppenheimer et al., 2019). Several typologies for SLR adaptation measures have been highlighted in the literature.

An interesting study by Harman et al. (2015) makes a comparison between the different SLR adaptation measures used by a number of countries around the world. The study recognizes that adaptation to SLR involves three main types of actions by governments with the aim of either protecting, accommodating, or retreating. *Protection* can be through hard defenses, such as concrete blocks and seawalls, or through soft defenses, such as beach nourishment and restoration of sand dunes, or a mixture of both. *Accommodation* involves extending the use of at-risk areas by introducing changes to building codes and design standards, such as elevated floors and improved drainage. It also involves requiring building owners to publicly disclose any information about their buildings being susceptible to flooding or being in or near an at-risk area. As for *Retreat* as an adaptation measure, it involves imposing a setback restriction between construction and the shoreline, prohibiting construction in flood-prone areas, and using a 'planned retreat' approach where citizens are allowed to build in land plots with the agreement that they will have to relocate, or move out, if the shore erosion reaches a threatening level. Each of these adaptation measures has its advantages and disadvantages and the decision rules applied relate to technical feasibility, political and social acceptability, and administrative capacity. Having said that, Harman et al. (2015) point out that hard coastal defense structures, coupled with soft beach nourishment techniques, are the most commonly used adaptation measures worldwide.

El Raey et al. (1999) came up with a possible typology for adaptation measures to SLR in coastal cities. Several adaptation methods were identified, covering: *Nourishing beaches* through adding sand and building artificial sand dunes, and/or building groins which are concrete structures laid perpendicular to the shores; *Constructing breakwaters* which are large walls built out into the sea to protect the coast from the force of waves, to reduce wave energy reaching the shore and therefore protect it from erosion; *Using legal measures* such as elevating taxes in areas where construction is not perceived as favorable, and/or imposing restrictions on developments along vulnerable coastal areas; *Changing land use* in the vulnerable areas, like for example prohibiting housing and construction and establishing fish farms; *Implementing an integrated coastal zone management system* (ICZM) through having a geographic information system (GIS), routine monitoring, decision support systems, and staff capable of coming up with decision to minimize degradation in resources; and finally, a fifth adaptation method is to *do nothing* and allow the land to get flooded.

Another typology for possible coastal protection measures was presented by Masria et al. (2015), whereby they listed four possible types of measures: hard, soft, a combination of both hard and soft, and a fourth more innovative type. Hard measures include building seawalls and breakwaters. Soft protection measures include beach fills and sand dunes construction. A combination of both hard and soft measures is also possible. Additionally, more innovative protection measures may include artificial mangrove root systems whereby mangrove trees that grow in salty water can be planted as a possible natural protection measure (Masria et al. (2015).

A yet more comprehensive typology of possible adaptation measures to SLR is that found in the Intergovernmental Panel on Climate Change report, whereby the following options are relayed:

- *Protection Adaptation Measures*: Hard protection measures covering: sea walls, embankments, surge barriers (moveable flood barriers which close when water levels are high), and dykes (walls built to prevent the sea from covering an area) are economically justifiable, especially in dense urban settings, but can lead to increased exposure in the long run;
- *Ecosystem-based Adaptation*: which is becoming more popular, is viable when enough space is available, but there is no agreement on their long-term effectiveness (EbA);
- *Advance Adaptation*: where new land is built into the sea. This is a useful method when land is scarce, but the option can lead to further increased exposure in the long term;
- *Accommodation Measures* which include floodproofing buildings, emergency planning, and early warning systems (EWS);
- *Retreat*, where developments are moved back away from the coastal line, or developments are avoided from the beginning in the space nearest to the coastline. The issue here is that there needs to be alternative land available to allow for relocation or avoidance (Oppenheimer et al., 2019).

The four different typologies include a great deal of overlap. They all mention hard and soft protection measures. They differ in how they bundle different measures together under one category or choose to emphasize a specific measure by recognizing it as a stand-alone category by itself. For example, 'using legal measures' was recognized as an independent category by El Raey et al. (1999) but was placed under the category of 'Accommodation' by Harman et al. (2015) and in the IPCC report (2019). Masria et al. (2015) recognized 'Innovative Measures' as a separate category, and similarly, the IPCC mentioned 'Ecosystem Measures' as a separate category. They are both giving the new 'Nature-Based Solutions' special attention by listing them as a separate category. The message is clear throughout this review of typologies for SLR Adaptation Measures, in that there is no one preferred adaptation method, and no one size fits all solution. Every locality needs to come up with a tailor-made mix that best suits its own needs, and that can be changed over time in response to anticipated SLR (Oppenheimer et al., 2019).

Good Practices for SLR Adaptation Policies

Coastal adaptation knowledge has evolved, and there are a lot of good practices being identified (Valente & Veloso-Gomes, 2020). The following section provides some insights from adaptation measures used in several regions from around the world known to be under severe levels of threat from SLR. We present here some brief findings from a select number of studies discussing and analyzing the problem of adapting to SLR in Australia, the Netherlands, the Maldives, and Singapore. The findings are relayed mostly in a jargon-free manner.

- A study about the Gold Coast city in *Australia* analyzes the amount of sand volume required to be added to the beach as an adaptation measure, how this should be synced with the speculated SLR over the years, and how much it would cost, plus putting in place new regulations to elevate base levels of new buildings and ensure they have proper flood protection guidelines (Cooper & Lemckert, 2012).
- In the *Netherlands*, there is acceptance of the fact that with climate change, there will be a continued trend of SLR and shoreline erosion. What they do is that there is a continuous flow of sand nourishment to limit erosion, and as a result, they have managed to protect the shores despite the constant SLR. The main factors leading to their success include something as simple as having sand available when needed, having a monitoring system based on science, having political support, and having an institutional setup capable of effective implementation (Keijsers et al., 2015). In the Netherlands, it is not only a problem with the yearly SLR but also with the annual increase in SLR, which requires flexibility to be integrated in the design of the adaptation measures. For example, the lifetime of solid protection infrastructure designed to last a century, as a result of the annual increase in SLR, ends up lasting a few decades and therefore needs different interventions to change the original plans (Alphen et al., 2022).
- In the *Maldives*, since the land is formed of small islands that are highly threatened by the SLR, the government has implemented lots of shore protection projects. It is reported that from 2013 to 2016 nearly 6 km. of shore protection projects were implemented. Additionally, the Ministry of the Environment and Energy is keen on reclaiming land from the sea, one example being the artificial island Hulhumalé raised about 1.8—2.0 meters above sea level and with an area of 400 hectares (equivalent to 952 feddans) (Gussmann & Hinkel, 2021).

- Because *Singapore* is a very small island state, therefore land is very scarce and costly and, at the same time, vulnerable to SLR (Cannaby et al., 2016). The majority of the land in Singapore is less than 15 meters above sea level (Bhullar, 2013). The expected cost of adapting to SLR on the island, through relying mostly on protection measures, is expected to rise over time as the country needs to meet the rising sea levels (NG & Mendelsoh, 2005). It is a continuous struggle. However, a diverse number of options are implemented, including hard stone embankments and walls, plus using natural adaptation mechanisms represented in the planting of coastal areas with Mangrove trees to protect the shoreline (Bhullar, 2013).

Challenges to Implementation of SLR Adaptation Policies

Every country chooses a mix of adaptation alternatives that suit its own needs. However, as much as there are good practices to emulate, there are also numerous challenges reported relating to the implementation of SLR adaptation policies in different parts of the world.

Abbott (2013) reports on how the adaptation policy in Maui, *Hawii*, focusing on determining setback measures in response to changes in SLR, faced serious hurdles during implementation. It was difficult to decide on setback areas when the land plot, for example, had cliffs, faced the ocean from more than one side, or was not regularly shaped (Abbott, 2013). Meanwhile, Wedin (2021), focusing on *Southern Sweden*, identifies several technical and ethical challenges related to the implementation of SLR adaptation policies, including lack of sufficient knowledge, incompetent human resources, problems with coordination, conflicting goals and inappropriate distribution of legal responsibilities. These results are asserted by Segge and Mauerhofer (2023), who point out to how Swedish municipalities may face challenges in SLR adaptation policies when they lack sufficient knowledge about solutions and when the legal frameworks do not adequately discern the division of responsibilities between the central and municipal levels.

Another study by Burger et al. (2023) discussed the challenge of determining the cost of 'relational values' when asking people to relocate in response to SLR and how relational values comprise the different ways people relate to their environment, build social capital and maintain customs and traditions; all may be lost when they are asked to move back. Additionally, Escudero and Mendoza (2021), focusing on coastal areas in *Mexico*, identified a number of critical challenges to the effective implementation of SLR adaptation policies, including, most importantly, insufficient communication with citizens and the lack of appropriate mechanisms that allow for their participation in the policymaking process, insufficient prioritization of climate change policies compared with pressing economic problems, and lack of proper information and monitoring systems.

SEA LEVEL RISE IN EGYPT

Meanwhile, the impact of climate change and rising sea levels on coastal zones in Egypt has been studied by various scholars. Some focused on using sophisticated modeling techniques in assessing the impact of the expected SLR, others focused on clarifying the potential negative impacts of SLR, while some focused on identifying the types of measures that can be used to adapt to SLR in Egypt.

Potential Negative Impact of SLR on Egypt

A detailed description and analysis of how SLR may impact the coastal zones in Egypt at large, including Alexandria governorate, was presented in an article by El-Nahry & Doluschitz, 2010). Here, it is not a question of flooding, but a detailed primary and secondary negative impact on coastal areas as a result of a minimum increase in sea level was discussed. Among the primary impacts predicted were increased erosion of shores, inundation of wetlands, increased flooding risks, coastal retreats, intrusion of salty water into underground water reservoirs, and possibly an increase in the salinity of the River Nile water. Meanwhile, the possible secondary impacts listed covered a threat to food production, whether through agriculture or fisheries, a decline in health standards as a result of a possible deterioration in the quality of drinking water, possible relocation of population groups, and possible threats to many economic and social activities in the affected regions, such as a decrease in the value of housing and land, rising cost of protection measures and them becoming unaffordable to a country like Egypt unless it receives international funding, plus threats to cultural values of displaced communities (El-Nahry & Doluschitz, 2010). The problem is complicated and has far-reaching effects on different aspects of economic, health, and social activities.

Potential Negative Impact of SLR on Alexandria

There seems to be a consensus among scholars that SLR in Alexandria is going to lead to negative effects. According to the Intergovernmental Panel on Climate Change (IPCC, 2022), the Nile Delta, where the city of Alexandria is located on its Northern West side, faces the threat of flooding for its low-lying coasts from SLR and from the expected increase in the intensity and frequency of storms; **a one meter SLR is expected to inundate 20% of the delta land by the end of the twenty-first century,** and that this will be coupled with increased salinization of land and water resources and will have negative impacts on fishing, agricultural and on the availability of fresh water. Mohamed (2023) mentions how coastal settlements in Egypt at large may be compromised as a result of melting polar icebergs and resulting SLR.

Assessing the Extent of the Damage

Modeling and satellite imaging techniques were used to assess the extent of the expected damage as a result of SLR in Alexandria. El-Masry et al. (2022) utilized a digital elevation model to assess the impact of climate change and sea level rises (SLR) on touristic activities in a specific section of the Northern shores of Egypt, extending from Alamein City to El Hammam City, both lying to the west of Alexandria. They concluded that the tourist activities in this area would be severely threatened by the SLR, particularly because of their location very close to the sea waters. Meanwhile, Kaloop et al. (2016), concurring that sea level change modeling is useful for studying the change in the shoreline, used two different types of models to assess the change in the shorelines of Alexandria city during the time period from 2008–2011and concluded that no 'extreme events' were detected during that period, but yet conceded that Alexandria is vulnerable to SLR and that this may have a negative impact on agriculture and human settlements. Mohamed (2020), through relying on satellite imaging, calculated a Climate Vulnerability Index (CVI), and concluded as well that despite the natural and artificial shore protection measures in place in Alexandria, there is still a risk of flooding through the lowest waterfront points in case of extreme storms or tsunamis. Abdrabo et al. (2023) developed and applied a Flood Vulnerability Index to Alexandria city and concluded that there is variability in flood vulnerability between different areas depending on the level of urbanization, and therefore recommended the tool to urban planners.

SLR Adaptation Policies in Alex

Effective adaptation to SLR in Alexandria is key. A study by Shaltout et al. from the Polish Institute of Oceanology (2015) concluded that Egypt's coastal lines along the Mediterranean can only be protected from the negative impacts of SLR and from flooding by 2100 if effective adaptation measures are put in place. Similarly, Sharaan et al. (2022,1) reiterate that: "without robust and effective adaptation, numerous coastal zones will be drastically affected." However, there were variations among scholars in their assessment of what would work and what would not amongst the different SLR adaptation options in Alexandria. In a study by Masria et al. (2015), they concluded that the use of the so-called hard measures and placing obstacles to block seawater from invading the coastline is not a sustainable solution for Egypt. The construction and maintenance costs for these constructions are relatively high, and they have negative side effects on the environment. Thus, Masria et al. (2015) recommended a combination of innovative and soft measures, where simply beaches are periodically nourished with sand and more mangrove trees planted along the shorelines.

Meanwhile, Kloos & Baumert (2015) looked into whether 'resettlement', as a future possibly unavoidable adaptation measure to future SLR in Alexandria, can be a viable option. The authors did a field study of SLR-vulnerable communities in Alexandria and tried to figure out what would work for them to convince them to relocate. They concluded that it would be quite difficult to convince people to relocate, especially those in urban households, and that the factors they would be looking into would include: the financial compensation package, housing options, job opportunities, income security, and social relations. Evidently, all these factors were related to offering the citizens a choice and convincing them to relocate but did not discuss the option of forced eviction.

El Raey et al. (1999), after assessing and evaluating the different SLR adaptation options focusing on Alexandria, concluded that a combination of hard and soft measures (groins, breakwaters, and sand nourishment) would be the best thing in the short run. However, they recommended on the long run adopting and implementing an ICZM plan that involves possibly not only hard and soft measures but also changes in land use when needed and using regulatory tools, and that effective ICZM implementation can only work if we engage more with the different groups of stakeholders and raise people's level of awareness. A more recent study by Sharaan et al. (2022), after reviewing different SLR adaptation measures used by the GOE in the Nile Delta coastal zone, concluded that: "the protection approach seems the most appropriate and effective coastal adaptation approach in Egypt, based on the Egyptian great historical experience, skills and best practices" (Sharaan et al., 2022, 9). Sharaan et al. (2022) also pointed out that traditional hard and soft protection mechanisms are in place and in use and that more experimental nature-based approaches are still being piloted.

Alexandria is under threat of SLR. Assessment is ongoing to figure out the extent of the expected damage. There is consensus amongst scholars on the need for adaptation and on using a mix of hard and soft measures. Voluntary relocation is perceived as a difficult option to resort to. ICZM is hinted at by some, and so is the need for more innovative and creative adaptation measures.

GOE POLICIES REGARDING CLIMATE CHANGE IMPACT ON ALEXANDRIA CITY

"Egypt is one of the most vulnerable countries when it comes to climate change", says Hani Sewilam, Egyptian Ministry of Water Resources and Irrigation (IMF YouTube Channel, 2022, June 16). "Alexandria is ranked number five worldwide amongst the cities threatened with drowning because of climate change", says the Governor of Alexandria (Abdelrehim, 2022, Nov. 14).

GOE Awareness of the Problem

It seems that the Government of Egypt is fully aware of the seriousness of the problem, not only of climate change impacts at large, but also of its specific potential impact on the Mediterranean coastal shores and specifically on the city of Alexandria. In its 2021 Egypt Human Development Report, it was clearly stated that several coastal cities are vulnerable to SLR and that there is a need to act proactively. Quoting the report:

> It is expected that the coastal areas of the Nile Delta will be highly flooded as a result of sea-level rises... [and] a sea-level rise above 0.5 meters will lead to permanent inundation of 1800 square kilometers of the agricultural lands in the low-lying areas of the Nile Delta. (EHDR, 2021, 229)

However, no exact dates were given for these speculations. Additionally, the report, citing a World Bank Study in 2012, mentions that several studies have predicted that in the event of a 0.5meter rise in sea level, 30% of Alexandria city will be inundated, 1.5 million individuals will have to relocate, and nearly 200 thousand jobs will be lost (EHDR, 2021). The World Bank 2012 cited is nearly a decade old.

SLR Governing Legislation

It is worth noting that the main governing law for environmental activities in Egypt is Law 4 for 1994 for the Protection of the Environment, amended by Law 9 for 2009. The Egyptian Environmental Affairs Agency was established in 1981, and the first Minister of the Environment was appointed in 1997 (The World Bank, 2017, Dec. 18). However, to date, there is no specialized law dealing with Climate Change.

Climate change policy formulation in Egypt speeded up in preparation for the COP27, which was held in November 2022 in Sharm El Sheikh in Egypt. A draft Climate Change Law was submitted to parliament whilst

COP27 was being held, but the law has not been passed. More importantly, Egypt National Climate Change Strategy (NCCS) 2050 was issued earlier during the same year of 2022. Although Egypt had ratified earlier in 2011 a National Strategy for Climate Change Adaptation and Disaster Risk Reduction, and later in 2018 a Low Emission Development Strategy (LEDS), the NCCS aimed at consolidating all climate change aspects in one updated plan (NCCS, 2022). The plan is well developed, with clear goals, objectives, key performance indicators, and a proposed monitoring and evaluation system. However, the monitoring and evaluation system is yet to be tested, and the needed financing for implementation is not yet secured.

The NCCS mentions SLR on several points. In the NCCS (2022, 11), a SWOT analysis is included that lists the strengths, opportunities, and threats of the climate change management landscape in Egypt, and among the threats mentioned are: "Severe weather phenomena, such as torrential rains, storms, high temperatures, and rising sea levels." This is again a clear indication of the GOE's awareness of the problem with SLR. As a result of this clear recognition of the threat of SLR, and other threats, Goal Number 2 is formulated, which calls for: "Enhancing resilience and adaptive capacity to climate change and alleviating the associated negative impacts" (NCCS, 2022, 11). Goal 2 cascades into more detailed objectives, so we find that Objective 2.d. calls for "Resilient infrastructure and services in the face of climate change impacts" (NCCS, 2022, 22), and a list of directives is listed as possibly contributing to the achievement of that objective. Amongst those directives are: "protecting coastal lowland and implementing integrated coastal zone management, implementing flood protection systems in areas prone to the phenomenon, strengthening sewage and rain drainage systems in cities and villages, ... [and] improving roads to be more resilient to the impacts of climate change such as high temperatures, floods, and sea level rise" (NCCS, 2022, 22).

Institutional Setup Responsible for Managing SLR in Egypt

The government organization directly responsible for dealing with the SLR problem is the Shore Protection Authority (SPA), affiliated to the Egyptian Ministry of Water Resources and Irrigation (MWRI). The Authority was established by virtue of presidential decree 261 for 1981. The main reason for the establishment of the Authority was the witnessed decline in shorelines along the Nile delta region due to both: the SLR due to climate change, and the decline in the rate of mud precipitation, after controlling the Nile flooding through the built dams (SPA, website). According to Prime Ministerial Decree 1599 for 2006, the 200 meters of land space inward from the shoreline is considered a special region to be regulated jointly by SPA and the Egyptian Environmental Affairs Agency (EEAA), and it is prohibited for any other entity to establish constructions without referring to both SPA and EEAA (Prime Ministerial Decree, 1599 for 2006, articles 1–3). Any touristic or maritime establishment wishing to start operations within the restricted 200 meters specified in the law needs to apply for needed licenses, accompanied by supplementary maps and studies, to the Supreme Committee established for that purpose at MWRI. The Committee is headed by the chairperson of SPA and has representation from five ministries—Tourism, Environment, Defense, Housing, and Culture–and the governorate in which the project lies. The Committee recommends either the approval or rejection of the needed license applications and the final decision is for MWRI (PM Decree 1599 for 2006).

Another important governmental entity supporting the work of SPA, EEAA, and MWRI is the Coastal Research Institute (CORI) which is a research center affiliated with MWRI and mandated with monitoring the evolution of coastal zones, predicting changes in the shoreline through the use of mathematical modeling, doing ICZM research and proposing the most efficient and effective protection measures, especially for the heavily populated areas (CORI, website, n.d.). CORI has its own international peer-reviewed journal called Water Science, published since the 1980s.

Examples of SLR Adaptation Projects Implemented in Alexandria

A number of projects were implemented by SPA in Alexandria for the purpose of protecting the shoreline, including:

- Protection and reinforcement of the historic sea wall for fish farms in Montazah: This is a sea wall project that started in December 2020, with a value of EGP 75 million and is marked as still ongoing on SPA's website (see Figure 15.3).
- Protection and development of the area in front of Qayetbay Citadel: This is a wave barricade about 520 meters long. The project started in January 2018 and ended in April 2021 with a total cost of EGP 235 million (see Figure 15.4).
- Reinforcement, development, and protection of Alexandria Corniche near *Mansheya* and *Mehattat El Raml*: This project involved building a sea wall 835 meters along the Corniche fence. It started in 2017 and ended in June 2021 with a total cost of EGP 104 million (see Figure 15.5).

- Protecting the shorelines of Alexandria, Phase 1 from Beer Massoud till El Mahroussa: This project was briefly described as immersed walls. It started in 2014 and ended in 2021, costing EGP 189 million (see Figure 15.6).

Figure 15.3   Sea wall for fish farms in Montazah. *Source:* Egyptian Shore Protection Authority (SPA) Website. https://www.mwri.gov.eg/spa/alex/

Figure 15.4   Wave barricade in front of Qayetbay Citadel. *Source:* Egyptian Shore Protection Authority (SPA) Website. https://www.mwri.gov.eg/spa/alex/

Figure 15.5   Sea wall near Mansheya in Alexandria. *Source:* Egyptian Shore Protection Authority (SPA) Website. https://www.mwri.gov.eg/spa/alex/

Figure 15.6   Sea wall near Beer Massoud in Alexandria. *Source:* Egyptian Shore Protection Authority (SPA) Website. https://www.mwri.gov.eg/spa/alex/

From what we can see in the listed projects implemented through SPA, the main focus is on hard adaptation measures through building sea walls and breakwaters, whether above and/or below water. These types of measures have been criticized in the literature as being costly and requiring a lot of maintenance, and without support from international development partners, Egypt may not have the needed resources to continue its battle against the continuing projected SLR. Another aspect pointed to in a study by Malm (2012) is that the chosen sites for building the sea walls are geared more toward protecting the rich touristic investments and gated communities, more than, for example, industrial areas in lowlands, or small agricultural plots by farmers; a matter that points out to existing "bias towards protecting capital rather than people" (6). This should not be the case.

ICZM in Egypt and National Initiatives

As a result of the excessive focus on using hard solutions as the prime adaptation measures for shore protection in Egypt, the Egyptian Minister of the Environment, Yasmine Fouad (2023, March 12), during a talk at the American University in Cairo, pointed out clearly that there is a need for ICZM that considers all social and economic aspects to avoid the continuation of such practices. In explaining the GOE response to the problem of SLR in Alexandria, she pointed out that a number of initiatives have been implemented over the past five years (2018–2023) and that these initiatives have improved the situation, including:

- Finalization for the first time ever of the *'Egypt Vulnerability Assessment Map'* that indicates how each part of Egypt may be affected by climate change over the coming 100 years. The plan was based on a model by the Intergovernmental Panel on Climate Change (IPCC);
- *A mathematical model to assess the SLR in nine governorates* in Egypt that are expected to be highly affected by the SLR, including but not limited to Alexandria;
- Development of an *Integrated Coastal Zone Management* (ICZM) *Plan* that has been finalized at the national level that realizes the overlap between economic and social dimensions in the vulnerability assessment for coastal zones;
- Establishment and construction of nature-based solutions in nine governorates under the umbrella of the so-called *'Nile Delta Adaptation Project'* funded by the Green Climate Fund and implemented by the Ministry of Water Resources in Egypt (Fouad, 2023, March 12).

Thus, according to the Minister of the Environment, Egypt has not only carried out evidence-based assessments of the extent of the problem of SLR in Alexandria, but also assessment of the extent of vulnerability of different geographic areas in Egypt to climate change, development of a coastal zone management plan, plus implementing a project for using nature-based ad-aptation solutions in a number of governorates including Alexandria.

More in-depth exploration of the *Nile Delta Adaptation Project,* as dis-cussed on the website of the Green Climate Fund, notes that its main aim is to develop an ICZM plan for the North Coast of Egypt and to introduce a paradigm shift in the traditional practices pursued by the GOE in coastal protection, by replacing the hard measures of protection with soft nature-based ones, or ecosystem-based approaches. The project was approved in 2017 and is expected to continue for seven years until 2025, with a total co-financed budget equivalent to USD 105.2 million (GCF, 2022, website). The latest available annual progress report for the project, dated 2021, men-tions that the project managed to achieve the following since it started in 2018: implementation of coastal protection measures in five hot spots in the Delta region, completion of the inception phase of the ICZM plan, plus the implementation of an ongoing capacity building program for staff in relevant government entities about the ICZM plan and the nature-based adaptation activities. Some delays were encountered with the onset of the COVID-19 pandemic resulted in delaying some of the field activities of the project, but real progress was reported to have been realized in *Kafr El Sheikh* governorate after completing 29 kilometers of protection work which has led to avoiding for the first time in decades inundation during Winter storms. Soft protection measures—including sand and reed fences–were used instead of hard protection measures (GCF, 2022). The same progress has not yet been reported for Alexandria.

ICZM is not a new thing in Egypt. Law 4 for 1994 for Environmental Protection, in detailing the mandates and responsibilities of the Egyptian Environmental Affairs Agency (EEAA), clearly mentions the responsibility of EEAA in: "participating in the development of the National Integrated Coastal Zone Management Plan for Coastal Zones in the Mediterranean Sea ad Red Sea in coordination with the relevant agencies and ministries" (Law 4/1994, article 5). A National Committee for Integrated Coastal Zone Man-agement was established and managed to issue a 'Framework Program for the Development of a National ICZM plan for Egypt' in 1996, and through this Framework Program, a number of projects were implemented in *Marsa Matrouh* governorate in Lake *Manzala* wetlands and in the Red Sea (The World Bank, 2017, Dec. 18). However, there were no specific initiatives that focused on Alexandria. Although the coordinating role for the development of the ICZM plans, and although there is a specialized department for Coast-al Zone Management within the EEAA, the implementation authority related to shore protection lies with the Ministry of Water Resources and Irrigation.

Although an ICZM plan is being developed through the Green Climate Fund for the North Coast of Egypt at large, a World Bank-funded project, with a total budget of USD 7.15 million, that ended in 2017 had reported the development for the first time of an ICZM for the Alexandria governorate and its adoption by the governor in place (The World Bank, 2017, Dec. 18). The main objective of the ICZM plan for Alexandria was to cater to the needs of the citizens by reducing their vulnerability to SLR, loss of habitat, degradation of water quality and other potential negative impacts related to climate change and to management of the coastal zones. An evaluation of the ICZM Plan project by the World Bank rated it as overall 'Moderately Satisfactory'. Despite the achievement of most project objectives there were a number of delays attributed to political events–the project duration overlapped with the Arab Spring and the years of political turbulence -, so a landfill developed as part of the project remained closed for two years, and the ICZM committee was not established except after the project ended with five months (The World Bank, 2017, Dec. 18). A number of lessons learned were mentioned in the ICZM Plan project evaluation by the World Bank (2017, Dec. 18), including the need for more consultation and engagement with citizens to guarantee awareness and buy-in, need for commitment from the central and local administrative levels in government, and for effective and sustainable implementation we need a strong implementing government entity, plus high-quality technical studies and monitoring indicators. The hope is that the ICZM plan for Alexandria that was adopted in 2017 is being implemented as needs be.

Egyptian Government officials seem to uphold a more positive and optimistic perspective on how the problem of SLR is being handled in Alexandria. Following the statement by the Former Prime Minister of England, Boris Johnson in 2021 about how Alexandria is drowning, a member of the Senate in Egypt requested an explanation by the Ministers in charge. Minister Yasmine Fouad, Minister of the Environment, stated to the Senate that: "not all public speeches and reports on climate change are correct, and many just aim to ring alarm bells" (Essam El-din, 2022, Jan. 6). Additionally, the Minister of Irrigation and Water Resources at that time added: "We have been taking actions that have rendered the submerging of Alexandria an impossibility" (Essam El-din, 2022, Jan. 6).

FINDINGS AND DISCUSSION: THE FIELD WORK— ENGAGING WITH ALEXANDRIA CITIZENS

Now to what extent are citizens of Alexandria aware of the seriousness of the problem of SLR, know about its potential impact on various aspects of life, have confidence in the GOE's policies, believe that the implemented projects will actually lead to proper management of the problem? This is what the fieldwork tried to achieve in order to identify any gaps between the formulated policies and their implementation.

Citizens' Awareness of the SLR Problem

When the sample of Alexandria citizens was asked if they were aware of the problem of Sea Level Rise in Alexandria, 90% of respondents stated that they were aware of the problem of the SLR in Alexandria, while 10% stated that they were not aware of the problem.

When asked about the extent they think that the SLR is a serious problem, 78%of the respondents stated that the problem of SLR in Alexandria is a 'very serious problem'; 16% stated that it is 'very serious'; while 4% were 'neutral', and only 2% stated that it is not an important problem at all.

Citizens' Perception of SLR's Negative Impacts

On asking the sampled citizens about what signs for SLR in Alexandria they have witnessed, they listed several tangible examples for its visible impact, including erosion of beaches, damage to the Corniche Road, flooding, and higher waves. This is how they expressed their concerns:

- Erosion and Reduction in Size of Beaches
    - "Most of our sand shores are totally or partially disappearing specially in the west area from *Raml* station to *Gleem*"
    - "Some beaches that I have witnessed as a child have disappeared . . . like in the areas of *Sidi Beshr* and *Rouchdi*"
    - "The beaches are shrinking in size."

The citizens were giving specific examples of beaches that have shrunk in size and some were reminiscing about childhood memories they had of these beaches when they were in better shape.

- Damage to the Corniche Road and Fence
    - "Erosion of the Corniche [coastal road]."
    - "The corniche fence has been worn out as a result of being pummeled by the Winter waves each year."
    - "Every year Alexandria has to put more sand . . . as the tide eats up *El Geish* road."
    - "Erosion of the wave blocks in the sea has been distinctly obvious."
    - "Repeatedly, the Corniche Road and some of the sea-shore buildings partially collapse."

The *Corniche Road,* which is the main seaside road extending for nearly 15 kilometers, is reportedly suffering from erosion and occasional collapse, especially during the Winter months. Even the government efforts represented with the hard adaptation measures, such as wave blocks, are getting eroded.

- Flooding
  - "Inundation of some of the cafes and restaurants along the coastline during the last storms [hitting Alexandria]."
  - "The Corniche is submerged with water during storms and especially in the *Montazah* area."
  - "Street flooding during storms."
  - "Some of the historical sites get flooded with water."
  - "In Winter with the heavy rains, the seawater crosses the fence, fills the streets, and this was not the case before and is intensifying over time."

Respondents reported the frequent flooding and inundation of streets, cafes, and restaurants, and even historical sites along the seaside. Flooding is also reportedly increasing compared to earlier times.

- Higher Waves
  - "Rise in water levels on the cement barriers."
  - "Increase in the intensity and height of waves."
  - "During storms, the waves reach the Corniche road."
  - "Weather extremes."

The waves were reported to be getting higher and more intense, and the storms striking Alexandria were much stronger than before.

Citizens' Perception of Government's SLR Adaptation Measures

On asking respondents about what the GOE is doing to protect the Alexandrian shores, the main observed interventions were focused on protection measures, mostly hard protection measures, and some soft protection measures. However, there were also a good number of negative observations, some even sarcastic about the government's limited efforts or lack of efforts. Examples of the respondents' observations included the following:

- Hard Structures:
  - "Increasing the wave barriers along the beach."
  - "They keep on protecting it using the big cement blocks."
  - "They are adding more concrete blocks in down town areas for the shore to be more resilient to the sea level rise."

- "Placing marble rock barriers."
- "Placing barriers in the middle of the Sea and I don't think this is right."
- "The [government] builds wave barriers, and puts concrete blocks."
- "Five miles into the sea they are putting up wave barriers."
- "Protecting *Qayetbay* Citadel and fighting the erosion of beaches through surrounding the citadel with concrete blocks to protect it from SLR."
- "Hard engineering . . . sea walls and groins."
- "[Implementing a] rainwater drainage project."
- "Building a concrete promenade into the sea."

Nearly half of the respondents, 51% focused on the hard protection measures and how the government is adding concrete blocks and wave barriers in different forms and shapes. One respondent mentioned a project for draining rainwater, one mentioned the groins, and another mentioned the concrete promenade or jetty as a way to protect the shores.

- Soft Structures:
  - "Building sand barriers to reduce the force of the waves before reaching the shores."
  - "Dumping sand"
  - "Throwing in more sand and widening beaches."

Very few respondents, 3% mentioned beach nourishment with sand as one of the tangible efforts by government to protect the shores against the problem of SLR.

Citizens' Perceptions of Effectiveness of Implemented SLR Policies

On asking the respondents to what extent they thought the government actions to manage the phenomenon of SLR was effective, 57% thought that the actions were 'not so effective' or 'not effective at all', 38% thought that the actions were 'somewhat effective', only 4% perceived the actions as 'very effective', and 1% perceived the government actions as extremely effective. These responses are in contrast with the earlier stated perceptions related to the gravity of the problem.

In the open comment section, the negative comments were a lot. Respondents commented on how the government was doing nothing, or not enough, or that the respondents did not know what the government was doing, or they stated negative perceptions citing different reasons. Either they said the government was spending too much money in a futile manner, was contributing to the problem rather than solving it, treating the symptoms and not the root cause, that they were just building cafes and restaurants by the sea, that the solutions were all temporary, that they were just focusing on money making, that they do not have a clear plan, and were even moving in the wrong direction. Some of the respondents mentioned the hard protection efforts in combination with negative, and sometimes even sarcastic, comments.

Examples of the citizens' comments included:

- "The government closes beaches and replaces them with cafeterias."
- "The government is not really interested in environmental issues but is utilizing these issues to receive foreign funding."
- "Nothing serious, only reaction to the action."
- "Spends a lot of money and I think [the actions are] not well studied."
- "There is no protection for the Alexandria beaches."
- "Insufficient solutions . . . just placing concrete blocks, but not enough and not in all beaches."
- "They are building cafes on the seaside and blocking the sea view."
- "Nothing."
- "Not much."

It is evident from the previous comments that despite all that the GOE is trying to do, the citizens lack trust in its abilities.

Citizens' Suggestions for Overcoming Implementation Challenges to SLR Adaptation Policy

On asking respondents about what more can be done and if they had any other suggestions for better management of the SLR problem in Alexandria, several very useful insights were shared, including suggestions for greater attention and prioritization of the problem, improved and more advanced hard and soft protection measures, more proactive planning, learning from other countries experiences, better law enforcement and regulation, and more importantly creating more awareness about the problem and implemented policies, and greater citizens' engagement. Citizens' responses were in line with the literature reviewed and the challenges identified with the implementation of SLR policies in other parts of the world.

- More Political Will and Attention to the Problem of SLR:
  - "Directing support and financial resources to the issue."

- "Giving the issue of SLR the same attention given to other issues like usage of plastic bags and other symbolic events organized to raise awareness about climate change."
- "I think we need [the government] to exert more efforts, so that we feel safe."

Amongst the main challenges to implementation was the lack of prioritization of the SLR issue and its placement on the top of the GOE's agenda and, therefore, the relatively limited financial resources directed towards alleviating the problem. For policies to get implemented, there is a need for political will, and there is a need for financial resources. If Egypt is facing a difficult economic situation, then funds will be directed to other more pressing priorities, like social protection and subsidy programs. This finding is in line with Escudero and Mendoza (2021) and them pointing to the same policy prioritization challenge in Mexico because of other more pressing economic problems. The NCCS has an ambitious list of adaptation and mitigation measures but no earmarked financing. They are waiting for international assistance.

- Better and more Advanced Hard and Soft Protection Measures:
  - "Putting in more blocks but at more sea depths...scientifically."
  - "Widening the breadth of the Corniche and increasing its elevation."
  - "Increasing the sand space on beaches and planting trees to preserve it."

Business as usual is not enough in dealing with a wicked problem like SLR and, therefore, the citizens are calling for more scientifically based solutions and more intensified efforts. Persistent protection efforts are called for. A lot of studies by Egyptian scholars suggest using a mix of protection and adaptation efforts to deal with the problem of SLR in Alexandria (Masria et al.; El Raey et al., 1999; Sharaan et al.). The results of those studies and others should be heeded and made use of.

- More Proactive Planning:
  - "Shifting the response from reactive to proactive through using early warning systems and equipment enabling the measuring of waves heights."
  - "Formulating urban development and urban extension plans that take into account the future SLR."
  - "Appoint true experts immediately to make proper studies."
  - "I haven't seen any proactive thinking by GOE so far."

Effective SLR policy implementation needs to be based on long-term planning and on a proactive, evidence-based approach. The GOE has an ICZM for Alexandria that was adopted in 2017. However, as directed by the World Bank Evaluation Study (2017, Dec. 18), there is a need for more consultation with citizens, more awareness creation, and, more importantly, effective implementation. The citizens do not seem aware of the existence of the ICZM plan. Additionally, the NCCS gives directives for how to protect coastal areas and enhance their resilience to climate change. The plans exist, but the problem is with effective implementation. Moreover, the governance structure for dealing with SLR in Egypt includes a Coastal Research Institute (CORI) that has its own peer-reviewed journal with rigorous studies in coastal zone management. However, the public seems to know little about this institute and does not see the impact of its research on improving the resilience of the coastal shores to the SLR problem.

- Better Regulation and Law Enforcement:
  - "Prohibiting the establishment of more restaurants [along the Corniche] and decreasing what is currently available."
  - "Prohibiting construction along the Corniche."
  - "Removing all violating buildings, constructions, clubs along the Corniche."

Citizens are aware that among the reasons for the challenges facing the implementation of the SLR adaptation policies are the regulatory environment and the weak law enforcement. So long as the government itself is committing violations and allowing for exceptions, including giving licenses to cafeterias and clubs along the Corniche, the adaptation measures will be incomplete. A clear Ministerial decree (1599 for 2006) requires a 200-meter setback from the shoreline, but it is the government, in many cases, that is violating this decree by giving licenses to cafes right at the shoreline.

- Learning from Other Countries Experiences:
  - "Learning from other countries' experiences, such as the Netherlands, the Maldives, and Seychelles."
  - "Better research and studies."

SLR is a global problem facing many coastal cities around the world and there are a lot of successful experiences and initiatives that are being developed and implemented. Again, the citizens are aware of the need to benefit from other international experiences, and they cited a number of good examples to model, such as the Netherlands and the Maldives. The literature reviewed also points out a mix of adaptation measures that can be used. In Australia, Cooper & Lemckret (2012) pointed out hot beach nourishment must be in sync with the expected SLR, and in the Netherlands, Keijsers et al. (2015) explain how the presence of a monitoring system guarantees a continuous flow of sand nourishment, and Alphen et al (2022) calls for the importance of heeding flexibility in design of adaptation methods to be used. In the Maldives, they are reclaiming land from the sea and building artificial islands (Gussmann & Hinkel, 2021), while in Singapore, they were using natural protection methods, such as Mangrove trees (Bhullar, 2013). A lot can be learned from the international experiences. In Egypt, so far, the concentration is on using either hard or soft protection methods, but there is a lot more that can be achieved from considering other options such as eco-based solutions, advance, retreat, and accommodation measures.

- Creating More Citizens' Awareness and Allowing for More Citizens' Engagement:
  - "Raising the awareness of citizens and communicating with them about the protection measures."
  - "Allowing the participation of civil society and the city dwellers in decisions and policies that deal with their city."

Citizens are asking for more room to be given to them in the public policy-making process, and for more information to be shared with them about the decisions taken by government in dealing with the SLR problem. The Minister of the Environment in Egypt is talking about a number of promising initiatives related to developing an Egypt Vulnerability Assessment Map, a Mathematical Model to Assess SLR, and a pilot of eco-based solutions in several governorates, but the general public may not be aware of those initiatives.

The citizens consulted have a good sense of the complexity of the SLR problem and have managed to propose several crucial elements needed for dealing with it, including giving the problem the attention it deserves from the policymakers, relying on evidence and knowledge, learning from other countries experiences, enforcing the laws and coming up with better regulations.

CONCLUSION

To conclude, what is more important than developing strategies and plans, is getting them implemented on the ground, or in this case, the sea, and achieving the needed results. There are a lot of efforts being undertaken by the government to try to protect Alexandria from SLR. The majority of the plans, programs, and implemented projects have been supported by international donor funding and focus mostly on a mix of hard and soft measures, building concrete walls and jetties, plus engaging in beach nourishment. Not all shores and beaches are covered, but the efforts target the most vulnerable spots. There seems to be a planned paradigm shift in the making, moving from a focus on hard adaptation structures to soft nature-based ones. Projects utilizing eco-based protection measures have started being piloted in other adjacent governorates, such as Kafr El Sheikh and them using reed fences to protect plantations, and planting mangrove trees along the Red Sea, but these innovative measures have not yet been seen in Alexandria. An ICZM plan has been developed for Alexandria through support from the World Bank and has been adopted by the governorate since 2017, but no major change in the type of protection or management has been witnessed since its adoption.

International experiences reviewed point out that there are a lot of other possible protection measures that can be used in protecting coastal areas under threat from the SLR, including revising building codes and guidelines, floodproofing buildings, imposing higher taxes on areas where construction is not favored, retreating developments from the coastline and imposing strict setback restrictions, changing land use for example to fish farms and restricting construction, or alternatively advancing into the Sea and building new artificial islands. For example, nourishing beaches with sand works well in the Golden Coast City of Australia because it is a well-studied process that regularly adds sand to compensate for the expected SLR. With all our deserts filled with sand, this seems like an affordable and feasible option.

The governance system for dealing with SLR in Alexandria, and similarly all the coastal areas in Egypt, is in the hands of the Shore Protection Authority (SPA) affiliated to the MIWR and in coordination with the EEAA and Ministry of the Environment, and the governing board of SPA has representation from six other government entities. There seems to be little or no involvement by civil society and no room for citizens' engagement in deciding on adaptation measures to deal with the SLR.

From the quick survey conducted with a sample of educated citizens in Alexandria, although the results cannot be generalized but it can be discerned that the citizens are aware of the problem of SLR and that it is serious. They can observe the tangible effects on the receding beaches, the increasing erosion of the Corniche, the higher waves, and harsher storms. They can observe the government's use of hard protection measures to protect the shores, but many believe that the efforts are not enough and that more is needed. What is more important is that the citizens were able to identify ways to overcome many of the policy implementation challenges and propose solutions that are in line with those discussed in the literature review. This makes a strong argument in favor of engaging citizens in the public policymaking process.

Each country has to choose the mix of adaptation measures that best suit its needs. The point is there needs to be a strong understanding and awareness about the magnitude of the problem and its ramifications, transparency from government, more engagement with the citizens, and sharing of information about what is happening to adapt to the SLR, what is working, and what is not, no denial about the seriousness of the problem, funding made available and a strong governance system in place whereby the implementing governmental entities are sufficiently empowered, and manage to coordinate their work effectively and efficiently.

Some take-aways for practitioners concerned with the problem of SLR in Alexandria:

- Strategies and plans are important, but useless if not implemented.
- Scientific based approaches should replace any ad hoc adaptation measures.
- There is a lot to learn from other countries' experiences in adapting to SLR, covering traditional and non-traditional options.
- Creating awareness about the magnitude of the SLR problem is key.
- Government should be transparent about both its successful adaptation measures and also about the challenges faced.
- Citizens of Alexandria should be consulted and engaged in SLR adaptation policies.
- No unwarranted exceptions to SLR adaptation policies should be allowed, such as the two-hundred meters shoreline setback restriction.
- Engaged and active citizens may be the way to ensure sustainability of government's efforts, by monitoring performance, and calling government to action when needed.

NOTE

1. For security reasons, the author refrained from doing face to face interviews with common citizens on the street.

REFERENCES

Abbott, T. (2013). Shifting shorelines and political winds—The complexities of implementing the simple idea of shoreline setbacks for oceanfront developments in Maui, Hawaii. *Ocean and Coastal Management, 73*, 13–21. http://dx.doi.org/10.1016/j.ocecoaman.2012.12.010

Abdelrehim, A. (2022, Nov. 14). *Alexandria governor warns that SLR is unprecedented and we have to be afraid* [in Arabic]. Akhbarelyom. Retrieved April 19, 2023 from, https://m.akhbarelyom.com/news/newdetails/

3938006/1/%D9%85%D8%AD%D8%A7%D9%81%D8%B8-%D8%A7%D9%
84%D8%A5%D8%B3%D9%83%D9%86%D8%AF%D8%B1%D9%8A%D8
%A9-%D9%8A%D8%AD%D8%B0%D8%B1-%D8%A7%D8%B1%D8%AA%
D9%81%D8%A7%D8%B9-%D9%85%D9%86%D8%B3%D9%88%D8%A8-
%D9%85%D9%8A%D8%A7%D9%87-

Abdrabo, K. I., et al. (2023). An integrated indicator-based approach for constructing an urban flood vulnerability index as an urban decision-making tool using the PCA and AHP techniques: A case study of Alexandria, Egypt. *Urban
Climate, 48*, 101426.

Amara, S. (2020, October 2). Climate change: Will Alexandria, the city of God be
swallowed by the sea? *BBC News Arabic.* https://www-bbc-com.translate.goog/
arabic/science-and-tech-54386263?_x_tr_sl=ar&_x_tr_tl=en&_x_tr_hl=en&
_x_tr_pto=sc Accessed on 16 Feb. 2023.

Bhullar, L. (2013). Climate change adaptation and water policy: Lessons from Singapore. *Sustainable Development, 21*(3), 152–159.

Burger, M. N., Nilgen, M., Steimanis, I., & Vollan, B. (2023). Relational values and
citizens' assemblies in the context of adaptation to sea-level rise. *Current Opinion in Environmental Sustainability, 62*, 101295.

C40 Cities. (2018). *Sea level rise and coastal flooding.* Retrieved April 17, 2023, from
https://www.c40.org/what-we-do/scaling-up-climate-action/adaptation
-water/the-future-we-dont-want/sea-level-rise/

Cannaby, H., Palmer, M. D., Howard, T., Bricheno, L., Calvert, D., Krijnen, J., Wood,
R., Tinker, J., Bunney, C., Harle, J., Saulter, A., O'Neill, C., Bellingham, C., &
Lowe, J. (2016). Projected sea level rise and changes in extreme storm surge
and wave events during the 21st century in the region of Singapore. *Ocean
Science, 12*, 613–632. https://doi.org/10.5194/os-12-613-2016

Central Agency for Public Mobilization and Statistics (2023). *The population clock
now.* Retrieved April 10, 2023 from https://www.capmas.gov.eg/HomePage
.aspx

Cooper, J. A. G., & Lemckert, C. (2012). Extreme sea-level rise and adaptation options for coastal resort cities: A qualitative assessment from the Gold Coast,
Australia. *Ocean & Coastal Management, 64*, 1–14.

Coastal Research Institute. (n.d.). *Coastal Research Institute.* Retrieved April 17, 2023,
from https://www.nwrc.gov.eg/CoRI.php

Egypt Human Development Report. (2021). *Egypt human development report 2021.*
Ministry of Planning and Economic Development & UNDP. https://www.undp
.org/egypt/egypt-human-development-report-2021?utm_source=EN&utm_
medium=GSR&utm_content=US_UNDP_PaidSearch_Brand_English&utm_
campaign=CENTRAL&c_src=CENTRAL&c_src2=GSR&gclid=Cj0KCQiAxbef
BhDfARIsAL4XLRqyOeM7Rk1yIEYjexqmfvBIDsSfmwJA-AZCKschPWU1aLL
2II_avzQaAlhREALw_wcB

Egyptindependent.com. (2022, November 15). Experts reveal shocking scenarios for
the sinking of Alexandria. *Al-Masry Al-Youm.* Retrieved February 16, 2023, from
https://egyptindependent.com/experts-reveal-shocking-scenarios-for-the
-sinking-of-alexandria/

El-Masry, E. A., El-Sayed, M. K., Awad, M. A., El-Sammak, A. A., & El Sabarouti, M.
A. (2022). Vulnerability of tourism to climate change on the Mediterranean

coastal area of El-Hammam-El Alamein, Egypt. *Environment, Development and Sustainability, 24,* 1145–1165.

El-Nahry, A. H., & Doluschitz, R. (2010). Climate change and its impact on the coastal zone of the Nile Delta, Egypt. *Environmental Earth Sciences, 59,* 1497–1506. https://doi.org/10.1007/s12665-009-0135-0

El Raey, M., Dewidar, K., & El Hattab, M. (1999). Adaptation to the impacts of sea level rise in Egypt. *Climate Research, 12*(2/3), 117–128. https://www-jstor-org .libproxy.aucegypt.edu/stable/pdf/24866006

Escudero, M., & Mendoza, E. (2021). Community perception and adaptation to climate change in coastal areas of Mexico. *Water, 13,* 2483. https://doi.org/ 10.3390/113182483

Essam El-Din, G. (2022, January 6). Saving Alexandria. *Ahramonline.* Retrieved April 17, 2023, from https://english.ahram.org.eg/News/454664.aspx

Fouad, Y. (2023, March 12). Minister of the Environment opening speech at the Research and Creativity Convention, RCC 2023. The American University in Cairo.

France24.com (2022, Nov. 2). Alexandria threatened with drowning as a result of Climate Change. *France24.com.* Retrieved February 15, 2023, from https://www.france24.com/ar/%D8%A7%D9%84%D8%A3%D8%AE %D8%A8%D8%A7%D8%B1%D8%A7%D9%84%D9%85%D8%B3%D 8%AA%D9%85%D8%B1%D8%A9/20221102-%D8%A7%D9%84%D8 %A7%D8%B3%D9%83%D9%86%D8%AF%D8%B1%D9%8A%D8%A9 -%D9%85%D9%87%D8%AF%D8%AF%D8%A9-%D8%A8%D8 %A7%D9%84%D8%BA%D8%B1%D9%82-%D8%A8%D9 %81%D8%B9%D9%84-%D8%AA%D8%BA%D9%8A%D8%B1 -%D8%A7%D9%84%D9%85%D9%86%D8%A7%D8%AE

Gbane, N. C. (2022, November 3). Egypt: Alexandria expected to sink by 2100. *Africanews.com.* Retrieved February 16, 2023, from https://www.africanews .com/2022/11/03/egypt-alexandria-expected-to-sink-by-2100/

Green Climate Fund. (2022). Projects and programs: FP053 enhancing climate change adaptation in the north coast and Nile delta region in Egypt. Retrieved April 14, 2023, from https://www.greenclimate.fund/project/fp053

Gussmann, G., & Hinkel, J. (2021). A framework for assessing the potential effectiveness of adaptation policies: Coastal risks and sea-level rise in the Maldives. *Environmental Science and Policy, 115,* 35–42. https://doi.org/10.1016/j.envsci .2020.09.028

Harman et al. (2015). Global lessons for adapting coastal communities to protect against storm surge inundation. *Journal of Coastal Research, 31*(4), 790–801.

IMF Channel (2022, June 16). *Egypt Adapts to Climate Change.* [Video]. YouTube. Retrieved February 16, 2023, from https://www.youtube.com/watch?v=i5B 6IzdbHYY

Kaloop, M. R., Rabah, M., & Elnabwy, M. (2016). Sea level change analysis and models identification based on short tidal gauge measurements in Alexandria, Egypt", *Marine Geodesy, 39*(1), 1–20.

Keijsers, Joep G. S. et al. (2015). Adaptation strategies to maintain dunes as flexible coastal flood defense in the Netherlands. *Mitigation Adaptation Strategies Global Change, 20,* 913–928.

Malm, A. (2012). Sea wall politics: Uneven and combined protection of the Nile delta coastline in the face of sea level rise. *Critical Sociology, 39*(6), 83–832.

Masria, A., Iskander, M., & Negm, A. (2015). Coastal protection measures, case study Mediterranean zone, Egypt. *Journal of Coastal Conservation, 19,* 281–294.

Ministry of Planning and Economic Development. (2023). *Alexandria.* Retrieved April 10, 2023, from http://www.alexandria.gov.eg/Alexandria/default.aspx

Mohamed, A. F. A. (2023). A study of strategic plans of sustainable urban development for Alexandria, Egypt to mitigate the climate change phenomena. *Future Cities and Environment, 9*(1), 1–14.

Mohamed, S. A. (2020). Coastal vulnerability assessment using GIS-based multicriteria analysis of Alexandria-northwestern Nile Delta, Egypt", *Journal of African Earth Sciences, 163*(2020) 103751. https://doi.org/10.1016/j.jafrearsci.2020.103751

NCCS. (2022). *Egypt national climate change strategy 2050.* Arab Republic of Egypt, Ministry of Environment.

Ng W-S., & Mendelsohn, R. (2005). The impact of sea level rise on Singapore. *Environment and Development Economics, 10*(2), 201–215. https://doi.org/10.1017/S1355770X04001706

Oppenheimer, M., Glavovic, B. C., Hinkel, J., van de Wal, R., Magnan, A. K., Abd-Elgawad, A., Cai, R., Cifuentes-Jara, M., DeConto, R. M., Ghosh, T., Hay, J., Isla, F., Marzeion, B., Meyssignac, B., & Sebesvari, Z. (2019). Sea level rise and implications for low-lying islands, coasts and communities. In H.-O. Pörtner, D. C. Roberts, V. Masson-Delmotte, P. Zhai, M. Tignor, E. Poloczanska, K. Mintenbeck, A. Alegría, M. Nicolai, A. Okem, J. Petzold, B. Rama, & N.M. Weyer (Eds.), *IPCC special report on the ocean and cryosphere in a changing climate* (pp. 321–445). Cambridge University Press. https://doi.org/10.1017/9781009157964.006.

Prime Ministerial Decree 1599 for 2006 for Protecting Egyptian Sea Shores. *The Official Gazette,* Issue 205, 9 September, 2006. https://manshurat.org/node/13167

Sadek, H. (2021, November 17). Boris Johnson warns that Alexandria will disappear. *Daily News Egypt.* Retrieved February 16, 2023, from https://dailynews-egypt.com/2021/11/17/opinion-boris-johnson-warns-that-alexandria-will-disappear/

Segge, S., & Mauerhofer, V. (2023). Progress in local climate change adaptation against sea level rise: A comparison of management planning between 2013 and 2022 of Swedish municipalities. *Urban Climate, 49,* 101555. http://doi.org/10.1016/j.uclim.2023.101555

Shaltout, M., Tonbol, K., & Omstedt, A. (2015). Sea-level change and projected future flooding along the Egyptian Mediterranean coast. *Oceanologia, 57,* 293–307. http://dx.doi.org/10.1016/j.oceano.2015.06.004

Sharaan, M., Iskander, M., & Udo, K. (2022). Coastal adaptation to sea level rise: An overview of Egypt's efforts. *Ocean and Coastal Management, 218*(2022) 106024.

State Information Service. (2022). *Alexandria governorate.* Retrieved April 10, 2023, from https://www.sis.gov.eg/Story/167045/Alexandria-Governorate?lang=en-us

Shore Protection Authority (n.d.). *Establishment of the Authority.* The Egyptian Shore Protection Authority. Retrieved April 11, 2023, from https://www.mwri.gov .eg/spa/about1/

Law 4 for 1994 for Environmental Protection. *The Official Gazette,* Issue 5, 3 February. https://manshurat.org/node/13071

The World Bank (2017, December 18). *Alexandria coastal zone management project—Report No: ICR00004075.* World Bank Document. https://documents1.worldbank .org/curated/en/717171513949741048/pdf/ICR00004075-12192017.pdf

Valente, S., & Veloso-Gomes, F. (2020). Coastal climate adaptation in port-cities: Adaptation deficits, barriers, and challenges ahead. *Journal of Environmental Planning and Management, 63*(3), 389–414.

van Alphen, J., Haasnoot, M., & Diermanse, F. (2022). Uncertain accelerated sea-level rise, potential consequences, and adaptive strategies in the Netherlands. *Water, 14*(10), 1527. https://doi.org/10.3390/w14101527

Wedin, A. (2021). Getting adaptation right—Challenges and ethical issues facing planners adapting to sea level rise in southern Sweden. *Local Environment, 26*(4), 504–516. http://doi.org/10.1080/13549839.2021.1901267

ENVIRONMENTAL EFFECTS ON GROUNDWATER AND URBAN DRINKING WATER QUALITY BY LANDFILLS LEACHATE INFILTRATION

A Governance Approach

Miriam A. García-Colindres
IITCA Universidad Autónoma del Estado de México

Ivonne Linares-Hernández
IITCA Universidad Autónoma del Estado de México

Luis Antonio Castillo-Suárez
IITCA Universidad Autónoma del Estado de México

Abraham David Benavides
University of Texas at Dallas

Carolina Alvarez Bastida
IITCA Universidad Autónoma del Estado de México

Monserrat Castañeda Juárez
IITCA Universidad Autónoma del Estado de México

Verónica Martínez-Miranda
IITCA Universidad Autónoma del Estado de México

Vanessa González-Hinojosa
IITCA Universidad Autónoma del Estado de México

Climate Governance in International and Comparative Perspective, pages 379–397
Copyright © 2024 by Information Age Publishing
www.infoagepub.com
379

ABSTRACT

This chapter first looks at the governance structure of Mexico and its inefficiencies in properly constructing and maintaining municipal landfills. Next it addresses some of the laws in place that decentralized the administration of landfills to the state and local governments. At this point, the chapter looks at groundwater contamination and the dangers of heavy metals in potable water. The implications for public health in Mexico are significant. Finally, the chapter looks at future trends and possible solutions to addressing the problem of leachate filtering into the countries source of drinking water.

Groundwater is a source of potable drinkable water for half the world's population; however, groundwater presents serious pollution and scarcity challenges worldwide. The main environmental impacts of groundwater contamination are related to significant effects on the health of human beings and the ecosystem (Chen et al., 2019; Li et al., 2021; Somani et al., 2019). In addition, the lack of adequate public policies that address the extraction of groundwater, the creation and maintenance of municipal landfills, and the governance structure to maintain systems that produce adequate sources of drinking water, are lacking.

In this chapter we observe that water quality is related to its geopolitical region, and contamination can be anthropogenic, and or naturally generated due to geology. For example, some landfills are considered a potential risk to the environment and human beings because of their lack of proper construction—for instance, lacking appropriate assembly liners and leachate collection systems. We review some of these potential threats to ground water by showing high concentrations of heavy metals, such as fluorine, nutrients, and organic matter found in groundwater near landfills. We look critically at the generation of municipal solid waste as an unsustainable lifestyle and question the economic, social, and environmental spheres that conform the triad of sustainable development.

Landfills are methods for the disposal of solid waste and are one of the oldest and most frequently used waste management methods in the world. Ninety-five percent of solid waste is deposited in landfills (Al-Ghouti et al., 2021). Archeologist often study ancient civilizations by going through "their trash" to find out about how they lived, worked, and played. In this sense, these ancient landfills are treasure troves of information for how our ancestors lived. Landfills play a significant role in our society, and for the present time will continue to be used as the preferred method for waste disposal. As the world population continues to grow, the amount of urban solid waste that is generated will increase. Therefore, proper construction, management and operation of landfills will help to minimize any potential negative impacts. For instance, this may be accomplished by locating, constructing, and operating landfills that take into account hydrological and geographic conditions. Additionally, assuring that appropriate leach-ate migration systems are in place to protect ground water from contami-nants (Najafi Saleh et al., 2020; Przydatek, 2021) are essential. If managed properly, landfills can continue to benefit society.

Unfortunately, many municipal landfills and illegal dumping sites world-wide are operated without the adoption of leachate segregation and chan-neling strategies such as geomembranes (Tenodi et al., 2020) and represent a current risk for groundwater contamination. The environmental impact of landfills that are improperly managed have the potential to cause great harm. One of the disadvantages of landfills is the formation of an aqueous effluent called leachate, which is the result of rainwater percolation mainly through waste, as well as the physicochemical and microbiological changes of the solid waste that is deposited, which generates a complex mixture of contaminants (Parvin & Tareq, 2021).

Leachate migration can be vertical and horizontal by advection, disper-sion and dilution, contaminating the soil and groundwater (Abiriga et al., 2020; Gonçalves et al., 2019). Landfills present serious pollution problems because of the accumulation of toxic compounds, especially from old mu-nicipal landfills, due to low biodegradability. The nature and load of pollut-ants in leachates change considerably between landfills and are influenced by various factors for example: waste composition, months of the year, site hydrology, type of compaction, waste age, and landfill design (Boateng et al., 2019; Gamar et al., 2018; Ganguly et al., 2019; Hussein et al., 2021; Mishra et al., 2019; Somani et al., 2019). The leachate may contain soluble and insoluble organic and inorganic products from physicochemical, hy-drolytic and biological processes (Jabłońska-Trypuć et al., 2021; Kapelewska et al., 2019; Peng, 2017). Therefore, this chapter first looks at the gover-nance structure of Mexico and its inefficiencies in properly constructing and maintaining municipal landfills. Next it addresses come of the laws in place that decentralized the administration of landfills to the state and local governments. At this point, the chapter looks at groundwater contamina-tion and the dangers of heavy metals in potable water. The implications for public health in Mexico are significant. Finally, the chapter looks at future trends and possible solutions to addressing the problem of leachate filter-ing into the countries source of drinking water.

THE ROLE OF GOVERNANCE

In Mexico as in several Latin American countries, municipal landfills were formally created in the 1950s as sites for the deposit of domestic solid waste. These were common in both large and small cities. As a point of comparison, developed countries such as the United States and England had modern landfills as those in use today since the 1920s. This disparity in the technological application of the modern landfill method in part is responsible for much of the contamination that is found in and around landfills in Mexico today.

In the 1980s through constitutional reforms in Mexico, the decentralization of federal public services was carried out in order to reduce the role of national federal government. This change made the state and municipal governments and administrations responsible for reactivating the political-administrative role of the municipalities. Specifically, local governments now had the responsibility for public services which included the demarcation of sanitary services including the gathering, transfer, treatment and final disposal of domestic solid waste (Art. 115 of the Mexican Constitution). This adjustment or devolution of power to local governments resulted in, for instance, Mexico City closing in 1983 two of the largest municipal landfills in the country—Santa Cruz Meyehualco and Santa Fe, both located in the then Federal District. The following year, two new landfill sites were selected—Bordo Poniente in the Federal Zone of the ancient Lago de Texcoco and Prados de la Montaña in the Delegation Álvaro Obregón.

Decentralization of public services in Mexico also applied to environmental public policy. As part of a comprehensive process, participation of the private and public-private sectors was now allowed. Joint ventures and other innovative ideas were introduced and pursued. For instance, in order to help solve the problem of handling domestic solid waste in landfills, ravines of more than two kilometers of surface area and one hundred meters of depth, were used as the final disposal sites in major cities. Unfortunately, this type of practice was extended throughout the country without consideration of local needs and as these new practices increased with population growth, a sense of complacency developed. Large tracts of vacant land needed to be used, bodies of water and underground cavities were compromised, and protected natural areas and ecological conservation areas were at risk. Therefore, the Protected Natural Areas (ANP) Act became one of the main tools of ecological conservation policies used by the National Commission of Protected Natural Areas (CONANP) which is a decentralized body of the Ministry of the Environment and Natural Resources in Mexico. It promoted the law that governs the ANP which is the Law of Ecological Balance and Environmental Protection (LGEEPA) which regulates the rules on ANP. By instituting these new environmental laws, the LGEEPA, published in 1988 with updated policies in 1996, various watershed environmental legislation which identified municipal solid waste as the main soil contaminants (art. 134, LGEEPA). This led to the additional closure of six municipal solid waste facilities in 1991, in the Federal District, and having the waste moved to the three newer landfills—Bordo Poniente, Prados de la Montaña, and Santa Catarina.

In the 1990s the comprehensive project on the management of solid waste considered the Federal District and the cities Monterrey, Puebla, and Guadalajara for its influence on the regional environment, as well as five border cities: Tijuana, Torreón, Monclova, Matamoros, and Reinosa. With investment from the World Bank and with the participation of the Secretariat of Natural Resources and Fisheries (SEMARNAP), the National Institute of Ecology and Secretary of Social Development (SEDESOL), concluded that there was institutional weakness at the federal level and that the state and local level lacked the staff with technical skills and financial and appropriate administrative know how to run their facilities. To add to these problems, with insufficient collection services and control of confinement sites, clandestine dumps appeared and surface and underground contamination began to occur more readily. Lack of service cost recovery—which has prevented an investment in this sector, ineffective rules, as well as poor monitoring and environmental impact analysis has prevented growth in this area. Additionally, another obstacle has been represented by the guild or the union of "the pepenadores" (garbage collectors) by hindering modern collection and disposal services. This is an ongoing issue that currently has no resolution. Recycling is really not very popular, since it works through informal processes of very low productivity. There is no federal policy or planning concerning recycling (León, 2005).

Since 2000, the Mexican Federal government has been responsible for the management of hazardous waste and the states have been responsible for industrial waste that are not considered hazardous (art. 7, frac. VI, LGEEPA). Local municipal city councils are responsible for implementing the legal provisions relating to the prevention and control of waste that is not considered hazardous (art. 8, frac. IV, LGEEPA). This delegation of functions involved the participation of the three levels of government for the management of the waste. However, in many cases this does not happen and most of the responsibility continues to lie with local municipalities (Galilea et al., 2011).

According to the General Law for the Prevention and Integral Management of Waste (Federacion., D.O. de la, 2003) and its regulations (DOF, 2014) it define waste in three categories, municipal solid waste which is mostly urban, hazardous waste from industrial plants, and special handling waste:

1. *Municipal Solid Waste (MSW):* those generated in the dwelling houses, resulting from the disposal of materials that are used in domestic activities, that come from any activity in establishments or on public roads.
2. *Hazardous waste (HW):* those whose characteristics make them risky (explosives, infectious medical waste, radioactive matter, toxic

materials, flammable components, and corrosive substances). This category includes agricultural residues.

3. *Special handling waste (SHW):* generated in the production processes, that are not hazardous HW or belong to MSW, or that are produced by large generators.

Looking at the composition of laws and practices from an operational level, the lack of agreements between the different governmental actors hinders the application of the laws causing social controversy about the scope and responsibility that each of the actors has in the management of solid waste in general. At the same time, responsibilities that are assigned at the local level are not always accompanied by financial resources from the federal government nor is the available technical and operational know how necessary to perform the waste management functions offered and, in most cases, nonexistent. In the case of Mexico, it is truly in need of a system of governance, as has been defined in this book, to unite the various actors to come together and find solutions to a common problem.

In the 2019, the State's strategy was aimed at encouraging the participation of non-governmental actors through private initiatives to improve public policy outcomes and enhance governance. Governance, in this case, referred to specific forms of administration aimed at social development at different governmental levels (global, national, regional, and local). In Mexico, the management of solid waste, from the ecological perspective of Martínez-Alier (Figueroa Sánchez & Cruz-Morales, 2019) divides the management of solid waste into three approaches:

1. *Conservationist ecology* this approach "romanticizes" the discourse of recycling. It appeals to young people and calls on them to promote the idea of saving the world through recycling, reuse y reduce of solid waste. This environmentalist perspective, however, does not recognize the diverse social context of solid waste and its relationship with nature.

2. *Eco-efficient or mercantilist ecology* proposes clean development mechanisms and energy efficient technologies combined with socially and environmentally responsible technologies in its discourses. Garbage then, becomes a commodity, where only materials that can be reinserted into the market are recovered. The objective is to generate the greatest amount of economic credits regardless of the social or ecological costs (incineration and other innovative solid waste handling technologies). Most are private initiatives to the detriment of public or community initiatives.

3. *Popular ecology or political ecology* posits that human beings are part of nature and coexist in an interdependent and inter-influential

environment. Thus, political ecology recognizes that "it is societal models and power relationships that determine the type and magnitude of appropriation, transformation, and consumption and with it the type of excretion."

In the political ecology perspective, governance is understood from a critical point of view, since it recognizes that communities affected by socio-environmental conflicts are part of the problem. Therefore, they must take part in decisions and possible solutions through consultation exercises and practice participatory democracy. Political ecology questions that the solutions for the management of solid waste (recycling, incineration and other technologies) are aimed at legitimizing the actions of the system, since they do not attack the system as the main cause of the generation of solid waste.

During the last decade in Mexico, 87% of the sites where garbage is deposited were either controlled or uncontrolled open dumps, and only 13% were in approved landfills (Instituto Nacional de Ecologia y Cambio Climático [INECC], 2022). This Institute goes on to indicate that there are:

1. 1,731 uncontrolled open dumps where only garbage is deposited without any treatment, or protection;
2. 116 controlled open-air garbage dumps, which have a public administration or public-private partnership that monitors the type of garbage deposited and separates it manually for recycling through personnel; and
3. 40 municipal landfills, of which the 97.5% are managed by the private sector and 2.5% are administered through the traditional bureaucracy.

Additionally, to provide a general perspective, in Mexico the states which generate the most garbage or better said that generate more than half of garbage in the country are: Mexico City (14.2%), State of Mexico (11.2%), Jalisco (7.2%), Veracruz (5.3%), Nuevo Leon (4.8%), Guanajuato (4.2%), and Puebla (3.7%). Each of these states have major urban centers with high populations evidencing Mexico's inadequate handling of solid waste. This lack of concern for a vital government function causes negative environmental impacts, in addition to generating significant public health issues. Contrary to countries such as Sweden, South Korea, and Germany (all in the global north) that have some of the best solid waste management systems in the world. Proper management of solid waste—through appropriate governance—has the advantage of making the most of a complex problem of high concern. The responsibility depends not only on the local state and federal government but also an awareness on the part of the citizens.

Although the public and private sectors have been involved in attempting to provide better management of the solid waste industry, a number of their efforts have failed because of their reliance on allies of convenience to cover their immediate objectives. In addition, each of these allies or actors responds to their own interest—mainly the international plastic recycling market. In this sense, there is no governance mechanism for solid waste, rather, there are hierarchical relationships where citizens are subject to rules promoted by "green" associations that are driven by a system of consumption and waste and government and private sector companies that use recycling to create legitimacy. Additionally, in most municipalities they do not have the technical or financial capacity to perform the functions of solid waste management. As noted earlier, many of these functions are delegated to them as part of the decentralization process, therefore a number of local governments concede their responsibilities and make agreements with recycling companies promoting private initiatives.

Finally, Mexico lacks a viable environmental policy that appropriately addresses the creation, execution, and management of solid waste facilities. Some of the current policies, laws, and city ordinances fall short of promoting appropriate actions to minimize its negative effects of illegal dumping. Resulting problems include public health and environmental concerns. Given the population growth and demand for natural resources, it is necessary to make efficient use of resources. Therefore, the public and private sectors must implement utilization and development strategies that create municipal landfills that are code compliant and serve the people.

GROUNDWATER CONTAMINATION

To highlight the consequences for the lack of appropriate governance in the Mexican system, we now turn to the effects of a lack of cooperation between the various levels of government and the public and private sectors. The migration and infiltration of leachates from the landfill into the groundwater depends on the physicochemical properties of the municipal solid waste and the flow pattern of the groundwater (Abiriga et al., 2020; Boateng et al., 2019; Mishra et al., 2019). The degradation of pollutants during the rainy seasons allows the infiltration of leachate with the precipitation in groundwater. Most groundwater affectation is observed around 1,000 meter radius of a landfill, while the highest concentrations of contaminants in groundwater is observed within a 200 meter radius, because of sulfate (SO_4^{2-}), nitrate (NO_3^-), iron (Fe), manganese (Mn), and chromium (Cr) which are all detected near the landfill (Turki & Bouzid, 2017; Mishra et al., 2019; Morita et al., 2021). Additionally, groundwater contamination is inescapable, when the bottom of the landfill is below the water level, or if the material separating the landfill from the aquifer is permeable or is damaged (Zeng et al., 2021). The main pollutants found in groundwater were heavy metals (Cr, Fe, Mn, Cd, Cu, Pb, Ni, Zn and Hg), arsenic (As), fluorine (F^-), organic matter, and nutrients (phosphorus and nitrogen). Therefore, leachate is characterized by inorganic compounds such as heavy metals, ions such as chlorides (Cl^-), sulfates (SO_4^{2-}), ammonium (NH_4^+), large amounts of organic matter such as, volatile fatty acids (Babaei et al., 2021; Luo et al., 2020), humic and fulvic acids (Castillo-Suárez et al., 2019; Kjeldsen et al., 2002; Gautam & Kumar, 2021; Przydatek, 2021), xenobiotic compounds, such as pesticides, and microorganisms (Luo et al., 2020; Tenodi et al., 2020).

Leachate, therefore, is generated by infiltration of rainwater and/or surface runoff. If leachate mixes with groundwater, a plume of pollution is created in the direction of flow, which is one of the main threats to groundwater quality (Guo et al., 2022). It is necessary to identify the mechanism of groundwater pollution surrounding the landfill and propose pollution remediation strategies. Most commonly water quality parameters, are pH, total coliform, conductivity, alkalinity, chloride, biochemical oxygen demand, chemical oxygen demand, temperature and nitrogen. These standards are determined to evaluate drinking water quality near landfills, during its exploitation and after the closure (Talalaj & Biedka, 2016).

TOTAL ORGANIC CARBON AND HEAVY METALS

By now it should be understood that leachate leaking into the water supply can be very harmful to humans. Unfortunately, those that govern many nations in the global south are not apt to take steps to mitigate this contamination because of a lack of understanding of the leachate pollution in the ground water, on the one hand, and the political will to spend the resources necessary to address the issue on the other. Therefore, in an effort to further show the harmful effects of leachate permeating into the water supply, we looked at the total organic carbon (TOC) results from oxidation of organic matter (humic and fulvic acids, amines, urea) and from chemical sources (detergents, pesticides, fertilizers, industrial chemicals, chlorinated organics). We found high levels of organic carbon that can increase the solubility of different pollutants in the leachate (Przydatek & Kanownik 2019).

Different case study's were analyzed in a worldwide context, for example, in Norway, TOC is considered a moderate pollutant because of its low levels, which remain stable regardless of the season. Their landfill operations started in 1958 and by 1974 continued functioning with neither a leachate collection system nor liners (Abiriga et al., 2021). Norway's maximum concentrations can range from 77 mg/L in winter to 82.7 mg/L in autumn

TABLE 16.1 Behavior of Total Organic Carbon in Groundwater Near Landfills in Different Countries

Country/ Sample	Landfill Features	TOC (mg/L)			Reference
		Min.	Med.	Max.	
Norway/ Well water	Without leachate collection system nor liners	1.3	10	47	(Abiriga et al., 2020)
Serbia/ Groundwater	No leachate control	1.22	5.04	13.69	(Kr et al., 2018)
South Africa/ Groundwater	Without leachate collection system	8.3	ND[a]	82.7	(Ololade et al., 2019)
Poland/ Groundwater	Peripheral drainage system	0.1	ND	20.2	(Kanownik et al., 2019)
Poland/ Groundwater	The operation started in 1981	10.92	112.2	616.60	(Talalaj, 2014)
Serbia / Groundwater	Drainage layer and geomembrane	2.24	ND	34.2	(Tenodi et al., 2020)
Iran/ Groundwater	Absence of an appropriate engineering	0.2	ND	26.9	(Vahabian et al., 2019)

[a] ND = not determinate

in groundwater, although in surface water, the change is more noticeable, ranging from 27 mg/L in autumn to 9.2 mg/L in winter (Table 16.1; Ololade et al., 2019).

In Poland, TOC was registered at 616.6 mg/L (Table 16.1), and a significant variation in TOC concentration was observed at their site (±122.7 mg/L; Talalaj, 2014). In South Africa, Ololade et al. (2019) also observed a significant variation with a minimum of 8.3 and maximum of 82.7 mg/L total organic carbon. In Iran Vahabian (2019) determined a minimum of 0.2 and a maximum of 26.9 mg/L (Przydatek & Kanownik, 2019), all these sites did not have adequate technology for the collection and treatment of leachate.

The permissible limit in South African according to their National Standard-241 is ≤10 mg/L (SANS 241:2015, 2015), in all cases, the maximum value was exceeded, possibly because their landfills do not have a leachate collection system, resulting in the penetration of runoff into the groundwater environment (Talalaj, 2014).

In the case of heavy metal pollutants, they affect human health and are very persistent, toxic, and they can remain for around 150 years, leached at a rate of 400 mm/year (Gworek et al., 2016; Haider et al., 2021; Hussein et al., 2021). Heavy metals in groundwater and drinking water from landfills may be due to the disposal of photographic products, pesticide manufacturing industries, plastics, and electronic wastes (Abiriga et al., 2020;

Foufou et al., 2017). They can also be from corrosion of soldered joints in galvanized pipes, vehicle brake waste, and waste disposal from the textile industry (Talalaj, 2014). The toxicity causes damage to multiple organs in humans—mainly in the kidneys—pulmonary emphysema, renal tubular damage and kidney stones, has high solubility and can destroy the protein structure and cell function, DNA damage, and teratogenicity (Mahajan & Kaushal, 2018). Different study cases reported high concentrations, of Cd 70 mg/L in Poland, followed by Egypt (51 mg/L) and India (42.39 mg/L). According to the World Health Organization (WHO), the permissible limit is 3 g/L and many countries are above the standards.

Lead (Pb) pollution can be produced by corrosion lead materials or galvanized pipes in landfills. Another source of lead is old lead-based paints and batteries, as well as insecticides (Abiriga et al., 2020). Lead contamination in humans is seen by ingesting of contaminated plants that can be absorbed into the bloodstream and accumulates in the body's tissues, blood, liver, kidneys, and bones, causing sublethal effects in high concentrations, which could cause death (Altarez & Sedigo, 2019; Krishan et al., 2021).

Zinc (Zn) is the fourth most abundant metal in the body and is an essential trace metal found in more than 300 enzymes and proteins. However, some adverse effects of high concentrations of Zinc, can be a risk factor in Alzheimer's disease in the elderly, and it has the risk of cognitive deterioration and adverse gastrointestinal effects. Excessive Zinc supplementation can result in toxic effects in children, adolescents, and younger adults with cognitive impairment (Chrosniak et al., 2006). Zinc pollution can result from the use of fertilizers, steel industry wastes, wood preservatives, waste fluorescent lamps and batteries (Ahamad et al., 2019; Boateng et al., 2019). Some research shows the concentration of Zinc in groundwater and drinking water at various levels in different countries. In India for instance, it exhibits a maximum concentration of 10364.5 mg/L (Ahamad et al., 2019), followed by Poland at 8893 mg/L (Talalaj, 2014) and Norway 5739 mg/L (Abiriga et al., 2020). The permissible limit according to the World Health Organization is 4000 mg/L.

Finally, it should be noted that there are a number of other contaminants that can affect the health and welfare of individuals that are found in ground water when polluted by leachate. For instance, arsenic, high levels of fluoride, nitrogen, excess potassium and sodium, chloride, sulfates, and bicarbonates. Excessive concentrations and long term exposure to any of these contaminants can cause damage to a number of the human organs including structural and functional damage to the nervous system, reproductive system, kidney, liver, and other organs (Alhassan et al., 2020; He et al., 2020; Pang et al., 2020).

FUTURE TRENDS

Sustainable development postulates in its doctrine the harmonization between the economic, environmental, and social dimensions, in order to consolidate societies in balance with nature. However, in practice, reality is exceeded by finding a clear mismatch in the biogeochemical processes of the cycles of nature—particularly water—in this sense. The perfect triad economy, environment, and society forges serious ruptures that far from being in a balance are presumed in severe contradictions in their interrelation by their own rationality. A criticism of this mismatch regarding municipal solid waste is the following.

ECONOMIC SPHERE

The inherent structure of the world economic-political system produces a consumer society, in which most of the population is subsumed towards an unsustainable lifestyle. The case of the generation of municipal solid waste is a clear example, of the excessive way in which products are discarded. Therefore, it is no coincidence that the ingrained habit of excessive consumerism derives from the rationality of the capitalist system that governs us. It entails for its execution various rules derived from the project of modernity that pursues as a goal maximum profit at the expense of whatever or whoever. This logic configures the world as it is known generating contradictions on the outside so that the harmonious relationship between the economy, the environment, and society seems almost impossible, under the ontology of capital.

The governmental sphere being the political expression of capitalism, requires the same utilitarian logic of the system, therefore, its fundamental objective, is to facilitate the conditions for the reproduction of capital. In this way, the generation of public policies, which could be in favor of society and the environment—in terms of the elimination, reduction and control of urban solid waste—is prioritized in the achievement of the objectives pursued by the economic-political system.

SOCIAL SPHERE

The social sphere is not detached from the economic sphere, since society is a functional element of the economic-political system, and as part of it must engage perfectly to pursue the goal that the accumulation of capital requires. In ancient times, civilizations maintained a close relationship with Mother Earth, respecting their vital cycles of regeneration. However, at present this vital communion with nature has been truly uprooted by the subjectivation of human beings towards unsustainable ways of life subtly nourished by the system.

The enormous generation of municipal solid waste accounts for the society in which we live, where the essential thing is to hedonically satisfy the desires of consumption, throw away and replace as soon as possible. Without stopping to reflect on the final disposal of waste, the pollution to the air, soil, and water that they can cause, and their geographical location in impoverished areas of peripheral countries.

ENVIRONMENTAL SPHERE

The reproduction of a system in crisis, chaotic and unsustainable consequently generates terrible environmental and health implications. The generation of municipal solid waste, which is mainly disposed of in landfills or open dumps, produces a serious impact on groundwater due to the infiltration of leachates whose composition results from low biodegradability, reducing in turn the quality of water for human consumption. These situations trigger severe impacts on the health of people due to exposure to pollutants generated in the disposal sites.

Given this bleak panorama, the proper functioning of landfills, in terms of the adoption of strategies of segregation and channeling of leachate, through the appropriate use of geomembranes, suggests a possibility in the face of the climatic chaos that is experienced and looming.

CRITICAL REFLECTION

In addition to the corrective actions that can be adopted in terms of the mitigation of leachate contamination through an adequate separation of waste prior to its disposal, contributes to reduce the infiltration of the leachate generated in landfills. At the same time, the use of landfill technologies, such as synthetic membranes or other engineering materials to prevent the infiltration into the soil and groundwater of heavy metals and other toxic compounds is encouraged. It is urgent in the face of the current climate and health crisis that is being experienced to promote preventive actions which as a society allow us to rethink the consumerist lifestyle to which we have arrived as humanity. This in turn, will reduce as much as possible the generation of municipal solid waste. Thus, this will enable the construction of sustainable societies in reconnection with nature and in harmony with the biogeochemical cycle of water.

CONCLUSION

The physicochemical quality of groundwater is modified by the infiltration of leachate from municipal solid waste landfills. This poses significant health risks, which limits the use of groundwater for human consumption. This chapter showed that the high concentrations of heavy metals, nitrates, phosphates, sulfates, organic matter and an increase of major ions in the groundwater near municipal solid waste landfills, confirms the migration of leachates into the groundwater, being more evident when the distance is shorter between the landfill and the source of water. In addition, designs, engineering, and inadequate technologies promote the continuous infiltration of the contamination plume in the soil strata. Therefore, it is recommended to avoid the use of groundwater and surface water at lengths of at least 500 meters from municipal solid waste disposal sites. The groundwater quality must be monitored periodically near landfills, at least twice a year, which will provide an evaluation of the changes that originate from the presence of landfills. The importance of public policies in minimizing the damage from the source of solid waste, would help mitigate the impacts on public health and on the environment. However, preventive action, in terms of the generation of municipal solid waste; as well as rethinking the consumerist lifestyle of society, will make it possible to address the environmental and health crisis that is experienced.

ACKNOWLEDGMENTS

This work was supported by the National Council of Science and Technology (CONAHCYT), CVU No. 266124 and financial support from project UAEMex 6738/2022CIB and COMECYT CAT2021-0016/CAT2022-0035.

REFERENCES

Abiriga, D., Jenkins, A., Vestgarden, L. S., & Klempe, H. (2021). A nature-based solution to a landfill-leachate contamination of a confined aquifer. *Scientific Reports, 11*(1), 1–12. https://doi.org/10.1038/s41598-021-94041-7

Abiriga, D., Vestgarden, L. S., & Klempe, H. (2020). Groundwater contamination from a municipal landfill: Effect of age, landfill closure, and season on groundwater chemistry. *Science of the Total Environment, 737*, 140307. https://doi.org/10.1016/j.scitotenv.2020.140307

Ahamad, A., Raju, N. J., Madhav, S., Gossel, W., & Wycisk, P. (2019). Impact of non-engineered Bhalswa landfill on groundwater from Quaternary alluvium in Yamuna flood plain and potential human health risk, New Delhi,

India. *Quaternary International, 507*, 352–369. https://doi.org/10.1016/j.quaint.2018.06.011

Al-Ghouti, M. A., Khan, M., Nasser, M. S., Al-Saad, K., & Heng, O. E. (2021). Recent advances and applications of municipal solid wastes bottom and fly ashes: Insights into sustainable management and conservation of resources. *Environmental Technology and Innovation, 21*, 101267. https://doi.org/10.1016/j.eti.2020.101267

Alhassan, S. I., He, Y., Huang, L., Wu, B., Yan, L., Deng, H., & Wang, H. (2020). A review on fluoride adsorption using modified bauxite: Surface modification and sorption mechanisms perspectives. *Journal of Environmental Chemical Engineering, 8*(6), 104532. https://doi.org/10.1016/j.jece.2020.104532

Altarez, R. D. D., & Sedigo, N. A. (2019). Existing land use and extent of lead (Pb) contamination in the grazing food chain of the closed Carmona sanitary landfill in the Philippines. *Heliyon, 5*(5), e01680. https://doi.org/10.1016/j.heliyon.2019.e01680

Babaei, S., Sabour, M. R., & Moftakhari Anasori Movahed, S. (2021). Combined landfill leachate treatment methods: An overview. *Environmental Science and Pollution Research, 28*(42), 59594–59607. https://doi.org/10.1007/s11356-021-16358-0

Boateng, T. K., Opoku, F., & Akoto, O. (2019). Heavy metal contamination assessment of groundwater quality: A case study of Oti landfill site, Kumasi. *Applied Water Science, 9*(2), 1–15. https://doi.org/10.1007/s13201-019-0915-y

Castillo-Suárez, L. A., Lugo-Lugo, V., Linares-Hernández, I., Martínez-Miranda, V., Esparza-Soto, M., & Mier-Quiroga, M. de los Á. (2019). Biodegradability index enhancement of landfill leachates using a Solar Galvanic-Fenton and Galvanic-Fenton system coupled to an anaerobic–aerobic bioreactor. *Solar Energy, 188*(March), 989–1001. https://doi.org/10.1016/j.solener.2019.07.010

Chen, G., Sun, Y., Xu, Z., Shan, X., & Chen, Z. (2019). Assessment of shallow groundwater contamination resulting from a municipal solid waste landfill—A case study in Lianyungang, China. *Water, 11*(12), 2496. https://doi.org/10.3390/w11122496

Chrosniak, L. D., Smith, L. N., McDonald, C. G., Jones, B. F., & Flinn, J. M. (2006). Effects of enhanced zinc and copper in drinking water on spatial memory and fear conditioning. *Journal of Geochemical Exploration, 88*(1–3 SPEC. ISS.), 91–94. https://doi.org/10.1016/j.gexplo.2005.08.019

DOF. (2014, October 31). Reglamento de la Ley General para la Prevención y Gestión Integral de los Residuos (RLGPGIR) [Regulations of the general law for the prevention and comprehensive management of waste]. *Diario Oficial de La Federación*, 1–63. https://www.diputados.gob.mx/LeyesBiblio/regley/Reg_LGPGIR_311014.pdf

Federación., D. O. de la. (2003). Ley General para la Prevención y el Manejo Integral de los Residuos [General law for the prevention and comprehensive management of waste]. *Periódico Oficial*, 1–52. https://www.gob.mx/cms/uploads/attachment/file/27266/Ley_General_de_Residuos.pdf

Figueroa Sánchez, J. C., & Cruz-Morales, J. (2019). ¿Gobernanza de los residuos sólidos? Estudio de caso sobre el ejido Los Ángeles, Reserva de la Biósfera La Sepultura, Chiapas, México [Solid waste governance? Case study on the Los

Ángeles ejido, La Sepultura Biosphere Reserve, Chiapas, Mexico]. *Sociedad y Ambiente, 20,* 79–102. https://doi.org/10.31840/sya.v0i20.1993

Foufou, A., Djorfi, S., Haied, N., Kechiched, R., Azlaoui, M., & Hani, A. (2017). Water pollution diagnosis and risk assessment of Wadi Zied plain aquifer caused by the leachates of Annaba landfill (N-E Algeria). *Energy Procedia, 119,* 393–406. https://doi.org/10.1016/j.egypro.2017.07.123

Galilea, S., Letelier, L., & Ross, K. (2011). Los casos de Brasil, Chile, Colombia, Costa Rica y México en salud, educación, residuos, seguridad y fomento [The cases of Brazil, Chile, Colombia, Costa Rica and Mexico in health, education, waste, security and development]. *CEPAL-Coleccion Documentos de Proyectos,* 292.

Gamar, A., Zair, T., El Kabriti, M., & El Hilali, F. (2018). Study of the impact of the wild dump leachates of the region of El Hajeb (Morocco) on the physicochemical quality of the adjacent water table. *Karbala International Journal of Modern Science, 4*(4), 382–392. https://doi.org/10.1016/j.kijoms.2018.10.002

Ganguly, R., Sharma, D., Sharma, A., Gupta, A. K., & Gurjar, B. R. (2019). Parametric evaluation of leachate generated from a non-engineered landfill site and its contamination potential of surrounding soil and water bodies. In R. A. Agarwal, A. K. Agarwal, T. Gupta, & N. Sharma (Eds.), *Pollutants from energy sources: Characterization and control* (pp. 307–322). https://doi.org/10.1007/978-981-13-3281-4_15

Gautam, P., & Kumar, S. (2021). Characterisation of hazardous waste landfill leachate and its reliance on landfill age and seasonal variation: A statistical approach. *Journal of Environmental Chemical Engineering, 9*(4), 105496. https://doi.org/10.1016/j.jece.2021.105496

Gonçalves, F., Correa, C. Z., Lopes, D. D., Vendrame, P. R. S., & Teixeira, R. S. (2019). Monitoring of the process of waste landfill leachate diffusion in clay and sandy soil. *Environmental Monitoring and Assessment, 191*(9), 577. https://doi.org/10.1007/s10661-019-7720-9

Guo, Y., Li, P., He, X., & Wang, L. (2022). Groundwater quality in and around a landfill in Northwest China: Characteristic pollutant identification, health risk assessment, and controlling factor analysis. *Exposure and Health, 14,* 885–901. https://doi.org/10.1007/s12403-022-00464-6

Gworek, B., Dmuchowski, W., Koda, E., Marecka, M., Baczewska, A., Brągoszewska, P., Sieczka, A., & Osiński, P. (2016). Impact of the municipal solid waste Łubna landfill on environmental pollution by heavy metals. *Water, 8*(10), 470. https://doi.org/10.3390/w8100470

Haider, F. U., Liqun, C., Coulter, J. A., Cheema, S. A., Wu, J., Zhang, R., Wenjun, M., & Farooq, M. (2021). Cadmium toxicity in plants: Impacts and remediation strategies. *Ecotoxicology and Environmental Safety, 211,* 111887. https://doi.org/10.1016/j.ecoenv.2020.111887

He, J., Yang, Y., Wu, Z., Xie, C., Zhang, K., Kong, L., & Liu, J. (2020). Review of fluoride removal from water environment by adsorption. *Journal of Environmental Chemical Engineering, 8*(6), 104516. https://doi.org/10.1016/j.jece.2020.104516

Hussein, M., Yoneda, K., Mohd-Zaki, Z., Amir, A., & Othman, N. (2021). Heavy metals in leachate, impacted soils and natural soils of different landfills in Malaysia:

An alarming threat. *Chemosphere, 267,* 128874. https://doi.org/10.1016/j. chemosphere.2020.128874

Instituto Nacional de Ecologia y Cambio Climático. (2022). Atlas Nacional de Residuos Sólidos Urbanos [National Atlas of urban solid waste]. *Secretaría de Medio Ambiente y Recursos Naturales,* 1–314. https://www.gob.mx/inecc

Jabłońska-Trypuć, A., Wydro, U., Wołejko, E., Pietryczuk, A., Cudowski, A., Leszczyński, J., Rodziewicz, J., Janczukowicz, W., & Butarewicz, A. (2021). Potential toxicity of leachate from the municipal landfill in view of the possibility of their migration to the environment through infiltration into groundwater. *Environmental Geochemistry and Health, 5.* https://doi.org/10.1007/s10653-021-00867-5

Kanownik, W., Policht-Latawiec, A., & Fudała, W. (2019). Nutrient pollutants in surface water-assessing trends in drinking water resource quality for a regional city in central Europe. *Sustainability (Switzerland), 11*(7). https://doi.org/10.3390/su11071988

Kapelewska, J., Kotowska, U., Karpińska, J., Astel, A., Zieliński, P., Suchta, J., & Algrzym, K. (2019). Water pollution indicators and chemometric expertise for the assessment of the impact of municipal solid waste landfills on groundwater located in their area. *Chemical Engineering Journal, 359,* 790–800. https://doi.org/10.1016/j.cej.2018.11.137

Kjeldsen, P., Barlaz, M. A., Rooker, A. P., Baun, A., Ledin, A., & Christensen, T. H. (2002). Present and long-term composition of MSW landfill leachate: A review. *Critical Reviews in Environmental Science and Technology, 32*(4), 297–336. https://doi.org/10.1080/10643380290813462

Kr, D., Tenodi, S., Grba, N., Kerkez, D., Watson, M., & Ron, S. (2018). *Science of the total environment preremedial assessment of the municipal land fill pollution impact on soil and shallow groundwater in Subotica , Serbia. 615,* 1341–1354. https://doi.org/10.1016/j.scitotenv.2017.09.283

Krishan, G., Taloor, A. K., Sudarsan, N., Bhattacharya, P., Kumar, S., Chandra Ghosh, N., Singh, S., Sharma, A., Rao, M. S., Mittal, S., Sidhu, B. S., Vasisht, R., & Kour, R. (2021). Occurrences of potentially toxic trace metals in groundwater of the state of Punjab in northern India. *Groundwater for Sustainable Development, 15*(July), 100655. https://doi.org/10.1016/j.gsd.2021.100655

León, J. M. N. C. (2005). El manejo de residuos sólidos municipales en México y la participación del Banco Mundial [The management of municipal solid waste in Mexico and the participation of the World Bank]. *Comercio Exterior, 55*(12848-me), 348–361. http://revistas.bancomext.gob.mx/rce/magazines/77/8/RCE.pdf

Li, P., Karunanidhi, D., Subramani, T., & Srinivasamoorthy, K. (2021). Sources and consequences of groundwater contamination. *Archives of Environmental Contamination and Toxicology, 80*(1), 1–10. https://doi.org/10.1007/s00244-020-00805-z

Luo, H., Zeng, Y., Cheng, Y., He, D., & Pan, X. (2020). Recent advances in municipal landfill leachate: A review focusing on its characteristics, treatment, and toxicity assessment. *Science of the Total Environment, 703,* p. 135468. https://doi.org/10.1016/j.scitotenv.2019.135468

Mahajan, P., & Kaushal, J. (2018). Role of phytoremediation in reducing cadmium toxicity in soil and water. *Journal of Toxicology, 2018.* https://doi.org/10.1155/2018/4864365

Mishra, S., Tiwary, D., Ohri, A., & Agnihotri, A. K. (2019). Impact of municipal solid waste landfill leachate on groundwater quality in Varanasi, India. *Groundwater for Sustainable Development, 9*(2352801X), 100230. https://doi.org/10.1016/j.gsd.2019.100230

Morita, A. K. M., Ibelli-Bianco, C., Anache, J. A. A., Coutinho, J. V., Pelinson, N. S., Nobrega, J., Rosalem, L. M. P., Leite, C. M. C., Niviadonski, L. M., Manastella, C., & Wendland, E. (2021). Pollution threat to water and soil quality by dumpsites and non-sanitary landfills in Brazil: A review. *Waste Management, 131*(June), 163–176. https://doi.org/10.1016/j.wasman.2021.06.004

Najafi Saleh, H., Valipoor, S., Zarei, A., Yousefi, M., Baghal Asghari, F., Mohammadi, A. A., Amiri, F., Ghalehaskar, S., & Mousavi Khaneghah, A. (2020). Assessment of groundwater quality around municipal solid waste landfill by using water quality index for groundwater resources and multivariate statistical technique: A case study of the landfill site, Qaem Shahr City, Iran. *Environmental Geochemistry and Health, 42*(5), 1305–1319. https://doi.org/10.1007/s10653-019-00417-0

Ololade, O. O., Mavimbela, S., Oke, S. A., & Makhadi, R. (2019). Impact of leachate from northern landfill site in Bloemfontein on water and soil quality: Implications for water and food security. *Sustainability (Switzerland), 11*(15). https://doi.org/10.3390/su11154238

Pang, T., Aye Chan, T. S., Jande, Y. A. C., & Shen, J. (2020). Removal of fluoride from water using activated carbon fibres modified with zirconium by a drop-coating method. *Chemosphere, 255,* 126950. https://doi.org/10.1016/j.chemosphere.2020.126950

Parvin, F., & Tareq, S. M. (2021). Impact of landfill leachate contamination on surface and groundwater of Bangladesh: A systematic review and possible public health risks assessment. *Applied Water Science, 11*(6), 100. https://doi.org/10.1007/s13201-021-01431-3

Peng, Y. (2017). Perspectives on technology for landfill leachate treatment. *Arabian Journal of Chemistry, 10,* S2567–S2574. https://doi.org/10.1016/j.arabjc.2013.09.031

Przydatek, G. (2021). Using advanced statistical tools to assess the impact of a small landfill site on the aquatic environment. *Environmental Monitoring and Assessment, 193*(2), 71. https://doi.org/10.1007/s10661-021-08850-4

Przydatek, G., & Kanownik, W. (2019). Impact of small municipal solid waste landfill on groundwater quality. *Environmental Monitoring and Assessment, 191*(3), 169. https://doi.org/10.1007/s10661-019-7279-5

SANS 241:2015. (2015). *South African national standard, 151*(1), 10–17.

Somani, M., Datta, M., Gupta, S. K., Sreekrishnan, T. R., & Ramana, G. V. (2019). Comprehensive assessment of the leachate quality and its pollution potential from six municipal waste dumpsites of India. *Bioresource Technology Reports, 6,* 198–206. https://doi.org/10.1016/j.biteb.2019.03.003

Talalaj, I. A. (2014). Assessment of groundwater quality near the landfill site using the modified water quality index. *Environmental Monitoring and Assessment, 186*(6), 3673–3683. https://doi.org/10.1007/s10661-014-3649-1

Talalaj, I. A., & Biedka, P. (2016). Use of the landfill water pollution index (LWPI) for groundwater quality assessment near the landfill sites. *Environmental Science and Pollution Research, 23*(24), 24601–24613. https://doi.org/10.1007/s11356-016-7622-0

Tenodi, S., Kr, D., Agbaba, J., Zrni, K., Radenovi, M., & Ubavin, D. (2020). *Assessment of the environmental impact of sanitary and unsanitary parts of a municipal solid waste landfill. 258,* 110019. https://doi.org/10.1016/j.jenvman.2019.110019

Turki, N. & Bouzid, J. (2017). Effects of landfill leachate application on crops growth and properties of a Mediterranean sandy soil. *Journal of Pollution Effects & Control, 5*(02). https://doi.org/10.4172/2375-4397.1000186

Vahabian, M., Hassanzadeh, Y., & Marofi, S. (2019). Assessment of landfill leachate in semi-arid climate and its impact on the groundwater quality case study: Hamedan, Iran. *Environmental Monitoring and Assessment, 191*(2), 109. https://doi.org/10.1007/s10661-019-7215-8

Zeng, D., Chen, G., Zhou, P., Xu, H., Qiong, A., Duo, B., Lu, X., Wang, Z., & Han, Z. (2021). Ecotoxicology and environmental safety factors influencing groundwater contamination near municipal solid waste landfill sites in the Qinghai-Tibetan plateau. *Ecotoxicology and Environmental Safety, 211,* 111913. https://doi.org/10.1016/j.ecoenv.2021.111913

CHAPTER 17

CLIMATE SELF-GOVERNANCE

Decentralized Solar Systems
in Kenya's Cities

Victor Ferreros
Walden University

Felix O. Vescovi
Walden University

Susan E. Baer
The University of Kansas

Angela M. Mai
Social Policy Research Consultant

ABSTRACT

Among the top five world renewable energy resource development leaders, Kenya's governance challenges continue to affect the country's renewable energy provision and distribution among the Kenyan population. This case study analysis used Ostrom's polycentric environmental governance approach

---

Climate Governance in International and Comparative Perspective, pages 399–422
Copyright © 2024 by Information Age Publishing
www.infoagepub.com
399

to explore Kenya's decentralized climate governance via systematic literature review and policy analysis regarding renewable solar energy development, provision, and distribution among citizens. Findings revealed practical climate governance efforts extended from multiple actors and incorporated vital contributing factors, including regulatory oversight, government-non-government collaboration, dynamic momentum generation, and political sway. Although no singular governing institution surfaced, the polycentric framework approach exposed and emphasized four significant contributors to Kenya's renewable energy governance endeavors: legal frameworks, private sector inclusion, capacity building, and political will. While the polycentric environmental governance model was integral to effectively analyzing Kenya's climate governance, the political will element needed increased focus. These elements indicated that cooperative efforts among government and community-based participants toward universal energy access goals are essential for sustainable renewable energy governance.

The population of the Sub-Saharan Africa (SSA) region will double to more than 2.1 billion people by 2050 (World Bank, 2023). It will need over USD 190 billion annually to meet energy goals (International Energy Agency [IEA], 2023). Despite ongoing global initiatives toward achieving the United Nations' Sustainable Development Goal 7 ([UN SDG 7]; 2015), the IEA reported that, in 2020, the number of SSA cities without electrical access rose for the first time since 2013. Cozzi et al. (2020) attributed some of the declines to COVID-19; nevertheless, the decreased energy availability left over 600 million people in the region without electrical access (IEA, 2023). Given population growth projections, urban concentration patterns, and the general SSA self-governance climate related to energy consumption (Burger & Weinmann, 2019; IEA, 2023; Michoud & Hafner, 2021), the growing problem requires systematic study toward identifying practical climate governance solutions.

Since the SSA's energy governance extends from self-governance endeavors, we framed this case study using Ostrom's (2009) polycentric environmental governance approach. The polycentric theoretical foundation was best for understanding the problem, identifying mitigating factors, and explaining the implications to the researchers, scholars, practitioners, and policymakers responsible for Kenya's energy needs. Climate self-governance is a continuous multilevel structure of rules, discussions, and negotiations involving a diverse group of national and local governments, the private sector, and other social actors responding to climate change risks and opportunities (Organization for Economic Cooperation and Development & Sahel West African Club [OECD/SWAC], 2020). Climate self-governance promotes opportunities and prompts actions addressing climate change, thereby contributing toward positive social values, albeit, Mattijssen et al. (2018) inferred local focus limited such contributions. Because reduced greenhouse emissions are considered a global public good, and most high-income developed governance approaches emphasize a collective, global response, Mattiissen et al. noted that past analysts focused on global-level institutional change. However, polycentric governance pioneers like Ostrom challenged the globally standardized approach with the theory of polycentric governance.

The purpose of this chapter was SSA climate governance analysis by examining Kenya's decentralized solar system governance development, deployment, and distribution, in a polycentric framework. Thus, Ostrom's (2009) polycentric model seems appropriate for evaluating the characteristics and features of Kenya's renewable energy governance. Therefore, our systematic case study literature analysis included the roles of the state, non-government organizations (NGOs), private sectors, local leaders, community associations, and individuals within the polycentric environmental governance framework to identify institutional and economic barriers affecting Kenya's solar energy system development. Chapter sections include (a) SSA renewable energy development, (b) Kenyan energy endeavors, (c) Kenya's current decentralized climate governance environment, (d) our polycentric methodology, (e) our Kenyan governance findings, and (f) lessons learned with implications thereof.

LITERATURE REVIEW

In recent years, Africa spent millions of dollars expanding electricity across the countryside; however, most Africans remain without connectivity (IEA, 2023). Since Africa's energy grid spans the country and the IEA considers the land energy-rich, we reviewed issues affecting SSA citizens' lacking energy access. Thus, we explored the SSA's decentralized renewable energy governance in a polycentric case study environment using multiple data source types: current scholarly literature and subsequent secondary data included within the empirical research discussed in this systematic literature review analysis. We conducted an associated in-depth policy review. We also utilized respected public government sources like AFSIA, EASAC, EPA, GOGLA, GPOBA, IEA, IIED, IRENA, KOSAP, NRDC, OECD, SDG-7, SE4ALL, SWAC, UN, and World Bank Group, among others wherein we also accessed their public posted secondary databases and dynamic tracking systems.

SUB-SAHARAN AFRICA RENEWABLE ENERGY DEVELOPMENT

Historically, decentralized solar systems first developed in Africa's rural areas and then expanded to their cities, becoming a regular part of SSA rural landscapes in the 21st century (Pillot et al., 2019). Small solar photovoltaic (PV) based mini-grids, energy charging centers, and other small-scale solar energy models served the remote settlements typically overlooked for electrification by full-scale grids or large conventional mini-grids (Ulsrud, 2020). Ulsrud (2020) and Pillot et al. (2019) demonstrated that the highly dispersed settlement patterns and meager incomes within rural SSA communities resulted in solar panel placement on the home and power plant roofs. Additionally, many users traveled to power plants to get electrical services, which included renting portable electric lanterns and other energy tools (Ulsrud, 2020). Thus, rural needs drove initial, ungoverned, decentralized applications.

In recent decades, the implementation of rural solar power provisions supported quality electrical access, befitting villagers' needs and economic situations, thereby driving a significant positive increase in decentralized village-scale electrical supply models across SSA cities (Ulsrud, 2020). Thus, those emerging off-grid solar systems became vital elements in the physical, institutional, financial, and digital infrastructures bringing electricity to people previously without power living in SSA cities, thereby improving their socioeconomic well-being but lacking needed oversight. However, scientific analysis, like that of IEA et al. (2020), Michoud and Hafner (2021), and IEA (2023), reflected that these experimental practices generated electrical knowledge among SSA leaders. Subsequently, that increased understanding resulted in expanding potential applications about technical and organizational models, service delivery systems, and integrating electricity into the social practices and everyday lives of low-income urban communities.

Consequently, two decades after initial implementation, studies show that off-grid solar systems became big business across SSA cities, driven by the alignment of off-grid electrification with SSA's socioeconomic, political, and environmental plans (Cross & Neumark, 2021; Winklmaier et al., 2020). Thus, solar systems in SSA acquired the characteristics of a "socio-technical regime." According to Kirshner et al. (2019), such regimes grow where co-evolving sets of social and technical developments build sufficient momentum to accept a particular technology as an established part of the energy provision system. Kirshner et al. posited that this privatized momentum spread across SSA and evolved into the region's current self-government model.

THE PRIVATE SECTOR

Due to private solar manufacturers, distributors, and installers, the private sector became a chief avenue for delivering access to clean, efficient, affordable energy for millions of people living without connection to a city's electricity grid (Shirley et al., 2019). By 2023, off-grid solar panels frequently appeared on rooftops of health centers, schools, and homes in cities throughout the SSA, wherein city residents use them to power mobile lighting, charging devices, generate individual household electricity (Boamah, 2020; Mukisa et al., 2022). Thus, SSA regions began aligning off-grid electrification with socioeconomic, political, and environmental plans giving renewable energy advocates unrestricted roles toward meeting the SDG-7 goal of inclusive growth through affordable, reliable, and sustainable energy for all (Cross & Neumark, 2021). Additionally, Cross and Neumark found that, since the last decade, international developmental aid organizations have increasingly promoted off-grid application expansion as the key to including impoverished citizens in SSA developments.

SSA's climate governance evolution toward its decentralized, polycentric-style practices was solidly reflected by funding allocations. For example, between 2010 and 2018, roughly USD 1.64 billion in equity and debt from government-operated and NGO sources went toward global energy access markets (Burger, 2019; Inshakov et al., 2019). Three-quarters of those funds were allocated to renewable energy sources in SSA cities (Cross & Neumark, 2021). According to Baker (2022) and the Global Off-Gid Lighting Association (GOGLA et al., 2022), these investment inflows connected SSA cities' low-income populations to clean, renewable energy sources and linked them to consumer-focused mobile financial infrastructures thereby instigating public–private partnerships (PPPs). Thus, PPPs became firmly integrated within the region's blossoming climate governance efforts.

PUBLIC–PRIVATE PARTNERSHIPS

Solar energy systems fueled PPPs falling costs, emerging mobile banking, and affordable financing across SSA cities, and soon, PPPs promoting solar energy infrastructure investments spanned SSA cities (Michoud & Hafner, 2021). In 2022, the Africa Solar Industry Association ([AFSIA]; 2023) reported over 7,600 PPP solar energy projects commissioned, awarded, or operational. Consequently, SSA's vast renewable resources made cities' solar energy projects an attractive market for international investment (GOGLA et al., 2018), with AFSIA reporting global investment in SSA cities' solar and wind projects exceeding USD 34.7 billion from 2011 through 2020. G7 Summit 2022 world leaders also pledged more than USD 80 billion to private sector SSA endeavors focusing on emerging off-grid pay-as-you-go (PAYG) systems and renewable energy technology (IEA, 2023). Thus, SSA PPPs cemented their power and authority within the region's climate governance system, naturally affecting most SSA citizens as energy consumers.

CONSUMERS

Accordingly, individual consumers' greater power autonomy progressed SSA electrical systems decentralization, directly impacting energy costs. For example, Finke et al. (2022) found that solar smart contracts replaced many third parties, reducing transaction costs. Macrinici et al. (2018) also demonstrated how such energy management platforms provided users with total business flow automation, improving buying and stop-buying power based on meeting specified conditions. Thus, smart contracts eliminated intermediaries, increased consumer roles, and reduced transaction costs.

Albeit, some PPP endeavors, including the structural relationship between solar power companies and consumers, adversely affected consumers, causing distrust, dissatisfaction, and hostility toward power companies (Cross & Neumark, 2021). Furthermore, energy privatization failures became catalysts driving consumers away from power companies (Burger & Weinmann, 2019). Additionally, Cozzi et al. (2020) discussed varying COVID-19 impacts negatively affecting PPPs, and Chengo et al. (2021) pointed out inclusivity problems within energy distribution channels. Considering the many factors influencing SSA climate governance, we delimited our analysis to Kenya's solar energy governance.

KENYAN ENERGY ENDEAVORS

Historically, Kenyans depended on traditional biomasses of firewood, charcoal, and agricultural waste, with seasonally dependent hydro-power reserves, for renewable energy (Mburugu & Gikonyo, 2019). In 2019, Kenyans harvested wood fuel and biomass (68%), oil (22%), hydro-power electricity (9%), and wind-power (1%) for their energy needs (Takase et al., 2021). The NRDC (2020) emphasized wood fuel as an environmental disaster and oil as a dirty, unsustainable resource. For example, the Environmental Protection Agency ([EPA], 2023) determined wood fuel increased greenhouse gas emissions, the NRDC documented related deforestation and degradation, Sterman et al. (2018) demonstrated it was not carbon neutral, and the European Academics Science Advisory Council ([EASAC], 2018) concluded that it worsened climate change. Stashwick (2019) posited wood fuel was not sustainable, and Koester and Davis (2018) demonstrated economic injustices resulting from its use. Thus, only 10% of Kenya's energy needs were sustainable in 2019. Kenya's climate governance needed to redress more than 90% of its energy sources and other consumer and infrastructure challenges to meet SDG7 goals.

KENYA'S CLIMATE GOVERNANCE CHALLENGES

Kenya's 2014 to 2018 increased focus on extending electrical access (Coffey, 2023) failed to deliver the electric grid to rural communities (Bercegol & Monstadt, 2018; Bonnell, 2021). Subsequently, rural populations labored gathering unsustainable wood fuel at the expense of adequate health care (Moner-Girona et al., 2021), water and food systems (Winklmaier et al., 2020), and essential economic activities (Mukisa et al., 2022; Takase et al., 2021). Kenya's climate governance failures also included exorbitant costs (Bonnell, 2021), fossil fuel competition, infrastructure issues, and interrupted power supplies (Burger & Weinmann, 2019). Pollutant concerns (EASAC, 2018; EPA, 2023; Sterman et al., 2018), degradation, deforestation (National Resources Defense Council [NRDC], 2020), unsustainability (Stashwick, 2019), and economic injustice (Koester & Davis, 2018) compounded the challenges affecting Kenya's climate governance.

Subsequently, these factors negatively affected consumer confidence in Kenya's energy grid management, causing widespread dissatisfaction (Bonnell, 2021), trust and credibility issues toward Kenya Power & Light Company (KPLC), reduced new power connections, and decreased consumer cooperation (Cross & Neumark, 2021). Cross and Neumark, and Bonnell posited that since the Kenyan government owns KPLC (https://www.kplc.co.ke/category/view/41/public-information), the publicly traded company's decline also negatively impacted community support in Kenya's climate governance and energy policy implementation. Further debasing the situation, the 2018 KPLC criminal indictments for metering, bribery, favoritism, procurement, substandard equipment, fraudulent billing, and mismanagement resulted in numerous company officials' arrests (Seii, 2018) and convictions (Kenya Power & Light Company [KPLC], 2020). KPLC reported transformer vandalism, component theft, and sub-standard installations widespread in urban settlement areas, likely caused by attempted illegal grid connections.

Illegal grid connection issues contributed to financial loss, safety issues, and misconceptions (Bonnell, 2021). Bonnell reported that Kenyan energy cartels profited from poorly managed KPLC projects by commandeering energy distribution using unlicensed, untrained intermediaries impersonating KPLC personnel, creating unsafe living situations. Bonnell explained that these dangerous conditions worsened after the grid connection as the intermediaries became violent cartel collectors. Thus, the problem increased Kenyan climate governance's difficulty, thereby deepening the rift between government-managed energy services and consumers.

Additionally, bureaucratic hurdles (AKA red tape) caused energy governance impediments. Boamah et al. (2021) reported prepaid connections took over 200 days to provide electricity and detailed frequent blackouts, exorbitant electricity bills, perceived corruption, and regular provision and maintenance delays increased the uptake of solar energy systems. Thus, Boamah et al. explained the frustrations that led some consumers to bribery. Bercegol and Monstadt (2018) proffered that few consumers afforded such increased costs to obtain and maintain their energy source. Still, most struggled because their low-income and rural populations frequently left without energy.

Kenya's electricity grid served densely populated urban areas with relatively cheap grid extension costs (GOGLA et al., 2018). Therefore, since affordability was a significant problem, Burger (2019) explained that Kenya began pursuing off-grid mini-grids; albeit, Cozzi et al. (2020) detailed how the COVID-19 pandemic exacerbated energy governance endeavors thereby increasing disorganization and further muddying political priorities. Even so, changing donor funding patterns and global renewable energy focuses stimulated Kenya's entrepreneurial hybridity among the national government, city officials, private sector, and consumers toward improved production and distribution of decentralized solar power to overcome existing climate governance challenges.

DECENTRALIZED SOLAR SYSTEM DEVELOPMENT

Kenya's decentralized solar-system development followed the same avenues as the larger SSA region in the early stages. However, Kenya's decentralized solar energy profitability led to a strong consumer chain for solar home systems (SHS) in populated cities, including Nairobi and Mombasa, and set a capitalizable energy-use behavior standard (Adwek et al., 2020). Kenya's low-income city residents, like those in other SSA cities, functioned within a normative energy behavior ideal that copied wealthier residents' houses and lifestyles (Haque et al., 2021). Haque et al. (2021) found that the low-income population's conformity to normative energy behavior ideals matched wealthier residents resulting in duplicate new appliance purchases and energy behaviors irrespective of household affordability or relevance. Takase et al. (2021) reported similar results wherein low-income households made substantial financial sacrifices to prove they fit into their perceived upward social mobility. These studies demonstrated how social norm conformity affected energy behavior and how perceived behaviors and lifestyles set a normality standard matching higher-income groups. Thus, the SHS stand-alone system became the most successful decentralized solar market in Kenya's cities (Wagner et al., 2021). Wagner et al. (2021) demonstrated that stand-alone SHS was Kenyan cities' most decentralized solar system development.

Albeit, decentralized solar systems in Kenya often consisted of grid-tied, hybrid (solar/diesel), and off-grid systems (Pedersen & Nygaard, 2018). Except for the diesel hybridization model, these systems allowed optimal renewable energy use, including PV systems, all external grid-connected solar platforms (Pedersen & Nygaard, 2018), and some experimental village-scale charging centers, portable rental outlets, and mini-grids (Finke et al., 2022). Initially, scholars did not believe these small-scale technologies could serve the massive national power grid scale; however, by 2019, Burger explained that many observers began considering microgrids as long-term solutions accelerating Kenya's rural electrification.

THE CURRENT STRUCTURE OF DECENTRALIZED SOLAR SYSTEMS IN KENYA

Kenya's climate governance increased urban and rural connectivity from 32% in 2013 to 75% in 2022 using a combination of grid-connected and off-grid systems (Coffey, 2023). Even so, while the urban access rate was 100%, rural Kenya was only 65% (IEA, 2023). The IEA reported Kenya's 2022 energy mix predominantly consisted of green energy with geothermal, hydro, wind, and solar accounting for roughly 81% generation and only 19% thermal, biomass, and imported. Thus, Kenya's decentralized climate governance practices increased sustainable clean energy provision by about 70% since 2018.

Despite impressive electrification rates, Kenya projected increased generation capacity on overzealous demand projections resulting in a demand constraint wherein the country's electric generation capacity exceeds demand Bonnell (2021). Alternatively, Cozzi et al. (2020) attributed the demand decline to the 2019 pandemic. Despite the increased generation, energy benefits remained out of reach for many of the nation's poorest and most vulnerable citizens (Chengo et al., 2021). Chengo et al. (2021) estimated more than 50% of the electrical infrastructure revenue came from over 3,000 high-volume industrial power users that increasingly moved off-grid and built private plants, increasing prices to average Kenyan consumers.

Subsequently, Kenya currently finds itself amid an electric equality crisis disproportionately affecting the rural and inner-urban poor (Bercegol & Monstadt, 2018; Bonnell, 2021). Thus, inadequate management, poorly planned outages, transmission, and distribution losses compounded Kenya's energy governance. Therefore, grid unreliability provided additional impetus for spreading solar systems to Kenya's cities as citizens were dissatisfied with Kenya's climate governance challenges. Urban Kenyans paid for a grid connection and an SHS as a backup system against widespread and unpredictable outages. Therefore, continual electrical infrastructure issues increasingly catalyze energy infrastructure decentralization, thereby expanding generation capacity investment. Albeit, the current structure of the Kenyan decentralized solar system remains insufficient as the economic, institutional, and political barriers within Kenya's climate governance persist unresolved.

KENYA'S CURRENT DECENTRALIZED CLIMATE GOVERNANCE SYSTEM

The Kenyan national government's broader policy landscape appreciation for decentralized solar systems focuses on shaping practical policy innovation and interventions, encouraging stakeholders' interaction (Chengo et al., 2021). Chengo et al. (2021) explained the devolvement of Kenya's national climate governance and transitioned decentralized energy planning responsibilities to local levels encouraging city officials' interactions with other non-state actors. Albeit, Takase et al. (2021) documented that the Ministry of Energy and Petroleum manages overall strategy and provides advice on the production and growth of energy sub-sectors, including power, petroleum, and renewable energies. Whereas the city level maintains no regulatory solar service policy, instead uses various initiatives supporting the distribution of SHS, and SE4ALL (Sustainable Energy for All, 2022) endeavors aid city officials' PPP interactions.

The combination of Kenya's Energy Act (2019) and the National Climate Change Action Plan (NCCAP, 2018) strengthened local entities' and communities' roles in developing solar energy systems. The Kenyan national government helped develop decentralized solar systems by moving toward climate government devolvement in 2010. This devolved system began shifting energy planning responsibility to the county government level. The Energy Act (2019) mandated that all cities develop energy plans to meet residents' energy needs (Mwendwa et al., 2022; Winther et al., 2018). Thus, urban boards manage communities on behalf of county governments. Additionally, with support from the World Bank, the Kenyan national government established the Kenya Off-Grid Solar Access Project ([KOSAP]; 2018) and the Kenya Electricity Modernization Project ([KEMP]; 2021). KOSAP's objectives supported solar technology use driving household electrification, and KEMP increased electricity access and service reliability and strengthened KPLC.

This climate self-governance directive gave city officials leverage and scope for developing energy plans considering city-wide development needs and working with other stakeholders for planning and implementation (Energy Act, 2019; NCCAP, 2018). For example, in 2013, the International Institute for Environment and Development (IIED), Loughborough

University, Caritas Kitui, and the Catholic Diocese of Kitui worked with Kitui City to develop the city's mandated energy plan (Mwendwa et al., 2022; Winther et al., 2018). Thus, this coalition strengthened Kitui's capacity for creating energy plans and designs. These actions demonstrated the role of the Kenyan national government as a loose oversight broadly managed by the Ministry of Energy and Petroleum, with the next actor level being each region's county governments.

COUNTY GOVERNMENT MANDATES

Under the Minister's general oversight, the county government was responsible for broad-range county-wide energy planning, strategies, and associated resource allocations (Energy Act, 2019; NCCAP, 2018). These regional area community planning objectives included strengthening smaller municipalities', supporting, developing, and maintaining solar system projects, and minimizing large-scale government involvement (Energy Act, 2019; Government of Kenya, 2018); thus, decentralization. County governments then mandated city officials to develop energy plans unique to their city's residential needs.

THE ROLE OF CITY OFFICIALS

City officials did not enforce policy dictating solar production, distribution, or access to service companies; instead, they worked in tandem with local stakeholders (providers and consumers) to meet residents' needs (Mwendwa et al., 2022; Winther et al., 2018). SE4ALL (Sustainable Energy for All, 2022) supported city initiatives aiding city officials' interactions with SHS stakeholders and other partners. Mwendwa et al. and Winther et al., among other noted scholars, describe the dynamic urban boards that city officials empowered to fulfill their responsibilities toward meeting their residents' needs.

URBAN BOARDS

Subsequently, urban boards typically consisted of a wide variety of invested residents, including (a) city officials and administrators; (b) donors from wide-ranging city constituents—power producers, suppliers, collection centers, and distributors; (c) licensed technical and safety service personnel; (d) city business representatives from assorted industries; and (e) NGOs, NPOs, and various other consumers and non-state actors from all income levels and sociocultural structures (Energy Act, 2019; Kenya Off-Grid

Solar Access Project [KOSAP], 2018; Mwendwa et al., 2022; NCCAP, 2018; SE4ALL, 2022; ...; Winther et al., 2018).

OTHER NON-STATE ACTORS

Non-state actor participation ensured consideration of all city residents' needs and the best feasible outcome (Energy Act, 2019; NCCAP, 2018). This climate self-governance supported residential needs and provided city officials with leverage and scope to develop energy plans centered on city-wide energy access (Energy Act, 2019; NCCAP, 2018). Chengo et al. (2021) suggested decentralized self-governance encourages working with other stakeholders during energy planning, implementation, distribution, and maintenance. Thus, Kenya's climate governance consisted of a continually evolving decentralized system designed to meet citizens' needs that were highly like polycentric governance.

A POLYCENTRIC ANALYSIS APPROACH: METHODOLOGY

Kenya's decentralized climate governance required a case study examination appropriate to the conditions and prospects of polycentric environmental governance. Thus, we described and explained Kenya's decentralized solar system development, including the key assumptions in a polycentric system necessary for efficient and effective solar energy governance. Polycentric governance requires strong, accountable, transparent, representative, and participative institutions (Ostrom, 2009). Ostrom emphasized that a poly-centric format is a complex, multi-scale system operating beyond formal policies, administrative boundaries, and multiple institutions; it is neither a binary nor static condition. Effective polycentric governance requires four general preconditions: legal framework, higher-level governance support, capacity building, and political will (Amaruzaman et al., 2022). Kenya's de-centralized solar system development aligned with this framework, which became apparent when analyzing the literature from a case study approach.

We examined climate governance in Kenya as a case study using the energy sector to illustrate opportunities and challenges. The case study method allowed us to dig deeper and discover factors affecting adaptation to climate governance and varying mitigating capacities. Using a case study approach isolated in Ostrom's (2009) polycentric environmental governance framework, we examined multiple contributing data exposing how Kenya connected its economic, environmental, political, and social-cultural factors facilitating transformation toward decentralized solar system development echoing a polycentric climate governance style.

OSTROM'S 2009 POLYCENTRIC ENVIRONMENTAL GOVERNANCE APPROACH: THEORETICAL FRAMEWORK

Climate change addressed by individuals, families, small groups, and voluntary associations actions is effectively employable by private firms and local, regional, and national governments (Ostrom, 1990). Ostrom's polycentric approach directly contrasted Buchanan and Tullock's (1962) ideology and Olson's (1965) classic collective action theory, which emphasized a top–down approach wherein external authority imposed enforceable rules and restrictions on actors, thereby curbing their behavior and changing their incentives. Ostrom posited that polycentric, bottom–up, small-scale efforts by small groups and communities without government involvement might produce collective action that cumulatively mitigates climate change.

Since 1990, Ostrom's polycentric environmental governance approach has resulted in many positive climate change endeavors and inspired aspects of the UN Framework Convention on Climate Change (UNFCCC) at the 21st Conference of Parties ([COP21]; 2015). In contrast to earlier international efforts, COP21 (AKA: The Paris Agreement (Paris Climate Accords) was a bottom–up approach wherein individual country policies aggregated around common global targets. By 2022, the polycentric model was a key analysis component of intergovernmental, trans-governmental, and cooperative endeavors among local and subnational governments and non-state actors' activities (Amaruzaman et al., 2022). Thus, climate change governance models evolved from cooperation between nation-states to multiple governing authorities at different scales. Amaruzaman et al. posited that such governance layers exercised considerable independence driving state norms and rules that included families, firms, local governments, and civic associations. Accordingly, such polycentric systems did not stand in hierarchical relationships with each other and frequently engaged in self-organization, cooperation, and mutual adjustment.

Initially, Ostrom's (1990) polycentric approach revolved around Sidgwick's (1874) common good theory, wherein common goods included non-excludable but rivalrous in consumption, including public, private, and club goods (Wolsink, 2020). Wolsink documented that the polycentric models expanded application to outside resource conservation, including construction, operation, and maintenance of production infrastructures. Considering distributed solar energy as a common good and electrical access as a public good, polycentric environmental governance becomes a viable governing system option. Additionally, Kenya's historic climate governance challenges are glaringly compared to devasting historical events like the Tragedy of the Commons, wherein individualistic behaviors dominated over collective interests resulting in ravaging resource depletion. Thus, we used Ostrom's (2009) polycentric environmental governance approach as

the base framework and critical lens in our case study style systematic literature analysis of Kenya's climate change governance.

POLYCENTRIC FINDINGS: KENYA'S DECENTRALIZED CLIMATE GOVERNANCE

We found that Kenya's decentralized solar system technology comprised a
wide actor range, including national and local governments, private sector
companies, NGOs, and donors, facilitating an even wider consumer variation.
An all-inclusive broad-scale polycentric governance approach would behoove
Kenya's climate governance practices. Institutional arrangements operating at
other governance scales, like national government agencies, international organizations, NGOs, private associations, and self-organized local institutions,
play critical roles in any scheme of climate governance (Andersson & Ostrom,
2008). Mattijssen et al. (2018) explained that these polycentric constituents
facilitated research and practical activities, projected installation, forecast
technical support, lobbied, communicated strategies, and pursued political
engagements creating positive perceptions of decentralized solar technology
nationwide. We further found that Kenyan climate governance challenges remained complex and involved cooperation between and across multiple governance levels and units. We discovered all four key polycentric governance
elements already exist within Kenya's decentralized climate practices: a legal
framework, the private sector, capacity building, and political will.

THE LEGAL FRAMEWORK

The legal framework, encompassing the governmental component, was a
critical element of Ostrom's (2009) polycentric environmental governance approach because it provided an essential structure fundamental to supporting
the governing system, stipulated policy, and ensured all parties had avenues to
redress grievances. To that end, the two most significant components of the
legal framework were Kenya's Energy Act (2019) and the NCCAP (2018). The
Energy Act was a comprehensive description of Kenya's policy framework and
energy sector detailing energy sector policies related to renewable energy, energy financing, and energy efficiency. The NCCAP was Kenya's 5-year nationwide
plan guiding action responding to climate change from 2018 through 2022.

The Energy Act

The Energy Act (2019) outlined the duties and responsibilities of the
principal agencies governing the traditional energy infrastructure. The

Energy Act included financial support for electrical access, subsidies, regulations, penalties, grid management best practices, and financial support for community electrical access. The Energy Act defined the Ministry of Energy and Petroleum's duties and responsibilities in overseeing general electricity policy formulation. It confirmed the Energy Regulatory Authority (EPRA) regulation of the electrical sector, including decentralized solar systems. Oversight checks and balances promoted fair governance and minimized corruption. For example, Adwek et al. (2020) discussed how the Energy Act allocated the Kenya Energy and Petroleum Tribunal (KEPT) for conflict resolution and appeals against the EPRA, regulation disputes, and policy interventions.

The Energy Act (2019) applied to all decentralized solar system technology actors serving, participating, or cooperating with any aspect of Kenya's climate governance practices. The Energy Act charged the Rural Electrification and Renewable Energy Corporation ([REREC]; https://www.rerec.co.ke/) with overseeing the rural electrification program implementation and renewable energy master plan development focusing on county-specific needs and equity. In tandem, Kenya's Energy Act (2019) created the Renewable Energy Resource Advisory Committee to advise ongoing renewable development and licensing criteria, utilizing subsidies and funding growth mechanisms. Through the Energy Act (2019), Kenya's climate governance removed renewable energy-related import duties and established the zero-rated value-added tax (VAT), and renewed the Feed-in-Tariff Policy (FiT Policy, 2008) promoting renewable energy source investments. The Consolidated Energy Fund financially supported development initiatives and necessary infrastructure in line with The Energy Act.

Kenya's Energy Act (2019) also regulated distributor responsibilities, thereby limiting illegal connections through strictly defining licensed electricity distributors, electricians' requirements and responsibilities, and delineating violation fines and punishments. The Energy Act further established penalties assessed against electricity providers allocating appropriate compensation for consumers suffering irregular or poor electric service quality. While Kenya's Energy Act (2019) detailed comprehensive climate governance legislation, the NCCAP (2018) provisions strengthened Kenyan climate governance grid management and expansion.

National Climate Change Action Plan

The NCCAP (2018) focused on goal-meeting projects addressing grid operating efficiency and reducing transmission and distribution losses, including grid management and expansion mandating increased renewable energy resource development based on evolving dynamic conditions within the Kenyan environment. Thus, the NCCAP mainly consisted of project management groups responsible for real-time strategic planning, goal-setting, and implementation support aligning the Kenyan government's national climate goals with county-level endeavors and local strategy implementation while maintaining compliance with The Energy Act. Notable NCCAP projects included implementing solar technology in Kenya's primary air- and seaports, Kenya Vision 2030, and the Kenya Slum Electrification Program ([KSEP]; Bercegol & Monstadt, 2018). Subsequently, the NCCAP portion of Kenya's climate governance legal framework typically projected significant hands-on contact with the private sector.

THE PRIVATE SECTOR

The private sector encompassed corporations, NGOs, NPOs, and community partners as critical keys to action in Ostrom's (2009) polycentric environmental governance approach providing crucial decision-making, active support, and ensuring equitable solutions to end-users' needs. We found the private sector was instrumental in structuring decentralized solar systems development and successful SHS expansion in Kenya's cities. We noted that private sector actors included small-size enterprises, engineers, businesspeople, sales, advertising, technical entities, donors, and NPOs like the Kenya Renewable Energy Association (KEREA), supporting sector training, awareness creation, funding, and decentralized solar system activities. We found that the private sector in Kenya's decentralized solar systems focused on transformations associated with PAYG technology, increased equitable SHS access, accelerated communication, and information knowledge distribution, thus demonstrating Ostrom's (2009) assumption of community equity via the private sector element.

Thus, we found that the private sector was the driving force behind economic equity and expanded socio-technical innovations furthering decentralized solar systems in Kenya's decentralized climate governance practices. We noted that such private sector agents added innovation and boosted economic conditions within Kenya's multilevel climate governance practices. Thus, the private sector element of Kenya's climate governance practices drove equitable solutions, so low-income households gained increased electric access and shared intellectual property within existing frameworks essential for capacity building.

CAPACITY BUILDING

Capacity building represents the essential fuel needed to maintain momentum in the face of complex natural resources systems' continual social and climate change enabling risk mitigation in Ostrom's (2009) polycentric environmental governance approach. We revealed capacity-building components within Kenya's decentralized climate governance, including adaptive management, collective capacity, diversity, fresh ideas, continual experimentation, productivity incentives, accountability, and redundancy, as posited by Ostrom, among other momentum-building factors. Subsequently, we discovered that collaboration between the Kenyan national government, county-level administrators, city officials, the private sector, and consumers significantly increased Kenya's transition to decentralized solar by continually building capacity.

We discovered that the national government approach supported capacity building and encouraged knowledge transfer which minimized the barriers preventing the forward progress of decentralized solar system initiatives. We identified power back-ups, alternative energy distributions, and diffusion providing redundancy and control, thereby increasing capacity through reliability. We found that self-empowerment energy access prompted expanding strategies toward impoverished citizens' energy needs, thus, building capacity through diversity, fresh ideas, and productivity incentives. Consumer involvement increased problem-solving capacity, and sensitized service providers increased momentum toward equitable energy products and delivery. Thus, we uncovered capacity building through adaptive management, collective ability, diversity, fresh ideas, continual experimentation, and accountability.

POLITICAL WILL

Political will must exist within an effective polycentric environmental governance system, as without the power of political will, a polycentric governance system fails (Morrison et al., 2019). Political will consists of several power factors integrated into a polycentric fashion which is the spearhead of Ostrom's (2009) polycentric environmental governance approach. Overlapping political units achieve greater efficiency, production, provision, and political clout and operate across multiple political jurisdictions (Morrison et al., 2019); thus, political will in a polycentric system comprises powerful actors holding legitimate power and authority. On September 13, 2022, President Ruto's demand that humanity's fossil fuel addiction end expressed political will. However, Kenya's fifth president demonstrated legitimate power and authority by reaffirming Kenya's commitment to reach 100% clean energy by 2030 through active legislation.

While not at the Kenyan president's power level, various local community groups, civil society groups, NGOs, and government personnel hold varying degrees of legitimate power and authority, thus, political will. Therefore, we determined that Kenya's decentralized climate governance practices likely comprised the political will necessary for a practical polycentric environmental governance approach, as modeled by Ostrom (2009). Albeit, while we noted that numerous political power levels existed, with Kenya's climate governance holding legitimate power and authority ranging from the Ministry of Energy and Petroleum to county governments, city officials, and private sector contributors, we were unable to analyze the actionable extent of that political will effectively.

LESSONS LEARNED AND IMPLICATIONS

Thus, we learned that Kenya's current decentralized climate governance practices possessed all the needed elements of a successful polycentric environmental government: (a) a comprehensive, practical legal framework; (b) robust, diverse, multilevel private sector; (c) ample capacity building endeavors; and (d) legitimate authoritative political will. Albeit, we discerned that Kenya's current climate governance still needs significant improvement despite the polycentric system's substantial recent successes and knowledge gaps preventing an adequate assessment of political will.

We ascertained that the substantial support provided by Kenyan legislation provided for effective governance management through regulation, provision, and accountability; therefore, we purport to build on The Energy Act and NCCAP policies. We also recognized that cooperative projects demonstrated the profound strength of polycentric environmental governance cooperation; thus, we strongly recommend capitalizing on those polycentric strengths. We also realized the robust impact of Kenya's capacity-building incentives toward increasing solar energy production, distribution, and consumption; we advocate exploring further momentum-building avenues, especially toward equitability. We further discovered the significant socio-economic benefits of Kenya's decentralized energy governance in increasingly providing electricity access to impoverished and underserved residents; accordingly, we certainly advise honing decentralized energy governance.

Conversely, we gathered that Kenya's current climate governance system remains inadequate in reaching significant numbers of rural residents. We discovered that mismanagement remains a sizeable challenge to Kenya's effective climate governance practices, as exampled by continuing KPLC issues. We ascertained that outlying decision-making centers often lacked comprehensive integration with the more extensive polycentric governance system, as current semi-autonomous channels were insufficient to characterize a polycentric governance arrangement alone.

While Ostrom's (2009) polycentric approach provided an appropriate lens for this micro-level systematic case study literature review into Kenya's climate governance, the model failed to adequately account for the power dynamic role of political will. Morrison et al. (2019) defined power as the uneven capacity of different actors to influence the goals, processes, and outcomes of polycentric governance. While that description is more insightful than Ostrom's, it remains inadequate for understanding effectiveness, nuances, and the complete role of power in polycentric environmental governance.

While descriptions of Kenya's decentralized solar energy implementation and climate governance practice were helpful, empirical studies for solar energy provision management in Kenya do not yet exist. That empirical knowledge gap hindered understanding the bigger strategic governing picture associated with medium to long-term sustainable development goals. Thus, future in-depth empirical research must (a) better account for the political will, legitimate power, and authority element in the polycentric environmental governance approach, (b) involve multilevel human-environment interaction, and (c) take a broader, longitudinal approach. For example, Kitui's cooperative 2013 city planning endeavors with NGOs, academic organizations, and the local governments presented a suitable candidate for empirical study. Such research requires a more detailed approach that identifies stakeholders, participating institutions, relative participants, and interaction outcomes assuming a more efficient, effective, equitable, and sustainable solar energy provision governance system. The next steps for such empirical study include extensive field study and surveys gathering the requisite information.

CONCLUSION

We delimited our case study analysis of Kenya's climate governance to focus on Kenya's climate governance specific to the nation's decentralized solar systems within the polycentric environmental governance framework. The discussion focused on polycentric governance system enabling conditions, stopping short of a complete examination of broader Kenyan governance features. We acknowledged natural resource governance research as complex and cross-level, wherein most human-environment interactions take place at multiple scales. We recognized that polycentricity is not a binary condition. Recent scholars recommend that future studies treat polycentric governance as a longitudinal evolutionary study exploring nuances and complexities associated with governance arrangement resiliencies.

Segueing to future studies, we recommend applying the locally-led adaptation approach (LLA) of solar technology policy designed, monitored, and evaluated by local communities. While the LLA approach is conceptually consistent with polycentric governance, it better accounts for the power dynamic. Proponents posit that an LLA approach enables a shift in power to local stakeholders, resulting in more effective governance, increased equitability, and inclusive justice considerations.

REFERENCES

Adwek, G., Boxiong, S., Ndolo, P. O., Siagi, Z. O., Chepsaigutt, C., Kemunto, C. M., Arowo, M., Shimmon, J., Simiyu, P., & Yabo, A. C. (2020). The solar energy access in Kenya: A review focusing on pay-as-you-go solar home system. *Environment, Development, and Sustainability, 22*, 3897–3938. https://link.springer.com/article/10.1007/s10668-019-00372-x

Amaruzaman, S., Hoan, D. T., Catacutan, D., Leimona, B., & Malesu, M. (2022). Polycentric environmental governance to achieving SDG 16: Evidence from Southeast Asia and Eastern Africa. *Forests, 13*(1), 68. https://doi.org/10.3390/f13010068

Andersson, K. P., & Ostrom, E. (2008). Analyzing decentralized resource regimes from a polycentric perspective. *Policy Sciences, 41*(1), 71–93. https://www.doi.org/10.1007/s11077-007-9055-6

Baker, L. (2022). New frontiers of electricity capital: Energy access in sub-Saharan Africa. *New Political Economy, 28*(2), 206–222. https://www.doi.org/10.1080/13563467.2022.2084524

Bercegol, R. D., & Monstadt, J. (2018). The Kenya slum electrification program: Local politics of electricity networks in Kibera. *Energy Research & Social Science, 41*, 249–258. https://doi.org/10.1016/j.erss.2018.04.007

Boamah, F. (2020). Desirable or debatable? Putting Africa's decentralized solar energy futures in context. *Energy Research & Social Science, 62*. https://doi.org/10.1016/j.erss.2019.101390

Boamah, F., Williams, D. A., & Afful, J. (2021). Justifiable energy injustices? Exploring institutionalized corruption and electricity sector "problem-solving" in Ghana and Kenya. *Energy Research & Social Science, 73*. https://doi.org/10.1016/j.erss.2021.101914

Bonnell, C. J. (2021). *Energy policy in Kenya: Barriers to electrification and policy solutions for more equitable and sustainable grid development* [Undergraduate honors thesis, University of Massachusetts Amherst]. https://icons.cns.umass.edu/sites/default/files/student-work/2021-12/bonnellthesis.pdf

Buchanan, J. M., & Tullock, G. (1962). *The calculus of consent: Logical foundations of constitutional democracy* (Vol. 3). Ann Arbor Paperbacks.

Burger, A. (2019, October 7). *Kenya continues rollout of off-grid mini-grids*. Microgrid Knowledge. https://www.microgridknowledge.com/microgrids/remote/article/11429380/kenya-continues-rollout-of-off-grid-minigrids

Burger, C, & Weinmann, J. (2019, June 26). *The failure of privatization in the energy sector and why today's consumers are reclaiming power*. Renewable Energy World. https://www.renewableenergyworld.com/baseload/geothermal/the

-failure-of-privatization-in-the-energy-sector-and-why-todays-consumers-are
-reclaiming-power/#gref

Chengo, V., Mbeva, K., Atela, J., Byrne, R., Ockwell, D., & Tigabu, A. (2021). Kenya: Making mobile solar energy inclusive. In A. Ely (Ed.), *Transformative pathways to sustainability* (pp. 109–124). Routledge. https://www.doi.org/10.4324/9780429331930-10

Cross, J., & Neumark, T. (2021). Solar power and its discontents: Critiquing off-grid infrastructures of inclusion in East Africa. *Development and Change, 52*(4), 902–926. https://doi.org/10.1111/dech.12668

Cozzi, L., Contejean, A., Samantar, J., Dasgupta, A., Rouget, A., & Arboleya, L. (2020, November 20). *The COVID-19 crisis is reversing progress on energy access in Africa.* IEA [Report, license No. CC BY 4.0]. www.iea.org/articles/the-covid-19-crisis-is-reversing-progress-on-energy-access-in-Africa

Electricity Modernization Project. (2021, December 31). *Project snapshot.* Climate Investment Funds. https://www.cif.org/projects/electricity-modernization-project

Energy Act, No. 1 of 2019. (2019, March 14). *The Energy Act, 2019.* Kenya. https://www.epra.go.ke/download/the-energy-act-2019/

Environmental Protection Agency. (2023). *Inventory of U.S. greenhouse gas emissions and sinks: 1990–2021.* US Environmental Protection Agency, EPA 430-R-23-002. https://www.epa.gov/ghgemissions/inventory-us-greenhouse-gas-emissions-and-sinks-1990-2021

European Academics Science Advisory Council. (2018, June 15). *Commentary by the European Academics' Science Advisory Council (EASAC) on forest bioenergy and carbon neutrality.* https://easac.eu/fileadmin/PDF_s/reports_statements/Carbon_Neutrality/EASAC_commentary_on_Carbon_Neutrality_15_June_2018.pdf

Finke, S., Velenderić, M., Severengiz, S., Pankov, O., & Baum, C. (2022). Transition towards a full self-sufficiency through PV systems integration for sub-Saharan Africa: A technical approach for a smart blockchain-based mini-grid. *Renewable Energy and Environmental Sustainability, 7*, 8. https://www.doi.org/10.1051/rees/2021054

GOGLA, Lightening Global, World Bank Group, ESMAP, & Dahlberg Advisors. (2018, January). *Off-grid solar market trends report 2018: Executive summary.* International Finance Corporation. https://www.lightingglobal.org/wp-content/uploads/2018/02/2018_Off_Grid_Solar_Market_Trends_Report_Summary.pdf

GOGLA, Lightening Global, World Bank Group, ESMAP, Efficiency for Success, & Berenschot. (2022). *Global off-grid solar market report: Semi-annual sales and impact data: July–December 2022 public report.* GOGLA The Voice of the Off-Grid Solar Energy Industry. https://www.gogla.org/wp-content/uploads/2023/05/gogla_sales-and-impact-report-h2-2022.pdf

Haque, A. N., Lemanski, C., & de Groot, J. (2021). Why do low-income urban dwellers reject energy technologies? Exploring the socio-cultural acceptance of solar adoption in Mumbai and Cape Town. *Energy Research & Social Science, 74.* https://www.doi.org/10.1016/j.erss.2021.101954

Inshakov, O. V., Bogachkova, L. Y., & Popkova, E. G. (2019). The transformation of the global energy markets and the problem of ensuring the sustainability of their development. In O. V. Inshakov, A. O. Inshakova, & E. Popkova (Eds.), *Energy sector: A systemic analysis of economy, foreign trade, and legal regulations* (pp. 135–148). Springer. https://www.doi.org/10.1007/978-3-319-90966-0_10

International Energy Agency. (2023, June). *African Energy Outlook 2022* [World energy outlook special report, license: CC BY 4.0]. *UpToDate.* https://www.iea.org/reports/africa-energy-outlook-2022

International Energy Agency, IRENA, UNSD, World Bank, & WHO. (2020). *Tracking SDG 7: The energy progress report 2020* [CC BY-NC 3.0 IGO]. International Bank for Reconstruction and Development/The World Bank. https://www.irena.org/-/media/Files/IRENA/Agency/Publication/2020/May/SDG7 Tracking_Energy_Progress_2020.pdf

Kenya Off-Grid Solar Access Project. (2018, November 23). *KOSAP project components.* Ministry of Energy. https://www.kosap-fm.or.ke/kosap-project-components/

Kenya Power & Light Company. (2020, March 18). *Kenya power staff and several other suspects arrested over various crimes undermining power supply.* Kenya Power. https://www.kplc.co.ke/content/item/3484/kenya-power-staff-and-several-other-suspects-arrested-over-various-crimes-undermining-quality-power-supply

Kirshner, J., Baker, L., Smith, A., & Bulkeley, H. (2019). A regime in the making? Examining the geographies of solar PV electricity in Southern Africa. *Geoforum, 103,* 114–125. https://doi.org/10.1016/j.geoforum.2019.04.013

Koester, S., & Davis, S. (2018). Siting of wood pellet production facilities in environmental justice communities in the Southeastern United States. *Environmental Justice, 11*(2). https://doi.org/10.1089/env.2017.0025

Macrinici, D., Cartofeanu, C., & Gao, S. (2018). Smart contract applications within blockchain technology: A systematic mapping study. *Telematics and Informatics, 35*(8), 2337–2354. https://www.doi.org/10.1016/j.tele.2018.10.004

Mattijssen, T., Buijs, A, Elands, B., & Arts, B. (2018). The 'green' and 'self' in green self-governance—A study of 264 green space initiatives by citizens. *Journal of Environmental Policy & Planning, 20*(1), 96–113. https://doi.org/10.1080/15 23908X.2017.1322945

Mburugu, P. N., & Gikonyo, N. (2019). Factors influencing utilization of solar energy in Kenya industries: The case of tea processing factories in Meru County. *International Academic Journal of Information Sciences and Project Management, 3*(4), 304–326. https://www.iajournals.org/articles/iajispm_v3_i4_304_326 .pdf

Michoud, B., & Hafner, M. (2021). *Financing clean energy access in Sub-Saharan Africa: Risk mitigation strategies and innovative financing structures.* Springer Briefs in Energy. https://link.springer.com/book/10.1007/978-3-030-75829-5

Moner-Girona, M., Kakoulaki, G., Falchetta, G., Weiss, D. J., & Taylor, N. (2021). Achieving universal electrification of rural healthcare facilities in sub-Saharan Africa with decentralized renewable energy technologies. *Joule, 5*(10), 2687–2714. https://www.doi.org/10.1016/j.joule.2021.09.010

Morrison, T. H., Adger, W. N., Brown, K., Lemos, M. C., Huitema, D., Phelps, J, Evans, L., Cohen, P., Song, A. M., Turner, R., Quinn, T., & Hughes, T. P. (2019). The black box of power in polycentric environmental governance. *Global Environmental Change, 57.* https://doi.org/10.1016/j.gloenvcha.2019.101934

Mukisa, N., Manitisa, M. S., Nduhuura, P., Tugume, E., & Chalwe, C. K. (2022). Solar home systems adoption in Sub-Saharan African countries: Household economic and environmental benefits assessment. *Renewable Energy, 189*, 836–852. https://doi.org/10.1016/j.renene.2022.03.029

Mwendwa, D. M., Tchouambe, J., Hu, E., Lanza, M. F., Brener, A. B., Hwang, G., Khanfar, L., Leonard, A. Hirmer, S., & McCulloch, M. (2022). Spatial data starter kit for OnSSET energy planning in Kitui County, Kenya. *Data in Brief, 45*. https://doi.org/10.1016/j.dib.2022.108691

National Climate Change Action Plan. (2018). Ministry of Environment and Forestry. https://climate-laws.org/documents/national-climate-change-action-plan-nccap-2018-2022-volume-i_7d48

National Resources Defense Council. (2020, May 8). *Our forests aren't fuel.* NRDC. https://www.nrdc.org/resources/our-forests-arent-fuel

Organization for Economic Cooperation and Development & Sahel West African Club. (2020, February 7). *Africa's urbanisation dynamics 2020: Africapolis, mapping a new urban geography.* West African Studies. OECD Publishing. https://doi.org/10.1787/b6bccb81-en

Olson, M. (1965). *The logic of collective action: Public goods and the theory of groups.* Harvard Economic Studies.

Ostrom, E. (1990). *Governing the Commons: The evolution of institutions for collective action.* Cambridge University Press. https://doi.org/10.1017/CBO9780511807763

Ostrom, E. (2009). *A polycentric approach for coping with climate change* (World Bank policy research working paper No. 5095). http://dx.doi.org/10.2139/ssrn.1934353

Pedersen, M. B., & Nygaard, I. (2018). System building in the Kenyan electrification regime: The case of private solar mini-grid development. *Energy Research & Social Science, 42*, 211–223. https://www.doi.org/10.1016/j.erss.2018.03.010

Pillot, B., Muselli, M., Poggi, P., & Dias, J. B. (2019). Historical trends in global energy policy and renewable power system issues in Sub-Saharan Africa: The case of solar PV. *Energy policy, 127*, 113–124. https://www.doi.org/10.1016/j.enpol.2018.11.049

Seii, J. (2018, May 31). *Power struggles: Unmasking the thieves behind the KPLC heist.* Elephant Politics. https://www.theelephant.info/features/2018/05/31/power-struggles-unmasking-the-thieves-behind-the-kplc-heist/

Shirley, R., Lee, C. J., Njoroge, H. N., Odera, S., Mwanzia, P. K., Malo, I., & Dipo-Salami, Y. (2019). Powering jobs: The employment footprint of decentralized renewable energy technologies in sub-Saharan Africa. *Journal of Sustainability Research, 2*(1). https://doi.org/10.20900/jsr20200001

Sidgwick, H. (1874). *The methods of ethics.* Macmillan & Co.

Stashwick, S. (2019, July 25). *How the biomass industry sent "sustainability" up in smoke.* NRDC. https://www.nrdc.org/bio/sasha-stashwick/how-biomass-industry-sent-sustainability-smoke

Sterman, J. D., Siegel, L., & Rooney-Varga, J. N. (2018). Does replacing coal with wood lower CO_2 emissions? Dynamic life cycle analysis of wood bioenergy. *Environmental Research Letters, 13*(1). https://www.doi.org/10.1088/1748-9326/aaa512

Sustainable Energy for All. (2022, July 28). *Sustainable energy for all annual report 2021.* SE4ALL. https://www.seforall.org/publications/sustainable-energy-for-all-annual-report-2021

Takase, M., Kipkoech, R., & Essandoh, P. K. (2021). A comprehensive review of energy scenario and sustainable energy in Kenya. *Fuel Communications, 7.* https://www.doi.org/10.1016/j.jfueco.2021.100015

Ulsrud, K. (2020). Access to electricity for all and the role of decentralized solar power in sub-Saharan Africa. *Norwegian Journal of Geography, 74*(1), 54–63. https://www.doi.org/10.1080/00291951.2020.1736145

UN Sustainable Development Goal 7. (2015, September). *7 affordable and clean energy.* https://www.un.org/sustainabledevelopment/energy/

Wagner, N., Rieger, M., Bedi, A. S., Vermeulen, J., & Demena, B. A. (2021). The impact of off-grid solar home systems in Kenya on energy consumption and expenditures. *Energy Economics, 99.* https://www.doi.org/10.1016/j.eneco.2021.105314

Winklmaier, J., Santos, S. A. B., & Trenkle, T. (2020). Economic development of rural communities in sub-Saharan Africa through decentralized energy-water-food systems. In N. Edomah (Ed.), *Regional development in Africa* (Chapter 7). IntechOpen. https://www.doi.org/10.5772/intechopen.90424

Winther, T., Ulsrud, K., & Saini, A. (2018). Solar powered electricity access: Implications for women's empowerment in rural Kenya. *Energy Research & Social Science, 44,* 61–74. https://www.doi.org/10.1016/j.erss.2018.04.017

Wolsink, M. (2020). Distributed energy systems as common goods: Socio-political acceptance of renewables in intelligent microgrids. *Renewable and Sustainable Energy Reviews, 127.* https://doi.org/10.1016/j.rser.2020.109841

World Bank. (2023). *DataBank: Population estimates and projections.* https://databank.worldbank.org/source/health-nutrition-and-population-statistics:-population-estimates-and-projections

CHAPTER 18

CLIMATE GOVERNANCE

Stakeholders Adapting
to Climate Change

N. S. Matsiliza
University of Fort Hare

ABSTRACT

In the post-apartheid era, climate change has been a challenge in South Africa. Hence, strategies are needed to mitigate environmental risks such as global warming, climate change, financial crisis, and post-COVID-19 adaptations. Even though there are significant deliberate commitments to operationalize the United Nations 2030 Agenda for Sustainable Development, the South African government still grapple with environmental policy reforms and legacy challenges. South Africa also faces severe climate change observed through the rise of temperature by 1.5 times higher than the expected global average of 0.65 degrees C. At the local level, climate change affects the quality of life, biodiversity and the ecosystem. The effects of climate change are observed through severe rainfalls that cause flooding, emerging sinkholes, drought, stunted forest growth, and changes in usual weather in most areas. Government and non-governmental organizations are mitigating and responding to climate change using advocacy, environmental justice strategies

Climate Governance in International and Comparative Perspective, pages 423–445
Copyright © 2024 by Information Age Publishing
www.infoagepub.com
423

and other shared activities organized as active movements. Stakeholders are willing to play active roles, move towards positive responsiveness, address climate change challenges, and frame the future success of the next generation in policy development and governance. Therefore, this chapter aims to explore the involvement of various stakeholders organizations and institutions in mitigating climate change in South Africa. In response to the aim, this chapter explores sustainable development, the impact of climate change, the responsiveness of stakeholders in mitigating and adapting to climate change, policy implications, conclusion and recommendations.

Despite the current environmental policies and adaptability strategies in South Africa, climate change (CC) still impacts living organisms and society. A paucity of literature is directed towards stakeholders' inclusion in mitigating climate change using a participatory approach. The current State of climate change is beyond what policies can provide regarding mechanisms to serve justice to those directly abusing the environment by emitting polluted gasses to destroy natural resources. Climate change is regarded as a disaster in South Africa. Hence there are policies to regulate complex issues and capability constraints during disasters (Nemakonde et al., 2021).

Stakeholders require an integrated and inclusive approach to address climate change in areas where vulnerable groups struggle to cope with life due to environmental degradation. Organizations like the United Nations (UN) have long been leading and mentoring other countries to comply with sustainable development's goal to reduce carbon emissions. Compliance and active involvement of nations on climate change can ease tensions in international relations and provide a better pace to deliver sustainable economic growth for the planet and improve citizens' quality of life (Bornemann & Weiland, 2021). For successful integration of policies with stakeholders, there is a need for strong relations among the stakeholders to be uncovered among policymakers and bureaucrats, consumers, industry and participatory forums and activists. This chapter will answer the critical research question, "What is stakeholders' involvement in climate change?"

Sustainable development is adopted to align its goals with the proposed strategies to address environmental challenges at the local level by building an inclusive community empowerment approach. Ozili (2022) believes sustainable development is relevant for research use since it can provide better solutions for use by policymakers and analysts. Scholars attended to diverse strategies to adapt and mitigate government failures of organizations and governments to meet the Sustainable Development Goals aiming at providing competent civil servants to advance the dialogues for climate change with other stakeholders through adaptive measures (Bornemann & Weiland, 2021). While civil servants are at the centre of adequate support of government ministers and other executives, it is detrimental for them to pursue the right path that will include all stakeholders. Data for this chapter is collected from qualitative secondary sources such as books and commissioned reports on various topics of environmental governance, climate change and sustainable development.

This chapter aimed to analyse the impact of climate change and unpack the role of stakeholders in addressing climate change issues in South Africa. Even though this chapter uses a South African case, there is recognition of international organisations that are involved in addressing and mitigating climate change. The secondary qualitative data used in this chapter is drawn from various sources such as books, and articles from accredited journals, reports of commissioned studies, and policy documents. The literature is reviewed from various sources focusing on knowledge areas such as environmental issues, climate change, governance, and public policy. Therefore, this chapter focuses on sustainable development and climate change, the impact of climate change, stakeholders' responsiveness to climate change, policy implications, conclusions and recommendations. The outcomes of this chapter will contribute to existing knowledge areas of governance and sustainable governance.

CONCEPTUAL AND THEORETICAL FRAMEWORKS

Climate Governance

The concept of climate governance is complex and involves the progress of nations, as well as the welfare of nature and people, which are all related to climate change governance. The negotiation by member states and parties in decision-making processes related to climate change that affect human well-being and determine the course of climate action are the main emphases of climate governance. Climate governance, according to political ecology and environmental policy, is the corrective action, diplomacy, and processes "meant to steer social systems towards mitigating and preventing, or adapting to the risks posed by climate change (Cote & Nightingale, 2012). The variety of political and social scientific traditions (such as development management, comparative politics, economics, and multilevel governance) involved in packaging analysis and conceptualising climate governance at diverse levels and arenas leads to the tough pronunciation of conclusive interpretation.

Scholars in diverse fields of study such as economics, anthropology, geography and business studies are paying attention to climate governance (see Figure 18.1). Although the management of the networks and forums on the global climate system is complex, it can add value if nations can upskill their capacity to manage climate change effectively since it is such an important issue in the world.

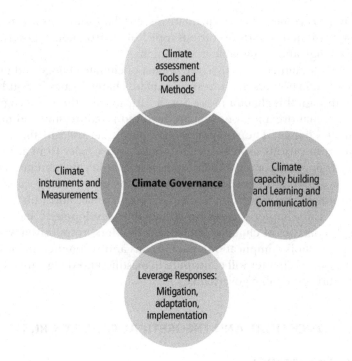

Figure 18.1   Governance conceptual framework.

Some nations are successful in managing and preventing some effects of climate change, but it is still necessary to form groups and networks to participate in the wider dialogues of climate governance and build string group mechanisms to regulate the impacts of climate change and address common difficulties, such as the ongoing refinement of scientific knowledge about our planet issues and global climate systems. This can add value to disseminating the knowledge on climate governance to the public and policymakers. Countries like South Africa can take recognisance of climate governance when they also participate at the global level. There is also a pressing need to address this issue at diverse governance levels; the Intergovernmental Panel on Climate Change (IPCC) can protect the international community—which has limited time to act to retain rising global temperatures at safe levels. Three common pillars—mitigation, adaptation, and means of implementation—form the framework for contemporary global climate governance.

Climate Change

Climate change has been a standing issue in South Africa's sustainable development for decades. In the post-apartheid era, the government formalized sustainable development and various policies and strategies to respond to environmental challenges associated with climate change. The genesis of sustainable development (SD) can be traced back to 1972 at the Stockholm Conference to discuss the future of the environment and the Earth's planet. The Stockholm Conference left a legacy of a complex network of global treaties, initiatives, agreements, and programmes aiming to address prevailing environmental problems and treat the lives of living organisms and humans (Farah, 2015). Several activities are official and intergovernmental, less formal and concern multiple stakeholders. Based on the advocacy of the SD, environmental issues are inseparable from international action. Diverse nations are battling with compliance with the UN Environment Assembly agreements on the Agenda 2030 ecological protection and have just begun discussions on a new plastic treaty (Janowski et al., 2018). The Basel Convention and partnerships were taking action to stem the plastic flood.

Sustainable Development

The DNA of Figure 18.2 on sustainable development (SD) is mainly on the interconnectivity of SD and the environmental factors (social, economic, political and physical). Ahmed et al. (2019) note the systematic

Figure 18.2 Sustainable development goals.

factors that affect sustainability and the environment with evidence from water stress and bioactivity, with the seasonal shift from spring to other seasons indicating increasing levels of phenolic compounds. Diverse scholars ushered on sustainable development with a shared focus on its nature of development that addresses present demands without jeopardizing future resources (Di Baldassarre et al., 2009).

A comprehensive action plan among partners committed to establishing a worldwide sustainable network to address climate change was met at the Rio de Janeiro Agenda 21 Earth Summit in June 1992. These stakeholders summited to create partnerships for sustainable development focusing on changing human lives and the protection of endangered species and the Earth's planet in more than 178 nations. Member States unanimously ratified the Millennium Declaration in September 2000 at UN Headquarters in New York. This summit developed eight millennium development goals (MDGs) to end extreme poverty by 2015.

South Africa's emerging movement can be traced through the genesis of the Johannesburg Declaration on Sustainable Development adopted in 2002 and the environmental policies that responded to the call for environmental protection and poverty eradication. In June 2012, at the Rio+20 United Nations Conference on Sustainable Development in Rio de Janeiro, Brazil, partners espoused the product document titled "The Future We Want," which they decided further resulted in the endorsement of the Sustainable Development Goals (SDG) to which a UN High-Level Political Forum was anchored on Sustainable Development. The SDGs are the main themes to be observed and complied with by member states in different countries, with the leading advocate of the 30-member Open Working Group of the General Assembly of 2013. to develop an SDG proposal. The General Assembly formed the post-2015 development agenda negotiations in January 2015, which passed some global policies and significant agreements such as:

- Framework for Sendai on Disaster Risk Reduction (March 2015)
- Addis Ababa Action Agenda on Development Funding (July 2015)

The UN SD Conference in New York in September 2015 further stamped the endorsement of the 2030 Agenda for Sustainable Development and its 17 Goals as changing the world. Another success was the Paris Climate Change Accord (December 2015). The UN now uses its annual High-Level Political Conference on Sustainable Development as its main forum for monitoring and reviewing the SDGs.

The broad goals of SDG 13 are impressive, even though other member states are reluctant to speed up the process of stopping gas emissions. According to the UN Report (2022), the increasing impact of climate change is experienced worldwide, with greenhouse gas emissions unambiguously wished to fall. According to the Emission Gap Report (EGR) (2022), the gap window is increasing instead of closing these calls for a rapid response to transformation in many societies that are failing the Paris Agreement of reducing to 1.5°C in place. There is a need for an urgent system-wide transformation to avoid disaster. South Africa also falls short of the bench-marked standards to reduce emissions. The following section will elaborate on the impact of climate change. The global climate is regarded as a global risk as it has reached a crisis beyond the single methods of nations can fit for the response, while it impacts the poor and marginalised people. The World Bank's response to the funding of climate change response strategies unveiled the Climate Change Action Plan, which will increase the funding for adaptation and mitigation in diverse nations. The United Nations also calls for nations to invest and contribute to multilateral funds and join the broader responses to climate change. Responses on adaptation to climate change must include communities and local networks in discussions and empowerment programmes aiming at mitigating climate change. Climate adaptation should cover all mechanisms aiming at exposing communities to be resilient to the impacts of climate change.

IMPACT OF CLIMATE CHANGE ON THE SOCIETY

Rising Temperatures

As greenhouse gas concentrations increase, so does the average global surface temperature. The decade from 2011 to 2020 was the warmest on re-cord. Since the 1980s, every decade has been more giving than the one before it. Nearly everywhere on land, there are more hot days and heat waves. Higher temperatures increase heat-related illnesses and make working out-side more difficult. When the weather is hotter, wildfires start more easily and spread faster. According to Wallace-Wells (2018), the Arctic warmed at least twice as quickly as the rest of the planet. Rising temperatures can cause ice caps to melt and affect the rising sea levels around the world. Rising temperature is a serious concern because these glaciers can cause flooding and beach erosion, severely damaging infrastructure along the coastline. As a result of natural and manufactured issues like burning fossil fuels, defor-estation, and industrial activities produce greenhouse gases such as meth-ane, carbon dioxide, and nitrous oxide.

The temperature shift is characterized by a long-term shift in global weather patterns to unpredictable temperatures (Mattern, 2017). These gases trap heat in the planet's atmosphere, raising the temperature and causing changes in weather patterns, rising sea levels, glacier melt, and oth-er environmental effects. Biodiversity, agriculture, water resources, coastal and low-lying areas, public health, and economies are all negatively impacted by climate change (Anderson et al., 2016).

Kweku et al. (2018) noted that climate change (CC) might prolong rising water demands and water scarcity brought on by population growth and economic development by modifying rainfall volume and variability. The imbalance between water demand and supply capacity in many places heightens competition among different users. Capolupo et al. (2023) note changes that affect water supply, especially regarding issues with the sustainable development of CC adaptation and resilience strategies. Water adaptation patterns later affect water access patterns, especially on its domestic use.

Scholars note climate change's impact on metal gasses that can alleviate drought and soil degradation (Mukherjee et al., 2018). The continued temperature changes affect environmental and soil conditions, such as the variation and acidification that can cause the mobilization of heavy metals from the solid to the liquid phase, and it can further allow contamination of the plants grown in these soils. The level of bioaccumulation can affect the food chain's contamination, known as biomagnification. The contamination can further affect the produce industry, like farmers' corporations that depend on crops' supply to barter them to make a living.

Intense drought exposes temperature changes as a natural disaster. Drought can be extreme through socioeconomic impacts and multiple eco-hydrological effects, including water scarcity, increased risk of wildlife, and loss of livestock and crops (Mukherjee et al., 2018). Due to the high intensity of climate change in Southern Africa, more drought patterns are speculated to occur due to global warming. As we move closer to the end of the century, there will still be more variety of drought in the sea levels and on the surface. These factors mainly hinder the region's climatology, making the speculation of temperature patterns complex.

Greenhouse gas generation and climate change are related. More greenhouses form in the atmosphere as the planet's mean temperature rises. The planet's expansion and warming eventually have an impact on many aspects of wildlife, ocean temperature, agriculture, melting sea ice and changes to marine life, increasing sea levels, and extreme weather events including drought and heavy rain. Burning fossil fuels releases greenhouse gas emissions that cover the earth, trapping solar heat and raising temperatures (Coelho et al., 2015). Examples of greenhouse gas emissions causing climate change include carbon dioxide and methane. The greenhouse gas carbon dioxide is one of them. One oxygen atom is coupled to each side of the one carbon atom involved (Fekete et al., 2021). Once the carbon dioxide molecule's atoms are tightly linked, the molecule can absorb infrared radiation and start vibrating. The radiation will eventually be released by the vibrating molecule and most likely be absorbed by a different greenhouse gas molecule (Smith, 2018). The heat is maintained on the surface through this cycle of absorption, emission, and absorption, successfully shielding it from the cold of space. The greenhouse gas carbon dioxide is one of them. Once the carbon dioxide molecule's atoms are tightly linked, the molecule can absorb infrared radiation and start vibrating. air pollution still exists today.

Impact of Climate Change on Biodiversity and Livelihood

Climate change affects the movement of living organisms and biodiversity when habitat loss or fragmentation occurs (Dejene, 2018). The rising temperatures disturb the ecosystem patterns and force certain species to relocate to other areas. The migration is opportunistic since living organisms, animals, and humans migrate to suitable positions that might not be ideal for their habitat. Moreover, when there is moisture in some regions, the land-atmosphere feedback progressions aggravate the condition by cumulative atmospheric temperatures, thus growing the atmospheric demand for moisture due to this heating of the land (Dirmeyer et al., 2021). The heating of land degenerates the growth of some plants, and living microorganisms that need to survive, thereby foremost enlarging the drying and heating of land surface concurrently, the impact of which is often alarming (such as wildfire risk) and further affects the propagation of drought under climate change. Since Africa is wealthy in natural resources wildlife has attracted many species. Some species are uncomfortable living in some areas due to climate change which affects biodiversity composed of species such as birds, plants, and animals.

These natural inhabitants need to be protected by the government and ordinary citizens since South Africa attracts tourists and other people to visit South Africa's natural attractions places due to its rich biodiversity heritage, characterized by its tropical climate and rich forests, marine and freshwater, wetlands, maritime, coral reef, and Mining land (Leshalagae & Buthelezi-Dube, 2023). Since tourism has the potential to induce the economy, it is significant for all stakeholders to advocate for sustainability and mitigation of climate change. The natural bio-diversified and protected areas provide an exciting outlook for tourists; therefore, all the stakeholders need to engage in advocacy and other initiatives to address the slow degrading of biodiversity and natural protected areas due to climate change. All the risks associated with climate change are threatening the future wealth of the earth planet, environmental harmony, and the loss of ecosystem function to service well.

Domestic Impact and Livelihoods

Climate change causes health risks to human beings that can lead to heat exhaustion, respiratory problems, and dehydration. Furthermore, it can stimulate health problems that can be opportunistic for spreading infectious diseases like dengue fever and malaria. Literature notes that climate change's effects on human health are long and short-term (Ha, 2022). The impact on health can be demonstrated by its ability to retard development on the ability to grow food, settlements and housing, safety and work. In some areas, people are more vulnerable due to natural opportunistic effects, like people living in informal settlements needing to be fully protected due to their poor housing infrastructure. Some already more vulnerable to climate impacts include those living in small island nations and other developing countries (Ziervogel et al., 2014). Conditions like sea-level rise and saltwater intrusion have advanced to the point where whole communities have had to relocate, and protracted droughts are putting people at risk of famine. The number of "climate refugees" is expected to rise.

STAKEHOLDERS AND NETWORKS RESPONDING TO CLIMATE CHANGE

International Experience

Diverse and efficient stakeholder participation at various geographic and governance levels is necessary to effectively respond to the rising impacts of climate change. Stakeholder involvement in climate change adaptation has been recognized as a supported strategy that encourages social learning, depolarizes beliefs, and promotes group action. However, the definition of stakeholder engagement involves organizations and institutions interested and involved in climate governance. However, at the International level, the stakeholders are more involved than at the national level, and the options for evaluating measures are few. This study uses social network analysis (SNA) to examine the relationship between stakeholder learning and changes in views of climate change, as well as the social ties among stakeholders that result from participation.

International organizations like the United Nations (UN) and the World Bank have been at the forefront of supporting victims of intended environmental degradation and climate change and sharing information on climate change adaptations. The UN offers support by providing funding, information through their expertise, and support to governments, the business sector, and communities where the problem is. Their scope depends on their available resources to offer support. Organizations like the UN first offer rescue support and allocate resources to member states and other countries in a disaster. Suiseeya et al. (2021) note the emerging social relations to mitigate climate change, which constitutes global climate governance, conflict management, decision-making, cooperation, service exchange, accommodation, commodities, and others. They aim to share ideas and cooperation, seeking trust to succeed. These relations are sensitive to compliance with the agreements on sustainability among the partners.

There is a strong movement of non-organizational organizations that demonstrated positive mitigating strategies to address the challenges of climate change in South Africa and globally. These non-organizations (NGOs) are willing to take the matter through advocacy forums where they can engage with governments to limit opportunistic factors that trigger dangers instigating climate change. They form social networks for communities to learn about nature conservation and climate change. A case in point is Birdlife South Africa, an organization that aims to protect the habitat of birds' conservation and their biodiversity. Birdlife South Africa uses its scientifically based programmes to preserve nature in specific Provinces by conserving estuaries, the Karoo and the forests. According to Leshalagae and Buthelezi-Dube (2023), the changing role of NGOs is based on their role to mitigate climate change by encouraging communities to adapt their lifestyles and join the more significant climate change movement. Through programmes like BRACED (building resilience, adaptation, and climate extremes and disasters) which is an NGO that is making progress in intermediary roles, assisting the users in having an understanding and worth of climate information to assist them in their decision-making processes (Leshalagae & Buthelezi-Dube, 2023).

In South and Southeast Asia, in the African Sahel and its neighbouring nations, and in South and Southeast Asia, BRACED is assisting communities in building their resiliency to climate extremes. BRACED aims to have an impact on practices and policies at the local, national, and international levels to enhance the integration of disaster risk reduction and climate adaptation technologies into development approaches.

However, it has surfaced that the risk where the NGOs can misuse the opportunity of leading the communities through uncoordinated efforts to move into the climate services sector. At the same time, national meteorological and hydrological agencies need to upskill their efforts and be better user-friendly.

The NGOs have been applauded for their participatory role globally, in Africa and locally. They provide experts to donate and to offer support as experts and as donors by first acknowledging the government's role, engaging the government to comply with the reduction of factors triggering climate change and depletion of the ozone layer, and adopting the adapting methods towards coping with climate change.

Global non-governmental organisations are known for their activism on issues related to emissions and pollution. For instance, the mission of 350. org, established in 2008 by author and activist Bill McKibben and a group of college friends, advocates for nations to keep the world's carbon dioxide concentration below 350 parts per million. To halt oil and gas production and switch entirely to renewable energy, they are mobilizing folks world-wide. There is also the Asia Pacific Adaptation Network (APAN), Asia and Pacific Region which focuses on developing a resilience strategy against the consequences of climate change; APAN supports adaptation across Asia and the Pacific Region by offering mechanisms for planning, financing, and building adaptive societies. The point in this case is the publications on Coastal Zone Management, which aim to reduce damages created at the sea-level rise in the coastal communities.

In addition, the Climate Alliance International is a vital nonprofit organization comprising regional and State governments, district municipalities and non-governmental organizations (NGOs), and other organizations; forums and Climate Alliance is one of the major European city networks dedicated to climate action. This large alliance seizes the opportunity to promote advocacy and actions to discourage the escalation of climate change costs and destructions and encourage compliance by European municipalities and the Amazon River basin. This network is also setting a good lesson for other structures willing to comply with sustainability commitments on the reduction of emissions and other types of pollution that are retarding the progress in the plight to address climate change. Also, the Climate Action Network (CAN), International CAN, is a global network of more than 1,300 environmental NGOs. This network is situated at the regional hubs in West Africa, South Asia, Latin America, and Eastern Europe, aiming to promote governmental and individual action on climate change. The CAN engages groups on diversity issues such as science policy impacts, agriculture, and technology.

The South African Government's Response

The government in South Africa endorsed several policies to respond to the environmental degradation and other effects of climate change. The Constitution of the Republic of South Africa 1996 (Constitution) has a statutory frame of reference in Section 24, which sets out the right to an environment that is not harmful to health or well-being and calls on the government to take the legislative approach to the following issues:

- To cease pollution and ecological deprivation.
- Promote conservation.
- Defend ecologically sustainable development.
- Preserve natural resources whilst advocating justifiable economic and social development.

The government in the post-apartheid era paid much attention to ending the green apartheid by formalizing and adopting several policies that can address all the ills and challenges associated with the environment to protect the citizens, industry and biodiversity without any favor and prejudice. It should be noted that law enforcement institutions have attended many cases dealing with the violation of this act and some due to non-compliance. There are also critical pieces of legislation (and related regulations) that have been enacted, which include the Hazardous Substances Act No. 15 of 1989.

Over and above the environmental regulation, there is the Environmental Management Inspectorate (Inspectorate) that is composed of the environmental enforcement officials from the three spheres of government and departments, who are chosen as either (*Chapter 7, National Environmental Management Act No. 107 of 1998 (NEMA)*, the Environmental Management Inspectors (EMIs) by the Minister of Environmental Affairs (and, in the case of Mining, by the Minister of Mineral Resources).

The members of a Provincial Executive Council have a critical role to play based on vested powers, which refers to the investigation of the violation of the National Environmental Act and other matters relating to climate change and disasters. They are known as green scorpions since they investigate incidental witnesses, inspect, and eliminate articles, shoot photographs and audio-visual recordings, detect samples, and remove waste. Inspectors investigate issues related to non-compliance on environmental issues by inspecting target areas and reporting premises to determine compliance with legislation and seek evidence of unlawful activity. Lastly, they can enforce legal prosecution after examining premises, vessels, vehicles, containers, aircraft, and pack animals, establishing roadblocks, seizing evidence, and making arrests. The administrate of these inspectors is based on compliance notices and admission of guilt fines.

Notably, EMIs are vested with search and seizure powers and can seize evidence and contraband associated with criminal activity. They can also choose the South African Police Service to bring the law brokers into books regarding an offence under NEMA or a specific environmental management act (section 310, NEMA). Therefore, EMIs function closely with the South African Police Services. EMIs have no power for prosecution; they need to hand over the cases by reporting the law brokers to the SAPS. Relevant courts have powers to prosecute cases; some cases can be handed over to the National Prosecuting Authority for prosecution. Aside from their policing powers, EMIs play an essential role at an administrative level, ensuring that individuals and entities comply by issuing compliance notices and fines. The relevant regulators also enforce the environmental requirements.

The South African government offers a climate change response strategy (CCRS) through the Department of Environmental Affairs and Tourism (DEAT) and the Department of Minerals and Energy (DME). According to Pasquini et al. (2015), as a result of the applied nature of adaptation, governments at all levels, development organizations, NGOs, businesses, households, communities, land users, and other groups that will need to respond to the negative impacts of a changing climate have begun to bear an increasing share of the responsibility for action. In this decade, the responsibility of municipalities to steer climate change adaptation has increased in importance. To address climate change adaptation, local governments, which are often the level of government most directly responsible for planning and management at the urban and local scale, are crucial. In the City of Cape Town, there is inter-departmental collaboration, for example, a Flood and Storms Planning Task Team (a multi-disciplinary team of city and supporting external agencies) that was established to accomplish flood risk reduction and response measures, established in 2008. In 2009, a Climate Change Think Tank was established to guide, advise, and guide the implementation of local-level climate change policies, programs, and interventions.

They also work with various organisations and stakeholders to network and share information while playing a role in addressing climate change. The following section will provide an analysis of how they are involved in adapting to climate change.

Institutions and Networks

Few organizations are part of a network responsible and have interests in climate change adaptations in South Africa.

The South African National Botanical Institute

The South African National Botanical Institute's (SANBI) role in responding to climate change is to conduct research by focusing mainly on ecological impacts, predictions, and the threat to biodiversity like the lives of birds. They explore the effects of climate change on South African endemic birds, which are important for conservation efforts worldwide, using the recently developed ensemble modelling approach. They evaluate the anticipated range shifts in the network of Important Bird Areas in southern Africa. Important Bird Areas (IBAs) are a network of locations around the world that BirdLife International has identified as being crucial for the preservation of the avifauna (Ziervogel et al., 2014). IBAs are recognized based on the presence of globally threatened, restricted-range, or biome-restricted bird species, as well as the presence of sizable bird species congregations. IBAs are made to overlap as much as feasible with the current reserve network.

South African National Parks (SANParks)

The SANParks also take stork in the game by identifying parks' threats and developing a plan to adapt to change (Mathivha et al., 2017). The role of SANParks includes capacity building and training gamers and park managers to ensure national parks incorporate climate change strategies into their management plans. South African National Parks (SANParks) have a beneficial impact on the economy of South Africa in addition to helping to preserve the country's natural environment. The tourist business is made up of a variety of distinct but connected service sectors, including lodging and catering, the production of food and beverages, transportation, as well as leisure and other support services. The Kruger National Park (KNP) is the single largest and most significant tourism product in the provinces of Limpopo and Mpumalanga, and visitor spending accounts for an estimated 5.72% of all tourism-related spending in the region. Therefore, it acts as a centre to distribute information to other parks about climate change and issues of droughts on how they can affect tourism in South Africa.

National Business Institute (NBI)

In addition, the NBI is involved in mitigation and energy efficiency about sustainable development (Annual Report 2020/2021) through the Energy Efficiency Accord. This Accord recognizes energy usage, which leads to greenhouse gas emissions. The Accord is significant since now industries and businesses in South Africa are now taking stock in integrating sustainability and adaptation to climate change in their core business and activities (Department of Minerals and Energy, 2005). In the case of an adaptation, there needs to be more progress in some institutions (and businesses in general), especially on limiting gas emissions. There is hope in the participation of the South African National Biodiversity Institute (SANBI), which is encouraged to leverage the advocacy of climate change in biodiversity and encourage the involvement in adaptation of other groups in South Africa. Bulkeley (2023) asserts that governing climate change has become one of the pressing issues in today's politics and government circles. The main problems relating to climate change governance are centred around the issue of many participants involved, too many decisions to take and various levels, lack of commitments by member states and stakeholders, and lack of legal prosecution when non-compliance.

Farmers and Cooperatives

Farmers and rural communities are also involved in mitigating and addressing the challenges of climate change. They are affected by climate change through drought seasons and pollution affecting their crop growth and livestock. Scholars investigated the effects of climate change, such as an increase in heat and landslides and how it affects farming and other activities. According to Gariano and Guzetti (2016), rain and floods that are not controllable can create unstableness and affect farmers' stock. When there is dry land, it will take more rain to create unstable conditions, which will enhance shear strength, soil suction, and cohesion. Farmers' cooperatives and communities form networks for advocacy and information sharing on climate change. Gcumisa et al. (2016) demonstrated the importance of farming in bringing positive livelihood to Thukela District in KwaZulu Natal. However, climate change threatens communities around Thukela District and the rest of the value chain in food distribution. It is believed that the greater participation of farmers in climate change adaptation depends on their understanding and sharing of current knowledge on climate change issues.

Popoola et al. (2020) argue that access to information on climate change and adaptation strategies is crucial for small farmers' holdings to survive climate change. The changing weather patterns can not only be determined through indigenous knowledge; advanced tools and technology are needed to detect the dangers of climate change. Also, it is imperative to impart agricultural information on adaptation and mitigation methods has increased considering the current State of global climate change. The need for effective communication, public outreach, and education to build support for the policy, collective action, and behaviour change is constant. It is likely essential in the context of anthropogenic climate change, as Popoola et al. (2020) alluded to. These scholars' calls for climate science by state agencies to promote climate change education demonstrate the need to enhance climate literacy. It has surfaced that there are concerns over the significant gaps in the understanding and dissemination of climate change knowledge. Farmers need to know the SDG goals agreed upon by the United Nations General Assembly adopted on the report of the Open Working Group on Sustainable Development Goals in 2014 as the source for aligning the SDGs into the post-2015 development agenda. SDG 2: 'End hunger, achieve food security and improved nutrition, and promote sustainable agriculture', like other SDGs, is prioritized to be implemented successfully in 2020 and 2030.

Small cooperatives have emerged to add value to sustainable solutions in some villages. Mudau Phathutshedzo, from Tshikuwi Village in the Vhembe district, is an entrepreneur passionate about sustainable solutions by developing recycled products that come from waste. His primary objective is to address the problem of youth unemployment in South Africa by creating jobs for the youth through sustainable solutions. Unemployment among young people in South Africa remains a significant concern. According to Reuters, youth unemployment is a staggering 43.4% which impedes young people's ability to actively participate in the economy by living sustainable lives, unleashing their potential, and being a source of hope for their families. Nevertheless, in contrast to this gloomy and disappointing State of affairs about youth unemployment in South Africa, Mudau Phathutshedzo's eagerness and ability to soar above the storms has meant that his company has created 12 jobs for youth in his local community. By repurposing waste plastic to create bricks for paving and building and generating outdoor furniture and fence poles, his business processes its products to protect the environment. It is interesting to note that the processing of the goods also uses biomass waste, including sawdust and cow manure, to create eco-briquettes for heating and cooking. Compared to charcoal, the briquettes endure longer and do not emit smoke. The eco-briquettes go by the name HASHA MULILO. Additionally, his business creates organic fertilizer using biochar based on waste biomass. The application of organic fertilizer to soil occurs in agriculture.

Indigenous Communities

Indigenous communities and groups have deep knowledge of natural and traditional methods that can be used to adapt to climate change in their local ecosystems; hence they are sometimes at the forefront of advocacy and adaptive measures on climate impacts. Some rural communities and farmers are exposed to increased climate variability. Seasons for rainfalls have changed, and continuous unpredictable rainfalls with flooding negatively impact crop vegetation and farming in some other areas. The increased instability in farming can depend on indigenous knowledge systems and biological and geographical indicators of seasonal forecasts for making decisions on crop production, such as social safety nets and food stock.

Researchers and development professionals must integrate with farmers' regional decision-making frameworks when decision-makers require technical production information. However, this indigenous knowledge's potential as a basis for decision-making is constrained because some of the indicators (such as biological ones) on which it is traditionally based are also negatively impacted by increased climate unpredictability (Klenk et al., 2021). Despite this, initiatives to increase the adaptability of these farming communities should still start with the existing indigenous knowledge base. Organizations like the Botanical Society of South Africa are known for their established natural heritage at Kirstenbosch Gardens, where eight National Botanical Gardens are established for nature conversation of a unique wildflower heritage and some educational programmes and projects initiatives on nature conservation. This place attracts millions of tourists locally and abroad, especially those interested in studying nature conservation. Infrastructure developers and business companies are also critical stakeholders that do not only support adaptive change but also resilience strategies that can address climate change impacts. The tourist industry, with non-governmental organisations, continues to form networks focusing on world tourist activities to update their clients on changing weather and rising temperatures in various destinations.

POLICY IMPLICATION

Government and some organisations have demonstrated success cases in mitigation and adaptation to climate change in diverse contexts. The ordinary people from communities are still left behind. Therefore, there is a need for improved policy advocacy on climate change. Policies must be revisited and updated continuously to adapt to the constant concern of climate change, which imposes risks and vulnerability to society. However, besides litigation by green advocacy groups, policy results imply that there is a keen interest and less accountability in cleaning up climate air pollution with costs. Due to rapidly rising gas emissions, there is a strong interest in climate change problems. Sharma (2020) highlighted the necessity to develop new ideas and appropriate strategies to design better, regulate, and optimize wastewater treatment plants (WWTPs) on a plant-wide scale. Bioremediation technology has emerged recently as one of the most affordable, cutting-edge, and promising options for reducing GHG emissions into the Earth's atmosphere (Singh, 2014). Bioremediation is known for using living microorganisms to reduce environmental contaminants into less toxic forms. The detoxification is reduced by the naturally occurring bacteria and fungi from plants to human health and the environment. Jabbar (2022) believes that a precise and necessary GHG emission quantification method needs to be improved because of measurement uncertainties and a lack of transposable data. Therefore, there is a need for scientists specialising in climate change to go deeper than what this mathematical formula predicts and focus on the practical side of solving air pollution.

To close the gap, Singh (2014) suggested that the Mathematical models, which provide practical tools for quantifying greenhouse gas (GHG) and evaluating various mitigation solutions before putting them into practice, are one suggestion. (Singh, 2014). For various WWTP systems, GHG modeling may improve the accurate estimation of GHG emissions and assess the consequences of varied operating circumstances. Several mathematical modelling studies have been created recently to incorporate GHG emissions. If these emissions are left unsolved for long, they can heavily impact natural organisms and society.

The National Climate Change Response White Paper (NCCRWP) outlines the nation's climate change policy and is the primary guide for South Africa's climate change response (RSA, 2011). Together with the NDP, the NCCRWP addresses the urgent and visible challenges climate change poses to the nation's society, economy, and environment. It also serves as the foundation for monitoring South Africa's transition to a society and economy that is more climate resilient and low carbon. The NCCRWP mandates that South Africa keep track of, evaluate, and report on its progress in combating climate change, as well as plan an effective national response to the phenomenon's unavoidable effects and reduce the country's greenhouse gas (GHG) emissions. However, there is a great need for government to assist networks and other local organisations and communities to understand its core focus and how to comply with the policy framework underpinned by it and other environmental policies.

The Department of Environmental Affairs (DEA) thus completed the National Climate Change Response Monitoring and Evaluation (M&E) framework in 2015 to guide the readiness of government departments and other non-government organizations to monitor the climate change scenario and compliance with the existing environmental policy and other agreements. The main obstacles to effective climate change monitoring and evaluation include the lack of funding, knowledge and monitoring expertise. The policy also says nothing about those who cause gas emissions and pollution. There is limited responsibility and accountability towards shaming and punishing perpetrators so far. It is critical to incorporate the understanding of all-inclusive stakeholders' needs and conditions, as they are affected by climate change, to prepare for the global and local future of sustainable adaptation to climate change. Therefore, this chapter advocates for inclusive policy implementation in climate change adaptation.

CONCLUSION

This chapter demonstrates how South Africa has responded to significant actions to retort to climate change impacts and risks. Diverse strategies to mitigate and respond to climate change include government interface with various groups and networks through dialogues, information sharing, social responsiveness, service offering, empowering communities with information and commitment to organizations and non-government organizations. The reporting at the international level and through the National Climate Change Annual Report conveys information that is crucial to the progress of the State of readiness and adaptability in undertaking actions aiming at recognizing other activities existing, calculating the climate change impact and costs, stimulating current actions to avoid wasting resources for the future generation. The sustainable development theory is aligned with climate change adaptability to understand the impact and indicate the actions adding to the national imperatives of addressing inequality and poverty and achieving continued economic growth. Activism and advocacy are stronger at the international level than at the local level. At the level of these organizations, the element of integration is still at the incipient stage to male climate change advocacy and action match sustainable goals at the local central planning level and service delivery, where most things are set to happen and impose risk and threats. Even though climate change is a necessity for the South African government to address as part of poverty alleviation, there is still a gap in the development of the knowledge area in social science. Also, there is a great need for the local government to assist social nets for food security and climate change to empower local communities with knowledge and strategies for adaptation to climate change.

RECOMMENDATION

It is crucial to integrate climate change networks with sustainability to strengthen advocacy at the local level. While all stakeholders are crucial in participating and mitigating climate change, the youth must be encouraged to participate in national, provincial, and local programmes aimed at empowering them on climate change issues by experienced change agents and facilitators. The youth involvement can also include topics such as sustainability and incorporate issues on culture and language barriers in sustainable development and climate change programmes.

Higher education, SETAs and other service providers can adopt an interdisciplinary approach to train and develop the capacities of trainers to understand and critically evaluate national and international policy planning processes and decision-making on climate change. Research institutions and universities can also adopt sustainable niche areas in topics such as climate change, and the youth should have the opportunity to study diverse specializations on aspects of climate change. Sustainable development can also be infused into university academic programs as a compulsory module—in both social science and natural science disciplines.

The Department of Environment and Higher Education must offer support and climate change resources to trainers, change agents and activists in the formal and informal sectors and education system to guarantee the success of the learning process where recipients can demonstrate an understanding of the outcomes of the programs and be able to plough back to the community what they have learnt.

REFERENCES

Ahmed, S., Griffin, T. S., Kraner, D., Schaffner, M. K., Sharma, D., Hazel M., Leitch, A. R., Orians, C. M., Han, W., Stepp, J. R., Robbat, A., Matyas, C., Long, C., Xue, D., Houser, R. F., & Cash, S. B. (2019). Environmental factors variably impact tea secondary metabolites in the context of climate change. *Frontiers in Plant Science, 10,* 939.

Anderson, T. R., Hawkins, E., & Jones, P. D. (2016). CO2, the greenhouse effect, and global warming: From the pioneering work of Arrhenius and Callendar to today's earth system models. *Endeavour, 40*(3), 178–187.

Bornemann, B., & Weiland, S. (2021). The UN 2030 Agenda and the quest for policy integration: A literature review. *Politics and Governance, 9*(1), 96–107.

Bulkeley, H. (2023). The condition of urban climate experimentation. *Sustainability: Science, Practice and Policy, 19*(1), 2188726.

Capolupo, A., Barletta, C., Esposito, D., & Tarantino, E. (2023, September). Earth observation data for sustainable management of water resources to inform spatial planning strategies. In A. Marucci, F., Zullo, L., Fiorini, & L. Saganeiti (Eds.), *Innovation in urban and regional planning.* INPUT 2023. Lecture Notes in Civil Engineering, vol 467. Springer. https://doi.org/10.1007/978 -3-031-54118-6_3

Coelho, L. M., Rezende, H. C., Coelho, L. M., de Sousa, P. A. R., Melo, D. F. O., & Coelho, N. M. M. (2015). Bioremediation of polluted waters using microorganisms. In N. Shiomi (Ed.), *Advances in Bioremediation of wastewater and polluted soil* (pp. 1–21). Interchopen Publishers.

Cote, M., & Nightingale, A. J. (2012). Resilience thinking meets social theory: Situating social change in socio-ecological systems (SES) research. *Progress in human geography, 36*(4), 475–489.

Dejene W. (2018). Impact of climate change on biodiversity and associated key ecosystem services in Africa: A systematic review. *Ecosystem Health and Sustainability, 4*(9), 225–239.

Department of Minerals and Energy. (2005). *National energy efficiency strategy of the Republic of South Africa.* Government Printers.

Di Baldassarre, G., Sivapalan, M., Rusca, M., Cudennec, C., Garcia, M., Kreibich, H.,..., & Blöschl, G. (2019). Sociohydrology: scientific challenges in addressing the sustainable development goals. *Water Resources Research, 55*(8), 6327–6355.

Dirmeyer, P. A., Balsamo, G., Blyth, E. M., Morrison, R., & Cooper, H. M. (2021). Land–atmosphere interactions exacerbated the drought and heatwave over northern Europe during summer 2018. *AGU Advances, 2*(2), e2020AV000283.

Farah, P. D. (2015, November 30). *Global energy governance, international environmental law and regional dimension.* World Scientific Reference on Globalisation in Eurasia and the Pacific Rim, Imperial College Press. https://papers.ssrn .com/sol3/papers.cfm?abstract_id=2701031

Fekete, H., Kuramochi, T., Roelfsema, M., den Elzen, M., Forsell, N., Höhne, N., Luna, L., Hans, F., Sterl, S., Olivier, J., van Soest, H., Frank, S., & Gusti, M. (2021). A review of successful climate change mitigation policies in major emitting economies and the potential of global replication. *Renewable and Sustainable Energy Reviews, 137,* 110602.

Gariano, S. L., & Guzzetti, F. (2016). Landslides in a changing climate. *Earth-Science Reviews, 162,* 227–252.

Gcumisa, S. T., Oguttu, J. W., & Masafu, M. M. (2016). Pig farming in rural South Africa: A case study of uThukela District in KwaZulu-Natal. *Indian Journal of Animal Research, 50*(4), 614–620.

Ha, S. (2022). The changing climate and pregnancy health. *Current Environmental Health Reports, 9*(2), 263–275.

Harvey, B., Jones, L., Cochrane, L., & Singh, R. (2019a, April 11). *NGOs are shaking up climate services in Africa. Should we be worried?* http://www.braced.org/news/i/NGOs-are-shaking-up-climate-services-in-Africa-Should-we-be-worried/

Harvey, B., Jones, L., Cochrane, L., & Singh, R. (2019b). The evolving landscape of climate services in sub-Saharan Africa: What roles have NGOs played? *Climatic Change, 157,* 81–98.

Jabbar, N. M., Alardhi, S. M., Mohammed, A. K., Salih, I. K., & Albayati, T. M. (2022). Challenges in the implementation of bioremediation processes in petroleum-contaminated soils: A review. *Environmental Nanotechnology, Monitoring & Management, 18,* 100694.

Janowski, T., Estevez, E., & Baguma, R. (2018). Platform governance for sustainable development: Reshaping citizen-administration relationships in the digital age. *Government Information Quarterly, 35*(4), S1–S16.

Jones, L., Harvey, B., & Godfrey Wood, R. (2016, September 28). *The changing role of NGOs in supporting climate services.* http://www.braced.org/resources/i/ngos-supporting-climate-services/

Klenk, N., Fiume, A., Meehan, K., & Gibbes C. (2017, May 31). Local knowledge in climate adaptation research: Moving knowledge frameworks from extraction to co-production. *Wiley Interdisciplinary Reviews: Climate Change, 8*(5), e475.

Kweku, D. W., Bismark, O., Maxwell, A., Desmond, K. A., Danso, K. B., Oti-Mensah, E. A., Quachie, A. T., & Adormaa, B. B. (2018). Greenhouse effect: Greenhouse gases and their impact on global warming. *Journal of Scientific Research and Reports, 17*(6), 1–9.

Lee, U., Han, J., & Wang, M. (2017). Evaluate landfill gas emissions from municipal solid waste landfills for the life-cycle analysis of waste-to-energy pathways. *Journal of Cleaner Production, 166,* 335–342.

Leshalagae, M., & Buthelezi-Dube, N. N. (2023). Impact of topsoil mining for un-fired mudbricks on soil quality in eastern KwaZulu-Natal, South Africa. *Land Degradation & Development, 34*(4), 1051–1066.

Mathivha, F., Tshipala, N., & Nkuna, Z. (2017). The relationship between drought and tourist arrivals: A case study of Kruger National Park, South Africa. *Jàmbá: Journal of Disaster Risk Studies, 9*(1), 1–8.

Mattern, S. (2017, November). The Big Data of ice, rocks, soils, and sediments. *Places Journal.* https://doi.org/10.22269/171107

Mukherjee, S., Mishra, A., & Trenberth, K. E. (2018). Climate change and drought: A perspective on drought indices. *Current Climate Change Reports, 4,* 145–163.

Nemakonde, L. D., van Kiekerk, D., Becker, P., & Khoza, S. (2021). Perceived adverse effects of separating government institutions for disaster risk reduction and climate change adaptation within the Southern African Development Community Member States. *International Journal of Disaster Risk Science, 12,* 1–12.

Ozili, P. K. (2022). *Sustainability and sustainable development research around the world.* Managing Global Transitions.

Pasquini, L., Ziervogel, G., Cowling, R. M., & Shearing, C. (2015). What enables local governments to mainstream climate change adaptation? Lessons learned from two municipal case studies in the Western Cape, South Africa. *Climate and Development, 7*(1), 60–70.

Popoola, O. O., Yusuf, S. F. G., & Monde, N. (2020). Information sources and constraints to climate change adaptation amongst smallholder farmers in Amathole District Municipality, Eastern Cape Province, South Africa. *Sustainability, 12*(14), 5846.

RSA. (2011). *White paper on the national climate change response.* Government Gazette No. 334695, Notice No. 757 of 19 October 2011. Government Printer.

Sharma, I. (2020). Bioremediation techniques for the polluted environment: Concept, advantages, limitations, and prospects. In *Trace metals in the environment—Approaches and recent advances.* IntechOpen.

Singh, R. (2014). Microorganism as a tool of bioremediation technology for cleaning environment: A review. *Proceedings of the International Academy of Ecology and Environmental Sciences, 4*(1), 1.

Smith, B. C. (2018). *Infrared spectral interpretation:Aa systematic approach.* CRC press.

Suiseeya, K. R. M., Elhard, D. K., & John Paul, C. (2021). Toward a relational approach in global climate governance: Exploring the role of trust. *Wiley Interdisciplinary Reviews: Climate Change, 12*(4), e712.

United Nations. (1998). *Kyoto protocol to the united nations framework convention on climate change.*

Wallace-Wells, D. (2018). *The uninhabitable earth: Life after warming.* https://www .crisrieder.org/thejourney/wp-content/uploads/2019/05/The-Uninhabitable -Earth-David-Wallace-Wells.pdf

Ziervogel, G., New, M., van Garderen, E. A., Midgley, G., Taylor, A., Hamann, R., Stuart-Hill, S., Myers, J., & Warburton, M. (2014). Climate change impacts and adaptation in South Africa. *Wiley Interdisciplinary Reviews: Climate Change, 5*(5), 605–620.

CHAPTER 19

CLIMATE GOVERNANCE

An Anatomy of Community Understanding and Awareness of Climate Change

Redemption Chatanga
University of the Free State, South Africa

Maréve Biljohn
University of the Free State, South Africa

ABSTRACT

Climate change is prioritized by the United Nations as sustainable development goal (SDG) number 13. Community participation in climate change adaptation and mitigation is considered fundamental to achieving SDG 13. Achieving this SDG 13 could however pose a challenge if factors such as communities' understanding and awareness of climate change are not addressed because they largely influence their participation. In the global South, the African continent is equally affected by climate change with dire consequences for its communities and its environment. Whilst literature on climate change is extensive, research analysing community understanding and awareness of climate change

Climate Governance in International and Comparative Perspective, pages 447–474
Copyright © 2024 by Information Age Publishing
www.infoagepub.com

as factors influencing their participation in climate governance is limited. Therefore, this chapter explores the anatomy of community understanding and awareness of climate change towards nudging community participating in climate governance. Using a qualitative research approach, a case study design was used to collect data from documents, and 9 focus group discussions in the Manonyane communities of Lesotho. The results of this study showed that a lack of community participation in the Manonyane communities is in part attributed to community members being unaware of climate change and not understanding its consequences for them and the environment. Therefore, this study argues that the Lesotho NCCP should delineate the conceptual meaning of community understanding and awareness of climate change. From this, the policy should inform the development of a contextual framework to operationalize community understanding and awareness of climate change to enhance community participation in climate change policy formulation and implementation. As a result, this study proposes principles of climate governance to enhance community understanding and awareness of climate change as a starting point for successful policy formulation and implementation.

The evident threat of climate change undermines the ability of countries to attain the 2030 sustainable development goals (SDGs; United Nations Environment Programme, 2016, pp. 6–7), and places climate change on the global agenda with an emphasis on community participation in climate governance. Climate governance is the coordinated effort, mechanisms, and response measures with the purpose of mobilizing stakeholder support to mitigate and adapt to risks posed by climate change (Jagers & Stripple, 2003, p. 388). This is attributed to climate change being one of the greatest challenges because of adverse impacts that exacerbate already existing problems faced by humanity, such as amongst other poverty, gender inequalities, natural disasters and unemployment (United Nations, 2015, p. 8). Thus, the term community participation in climate change policy formulation and implementation in this context will refer to direct or indirect involvement in climate change decisions and implementation of such decisions. Achieving this SDG 13 through community participation in climate governance could be a challenge if community understanding and awareness of climate change is not addressed because of their influence on community participation. Communities are more likely to act voluntarily based on something that is known and understandable.

In the global South, the African continent is affected by climate change with dire consequences for its communities and its environment (African Union, 2014). Research analyzing community understanding and awareness of climate change as factors influencing their participation in climate governance is limited. Inadequate understanding and awareness of climate change issues amongst communities contribute to a lack of participation in climate governance and hinder the implementation of climate change policies in the Kingdom of Lesotho (referred to as Lesotho henceforth; Lesotho Government and World Food Programme [WFP], 2019). If this is not addressed, communities' poor agricultural practices and activities, deforestation, and anthropogenic wildfires could persist (Lesotho Meteorological Services 2013, p. 42). Therefore, this chapter explores the anatomy of community understanding and awareness of climate change towards enhancing community participation in climate governance, particularly on policy formulation and implementation. The research question answered is: "How does community awareness and understanding regarding climate change influence their participation in climate change policy formulation and implementation?"

Using a qualitative research approach, a case study design was used to collect data from documents and 9 focus group discussions in the Manonyane communities of Lesotho. The selection of this research design was based on three reasons. Firstly, to obtain a contextual understanding regarding how the Manonyane communities participated in climate change policy formulation and implementation (Hafiz, 2008, p. 545). Secondly to focus on the Manonyane communities in-depth as guided by the aim of the study (Blatter, 2008, p. 68). Thirdly to minimize bias through data, investigator, methodological and theory triangulation on one case study—Manonyane communities. Thus, this chapter proceeds by discussing and presenting a literature review, case of Manonyane communities, methodology, findings and discussion, lessons learned, and implications for policy and theory.

LITERATURE REVIEW

The literature review focuses on how community awareness and understanding regarding climate change influence their participation in policy formulation and implementation as part of climate governance. Climate governance highlights community participation as a success factor in the implementation of climate change policies. Community participation is seen as important in bringing sustainable development through policies that consider community needs when addressing climate change (United Nations Economic Commission for Europe [UNECE], 2015). Two key factors emerge from the approach used by UNFCCC to tackle climate change. These are community involvement in climate change policy formulation and implementation and emphasis on context-based participation, which is in line with the aim, and research question of this study.

Community Understanding of Climate Change

Generally, the word "understanding" means having knowledge about a situation or subject, the ability to comprehend something, sympathetic awareness or tolerance, an informal agreement between people, and having reason to believe in something (Cambridge Dictionary, n.d.). One of the distinguished conceptions for community understanding of climate change is the scientific understanding of climate change (Hegger et al., 2017). This is having the knowledge of the definition of climate change and its causes, the likelihood and severity of expected impacts, the risks, and planned future responses through the accumulation of observational data, construction of theories and models to interpret climate variability and empirical testing of theories and models (Intergovernmental Panel on Climate Change [IPCC], 2014, p. 9). This section discusses community understanding of climate change as a factor that influences their participation in climate change policy formulation and implementation.

The literature on community understanding of climate change highlighted that participation during policy formulation and implementation is influenced by the extent to which a community understands the likelihood and severity of expected climate change impacts (IPCC, 2014, p. 11). Noteworthy is that impacts vary from place to place, depending on the climate and human activities (Cai et al., 2017). The impacts include compromised agricultural production, floods, water insecurity problems, health impacts, droughts, heatwaves, wildfires, cyclones, and significant loss of biodiversity and ecosystems (IPCC, 2007, pp. 9–18; IPCC, 2014, pp. 12–14).

What is also known about community understanding of climate change as it relates to their participation during policy formulation and implementation is its description in terms of understanding climate change causes (Cai et al., 2017; Lorenzoni et al., 2007), which are mainly caused by human activities. It may be important for communities to know that naturally occurring greenhouse gases, including carbon dioxide, methane and nitrogen oxide, keep the earth warm enough for humans to live (IPCC, 1990, p. xiv). Human activities such as deforestation, wildfires, air pollution, urbanization, industrialization and poor agricultural practices increase the concentrations of these gases in the atmosphere, thereby causing climate change. Despite human activities, natural causes such as climate variability, temperature changes, weather changes and ozone depletion also cause climate change. It is when communities have an understanding that human activities increase greenhouse gas concentrations in the atmosphere that their behavior towards participation in climate change policy formulation and implementation is expected to be positive (IPCC, 1990, p. xiv; IPCC, 2014, p. 9).

The studies presented thus far provide evidence that community understanding of climate change that influences participation in policy formulation and implementation is described in terms of having knowledge about climate change policy processes. The communities should have an understanding of how governance processes in terms of policy formulation and implementation work. Community understanding of climate change that influences participation during climate change policy formulation and implementation is described in terms of understanding climate change risks (Larsen & McGuinness, 2016, p. 402). Risks faced by communities vary from community to community and include but are not limited to floods, food insecurity because of drought, land degradation and loss of biodiversity (Hegger et al., 2017). Community understanding of climate change that influences participation during climate change policy formulation and implementation is also described in terms of understanding the planned future adaptation and mitigation responses (Smit et al., 2000, p. 747). Climate change adaptation occurs when natural or human systems adjust in response to expected actual climatic factors or their effects, which reduces harm or exploits beneficial opportunities (IPCC, 2001, p. 982). Mitigation is defined by the IPCC (2001, p. 990) as an anthropogenic (originating from human activity) intervention to reduce the sources or enhance the sinks of greenhouse gases. Adaptation actions have consequences for mitigation and vice versa (IPCC, 2007, p. 747; Smit et al., 2000, p. 245). Understanding climate change becomes important.

Against this background, the anatomy of community understanding is outlined in Figure 19.1 in terms of its dissected defining attributes. "Anatomy" is defined as a branch of science concerned with bodily structure of humans and other living organisms, focusing on the dissection and separation of parts (Cambridge Dictionary, n.d.; Larsen & McGuinness, 2016). Based on the preceding discussion, the defining attributes of community understanding of climate change, as noted in literature, are many. Attributes are defining characteristics that best describe a concept, without which the concept cannot be said to occur (Von Colln-Appling & Giuliano, 2017, p. 107). Thus a community that understands climate change would demonstrate the following:

1. having reason to believe in climate change causes, impacts, policy process, risks and planned mitigation and adaptation measures (Wi & Chang, 2019);
2. being aware of climate change (Lorenzoni et al., 2007);
3. have knowledge about climate change; and
4. have the ability to comprehend (grasp the definition of climate change mentally) (Wi & Chang, 2019).

Without these attributes, community understanding of climate change cannot be said to have occurred.

Figure 19.1 provides a holistic overview of attributes that should be present to enhance community understanding of climate change. In turn, it also becomes synonymous with a cyclical process of community understanding. From this it could be deduced that if one of these attributes are missing

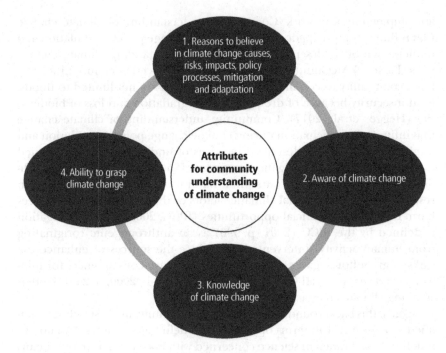

Figure 19.1 An anatomy of community understanding of climate change.

it might affect the ability of the community to meaningfully participate in climate change policy processes and planned mitigation and adaptation measures that requires their participation. Moreover it could present communities with a deficit in fully comprehending the importance of their participation in climate change policy processes and implementing policy solutions. It is, therefore, imperative that governments adopt contextual strategies underpinned by these attributes to support community participation in climate governance.

Community Awareness of Climate Change

Awareness refers to the understanding of a current subject based on information or experience (Shepard et al., 2018). Community awareness of climate change is defined as when information about environmental issues is available and accessible to community members (Wi & Chang, 2019, p. 10). Communities are, at least, aware of the 'normal' hazards caused by changes in weather, but are unaware of more dangers as a result of climate change (Samaddar et al., 2019, p. 352). In this study, community awareness of climate change is defined as the community's understanding of climate change in terms of its causes, impacts, risks and planned responses (mitigation and adaptation) based on information or experience (Shepard et al., 2018). Having information and experience will lead to communities being well informed and critically conscious about climate change (Friis-Hansen, 2017, p. 15; Lorenzoni et al., 2007, p. 447).

Similar to community understanding, community awareness of climate change is described in relation to having knowledge about climate change causes (Shepard et al., 2018). The causes of climate change are mainly human activities, which increase the greenhouse gas concentrations in the atmosphere (IPCC, 2007, pp. 82–85). Raising such awareness of climate change causes should aim at giving information to influence behavior change towards active participation during climate change policy formulation and implementation (Shepard et al., 2018; Wi & Chang, 2019).

Another description for community awareness of climate change is in relation to having knowledge of climate change impacts (Dhungana et al., 2018). The impacts of climate change vary from context to context but also include, among others, floods, loss of livelihoods, hazards, heat waves, exacerbated water related problems and health problems (IPCC, 2007, pp. 11–18; Shepard et al., 2018). For communities to influence climate change policy formulation and implementation, access to adequate information about climate change impacts should be ensured (Cai et al., 2017; Shepard et al., 2018). Nevertheless, the way information is disseminated and collected plays a role in ensuring that communities input their ideas in climate change policy formulation and implementation (Dhungana et al., 2018).

Having information about climate change risks is also a key aspect of community awareness of climate change observed in the literature (Larsen & McGuinness, 2016). Depending on the context, the risks associated with climate change are many and include, among others, extreme weather events such as drought, floods, wildfires and less rainfall (Larsen & McGuinness, 2016, p. 396). Usually, such effects have a huge impact on human beings and their environment.

Having knowledge about planned adaptation and mitigation responses have also proven to be a dominant feature of community awareness of climate change that increases participation in policy formulation and implementation (Hegger et al., 2017). In this regard, informed consent must be sought in advance by the policymakers prior to any direct community involvement in climate change policy formulation and implementation. Communities should know the aims, procedure and expected mitigation and adaptation responses in their context (Hegger et al., 2017, p. 13). Based on the preceding discussion, the anatomy of community awareness of climate change is summarized in Figure 19.2. Subsequently, similar attributes between community awareness and understanding and deductions are presented in Table 19.1.

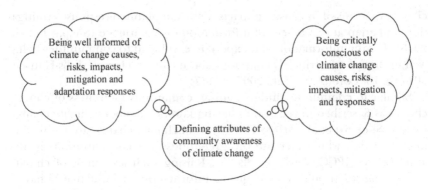

Figure 19.2 An anatomy of community awareness of climate change.

In concluding the discussion on the influence of understanding and awareness of climate change and climate change policy on community participation, Table 19.1 illustrates similar attributes between community awareness and understanding and deductions based on the literature discussed. The first similarity is that community awareness and understanding of climate change do not always change lack of participation to participation or communities from being not concerned to concerned (Tannebaum et al., 2017; Shepard et al., 2018). Therefore, community awareness and understanding of climate change do not always positively influence their participation during climate change policy formulation and implementation (Schmidt, 2017, p. 1026). This is because the influence of community understanding and awareness of climate change on their participation during policy formulation and implementation is debatable. The second similarity is that community understanding and awareness may result in conflicts between communities and policymakers. The third similarity suggests that community's understanding of the policy process may not necessarily lead to building trust, co-production, and collaboration during policy formulation and implementation. In fact, community understanding and awareness of climate change is not something tangible or visual, but an expression of an individual in the mind. Nevertheless, it can still be argued that understanding and awareness of climate change largely increase community participation during policy formulation and implementation because a person is more likely to act voluntarily based on something that is known than on an unknown subject (Brink & Wamsler, 2019, p. 94; Shepard et al., 2018, p. 5). Against this background, the analysis of community understanding and awareness of climate change is inevitable to come up with a theory of participation in climate change policy formulation and implementation in future research.

TABLE 19.1 Similar Attributes Between Community Awareness and Understanding and Deductions		
	Community understanding of Climate change	**Community awareness of climate change**
Similarities	<ul><li>Do not always change lack of participation to participation.</li><li>May results in conflicts between communities and policymakers.</li><li>Policy process—may not necessarily lead to building trust, co-production, and collaboration during policy formulation and implementation.</li></ul>	
Deductions	Community understanding should result in the following:<ul><li>Community understanding of their relationship to the environment and climate change.</li><li>Community establishing and understanding information on trees as resources, crop patterns and current level of livelihood production.</li><li>Community understanding the importance of terracing, contour ploughing and water diversion furrows.</li><li>Community understanding the need to change behavior concerning deforestation, wildfires and the use of poor agricultural practices.</li><li>Community understanding about their natural resources, soils, topography, vegetation and farming practices in relation to climate change.</li><li>Community understanding how to interpret climate change associated risks.</li><li>Community understanding the climate change policy process.</li><li>Community developing trust in the policymakers' leadership on climate change policy formulation and implementation.</li></ul>	Community awareness should result in the following:<ul><li>Community having a voice in climate change awareness program.</li><li>Increased awareness of climate change threats.</li><li>Community awareness of climate change impacts.</li><li>Raised awareness on the existing natural resources.</li><li>Development of a basis for comparing different activities.</li><li>Visual understanding through mapping.</li><li>Helping communities to create a basis for impact analysis or for monitoring and evaluation of climate change.</li><li>Raised awareness on the state of depletion of natural resources as a result of deforestation and poor agricultural methods.</li><li>Community's ability to comprehend rainfall and temperature trends.</li><li>Community being able to articulate indigenous seasons, crop diseases and pests.</li></ul>

CASE OF MANONYANE COMMUNITIES IN LESOTHO

Lesotho is a Southern African country in the global South with a population of 2.2 million where 80% of this population resides in the rural areas (World Bank, n.d.). The country is exposed to climatological patters from both the Atlantic and Indian oceans because of its location and geography

(Climate Change Knowledge Portal, n.d.-a). The impacts are more felt in rural communities than in urban areas. One such rural community is the Manonyane communities situated in the Manonyane Community Council area (foothills), in Maseru District. The Manonyane communities experience vulnerabilities such as water shortages, environmental problems and extreme weather conditions including droughts, strong winds, prolonged cold-weather seasons, and wetlands degradation and land (Climate Change Knowledge Portal, n.d.-b). Consequently, the Manonyane communities are affected by climate change impacts such as sudden snowfalls, reduced agricultural production, floods, strong winds and frequent rainstorms (UNDP, 2013, p. 8). Figure 19.3 shows the map of Lesotho and where the Manonyane communities are located. The Manonyane communities comprise 108 villages and a population of 25,143 people (Bureau of Statistics, 2018). Predominantly rural, these communities rely on subsistence farming and animal rearing (e.g., cattle, sheep, & goats) as a source of livelihood (Bureau of Statistics, 2018).

Regarding climate governance, Lesotho is a signatory to a number of international treaties in support of climate governance (Department of Environment, 2014, p. 4). Lesotho's adopted National Climate Change Policy (NCCP) acknowledges inadequate awareness of climate change issues amongst its communities at all levels (Ministry of Energy and Meteorology, 2017, p. 35). In order to increase awareness, the policy aims at raising

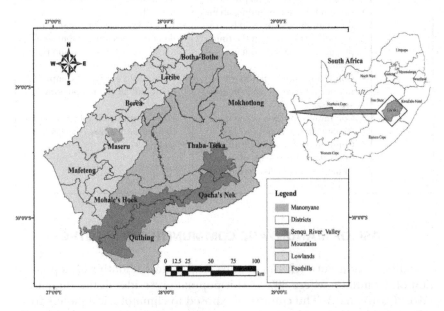

Figure 19.3 Lesotho Map with administrative districts and agro-ecological zones showing location of Manonyane communities.

awareness and building capacities to mitigate and adapt to climate change at all levels, however, it is not clear which levels are being referred to regarding community participation (Ministry of Energy and Meteorology, 2017, p. 35). The Manonyane communities cited lack of awareness of the NCCP process, thus, they did not participate during the NCCP policy stages. Limited understanding and awareness of climate change and the decision-making processes emerged as part of their reasons why the Manonyane communities did not participate in the NCCP (Chatanga & Biljohn, 2023).

METHODOLOGY

The study employed a qualitative research approach to collect and analyze data. The research is part of a doctoral study. Ethical clearance was obtained from the university where the student registered with ethical clearance number UFS-HSD2021/1838. The research question answered was: "How does community awareness and understanding regarding climate change influence their participation in climate change policy formulation and implementation?" The justification for using a qualitative research approach was that it allowed the researcher to collect data at the site where participants experience the problem under study (Creswell & Creswell, 2018, p. 257). Beyond the general qualitative research approach are more specific designs in conducting qualitative research (Creswell & Creswell, 2018, p. 259). Many designs exist, but this study used a case study design. A case study design focuses on one or a few instances of a phenomenon that is studied in-depth (Blatter, 2008, p. 68). The researcher had an opportunity for face-to-face interaction with the Manonyane communities. Therefore, a single research approach was deemed appropriate to answer the research questions using a case study design. The case study research design followed three stages—conceptual, theoretical and empirical stage. A conceptual stage included a literature review and analysis of data sources. At this stage, a conceptual demarcation of the concepts community understanding and awareness of climate change was done. The theoretical stage formed the foundation of knowledge construction for this study to provide a basis for the literature supporting the analytical framework for the study. Data collection was done during the empirical stage.

The study used a research objective and question, not hypotheses. Thus, no quantitative measurement was done but the interpretation of data (Creswell & Creswell, 2018, p. 259). Subsequently, the selection of the study's population and sample applied the purposive and non-probability sampling to conduct 9 focus group discussions with 108 Manonyane community members comprising 8–12 research participants per focus group. The identification of this population was in line with the case study design of the study that supports an in-depth exploration of the research problem and the phenomenon under investigation. The general inclusion criteria for the research participants were to first provide in-depth and relevant information regarding the phenomenon under investigation and the research question. Secondly, only the people living in the Manonyane communities were selected.

This research was conducted in a community governed by traditional leadership such as chiefs and a community council. Permission to collect data was thus obtained from the Manonyane principal chief and Manonyane community council. A research study information leaflet and consent form was also sent out to participants before participation. Notable from this data is that the participants' views did not require approval by the community representatives. Approval to conduct research in the Manonyane communities was granted by the principal chief before the individuals were invited for a focus group discussion. Letters explaining the purpose, benefits, risks, confidentiality and publication of results were hand delivered by the researchers asking for permission to interview individual community members. Research proposal and data collection instruments were tabled before the scientific and ethical committees for approval to foster scientific integrity in the current study. The researchers collected data from December 2021 to March 2022. Participants were assured that no personal information was to be disclosed and informed of how data would be used and stored. The researchers will store hard copies of participants' responses for 5 years in a locked cupboard/filing cabinet at the researchers' homes for future research or academic purposes; electronic information will be stored on a password-protected computer. Future use of the stored data will be subject to further research ethics review and approval if applicable. After 5 years elapsed, hard copy documents will be destroyed, and electronically stored information deleted from the computer. Furthermore, individuals interviewed were glad because they knew they had the permission of their chiefs and councillors to respond to interview questions. This helped participants to give their views voluntarily and honestly after understanding how the researchers would collect, use, and store data.

The qualitative research approach applied in this study, explores and understands the meaning individuals ascribe to a human social problem (Creswell & Creswell, 2018, p. 41). Therefore, a single research approach answered the research question noted during the introduction. This study applied content analysis of qualitative data which entails descriptions of manifest content and latent content (Graneheim et al., 2017, p. 30; Graneheim & Lundman, 2004). Manifest content are interpretations that are close to the text being studied, whereas latent content is the interpretation of the underlying meaning (Graneheim et al., 2017, p. 30; Graneheim & Lundman, 2004). Content analysis enabled systematic coding and thematic analysis of textual information from the transcribed interviews to address the research questions (Vaismoradi, Turunen, & Bondas, 2013, p. 400). The content analysis in the current study entails a deductive approach or concept driven approach to test the implications of community awareness and understanding of climate change. Data analysis process included transcribing the data from the audio recordings of the focus group discussions. The transcribed data were labeled according to themes emanating from the study's theoretical framework. The themes were (a) community understanding of climate change, and (b) community awareness of climate change.

FINDINGS

The findings emerged from the focus group discussions according to themes noted in theoretical framework.

How community understanding regarding climate change influenced participation in policy formulation and implementation of the Manonyane communities.

Question 1 probed Manonyane communities to get their view of climate change impacts. The focus group participants in all the Manonyane communities (Hata Butle, Ha Lebamang, Ha Maliele, Ha Pasane, Ha Tsunyane (Mafefooane), Nyakosoba, Popanyane, Thoteng, & Tloutle) confirmed persistent drought years because of climate change. Although not related to climate change, the participants pointed out to a lack of roads, lack of jobs and poverty. The participants concerted that "there is poverty. We are aware that due to heavy rainfall, agricultural productivity is poor due to climate change. We are already lacking food. The fields harvest is negatively affected," yet the communities depend on agriculture. The participants agreed that the "hail is too much and destructive. This had been happening for years. It affects the peach trees. There are no jobs, we depend on crops." In the Ha Pasane focus group, some participants believed that "God is punishing us because of our actions." To these participants, "people are out of control. Girls and boys are not afraid to walk naked in front of elders. God is angry with us." The Sesotho New Year begins in August, but participants in the Nyakosoba focus group indicated that planting time has changed to October towards the months of December. The winter was also noted to be beginning in February instead of April towards the end of the month to July with cold air blowing from the East of Nyakosoba villages. "Coldness comes faster (in February), so you will learn that the air comes from the East. The winter season is prolonged to the month of November," said a concerned participant.

Still responding to Question 1, all the participants reported compromised agricultural production, drought, floods, health impacts, heatwaves and roofs (on houses) blown away by strong winds in their communities which is something that was not happening before. Many participants testified that "growing up, we had regular rainfall, but this year, I got shocked by the change in rainfall patterns and the drought we experience." Trees were also drying up and the reasons for drying up were not clear to participants. The Nyakosoba focus group participants noted recurring thunderstorms and lightning in one of their communities. Thoteng focus group discussion participants reported unusual heat, dongas and unusual coldness. The Tloutle focus group participants reported insufficient drinking water. "Drinking water in this village is seasonal." There is no "water in winter. It does not rain in a normal way . . . a lot of destruction in the environment."

The purpose of the second question was to get the Manonyane community's understanding of the causes of climate change through a focus group discussion. The responses varied accross different communities. The Hata Butle focus group discussion participants mentioned environmental pollution in their communities because of pampers. The participants said that pampers polluted the environment and caused diseases. Therefore, the participants pleaded the government of Lesotho to address such problems to clean the environment. In Ha Lebamang, participants indicated that they have been receiving regular rainfall except in the year 2022. "The heavy rains washed away all our crops in the fields." This is a problem we encountered. We were not able to cultivate due to heavy rains. The participants in Ha Maliele agreed that "there are factories in South Africa that emit smoke, that's climate change. We also have dams here in Lesotho and it is said that they are the ones that cause the rain to fall." Some were in agreement, some participants looked unsure. However, many participants in the Ha Maliele focus group discussion blamed climate change on COVID-19 and claimed that the educated people (no description was given) constructed dams in Lesotho. Therefore, according to these participants' view, the educated people have the power to purposefully stop rain or make it rain. The Ha Maliele focus group participants holding this view claimed that the formally educated people "threw something knowingly in the ocean for something to rise up." Afterwards, "the rain will fall." "I am saying that because we have never experienced the rain like this."

The Ha Pasane focus group participants associated climate change causes to not following traditional practices. According to culture, the mourning attire is put on in January and be removed in June. However, it seemed some community members "choose to remove it in March that is not it." Largely, the members who participated in Ha Pasane focus group acknowledged that they "don't have any idea" of what causes climate change. They are dumbfounded because of climate change and "think it is nature"—The participants in Ha Tsunyane focus group discussion spoke of air pollution, especially chemical emissions and smoke, agreeing that they cause people to experience drought and unusual rainfall. Nevertheless, the focus group participants emphasized that they needed information from the experts to adapt to climate change. The participants in Nyakosoba focus group showed that they did not have information on what causes climate change. They pleaded in their response that, if possible, they would be glad to have experts guiding them on climate change to adapt "and not blame everything on people who were born in the years from 2000." Participants also indicated that some believed that "the world is coming to an end, yet it is the issue of climate change."

The response from the Popanyane focus group discussion indicated that ever since the cultural practices have stopped, the climate conditions have changed too. According to Popanyane focus group participants, the crops such as sorghum, wheat and beans are no longer yielding high production. The researchers noted that about two participants in Thoteng group explained about emitted smoke from industries which has chemicals causing climate change by "drying the ozone layer." Some participants in the Thoteng focus group discussion kept quiet and did not respond to this question. The researchers perceived that the participants did not have knowledge of climate change or they completely had a different view which they did not, however, share because after two participants contributed to this question, the other participants kept quiet. It was not clear to the researchers if they disagreed or agreed with the two participants who had contributed. Lastly, Tloutle participants said, "Climate change is caused by the building of dams." The Tloutle focus group participants agreed that constructed dams had caused prolonged cold weather conditions up to the month of December.

In the third question, communities were asked how they get informed of climate change decision-making processes. The Hata Butle, Ha Lebamang, Ha Maliele, Ha Pasane, Ha Tsunyane, Nyakosoba, Popanyane, Thoteng, and Tloutle focus group participants agreed and indicated that "no government official or anyone was facilitating community participation in climate change policy formulation and implementation." There was "no such a thing." The participants "agreed that they have never gotten such an opportunity." The Tloutle focus group participants claimed that "government officials consult unconcerned people, yet the people who have problems in the rural areas are not asked to hear their perspectives." The participants further said that they go to "the community council to ask two councelors, "Do they say two councelors and three chiefs through the workshops and having nice food represent community members?" One participant rhetorically asked, expressing that it is not possible for a few individuals to give their views on climate change and claim to represent other community members. The Tloutle focus group participants reiterated that the chiefs and councelors are not the communities.

The fourth question asked about the community's understanding of climate change risks for themselves and the environment. Hata Butle and Ha Pasane focus group participants said that no one was giving information to the community members on climate change. The participants agreed that they observed the climate changes on their own and also heard from Lesotho radio and television. Participants in the Ha Lebamang focus group discussion reported that the risks of climate change problems that are brought by climate change were that after planting their fields, heavy rains fell and washed away crops in the fields. "This is a problem we encountered. We were not able to cultivate due to heavy rains. There will be drought." Ha Maliele focus group discussion participants reported with concern that whenever they receive a lot of rain, the grass will ultimately be unavailable such that the animals eat soil to survive. Animal rearing forms part of a source of livelihood and the participants were worried, equating losing an animal to "losing a child." The participants in the Nyakosoba focus group discussion agreed that "we do not have any information regarding climate change." Participants who participated in Thoteng focus group discussion noted climate change risks such as dongas, unusual heat and unusual coldness. The participants for this focus group discussion noted with concern that "as long as the government doesn't want anything to happen, there is nothing we can do." The main concern of participants in the Tloutle focus group discussion was the house and infrastructural destruction as well as washed away fields because of unusually heavy rainfall.

How community awareness regarding climate change influence the Manonyane community participation in policy formulation and implementation

The first question for this section sought to get the community's knowledge about how they get informed of climate change causes such as air pollution, deforestation, climate variability, industrialisation, poor agricultural practices, ozone depletion, temperature changes, urbanisation, weather changes, wildfires and others. Participants in all the 9 focus group discussions consensually said that there is "no other person informing them of climate change causes," but they see it for themselves. "We look at it (observe) ourselves" to predict "hunger and drought." Asked what they see for themselves, participants mainly talked about climate variability, temperature changes, weather changes and wildfire. On wildfire, participants said that it is community knowledge that one should not burn grass. Again, participants testified that they have also seen on the Lesotho television when "the Minister of Agriculture would inform them about the upcoming heavy rainfalls. However, community members without radios and televisions reported that they hear the information from their fellow community members "when they pass it through the word of mouth." Of note were participants in a focus group discussion of Ha Tsunyane and Tloutle where many participants in these groups said that they hear about climate change causes on television and radio. "Mostly, they speak of air pollution, especially chemical emissions and smoke. We can see that they are the causes as we experience drought and unusual rainfall," said one of the participants with other participants nodding their heads in agreement. The Tloutle community members who watched television and listened to radio said that they "learn about climate change causes through announcements by the Ministry of Agriculture in South Africa." In other words these community members watched South African television and listened to different radio stations. In Popanyane, the participants believed that the government was making announcements regarding the climate change issues, but the challenge was that many community members who participated in this focus group had no television.

The third question aimed at getting knowledge on how communities get informed of extreme weather events, floods, food insecurity, insufficient access to drinking water, land degradation, loss of biodiversity, loss of rural livelihoods, reduced agricultural production and other. Regarding these risks, all the focus group discussions (Hata Butle, Ha Lebamang, Ha Maliele Ha Pasane, Ha Tsunyane, Nyakosoba, Popanyane, Thoteng, & Tloutle) emphasized that they were not getting information from different sources other than from television, different radio stations, and community members through word of mouth. However, some community members who participated in the focus group discussions agreed that they only got information from other community members.

The fourth question asked how communities get informed of compromised agricultural production, cyclones, drought, floods, health impacts, heatwaves, significant loss of biodiversity and ecosystems, weather insecurity problems, wildfires and other climate change impacts for them and the environment. Generally, across the 9 focus groups (Hata Butle, Ha Lebamang, Ha Maliele Ha Pasane, Ha Tsunyane, Nyakosoba, Popanyane, Thoteng, & Tloutle) conducted, community members unanimously agreed that information was passed on from one community member to the other. The participants said that there was "no one to explain to us about the changes that we observe after certain years and also there are no people who tell us the measures that we have to take because we are dependent on the crops." This implied that there was no one to assist communities with adaptation because they depended on agriculture. Some participants claimed that they get "information when we are already in the situation," citing an example of El Nino drought. The participants who listened to radios, watched television and read on the internet complained that information about climate change impacts would just say, "take care, you should expect rain or drought" without giving measures of what to do. Although COVID-19 was not the subject of discussion, some participants were concerned about COVID-19 causes, impacts and risks.

DISCUSSION

The findings in this section are discussed in relation to the identified themes

Influence of community awareness of climate change on community participation in policy formulation and implementation

The findings from this study established that the Manonyane communities' lack of participation in climate change policy formulation and implementation was because they lacked scientific knowledge of climate change and climate change policy processes. These findings suggest that the Manonyane communities' awareness of climate change was affected by the system of governance, which was closed. Based on the theoretical framework, community-based climate change solutions require the Lesotho government to be open and accessible to communities which was not the case in the Manonyane communities (Dang, 2018). The Manonyane communities were unaware of the NCCP. Consequently, these results are significant in two major respects (a) the need to have an open system of governance and (b) the need to reach out to the Manonyane communities to make them understand the processes surrounding the debating of a bill or policy to enhance their awareness of the climate change policy formulation and implementation.

Disseminating information on climate change causes, risks and impacts

Repeatedly, the findings that the Manonyane communities consensually refute that there was no formal way they received information on climate change causes, risks and impacts illustrate that information on climate change causes, risks and impacts was not readily available. The results indicated that the Manonyane communities observed climate change on their own. Despite acknowledging that wildfires were discouraged by law, the case of the Manonyane communities demonstrated that wildfires were still a problem because community members secretly burnt grass. This behavior (of burning grass) by the Manonyane communities showed that it could benefit if information about climate change was shared on Lesotho radios and television and through any other accessible platforms.

The literature premised that for communities to influence climate change policies, access to adequate information on climate change causes, risks and impacts was required for active community participation (Cai et al., 2017; Shepard et al., 2018). The way information was disseminated and collected in the Manonyane communities had a huge bearing on their lack of participation in the NCCP formulation. Thus, the term 'well informed' as defined by Schütz (1946, p. 465) could not apply to the Manonyane communities. The findings confirm that the Manonyane communities were not well informed of climate change because they did not understand climate change causes, risks and impacts technicalities and implications from a scientific point of view as they experienced it in their communities. An implication of these findings from the Manonyane context is that critical consciousness through dialogue, participatory action and collective empowerment is key to enhancing community participation.

Influence of community understanding of climate change on community participation in policy formulation and implementation

Contrary to how the literature had defined community understanding, the Manonyane communities' understanding of climate change could not be measured against the proposed definition. The Manonyane communities' understanding of climate change established that climate change can be defined using indigenous knowledge systems. Even so, the Manonyane communities' case demonstrated that indigenous knowledge systems should be complemented with scientific information on issues of climate change. This is possible if participation procedures are availed to communities. Further, the lack of an operational definition for the concept (community understanding of climate change) may account for fragmented community participation during climate change policy formulation and implementation in the Lesotho NCCP formulation and implementation (Ministry of Energy and Meteorology, 2017, p. 24).

Community's understanding of climate change impacts

This research found that the Manonyane communities understood climate change impacts as follows: reduced agricultural production, punishment by God because of moral decadence, change of seasons, heavy destructive rainfall, hailstorms, destroyed infrastructure and property by strong winds or rain and changes in seasons of the year. Comparison of the findings from the Manonyane communities, three issues emerged from the Manonyane communities' knowledge about the likelihood and severity of expected climate change impacts. First, the results showed that rural communities were differently affected compared to communities which are like townships, for example, the Roma Valley communities surrounding the NUL campus. Second, Manonyane communities made it clear that their communities did not understand climate change through the facilitation of the government. Some community members heard about climate change through the radio but that did not translate into understanding. The Manonyane communities concurred that the community members were not knowledgeable about the likelihood and severity of expected climate change impacts because no one was providing such information. A possible explanation for this finding could be that the Manonyane communities believed that being aware of climate change does not mean understanding.

Manonyane communities' understanding of climate change causes

The findings of this study regarding the Manonyane communities' understanding of the causes of climate change were that communities did not understand the causes of climate change from a scientific point of view. Noteworthy is that the Manonyane communities did not have the ability to comprehend climate change outside the impacts they were witnessing. Using their indigenous knowledge systems, many themes were coming out of the Manonyane communities' responses on how they understood the causes of climate change. It seems possible that the Manonyane communities were not considering human activities such as deforestation, wildfires, air pollution, urbanization, industrialization, and poor agricultural practices as causing climate change. Despite noticing natural causes such as climate variability, temperature changes and weather changes, the Manonyane communities' understanding of climate change causes was attributed to many different things stemming from lacking adherence to indigenous knowledge systems, pampers, built dams and South African factories. Consequently, consistent with the (IPCC, 1990, p. xiv; 2014, p. 9), the Manonyane communities' case demonstrates that it is when communities understand how human activities increase greenhouse gas concentrations in the atmosphere that there is a chance of behavior change in motivating their participation in climate change policy formulation and implementation.

Enhancing community participation through understanding climate change decision-making processes

The study found that the Manonyane communities were not getting official information about climate change decision-making processes. Another issue, which emerged from the findings, was that the Manonyane communities disproved the idea of asking councillors and chiefs about community perspectives arguing that a few individuals will not represent their communities, particularly on climate change issues.

Communities' understanding of climate change risks for themselves and the environment

Another notable finding from the Manonyane communities' response was the inability to separate climate change impacts and climate change risks. Generally, community members narrated the climate change impacts as their risks. Consistent with the literature, the Manonyane communities expressed willingness to participate in climate change policy process because climate change risks placed their communities on the risk of food insecurity, loss of rural livelihoods, insufficient access to drinking water, reduced agricultural production, strong winds and floods (IPCC, 2014, p. 15; Tan & Xu, 2019). Owing to this explanation, it is possible to assume that community understanding of climate change risks enhances community participation.

LESSONS LEARNED

The results from the Manonyane communities indicate that the use of community understanding and awareness of climate change by Lesotho policymakers is abstract regarding their influence on community participation in climate change policy formulation and implementation. This is because of the absence of a contextual general meaning of community understanding and awareness, which could contrast with the UNFCCC's expectation to see governments increasing community participation in climate change policy formulation and implementation. The Lesotho context raises a question regarding what the communities could do to increase their understanding and awareness of climate change. Moreover, it highlights the pivotal role of policymakers in adopting contextual strategies to enhance community participation in climate change governance. As such, this study proceeds by reflecting on and discussing the two lessons learned from community understanding and awareness of climate change, arguing that this requires a net balance of support, where policymakers support the perspectives of the communities and the communities supporting government policies on climate change (Reed et al., 2018, pp. 1–2).

First, the communities can initiate bottom-up deliberation or co-production approach with policymakers to increase their understanding and awareness of climate change. Hence this study uses the case of Manonyane communities to demonstrate what communities could do differently to enhance their understanding and awareness of climate change. Using their indigenous knowledge systems that has been developed over time communities have valuable knowledge to share with scientists to address climate change risks and impacts. Thus, communities can initiate climate change dialogue through the bottom–up/co-production and bottom–up one way or as noted by Reed et al. (2018). For this, the chiefs as the leaders of the community should lead such dialogues through pitsos (traditional public gatherings). During the application of bottom–up/co-production approach, stakeholders initiate engagements with other stakeholders (Reed et al., 2018, p. 10). The bottom–up one-way communication or consultation occurs when stakeholders via grassroots networks and social media, initiate participation by persuading decision-makers to include the communities in decision-making processes (Reed et al., 2018, p. 9).

Second, the communities can generate ideas together, observe the climatic changes and note their changes over a period by constructing the meaning of climate change through their indigenous knowledge systems. The Manonyane communities were aware of climate change for decades using indigenous knowledge systems. They have, however, defaulted in applying this knowledge from a communal perspective to address climate risks and impacts. Observing climate changes and adapting their own agricultural practices to mitigate further climate risks is part of the dialogue that should be driven by communities. This should facilitate their use of indigenous knowledge systems to enhance their understanding and awareness of climate change (Wamsler et al., 2020). What makes one curious about the Manonyane communities' case regarding indigenous knowledge systems is that there was a lot of joy and excitement during the focus group discussions in all the communities when community members were sharing their knowledge on the subject matter. It felt good for the Manonyane communities to talk about something they knew, lived and experienced. Community elders through the mobilization of the chiefs can record community understanding and awareness of climate change through indigenous knowledge systems. Such processes should include the chiefs. No plan can be accepted without the chief's consent because chiefs mobilized their community members for collective action (Samaddar et al., 2019).

CONCLUSION

This chapter explored the anatomy of community understanding and awareness of climate change towards nudging community participating in climate governance. This study found that the Manonyane communities understand climate change from an indigenous knowledge systems perspective. Scientific knowledge of climate change and climate change policy processes was lacking. Against this background: "How should the current NCCP be adjusted to improve community understanding and awareness of climate change?" and "How does the research contribute to advance knowledge and practice of climate governance?" These questions are answered in this section by discussing the implications for (a) climate governance through policy and (b) theory.

Policy Implication

This study confirmed that a lack of community understanding of climate change might seriously hamper community participation in climate change policy formulation and implementation. The implications for policy and theory are in relation to the importance of translating the theoretical conceptualising of community understanding and awareness of climate change into practice for communities. Whilst it is critical to conceptualize community understanding of climate change from a theoretical perspective its practical application should be located in enhancing community awareness and understanding through pragmatic policy approaches and solutions. Framing implementation of the NCCP requires viewing environmental problems with human activities which causes climate change. Despite the natural causes over a long period, community understanding, and awareness should be measured per community. Community understanding and awareness of climate change are complementary and not independent. As a result, incorporating the conceptualizing of community understanding and awareness of climate change in the NCCP could specify areas which should be worked on to ensure a standardized use of community understanding and awareness of climate change. The operationalizing of the concepts understanding and awareness of climate change in the NCCP should not be broad and vague, considering the complexity of community participation in climate change policy formulation and implementation. A lack of communities' understanding of climate change may seriously hamper their participation in climate change policy formulation and implementation, which could result in failed policy implementation. It will therefore be more useful if the operationalizing of the meaning of community understanding and awareness of climate change are standardized in the NCCP.

Thus, from a policy perspective, such approaches and solutions in the NCCP should be located in a policy praxis that fosters communities' awareness and understanding considering the unique context of communities. Consequently, this could entail fostering community participation in the respective policy stages of formulating and implementing climate changes policies. Secondly, such a policy praxis could also apply nudging to community participation to enhance their awareness and understanding of climate change risks, mitigation and adaptation. Particularly carving a more participative role for communities in devising mitigation and adaptation activities is essential. Thirdly, climate change terminology related to articulating risks, adaptation and mitigation should be communicated in a language compatible with the local vernacular of communities and indigenous knowledge systems.

Effective climate education and literacy should target selected groups in the NCCP. It may be better if the policy encourages compulsory climate literacy for politicians first before any other groups because policies detached from community realities may not be successful during implementation. The other group of people is the media. Participants interviewed for this study perceived that the media misrepresented scientific results. In some cases, communication was done through the media and the communities were unaware that the climate change message was being communicated to them. Because of the growing gap between what is known about climate change by the communities and what is understood by communities, urgent attention is needed to define climate literacy in the NCCP and start climate education which is currently low, as proven by this study. The aim of climate literacy in the NCCP should be to make learners understand that human activities are the primary cause of climate change. The following questions about climate literacy should be addressed in the NCCP: "How do you intend to address misconceptions?"; "Which language to use?"; "What time should interventions be done in the curricula?"; "Who are the formal educators to carry out the task?"; and "What is the role of life experiences such as indigenous knowledge systems in understanding climate change?" This is critical in changing the community's mindset towards climate change (Wamsler et al., 2020).

Theoretical Implication

This section elucidates how this research contributes to advance knowledge and practice of climate governance. Community awareness that enhances community participation relates to climate governance which considers the following principles when enhancing community understanding and awareness of climate change (see Figure 19.4). These are (a) having a clear policy process and planned mitigation and adaptation measures; (b) engaging communities not individuals, seeking community approval and agreement regarding a negotiated meaning of community understanding and awareness of climate change; (c) provision of relevant information (the NCCP should explain what is known about climate change to communities and correct misconceptions); (d) demystifying misconceptions about climate change by giving relevant scientific information on the causes, risks, and impacts of climate change. Balanced information dissemination on local weather conditions and climate change using known community scenarios (e.g., an extreme weather event) could enhance community understanding and awareness of climate change. shows an overview of what

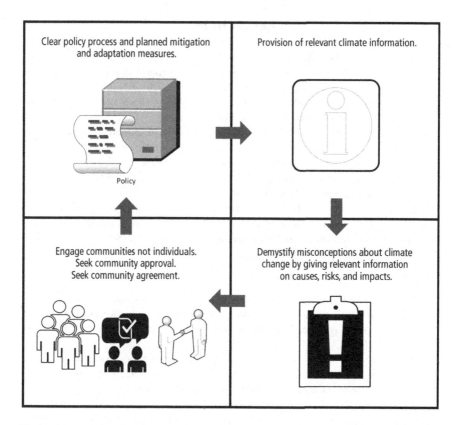

Figure 19.4 Enhancing community understanding and awareness of climate change.

could be done by policymakers and researchers to enhance community understanding and awareness of climate change.

Against the background of this chapter, it can be concluded that future research aimed at enhancing community awareness and understanding of climate change should be embedded in policy processes that advance social learning, develop place-based knowledge of climate change, and empowers communities to become social experts and community advocates of climate change. It should also be clear how policymakers can incorporate indigenous knowledge systems in the climate change policy process to ensure community understanding and awareness of climate change that leads to their participation in policy formulation and implementation.

REFERENCES

African Union. (2014). *African strategy on climate change.* European Parliament. Retrieved July 20, 2019, from https://www.europarl.europa.eu/RegData/etudes/BRIE/2022/738201/EPRS_BRI(2022)738201_EN.pdf

Blatter, J. K. (2008). Case study. In L. Given (Ed.), *The SAGE encyclopedia of qualitative research methods* (pp. 68–71). SAGE Publications.

Brink, E., & Wamsler, C. (2019). Citizen engagement in climate adaptation surveyed: the role of values, worldviews, gender and place. *Journal of Cleaner Production, 209,* 1342–1353. https://doi.org/10.1016/j.jclepro.2018.10.164

Bureau of Statistics. (2018). *2016 population and housing census analytic report: Population Dynamics, Vol. III.* A. Maseru, Government of Lesotho.

Cai, X., Haile, T. A., Magidi, J., Mapedza, E., & Nhamo, L. (2017). Living with floods household perception and satellite observations in the Barotse floodplain, Zambia. *Physics and Chemistry of the Earth, 100,* 278–286. https://doi.org/10.1016/j.pce.2016.10.011

Cambridge Dictionary. (n.d.). *Anatomy.*

Cambridge Dictionary. (n.d.). *Understanding.*

Chatanga, R., & Biljohn, M. (2023). New public governance theory: A framework for lesotho policymakers to enhance community participation during climate change policy formulation and implementation. *Administratio Publica, 31*(2), 1–24.

Climate Change Knowledge Portal. (n.d.-a). *Lesotho: Climate change overview > Country summary.* World Bank Group. Retrieved October 19, 2022, from https://climateknowledgeportal.worldbank.org/country/lesotho.

Climate Change Knowledge Portal. (n.d.-b) *Lesotho: Risk > Historical hazards.* Retrieved October 19, 2022, from https://climateknowledgeportal.worldbank.org/country/lesotho/vulnerability.

Creswell, J. W., & Creswell, J. D. (2018). *Research design: Qualitative, quantitative, and mixed methods approaches.* SAGE Publications.

Dang, W. (2018). How culture shapes environmental public participation: Case studies of China, the Netherlands, and Italy. *Journal of Chinese Governance, 5*(3), 1–23. https://doi.org/10.1080/23812346.2018.1443758

Department of Environment. (2014). *Lesotho environment outlook report: Environment for national prosperity.* Ministry of Tourism, Environment and Culture, Government of Lesotho.

Dhungana, N., Silwal, N., Upadhaya, S., Regmi, S. K., & Adhikari, S. (2018). Local people's perception and awareness of climate change: A case study from community forests in Lamjung District, Western Nepal. *Banko Janakari, 2018, 28*(2), 60–71. https://doi.org/10.3126/banko.v28i2.24189

Friis-Hansen, E. (2017). Implementing African national climate change policies. In E. Friis-Hansen (Ed.), *Decentralized governance of adaptation to climate change in Africa* (pp. 11–25). CABI.

Graneheim, U. H., Lindgren, B.-M., & Lundman, B. (2017). Methodological challenges in qualitative content analysis: A discussion paper. *Nurse Education Today, 56,* 29–34. http://dx.doi.org/10.1016/j.nedt.2017.06.002

Graneheim, U. H., & Lundman, B. (2004). Qualitative content analysis in nursing research: Concepts, procedures and measures to achieve trustworthiness. *Nurse Education Today, 24*(2), 105–112. https://doi.org/10.1016/j.nedt.2003.10.001

Hafiz, K. (2008). Case study sample. *The Qualitative Report, 13*(4), 544–559.

Hegger, D. L.T, Mees, H. L.P, Driessen, P. P. J., & Runhaar, H. A. C. (2017). The roles of residents in climate adaptation: A systematic review in the case of the Netherlands. *Environmental policy and governance.* http://dx.doi.org/10.1002/eet.1766

Intergovernmental Panel on Climate Change. (1990). *Climate change: The IPCC scientific assessment* (0003-0996). Cambridge University Press.

Intergovernmental Panel on Climate Change. (2001). *Climate change 2001: Impacts, adaptation, and vulnerability.* Cambridge University Press.

Intergovernmental Panel on Climate Change. (2007). *Climate change 2007: Impacts, adaptation and vulnerability. Contribution of Working Group II to the Fourth Assessment Report of the Intergovernmental Panel on Climate Change.* Cambridge University Press.

Intergovernmental Panel on Climate Change. (2014). *The IPCC's fifth assessment report: Whats in it for Africa.*

Jagers, C. S., & Stripple, J. (2003). Climate governance beyond the state. *Global Governance, 9*(3), 385–399.

Larsen, C., & McGuinness, S. (2016). Climate change adaptation planning with peri-urban local government in Victoria, Australia. In B. Maheshwari, V. P. Singh, & B. Thoradeniya (Eds.), *Balanced urban development: Options and strategies for liveable cities, water science, and technology* (pp. 395–407). Springer.

Lesotho Government, & World Food Programme. (2019). *Improving adaptive capacity of vulnerable and food-insecure populations in Lesotho.* https://www.adaptation-fund.org/project/improving-adaptive-capacity-vulnerable-food-insecure-populations-lesotho-2/

Lesotho Meteorological Services. (2013). *Lesotho's second national communication to the conference of parties (COP) of the United Nations: Framework convention on climate change.* Maseru. https://unfccc.int/resource/docs/natc/lsonc2.pdf

Lorenzoni, I., Nicholson-Cole, S., & Whitmarsh, L. (2007). Barriers perceived to engaging with climate change among the UK public and their policy implications. *Global Environmental Change, 17*(3), 445–459. http://dx.doi.org/10.1016/j.gloenvcha.2007.01.004

Ministry of Energy and Meteorology. (2017). *The kingdom of Lesotho national climate change policy 2017–2027.* Government of Lesotho.

Reed, M. S., Vella, S., Challies, E., de Vente, J., Frewer, L., Hohenwallner-Ries, D., Huber, T., Neumann, R. K., Oughton, E. A., & Sidoli del Ceno, J. (2018). A theory of participation: What makes stakeholder and public engagement in environmental management work? *Restoration Ecology, 26,* S7–S17.

Samaddar, S., Ayaribilla, A. J., Oteng-Ababio, M., Dayour, F., & Yokomatsu, M. (2019). Stakeholders' perceptions on effective community participation in climate change adaptation. In A. Sarkar, S. R. Sensar, & G. W. VanLoon (Eds.), *Sustainable solutions for food security* (pp. 355–379). Springer.

Schmidt, A. T. (2017). The power to nudge. *American Political Science Review, 111*(2), 404–417. https://doi.org/10.1017/S0003055417000028

Schütz, A. (1946). The well-informed citizen: An essay on the social distribution of knowledge. *Social Research, 13*(4), 463–478.

Shepard, S., Boudet, H., Zanocco, C. M., Cramer, L. A., & Tilt, B. (2018). Community climate change beliefs, awareness, and actions in the wake of the September 2013 flooding in Boulder County, Colorado. *Journal of Environmental Studies and Sciences, 8*(3), 1–14. https://doi.org/10.1007/s13412-018-0479-4

Smit, B., Burton, I., Klein, R. J. T., & Wandel, J. (2000). An anatomy of adaptation to climate change and variability. *Societal Adaptation to Climate Variability and Change, 45*, 223–251. https://doi.org/10.1007/978-94-017-3010-5_12

Tan, H., & Xu, J. (2019). Differentiated effects of risk perception and causal attribution on public behavioral responses to air pollution: A segmentation analysis. *Journal of Environmental Psychology, 65*, 1–7. https://doi.org/10.1016/j.jenvp.2019.101335

Tannenbaum, D., Fox, C. R., & Rogers, T. (2017). On the misplaced politics of behavioural policy interventions. *Nature Human Behaviour, 1*(7), 1–7. https://doi.org/10.1038/s41562-017-0130

United Nations Development Programme. (2013). *Reducing vulnerability from climate change in the foothills, lowlands and the lower Senqu River Basin Maseru, Lesotho.* UNDP Lesotho.

United Nations Economic Commission for Europe. (2015). *Maastricht recommendations on promoting effective public participation in decision-making in environmental matters: Prepared under the Aarhus Convention.* https://cir.nii.ac.jp/crid/113000 0795238636416

United Nations Environment Programme. (2016). *The rise of environmental crime: A growing threat to natural resources, peace, development ande security.*

United Nations. (2015). *Transforming our world: The 2030 agenda sustainable development* (A/RES/70/1). Retrieved February 4, 2018, from https://sdgs.un.org/2030agenda

Vaismoradi, M., Turunen, H., & Bondas, T. (2013). Content analysis and thematic analysis: Implications for conducting a qualitative descriptive study. *Nursing & Health Sciences, 15*(3), 398–405. https://doi.org/10.1111/nhs.12048

Von Colln-Appling, C., & Giuliano, D. (2017). A concept analysis of critical thinking: A guide for nurse educators. *Nurse Education Today, 49*, 106–109. https://doi.org/10.1016/j.nedt.2016.11.007

Wamsler, C., Alkan-Olsson, J., Björn, H., Falck, H., Hanson, H., Oskarsson, T., Simonsson, E., & Zelmerlow, F. (2020). Beyond participation: When citizen engagement leads to undesirable outcomes for nature-based solutions and climate change adaptation. *Climatic Change, 158*(2), 235–254. https://doi.org/10.1007/s10584-019-02557-9

Wi, A., & Chang, C.-H. (2019). Promoting pro-environmental behaviour in a community in Singapore—From raising awareness to behavioural change. *Environmental Education Research, 25*(7), 1019–1037.

World Bank. (2022). *The World Bank in Lesotho.* Retrieved from October 19, 2022, from https://www.worldbank.org/en/country/lesotho/overview

PART V

CLOSING

CHAPTER 20

ASSESSING THE QUALITY OF CLIMATE GOVERNANCE

Lessons Learned and the Way Forward

Peter F. Haruna
Texas A&M International University

Abraham David Benavides
University of Texas at Dallas

ABSTRACT

Climate governance is the term that scholars, researchers, and policy makers have used to better understand and inspire thinking, development, and practice in the era of climate change. What does climate governance look like in the Global South countries which are disproportionately affected by climate change and severely understudied? To address this question, the book examined Africa, Latin America, and the Caribbean (ALAC) to gain a better understanding of how these regions understand climate change, how they organize themselves, and how they respond to it with existing limited capacities. Synthesizing ideas, empirical findings, and policy recommendations, this closing chapter broadens the understanding of climate governance by sum-

Climate Governance in International and Comparative Perspective, pages 477–499
Copyright © 2024 by Information Age Publishing
www.infoagepub.com
477

marizing their collective experience and evaluating their performance from a pragmatic administration and governance perspective based on what works and what doesn't. In addition, it specifies lessons learned and highlights policy pathways even as knowledge of climate change and climate governance norms, processes, and facts evolve and remain uncertain and contested.

Climate governance is the term that scholars, researchers, and policy makers have used to better understand and inspire thinking, development, and practice in the era of climate change—operationalized as the level of global warning. Because climate change is a negative collective externality, knowledge of what climate governance is and/or could be is evolving (Intergovernmental Panel on Climate Change [IPCC], 2023). However, and as Huitema et al. (2016) argue, current knowledge and understanding basically revolve around coordination and collaboration among groups, communities, organizations, governments, and transnational institutions to achieve the mutually beneficial goal of minimizing vulnerabilities to climate change impacts. In essence, climate policy and climate action require governance encompassing "the patterns that emerge from the governing activities of social, political, and administrative actors" (Kooiman, 1993, p. 2). But governance, even in the best-case scenario, is debatable and it is reasonable to ask how the Global South engages with diverse actors and negotiates solutions within unprecedented climate change impacts.

As stated in the introductory chapters, the primary goal in writing this book is two-fold. The first goal is to tell the story of climate governance through the prism of the development process, focusing on the "Global South's particularities" (Sapiains et al., 2021, p. 46). Likewise, the goal was to illustrate the unique experiences inadequately and/or inaccurately portrayed by the current literature in the field. Consumed by Eurocentric epistemic claims, the consideration of alternative legitimate ways of knowing and being remains stifled amid calls for decolonial studies and epistemic justice (Sapiains et al., 2021; Yacob-Haliso et al., 2021). Additionally, the purpose of the book was to distill lessons and contribute to emerging "Southern theorizing" about the texture of life based on non-European ways of knowing and acting on public problems. In sum, the book evaluated climate governance based on the difference that contextualized knowledge can make in the fight against climate change.

In identifying these goals, we relied on literature that offers conceptual/theoretical clarity, and sheds light on empirical observation of the struggles and stresses that the Global South countries are experiencing with climate change action required by the UN's SDG13. Such struggles and stresses originate from what we see as the paradox of climate change. Within this context, these counties contribute least to GHG emissions, respond with scant resources, and apply technocratic knowledge incompatible with their lived experience. Nevertheless, we found two literature reviews with structured classification schemes that enlighten the current state of knowledge and pinpoint gaps that need to be filled. For example, Sapiains et al. (2021) uncover "an absence of perspectives from the Global South" and argue for "dedicated efforts...to build a Southern perspective of climate governance" (p. 56). Likewise, Baninla et al. (2022) uncover sparse and uneven research, arguing for more work "to improve our understanding of the status and trends of research on climate change adaptation and mitigation in Africa" (p. 1). Both sources helped us to frame the call for proposals and evaluate the quality of climate governance research and scholarship in regions that in our view are burdened most and yet are united by their mutual vulnerability to climate change impacts. In addition, Filho's (2015) authoritative and interdisciplinary book volume inspired the current book in more ways than one.

By leveraging these sites of discourse along with the UNFCCC (1992) principles and IPCC (2023) predictions about global warming, the book contextualized and highlighted climate-system responses in the world's poorest but least emitting ALAC regions. While these discourses along with insights from the World Bank and regional protocols need unpacking, they serve as the foundational knowledge and a springboard for examining climate governance systems. Based on the contributions to this volume and the extant literature, we provided context and conclude by assessing the quality of climate governance as measured by the extent to which climate initiatives close gaps between climate governance thinking, practice, and development in a complex and interconnected world. In sum, climate governance is being examined from varied perspectives in the continued quest for knowledge and understanding, but they must be brought into meaningful dialogue with one another.

To synthesize perspectives and enhance coherence in the context of literature, we developed and applied a five-part assessment framework (Table 20.1) with the goal of making sense of the complex and dynamic phenomenon of climate governance. Included in the framework were structuring questions (Column 1), chapter focus (Column 2), assessment tasks (Column 3), concepts and theories anchoring the chapter (Column 4), and methodologies for summarizing and assessing the issues and themes (Column 5). Both the structuring question and chapter focus facilitate the identification of the contours of climate governance in a snapshot view. In addition, the assessment tasks highlight and elaborate on theories and concepts such as coordination, collaboration, vulnerability, food, water, and energy in the three regions (Baninla et al. (2022). The appropriate methodology for assessing the quality of climate governance included synthesizing ideas from across disciplines such as politics, economics, and ecology as a way of building on and contributing to existing knowledge (Appleby, 2019).

TABLE 20.1 Framework for Assessing the Quality of Climate Governance in Africa, Latin America, and Caribbean

Question	Chapter Focus	Assessment Tasks	Theoretical tools guiding the synthesis	Methodologies for organizing and synthesizing the chapter
How does climate governance look like in Africa, Latin America, and Caribbean regions?	Describing, comparing, and connecting climate governance patterns, systems, institutions, processes, and outcomes to development policy in ALAC regions for protecting marginalized and vulnerable populations within the UN's Framework Convention on Climate Change.	Assessing the design and implementation of climate governance systems, institutions, processes, adaptation capacities, and outcomes related to development policy for protecting vulnerable populations in the ALAC regions.	Development, decolonization, democratization, transnationalism, new public service, new public passion, geo-strategic governance, adaptive capacity, resiliency, collaboration, social equity, social justice, eco-systems.	Comparative analysis, empirical analysis, case study analysis, documentary analysis, historical analysis, explorative, and conceptual analysis.

Source: Compiled by authors based on review of climate governance literature.

The remainder of this chapter is structured as follows: In the next section, the question of how climate governance looks in these regions is addressed by summarizing and specifying the contours of global and regional climate governance systems as they reflect in the three regions. Emphasizing institutions, organizations, actors, policies, norms, and processes, examples from different chapters were cited to ground and connect them to literature. Consistent with the IPCC's (2023) recommendation, the quality of climate architecture is assessed based on the extent to which it connects to society, promotes collaboration across sectors, reflects life and lived experience, and advances sustainable development. As the IPCC argues, the climate impacts people, economies, ecological, and ecosystems, which can be widespread and severe. Hence, because future risks are likely to escalate with every increase in the degree of warming, collaboration and coordination among people and across boundaries are needed to minimize vulnerability.

The final section considers outcomes, challenges, and lessons learned, arguing that there is a need to better align strategies and institutions to the development goal of achieving inclusion, participation, improvement, and transformation. Framed and based on Eurocentric governing assumptions, climate governance design and implementation follow the coloniality of western conceptualization of administration and governance without considering contextual realities of the Global South (Dadze-Arthur, 2022). Improving the quality of climate governance requires assessing the degree of fit of national, provincial, and local adaptation and mitigation initiatives with local social, cultural, political, and economic conditions and values. As a human enterprise, climate governance must respond to change as and when situated knowledge and technologies become available. Ultimately, the quality of climate governance will depend on how microlevel governance actors, especially communities view their roles relative to macrolevel goals and the extent of the investment in research, education, and training to prepare individuals for their interdisciplinary, transdisciplinary, and cross-cultural roles in climate governance.

ASSESSING CLIMATE GOVERNANCE APPROACHES

This section begins the assessment of the quality of climate governance in the ALAC regions by examining the approaches that the countries have adopted and how well those approaches align to their experience and to current knowledge conveyed through research and scholarship. Literature reflects two broad climate governance regime types—adaptation and mitigation. While it is unrealistic to dichotomize them, UN's climate action is based on either plans and measures to adapt to climate impacts or to prevent or reduce GHG emissions. Since 2010, the UN's Adaptation Committee has promoted enhanced action on adaptation and highlighted ways for Parties and actors to increase adaptation ambition (United Nations, 2019). On the other hand, the UNEP takes a multifaceted approach to mitigation, including using recent technologies and renewable energies, changing management practices and consumer behavior, among others (https://www.unep.org/explore-topics/climate-action/what-we-do/mitigation). How do ALAC policies align and respond to these global and international frameworks? And do they apply cultural sensitivity to define, design, and manage trade-offs among climate action goals?

Most of the chapters in this volume deal directly with climate adaptation governance. In this sense, adaptation concerns minimizing climate change vulnerabilities, capacity development to withstand extreme weather and climatic conditions, and addressing socioeconomic inequities and injustice, among others (Huitema et al., 2016). A couple of chapter examples will suffice: "Climate Governance in the East African Community," and "Trouble in Paradise" confront vulnerabilities and adjustments to the consequences of climate change impacts. But this is not to say that climate change risks and responses are the same in both cases. Nor does it underestimate the importance of mitigation risks and responses in those countries. To the contrary, chapter descriptions reflect the interconnectedness between adaptation and mitigation responses in multiple sectors, including transportation, agriculture, forestry, and other land use sectors.

In contrast, a handful of chapters engage directly with mitigation governance or efforts to understand and control global warming such as in the chapter on "Decentralized Solar Systems in Kenya's Cities," which explores the steady spread of micro solar systems to address the energy needs of urban dwellers in a cost-effective fashion. But this initiative has both adaptation and mitigation dimensions. As well, the chapter on "Forestry and Evolving Climate Governance of Costa Rica and Panama, 1969–2022," focuses on reforestation and agroforestry to meet commitments to reducing anthropomorphic greenhouse gas emissions. But the issue at the core of climate change action—an obstacle to meaningful action—is the question of "who is responsible for the problem and who must bear the cost of the solution" (Grunstein, 2021, p. 3). In short, the issue is one of climate justice in which the most emitting nations must bear the cost of both adaptation and mitigation through the "Loss and Damage" Fund. Nonetheless, the cause of the problem requiring either adaptation and/or mitigation governance is human-induced climate change impacts (Cook et al., 2016). To this extent, both adaptation and mitigation have a common history, and it is reasonable to frame them as collective or public goods for which global communities and citizens are responsible.

The disparity between adaptation and mitigation research and scholarship in these regions is not surprising. First, and as Adams (2009) argues, the most common response to environmental risk is for individuals and communities "to adapt their lives and systems of production to cope with it" (p. 364). This is the basis for recognizing the adaptability of developing societies to the scarcities of shelter, food, water, and security, as well as natural disasters such as in Haiti (2010). Baninla et al.'s (2022) review of literature reveals that climate research in Africa is more focused on adaptation. Second, the disparity lies in elevated vulnerability of the regions due to population growth, urbanization, and reduced average precipitation (IPCC, 2023). For that reason, the African Capacity Building Foundation's (2022) is implementing plans and programs to improve the performance of institutions working on sustainable climate adaptation eco-systems in Africa (https://ruforum.wordpress.com/2022/06/03/). Likewise, the World Bank Group's (2022) "A Roadmap for Climate Action in Latin America and the Caribbean" prioritizes adaptation while recognizing that the region accounts for only eight percent of global GHG emissions.

As alluded to above, chapter descriptions of climate governance patterns comprising of institutions, organizations, actors, strategies, policies, norms, resources, and processes at various levels of government tend to converge rather than diverge from the literature. Most of the chapters highlight climate governance based on adaptation principles and practices ranging from self-governance by localities to society-wide responses through national planning for change and transformation. For example, "Participatory Climate Change Governance in Enabling Leadership for Climate Action" argues that "the lowest existing local participatory legislated structure is the most suitable space to foster climate change understanding and debate." This explains in part why adaptation governance is associated with localities, where coordination between private entities and communities facilitates the achievement of collective goals. Thus, adaptation governance benefits from fostering community-based management and decision-making systems that tap into native wisdom and indigenous knowledge and experience (Huijsman & Savenije, 1991). But native wisdom tends to conflict with the current utilitarian civilization (Suzman, 2021), and its nonutilitarian and humanizing nature often gets lost in development discourses.

On the other hand, adaptation governance through national planning for change and transformation reflects in literature and features substantially in the chapter descriptions. For example, national adaptation planning is not only common in the regions but also, it is based on the principle of "nationally determined contributions" (NDC) embedded in the UNFCCC under the Paris Agreement (2015, Article 4, para 2). Empowered by this principle, ALAC countries articulate and communicate both adaptation and mitigation efforts. In accordance with this principle, the responsibility for designing and implementing adaptation and mitigation policies resides in national governments. For example, 53 African countries submitted intended NDCs ("Number of Countries," n.d.).

In Latin America and the Caribbean, 28 countries either submitted or planned to submit NDCs (https://unfccc.int/sites/default/files/resource/NDC%20Survey%20Report%202020-Caribbean_Jan2021.pdf). The significance of the NDCs is that they integrate efforts both to reduce GHG emissions and adapt to climate impacts. Moreover, coordinated global climate actions can determine whether the world achieves the long-term goals of the Paris Agreement. But in the absence of adaptation and mitigation funding—there is no guarantee that the Global North's promises will come through—NDCs remain but meaningless rhetoric.

Several chapter descriptions exemplify national adaptation planning coordinated by governments, encompassing complex interactions of public, private, and civic actors (Brink & Wamsler, 2018). For example, Suriname embarked on an elaborate national adaptation plan for the near to long term, culminating in the creation of a new national climate institute. In addition to administering the national adaptation plan involving actors, norms, and rules, Suriname's National Climate Institute serves as a resource and plays a lead role in adaptation implementation efforts. Likewise, Egypt, described as "highly vulnerable to climate change impacts" just as other Mediterranean countries suffering from heatwaves, has a broad-based

National Climate Change Council (NCCC) responsible for coordinating policies and plans. Headed by the Minister of the Environment, Egypt's NCCC draws representatives from NGOs and Think Tanks as co-actors and partners in adaptation and mitigation planning.

Such descriptions indicate that the national level framework is common in the selection of appropriate authorities for climate action. In short, the approach in developing climate governance in ALAC is consistent with literature indicating that adaptation plans and practices typically include selecting appropriate levels of authority for climate action, creating new institutions by adopting norms and rules, and specifying measures to gauge policy impacts (Cox, 2016; Sapiains et al., 2021).

However, national adaptation planning presents both an opportunity and a challenge in these regions. On the one hand, national governments have power to decide national priorities and determine the role that various levels of government and actors can play in planning and implementing adaptation policies. As Jordan et al. (2010) and Massey et al. (2015) argue, the ability to decide at which level to act in the context of adaptation is important because it has implications for effectiveness and distributional impact. On the other hand, nationally planned large-scale projects do not have a good reputation in these regions. For example, resettlement schemes, dams, and irrigation systems are the source of unsustainability there because they are implemented without environmental, social, and economic appraisals (Adams, 2009, pp. 322–324). In addition, national-level planning runs counter to the paradigm of bottom-up planning, decentralization and participation, and community development (Agrawal & Gibson, 1999). In short, national adaptation plans entail dilemmas and raise questions for climate governance.

The dilemmas about national adaptation planning addressed in literature and reflected in book chapter descriptions concern the emergence of the multi-level and multi actor environment. Comprising of global, regional, national, and subnational jurisdictions involved in co-creating and co-implementing climate policies, multi-level and multi actors originate from literature on international cooperation and its challenge of the authority of nation states (Rhodes, 1996; Rosenau & Czempiel, 1992). This literature emphasizes how organizations outside of nation states manage international cooperation and how actors outside the state realm provide services either as complements to or substitutes for government services. In recent governance literature, Sapiains et al. (2021) identify multi-level governance, among several others, arguing that the state must not go it alone and that "different disciplines and perspectives must be integrated" (p. 56), while Massey et al. (2015) suggest that adaptation should be implemented across multiple levels.

The multi-level and multi actor perspective reflected in chapter descriptions such as "Climate Governance in Development Perspective" that discusses the experiences of Ghana and Nicaragua in developing and managing natural resources. For example, the chapter identified and discussed the roles of multinational, regional, and civil society organizations in the development of Ghana's Climate Change Action Strategy. Likewise, "Environmental Vulnerability and Disaster Prevention" emphasizes social participation involving civil society organizations, communities, and academics in different spaces and levels as a critical factor for improving local government capacity in reducing vulnerability and mitigating disaster impacts. In "Ocean Health: Living Resources and Blue Economy Governance in the South African Context," the author specifies the primary international governance framework, including the UNFCCC, UN Convention on Biological Diversity, Convention on Wetlands, and Regional Seas Conventions and Action Plans discussed in literature. While the multi-level model is prevalent, it poses challenges, and its degree of influence varies by region/country as discussed below.

In the ALAC regions, chapter descriptions show how international organizations such as the United Nations, World Bank, International Monetary Fund, Organization of American States, and the African Union pressure national governments to take climate-related action (UN's SDG13; AU's Draft Africa Climate Change Strategy 2020–2030). For example, national governments abide by the Paris Agreement (2015), Helsinki Principles (2019), and UN Convention on the Law of the Sea (1982), aligning policies and practices and requiring them to promote adaptation and mitigation. At the same time, global networks such as the United Nations' (2015) "Sendai Framework for Disaster Risk Reduction (2015–2030)" and The National Academies of Sciences, Engineering, and Medicine (2020) do not only facilitate knowledge exchange as is presumed but also, they perpetuate hegemonic knowledge from western society to the rest of the world.

In these instances, the role of the global and international level and actors is essential and policy choices remaining for national authorities are the degree of involvement as coordinating mechanisms. In principle, regions learn from each other through these networks, as Smith (2007) argues, and municipalities can create their own action plans (Schreurs & Tiberghein, 2010). And at subnational levels, where, adaptation action occurs, businesses, NGOs, and citizens groups influence decision makers through strategies such as media campaigns, advocacy, and/or protests. Thus, in principle, adaptation outcomes are not necessarily the product of international and/or central government decisions alone, but also they are shaped by local government (Rhodes, 1996).

However, this should not be construed to mean that the multi-actor and multi-level paradigm is uniquely in the Global South. In fact, it plays a significant role globally in what Sapiains et al.'s (2021) literature review reveals as "adaptive governance" (p. 52), focusing on local experience with natural resources management. The chapter case studies on "Environmental Vulnerability and Disaster Prevention," "Environmental Effects on Groundwater and Urban Drinking Water Quality," and "Public–Private–Community Partnerships as Pathways for Sound Climate Governance" serve as examples only. As Marquardt (2017) argues, environmental governance is "the second most studied policy in connection with multi-level governance" (p. 168). And Sattler et al. (2016) asserts that such multi-actor governance is necessary for addressing climate change vulnerabilities. The argument is that multi actor perspectives create opportunities to do things differently by promoting public participation, recognizing and integrating knowledge, and developing context-relevant education and learning. Thus, climate multilateralism provides an enabling environment for social and environmental innovation and experimentation. Marks and Hooghe (2004) insist that "the dispersion of governance across multiple jurisdictions is both more efficient than, and normally superior to, central state monopoly" (p. 16).

But the problem is that multi-levels and multi actors do not exist in a vacuum. Neither do policy making and policy implementation occur in a vacuum. They exist in complex political, economic, social and cultural settings involving diverse individuals and groups with unequal power relations and conflicting interests. Benz et al. (2009) highlight the complexity and connects it to questions of power. The complexity occurs in ALAC nations, where environmental causes are hardly unifying. In such a complex environment, the mode of governance, willingness to commit to norms, and ability to cooperate are key to achieving desirable goals (Jordan et al., 2010). As literature suggests, adaptation governance requires legitimacy through consensus about the public interest (McHarg, 1999). In this light, several chapter descriptions show that the preferred mode of climate governance across ALAC is hierarchically structured through policy and regulation with little room for bottom up or networked arrangement as discussed below.

In Climate Governance in Development Perspective, co-authors discuss the centralized and hierarchical governing system of Ghana and Nicaragua to show how it mismatches with societies there and disconnects policy making largely from rural and marginalized populations. At the top of the hierarchy are multinational institutions such as the United Nations followed by regional bodies such as the African Union or InterAmerican Development Bank, national governments, and subnational jurisdictions in that order. Similarly, Forestry and Evolving Climate Governance of Costa Rica and Panama shows that reforestation programs are being implemented within a hierarchical framework informed by neo-liberalism as governing theory. While both nations are relatively successful in reducing carbon emissions and absorption of carbon through a combination of public-sector initiatives such as legislation and conservation, they emphasize markets, competition, and growth at the expense of public value frames such as participation, community, equality, ethics, and inclusion (Blessett et al., 2019).

With hierarchical institutional frameworks inherited from their respective colonial eras, the post-colonial states have neither the political will nor the capacity to create climate adaptation governance that is grounded in real life experience. Reinforced by decades of military dictatorship and authoritarian rule, the community participation needed by and for effective adaptation governance is lacking. As a result, adaptation policies and plans implemented within hierarchical frameworks are at best ineffective because they receive little to no cooperation and collaboration from citizens and communities. The best that the framework offers, as Sapiains et al. (2021) argue, is a "focus on preparedness for situations of stress and change" (p. 54) rather than changes in behavior. And as Marquardt (2017) observes, the lack of community involvement translates to differential opportunities to participate and resource disparities among multi-actors.

In the end, climate governance legitimacy—how actors perceive their roles, especially how citizens perceive institutional performance—remains in question (Ford & Ihrke, 2019).

As ALAC nations make challenging transitions to build resiliency, minimize inequities, and foster decarbonized and circular economies, the quality of climate governance remains in question. The chapters indicate that there is evidence of a shift toward integrating innovation and inclusive knowledge and emphasizing cooperation and collaboration among varied entities, but substantive challenges remain, most certainly on the question of cooperation and collaboration among multi-actors at multi-levels. In the next section, we draw insights from chapters to explore adaptation and mitigation conceptual, empirical, and normative challenges and outcomes in the face of shifting geo-political alliances before considering lessons learned and the way forward. The ongoing climate crisis and the responses to it create dilemmas and raise the question of whether commitments that ALAC nations make will necessarily survive political, economic, and social costs and trade-offs that come with formulating and implementing decarbonized policies.

CHALLENGES, OUTCOMES, LESSONS, AND THE WAY FORWARD

The foregoing discussion indicates, first, that there is convergence between chapter descriptions and literature on climate governance in ALAC regions with a focus on multi-levels and multi actors as flagship. This is based on the notion that environmental issues must not be left to national level planning but rather engage with global, subnational and non-state levels to facilitate the credibility of policy commitments (Piatoni, 2010; Pollack, 1997). But multi-level governance suffers from potential policy fragmentation and power imbalances between powerful Global North nations with financial and technical leverage and weak Global South nations. Second, the discussion highlights conceptual tools and principles such as coordination, cooperation, and collaboration reflected in literature and UNFCCC frameworks. While ALAC nations mostly rely on multilateral institutions such as the African Development Bank and Economic Commission for Latin America and the Caribbean, climate action derives from national planning. Given these complex interactions, we explore conceptual, empirical, and normative challenges and outcomes before considering lessons learned and the way forward. One of the most important aspects of multi-level governance is the annual UN Climate Change Conference held within the UNFCCC framework both to negotiate obligations to reduce GHG emissions and to increase climate adaptation finance for vulnerable nations.

Conceptually, it is hard to distinguish the state-specific role and function from the transnational role at this level. For example, previous hosts of these conferences such as Kenya (COP12), Mexico (COP16), South Africa (COP17), and Egypt COP27) played important roles both as global actors in climate governance mainly dominated by the Global North. At COP12, Kenya advocated in vain for the adoption of a five-year plan to support climate change adaptation in Global South nations and also for the adoption of the procedures and processes to create an adaptation fund. Likewise, at COP16, Mexico advocated for the parties to adopt the Cancun Framework, including USD100 billion per annum to create a green climate fund. While COP17 moved the ball afield, South Africa also advocated for the adoption of a management framework for a green climate fund of USD100 billion per annum. Among the successes of COP27 in Egypt as described in *Climate Change Governance and Institutional Structures in Egypt Pre and Post COP27* is symbolically influencing the global climate agenda and promoting climate awareness through domestic media. But the historic achievement is the breakthrough in negotiations to create a "loss and damage" fund and compensate Global South nations as they struggle to address climate change impacts (Ògúnmódèdé, 2022). This move recognizes the "widespread and rapid changes in the atmosphere, ocean, cryosphere, and biosphere" (IPCC, 2023, p. 5) and supposedly supports the call for "just transition declaration" adopted at COP26 (IndustriAll, 2021).

However, to reach the milestone achieved at COP27 required decades of solidarity work involving Global South nations, G77+China, and other stakeholders, highlighting exclusionary practices and inequities in global governance. As Shoukry (2022) describes it, it took calls of "anguish and despair" to establish the fund, and much remains to be done to make it operational.

While the creation of a "loss and damage" fund is a positive step toward achieving meaningful climate action, it has taken nearly three decades of hard-fought negotiations between the Global North and Global South nations to get there. They have to navigate geopolitical fights and a hierarchy of Global North v. Global South to build compromises (Quijano, 2000). But some of the outcomes show that international cooperation is possible even though logistics are unclear—who pays into the fund and how much do they pay? The IPCC (2023) and Guterres (2022) are concerned that progress is short of what Global South nations need to address climate impacts.

The uneven interactions between the Global North and Global South nations in global governance and their impacts on how to implement climate policies are the core of literature advocating a shift away from "nation states" as analytical units toward global policy and transnational administration (Bauer et al., 2018; Moloney & Stone, 2019). Among the crowd of global levels and actors are varied international organizations, civil society, and corporate actors. Because, as Moloney and Stone (2019) argue, the state does not exist in a vacuum but within a global community, global governance impacts domestic administration, questioning the inside/outside dichotomy. But while Moloney and Stone ironically caution against the "dominance of Western centric conceptualizations of 'global policy' and 'transnational administration,'" book chapter descriptions show that administration and policy frameworks in ALAC nation states are deeply implicated in Western mindset and culture, a phenomenon Grosfoguel (2007) describes as "global coloniality" (p. 221). According to Grosfoguel, the Global South's economic and political systems are visibly tethered to and shaped by "their subordinate position in a capitalist world system organized around a hierarchical international division of labor" (pp. 220–221).

Based on environmental laws, protocols, and organizational structures described and analyzed in *Climate Change Governance and Institutional Structures in Egypt Pre and Post COP27, Climate Governance and Sea Level Rise: A Case Study of Alexandria, Egypt,* and *The South African Policy Experience and Lessons learned in Sustaining Ocean Life,* climate governance architecture follows Euro-American governance traditions both in substance and process. Each of these case studies reflects interventions using formal state infrastructure and centralized authority to seek desirable outcomes. Informed by Western universalistic, objective, rational, technical, and institutionalized constructs (White & Adams, 1995), these interventions assume that climate change impact is subject to both negative externality and market failure for which "government know-how" action is needed. Scattered so-called infrastructural projects across ALAC landscapes are disconnected from life-world experience and controlled by technical experts and professionals with narrow disciplinary orientation. Technical and rational approaches dominate the UN's SDGs, World Bank's low carbon transition programming, and Africa

Agenda 2063, focusing on growth rather than resilience and risk-spreading. As Brooks et al. (2009) argue, projects' costs and benefits are unequally and inequitably distributed while the environmental costs include degradation (Millennium Ecosystem Assessment, 2005).

In refocusing the development narrative on integrating human wellbeing with economic growth and environmental protection, the UN SDGs (2015) present Global South nations with the opportunity to anchor climate-based development to the principle of "Leave no one behind." A blueprint aimed at ending poverty, achieving gender equality, and confronting climate change, SDGs require that implementation must be "participatory and inclusive, including all levels of sectors of government, civil society and the private sector, members of Parliament, national human rights institutions" (United Nations, 2023, p. 11). This is a tall order for the Global North governments, not to mention the Global South. Given their centralized and authoritative frameworks, this multi-stakeholder participation and inclusion requirement is unrealistic in ALAC nations. In fact, mainstreaming adaptation and low carbon transitions are only incremental responses in what appears as a business-as-usual development paradigm.

As the UN Voluntary National Assessment Reviews (United Nations, 2022) indicate, Global South nations are still struggling to address questions such as how the SDGs align with national policy frameworks, and how to address a gap analysis of SDGs and national frameworks. Most of the changes in investments, increased finance, and expanded markets fail to challenge the growth paradigm defining how ALAC societies interact with the environment. As Sabatini (2023) argues, Global South nations are trying to build formal economies with jobs that come with pensions and unemployment insurance. In sum, there is much frustration to go around. ALAC nations are frustrated with delays in adaptation and mitigation funding flows. Communities are frustrated with the lack of self-determination in and control over environmental affairs. On the other hand, climate scientists are frustrated with the slow pace of climate action while activists and human rights defenders are frustrated with the lack of action to protect them (United Nations, 2022). In the section below, we summarize and reflect on lessons learned and consider the way forward.

Lessons Learned and the Way Forward

The topics covered in this book volume are diverse including land use, sea level rise, and ocean health, but the themes and trajectories reveal that in the fight against climate change impacts, ALAC nations are losing if they have not lost already. The anti-statist fervor of governance reform based on decentralization and democratization (Pierre, 2000) has fizzled. And literature on climate governance hardly recognizes the deep-seated roots of the developmentalist approach (Brooks et al., 2009). ALAC nations find themselves in a quagmire that Quijano (2000) and Grosfoguel (2007) describe as "coloniality of power." Grosfoguel argues that in such an environment there exists a "continuity of colonial forms of domination" perpetuated by "colonial cultures and structures in the modern/colonial capitalist/patriarchal world system" (p. 219). He claims that while "classical colonial administrations" have been eradicated, the Global South nations live under a "postcolonial world" myth. Indeed, it is hard to argue against these claims, given the struggles and stresses over adaptation and mitigation funding and recent doubts about global ability to meet SDGs' targets by the 2030 deadline (https://www.devex.com/news). What lessons can one extrapolate from these experiences and developments in ALAC nations?

Based on the issues arising from and experiences with climate governance, the lessons that one can extrapolate are two-fold both triggered by governance and intellectual failures. On the issue of governance, it is obvious that there is no world authority to fight climate change, just as there is no world authority to fight COVID-19. By over-relying on Global North nations for adaptation and mitigation funding support, ALAC nations put all their eggs in one basket. Even as scholars write "liberal capitalism is bust" (Gray, 2020, p. 3; Zakaria, 2020), the bet on governance reform based on the assumptions of economic growth, trade liberalization, and electoral democracy gained momentum. In fact, a "second liberation of Africa" was proclaimed even as "personalization and concentration of government power remain stubborn realities in Africa" (Diamond, 2008, p. 3). By engaging less with, and seeking less input from, and building less trust with people to develop home-based solutions, ALAC nations missed opportunities. While they create portals and use the media to propagate information, they do so sporadically for political gain (Afrobarometer, 2022; Cruz et al., 2017). Thus, the consequence is governance failure manifested in and by lack of trust for public institutions and governing systems.

This instance of governance failure should not be construed to mean that ALAC nations should not engage with the Global North, nor should it imply that they must withdraw from bilateral/multilateral agreements. Since the UN Conference on the Human Environment (1972), they have been signatories to major environmental conventions and protocols that connect them to the global community, and it will be unreasonable to expect them to disengage. But some scholars such as Park (2022) argue that despite these structural arrangements, the outcomes are disappointing and climatic conditions either continue and/or are worsening. The reasons for such nonperformance are varied, including overemphasis on economic growth inherent in strategic sustainable development principles (Rasworth, 2017; Redclift, 1987; Schmelzer et al., 2022).

If the neoliberal state form as currently configured is failing ALAC nations as discussed above, what governance option is available to them in the midst of the climate crisis? In the recent fight against COVID-19, the sentiment reflects confidence in and the ability of the "postcolonial" state to beat back pandemics and restore countries to full health without resorting to market mechanisms (Amoah, 2020; Oloruntoba & Falola, 2020; Zondi, 2020). Amoah (2020) proposes "re-envisioning the African state as a deliberately interventionist one" (p. 9) and the "main vehicle" for responding to pandemics, while Zondi (2020, p. 190) suggests the possibility of "a return of the state" and the potential reversal of "neoliberal designs in favor of a lean and mean state in Africa." And there are calls elsewhere for the role of the state to be prioritized in addressing environmental governance (Haque, 2023; Heinrichs, 2022). In fact, Harrell and Haddad (2021) argue that the state can enhance its legitimacy by adopting what they describe as the eco-developmental state model through pro-environment policies and institutions.

The debates about how to address governance failure whether focused on an "emergent global administrative order" with a pro-market slant (Moloney & Rosenbloom, 2020, p. 227) or a return of the state with pro-government stance (Zondi, 2020, p. 190) is needlessly binary. Both can co-exist within a framework that embraces the value of indigenous and contextualized wisdom and knowledge (Grosfoguel, 2007; Ndlovu-Gatsheni, 2020; Yacob-Haliso et al., 2021), engages communities (Cheeseman & Sishuwa, 2021; Muthomi & Thurmaier, 2020), and creates an entrepreneurial-type state (Mazzucato, 2015) while holding public officials accountable. This stream of literature both challenges and shifts the discourse away from a parochial conceptualization of governance and administration based on Euro-centric hegemonic ontology, epistemology, and methodology (Afolayan et al., 2023; Haque, 2023). And yet the policy frameworks and programs exemplified in natural resources policy, for example, indicate that the administrative state in ALAC nations lacks the political will to develop relational governance grounded in experience and public service ethos that enhances compassionate treatment of the most vulnerable and underprivileged in society. In the next section, we conclude by exploring the way forward and considering policy, theory, and research implications.

SUMMARY AND CONCLUSION

The chapter focuses attention on ALAC nations' experience with institutionalizing climate governance in the context of fragile democracy and thinly structured governance and public administration systems. Of the multiple climate governance models identified in literature, the multi-level and multi actor model roughly exemplifies what exists in the ALAC nations covered in this book volume. While theoretically speaking decision making based in this model should reflect coordination between different actors at multiple jurisdictional levels, chapter descriptions mirror dominance by supranational bodies where climate agreements and protocols are framed and adopted. In terms of climate change impacts with cross-national and transnational concerns global governance makes sense and may in fact contribute to enhance the quality of climate governance if fairly conducted. But the question is, at what cost? To what extent do external influences affect and are affected by state sovereignty. This empirical question requires a case-by-case analysis of sectors by global governance scholars (e.g., Anderson, 2019).

It is true that national governments are responsible for preparing NDCs, formulating, and implementing adaptation and mitigation policies based on the latest technologies and scientific facts, especially IPCC periodic reports. But they are subject to monitoring and evaluation at the international level necessitated by IPCC recommendations and donor support conditionalities that may not align well to domestic social, economic, and cultural circumstances. For example, the push for renewables and green jobs to replace fossil fuels is reasonable but they do not have the current technology to support implementation. And with weak institutional, financial, and technical capacities, ALAC national governments are at the mercy of and tethered to the priorities of supranational bodies. Thus, contrary to claims that modern governance is dispersed across multiple centers of authority (Hooghe & Marks, 2004), centralized authority resides in and through multinational institutions, substantially influencing the quality of climate governance in ALAC nations.

In addition, continental institutions play roles and regional initiatives influence state-level climate decision making in ways that determine climate action outcomes and effectiveness. For example, the *Africa Agenda 2030* and *Draft Africa Climate Change Strategy 2020–2030* both demand concerted efforts to adapt to and mitigate climate change impacts. Also, the World Bank Group's (2022) regional initiative, *A Roadmap for Climate Action in Latin America and the Caribbean 2021–2025*, galvanizes climate action to deliver resilient, sustainable, and equitable growth. Owing to these structural arrangements and dilemmas entailed, much of the coverage takes a national perspective "in which methodological nationalism is upheld as the primary, or most appropriate policy and administrative foci" (Moloney & Stone, 2019, p. 12). Granted that a national perspective is narrow, the green transition is turning into "an electoral minefield" (Clarkson, 2023) that makes such a perspective useful for approaching climate governance. As human rights defenders face persecution, wildfires worsen, and loss of cultural heritage increases, methodological nationalism remains a reasonable approach to climate studies.

However, the challenge facing climate governance policy, research, and scholarship in ALAC nations is the prevalence of Western hegemonic knowledge claims that relegate or even ignore indigenous alternative pathways of knowing and becoming. Researchers and scholars opt for a universal objective truth in place of knowledge consciousness arising naturally and locally and in association with a long-term occupancy of a place (Dei, 2000, p. 72). As a result, much of the research and scholarship decontextualizes climate action and ignores what Haque (2023) describes as embeddedness in philosophy, history, and culture. And yet there is literature documenting indigenous knowledge, emphasizing consensus, participation, decentralization, community, and freedom of expression (Ayittey, 1991; Tiky, 2014; Williams, 1987). Combined with Anglo-American perspectives of collaboration (Ansell & Gash, 2008; King & Stivers, 1998), such culture-sensitive ideas can broaden climate governance and achieve sustainable solutions. As Francis et al. (2020) argue, technocratic solutions seem good, but they may not be feasible unless scholars address contextual, social, and political forces. Consistent with this thinking, Bryson et al. (2014) foresee an emerging perspective, emphasizing "public value, public values, and the public sphere" (p. 451). They assume that "public value involves producing what the public values or is good for the public."

The above suggestion builds on post New Public Management reform ideas and criticism of neoliberal-based development. It moves the discourse in the direction of public value governance toward theory development while encouraging reflective policy making. Such an approach considers governing that draws on networks of local actors. The epistemological grounding focuses on democratic theory, making it appealing to scholars with diverse orientation. While a composite approach may be less than ideal, a community-based model is more viable and timelier than current capacity building initiatives and is closer to what the African Capacity Building Foundation (2019) considers as "inclusive multistakeholder collaboration" (p. iii). Thus, to fight climate change, climate governance designs, especially in ALAC nations must rediscover and rebuild their communities. As Nickels and Rivera (2018) assert, engaged communities can lead to substantial and sustainable outcomes. These are incremental responses within what is essentially a business-as-usual approach to development. Proposed changes in investment patterns, increased finance and expanded markets (including global carbon markets) do not challenge fundamental ideas of growth and progress that define the way that "modern" human societies interact with their environment. Climate governance appears to be a solid solution to addressing some of the major challenges that countries in the global south experience. Increased attention to training governments in the global south in climate governance techniques will provide them another tool to addressing the challenges of climate change.

REFERENCES

Adams, W. (2009). *Green development: Environment and sustainability in a developing world.* Routledge.

Afolayan, A., Yacob-Haliso, O., & Oloruntoba, S. (2023). Pathways to alternative epistemologies in Africa. *Mind, 132*(527), 861–871.

African Capacity Building Foundation. (2019). *Capacity imperatives for the SDGs.* The African Capacity Building Foundation.

African Capacity Building Foundation. (2022, June 3). *Strengthening African leadership for climate adaptation.* Retrieved July 21, 2023, from https://ruforum .wordpress.com/2022/06/03/

Afrobarometer Survey. (2019). Retrieved July 16, 2023, from https://www.afro barometer.org/countries/ghana/

Agrawal, A., & Gibson, C. (1999). Enchantment and disenchantment: The role of community in natural resource management. *World Development 27,* 629–649.

Amoah, L. (2020). COVID-19 and the State in Africa: The state is dead, long live the state. *Administrative Theory & Praxis,* 1–11. https://doi.org/10.1080/108418 06.2020.1840902

Anderson, L. (2019). The future of the public policy school in a world of disruptive innovation. *Global Policy, 10*(1), 84–85.

Ansell, C., & Gash, A. (2008). Collaborative governance in theory and practice. *Journal of Public Administration Research & Theory, 18*(4), 543–571.

Appleby, M. (2019). *What are the benefits of Interdisciplinary Study?* Retrieved from https://www.open.edu/openlearn/education-development/what-are-the -benefits-interdisciplinary-study

Ayittey, G. (1991). *Indigenous African institutions.* Transnational Publishers.

Baninla, Y., Sharifi, A., Allam, Z., Tume, S., Gangtar, N., & George, N. (2022). *An overview of climate change adaptation and mitigation research in Africa.* Frontiers in Climate. https://doi.org/10.3389/fclim.2022.976427

Buaer, M. W., Ege, J., & Schumaker, R. (2018). The challenge of administrative internationalization: Taking stock and looking ahead. *International Journal of Public Administration.* https://doi.org/10. 1080/01900692.2018.1522642

Benz, A., Breitmeier, H., Schimank, U., & Simonis, G. (2009). *Politik in Mehrebenensystemen, VS Verlag fur Sozialwissenschaften* [Politics in Multi-level Systems]. Wiesbaden.

Blessett, B., Dodge, J., Edmond, B., Gourde, H. T., Gooden, S. T., Headley, A. M., DiCicco, N. M., & Williams, B. N. (2019). Social equity in public administration: A call to action. *Perspectives on Public Management and Governance,* 283–299. doi:10.1093/ppmgov/gvz016

Brink, E., & Wamsler, C. (2018). Collaborative governance for climate change adaptation: Mapping citizen–municipality interactions. *Environmental Policy and Governance, 28,* 82–97.

Brooks, N., Grist, N., & Brown, K. (2009). Development futures in the context of climate change. *Development Policy Review, 27*(6), 741–765.

Bryson, J., Crosby, B., & Bloomberg, L. (2014). Public value governance: Moving beyond traditional public administration and the new public management. *Public Administration Review, 74*(4), 445–456.

Cheeseman, N., & Sishuwa, S. (2021). African studies keyword: Democracy. *African Studies Review, 64*(3), 704–732.

Clarkson, A. (2023). *The green transition is becoming an electoral minefield.* Retrieved August 7, 2023, from, https://www.worldpoliticsreview.com/policies-climate -change-europe-us-green-cities-transition/

Cook, J., Oreskes, N., Doran, P., Anderegg, W., Verheggen, B., Maibeck, E., & Nuccitelli, D. (2016). Consensus on consensus: A synthesis of consensus estimates on human-caused global warming. *Environmental Research Letters, 11*(4), 04 8002.

Cox, M. (2016). The pathology of command and control: A formal synthesis. *Ecology and Society, 21*(3), 33.

Cruz, J., Marenco, E., Rodriguez, M., & Zechmeister, E. (2017). *The political culture of democracy in Nicaragua and in the Americas, 2016/2017: A comparative study of democracy and governance.* USAID, LAPOP. Retrieved August 5, 2023, from, https:// www.vanderbilt.edu/lapop/nicaragua/AB2016-17_Nicaragua_Country _Report_V6_English_V1_04.23.19_W_04.25.19.pdf

Dadze-Arthur, A. (2022). Democracy, governance, and participation: Epistemic colonialism in public administration and governance courses. In K. Bottom, P. Dunning, J. Diamond, & I. Elliot (Eds.), *International handbook on the teaching of public administration and management* (pp. 218–226). Routledge.

Dei, G. (2000). African development: The relevance and implications of indigenousness. In G. Dei & D. Rosenberg (Eds.), *Indigenous knowledge in global contexts: Multiple readings of our world* (pp. 70–80). University of Toronto Press.

Diamond, L. (2007). The state of democracy in Africa. In *Democratization in Africa: What progress toward institutionalization?* (pp. 1–14). National Intelligence Council. Retrieved July 31, 2023, from https://fas.org/irp/nic/african _democ_2008.pdf

Filho, W. L. (2015). *Handbook of climate change adaptation.* Springer.

Ford, M., & Ihrke, D. (2019). Perceptions are reality: A framework for understanding governance. *Administrative Theory & Praxis, 41*(2), 129–147.

Francis, D., Valodia, I., & Webster, E. (2020). *Inequality studies from the global south.* Routledge.

Gray, J. (2020). *The new leviathans: Thoughts after liberalism.* Macmillan

Grosfoguel, R. (2007). The epistemic colonial turn: Beyond political economies paradigms. *Cultural Studies, 21*(2–3), 211–223. https://doi.org/10.1080/095 02380601162514

Gunstein, J. (2021). *The climate crisis in rooted in the human condition.* Retrieved August 12, 2023, from https://www.worldpoliticsreview.com/despite -mitigation-climate-change-is-now-locked-in/

Guterres, A. (2022). *Secretary-General's remarks at COP27 stakeout.* Retrieved March 27, 2024, from https://www.un.org/sg/en/content/sg/speeches/2022-11-17/ secretary-generals-remarks-cop27-stakeout

Haque, S. (2023). Achieving sustainable development through developmental states in the 21st century. *International Review of Administrative Sciences, 89*(2), 315–329.

Harrell, S., & Haddad, M. (2021). The evolution of East Asian eco-developmental state. *Asian-Pacific Journal: Japan Focus, 19*(6), 1–22.

Heinrichs, H. (2022). Sustainable statehood: Reflections on critical (pre-)conditions, requirements, and design options. *Sustainability, 14*(15), 1–14.

Hooghe, L., & Marks, G. (2004). Contrasting visions in multi-level governance. In I. Bache & M. Flinders (Eds.), *Multi-level governance* (pp. 15–30). Oxford University Press.

Huijsman, B., & Savenije, H. (1991). Making haste slowly. In H. Savenije & B. Huijsman (Eds.), *Making haste slowly: Strengthening local environmental management*. Royal Tropical Institute, Amsterdam.

Huitema, D., Adger, W., Berkhout, F., Massey, E., Mazmanian, D., Munaretto, S., Plummer, R., & Termeer, C. (2016). The governance of adaptation: Choices, reasons, and effects. *Ecology and Society, 21*(3), 37.

IndustriAll. (2021). *Just transition declaration adopted at COP26.* Retrieved July 31, 2023, from https://www.industriall-union.org/just-transition-declaration-adopted -at-cop26

Intergovernmental Panel on Climate Change. (2023). *Climate change 2023: Synthesis report.* Retrieved July 13, 2023, from https://www.ipcc.ch/report/ar6/syr/ downloads/report/IPCC_AR6_SYR_SPM.pdf

Jordan, A., Huitema, D., van Asselt, H., Rayner, T., & Berkhout. (Eds.). (2010). *Climate change policy in the European Union: Confronting the dilemmas of mitigation and adaptation.* Cambridge University Press.

King, C. S., & Stivers, C. (Eds.). (1998). *Government is US: Public administration in the anti-government era.* Sage Publications.

Kooiman, J. (1993). *Governing and governance.* SAGE Publications.

Marks, G., & Hooghe, L. (2004). Contrasting visions in multi-level governance. In I. Bache & M. Flinders (Eds.), *Multi-level governance* (pp. 15–30). Oxford University Press.

Marquardt, J. (2017). Conceptualizing power in multi-level climate governance. *Journal of Cleaner Production, 154,* 167–175.

Massey, E., Huitema, D., Garrelts, H., Grecksch, K., Mees, H., Rayer, T., Storbjork, S., Termeer, C., & Winges, M. (2015). Handling adaptation policy choices in Sweden, Germany, the UK, and the Netherlands. *Journal of Water and Climate Change, 6*(1), 9–24.

Mazzucato, M. (2015). *The entrepreneurial state: Debunking public v. private myths.* Public Affairs.

McHarg, A. (1999). Reconciling human rights and the public interest: Conceptual problems and doctrinal uncertainty in the jurisprudence of the European court of human rights. *Modern Law Review, 62*(5), 671–696.

Millennium Ecosystem Assessment. (2005). *Ecosystems and human well-being: Synthesis.* Island Press.

Moloney, K., & Rosenbloom, D. (2020). Creating space for public administration in international organization studies. *The American Review of Public Administration 50*(3), 227–243.

Moloney, K., & Stone, D. (2019). *Beyond the state: Hlobal policy and transnational administration.* (online). https://doi.org/10.4000/irpp.344

Muthomi, F., & Thurmaier, K. (2020). Participatory transparency in Kenya: Toward an engaged budgeting model of local governance. *Public Administration Review, 81*(3), 519–531.

Ndlovu-Gatsheni, S. (2020). *Decolonization, development and knowledge in Africa: Turning over a new leaf.* Routledge.

Nickels, A., & Rivera, J. (2018). *Community development and public administration theory: Promoting democratic principles to improve communities.* Routledge.

Number of Countries That Submitted Nationally Determined Contributions (NDCs) in Africa as of May 2022, by Status. (n.d.). Retrieved July 22, 2023, from https://www.statista.com/statistics/1311387/status-of-ndc-submissions-in-africa/

Ògúnmọ́dẹdé, C. (2022, November 23). *For Africa, 'loss and damage' funding is a step toward climate justice.* World Politics Review.

Oloruntoba, S., & Falola, T. (2020). The political economy of Africa: Connecting the past to the present and future development in Africa. In S. Oloruntoba & T. Falola (Eds.), *The Palgrave handbook of African political economy* (pp. 1–28). McMillan.

Park, S. (2022). The role of the sovereign state in the 21st century environmental disasters. *Environmental Politics, 31*(1), 8–27.

Piatoni, S. (2010). *The theory of multi-level governance: Conceptual, empirical, and normative challenges.* Oxford University Press.

Pierre, J. (2000). Introduction. In J. Pierre (Ed.), *Debating governance: Authority, steering and democracy* (pp. 1–12). Oxford University Press.

Pollack, M. (1997). Delegation, agency, and genda setting in the European community. *International Organization, 51*(1), 99–134.

Quijano, A. (2000). Coloniality of power, ethnocentrism, and Latin America. *NEPANTLA, 1*(3), 533–580.

Rasworth, K. (2017). *Doughnut economics: 7 ways to think like a 21st century economist.* Chelsea Green Publishing.

Redclift, M. (1987). *Sustainable development: Exploring the contradictions.* Methuen & Co.

Rhodes, R. (1996). The new governance: Governing without government. *Political Studies, 44*(4), 652–667.

Rosenau, J., & Czempiel, E. (1992). *Governance without government: Order and change in world politics.* Cambridge University Press.

Sabatini, C. (2023). '*Green jobs' alone are no magic bullet for Chile's informal economy.* Retrieved May 23, 2023, from https://www.msn.com/en-us/money/markets/green-jobs-alone-are-no-magic-bullet-for-chiles-informal-economy/ar-AA1bAgpc

Sapiains, R., Ibarra, C., Jimenez, G., O'Ryan, R., Blanco, G., Moraga, P., & Rojas, M. (2021). Exploring the contours of climate governance: An interdisciplinary systematic literature review from a southern perspective. *Environmental Policy and Governance, 31*(1), 46–59.

Sattler, C., Schroter, B., Meyer, A., Giersch, G., Meyer, C., & Matzdorf, B. (2016). Multi-level governance in community environmental management: A case study from Latin America. *Ecology and Society, 21*(4), 24.

Schmelzer, M., Vetter, A., & Vansintjan, A. (2022). *The future is degrowth: A guide to a world beyond capitalism.* Verso.

Schreurs, M., & Tiberghien, Y. (2010). Multi-level reinforcement: Explaining European union leadership in climate change mitigation. *Global Environment Politics, 7*, 19–46.

Shoukry, S. (2022). *Cop27 agrees historic 'loss and damage' fund for climate impact in developing countries*. Retrieved July 29, 2023, from https://www.theguardian .com/environment/2022/nov/20/cop27-agrees-to-historic-loss-and-damage -fund-to-compensate-developing-countries-for-climate-impacts

Smith, A. (2007). Emerging in-between: The multi-level governance in renewable energy in the English regions. *Energy Policy, 35,* 6266–6280.

Suzman, J. (2021). *Work: A deep history, from stone age to the age of robots.* Penguin.

The National Academies of Sciences, Engineering, and Medicine. (2020). Public private partnership responses to COVID-19 and future responses. In *Proceedings of a Workshop in Brief.* Retrieved July 2023, from https://www.national academies.org/event/40346_07-2023_challenges-and-opportunities-towards -a-just-transition-and-sustainable-development-a-workshop

Tiky, L. (2014). *Democracy and democratization in Africa.* Common Ground Publishing.

United Nations. (2015). *Sendai framework for disaster risk reduction 2015–2030.* Retrieved July 24, 2023, from https://www.undrr.org/publication/sendai-framework -disaster-risk-reduction-2015-2030

United Nations. (2019). *25 years of adaptation under the UNFCCC: Report by the adaptation committee.* Retrieved July 27, 2023, from https://unfccc.int/sites/default/ files/resource/AC_25%20Years%20of%20Adaptation%20Under%20the%20 UNFCCC_2019.pdf

United Nations. (2022). *Human rights and voluntary national reviews.* Retrieved August 2, 2023, from https://www.ohchr.org/sites/default/files/documents/ issues/sdgs/2030/2022-07-01/HRandVNRs_Guidance_Note2022.pdf

United Nations. (2023). *Handbook for the preparation of voluntary national reviews.* Retrieved August 2, 2023, from https://hlpf.un.org/sites/default/files/vnrs/ hand-book/VNR%20Handbook%202023%20EN_0.pdf

United Nations. (1992). *United Nations framework convention on climate change—Multilateral environmental agreement.*

White, J., & Adams, G. (1995). Reason and postmodernity: The historical and social context of public administration research and theory. *Administrative Theory & Praxis, 17*(1), 1–18.

Williams, C. (1987). *The destruction of Black civilization.* Third World Press.

World Bank Group. (2022). *A roadmap for climate action in Latin America and the Caribbean 2021–2025.* World Bank.

Yacob-Haliso, O., Nwogwugwu, N., & Ntiwunka, G. (2021). *African indigenous knowledges in a postcolonial world: Essays in honor of Toyin Falola.* Routledge.

Zakaria, F. (2020). *Ten lessons for a post-pandemic world.* W. W. Norton & Company.

Zondi, S. (2020). COVID and the return of the state in Africa. *Politikon, 48*(2), 190–205. https://doi.org/10.1080/02589346.2021.1913805

ABOUT THE EDITORS

Peter Fuseini Haruna, PhD, is professor of public administration & government at Texas A&M International University. Founding chair of ASPA-SAPA (2018), senior Fulbright scholar to Ghana (2010–11), and Texas A&M University System Chancellor's Diversity, Equity, and Inclusion Awardee (2020), Dr. Haruna is the recipient of the ASPA-SICA Jeanne Marie Col Leadership Award (2022). He is also the recipient of the Senator Judith Zaffirini Award for leadership and Scholarship and the College of Arts & Sciences Distinguished Teacher of the Year Award (2022) at his University. His previous awards include Texas A&M International University's International Scholar of the Year (2010), College of Arts & Sciences Scholar of the Year (2016), and the College of Arts & Sciences Scholar of the Year (2005). His research focuses on international and comparative administration, governance structures, public leadership, public administration education, and public service training. His work has appeared in *Public Administration Review, Public Integrity, International Journal of Public Administration, Administrative Theory & Praxis, African Studies Review, and Journal of Public Affairs Education*. In addition, Dr. Haruna is the lead co-editor of *Public Budgeting in African Nations* (2016) and *Public Administration Training in Africa* (2015) both published by Routledge. Email: pharuna@tamiu.edu

Laila El Baradei, PhD, professor of public administration at the American University's Global Affairs & Public Policy and Administration Department, Cairo, Egypt. Previous acting dean and associate dean of global affairs & public policy and administration, associate dean of graduate studies & research, and member of Egypt's Human Development Report Team, Dr. El Baradei serves as the director of the 'public policy hub' at the School of Global Affairs and Public Policy. Dr. El Baradei has provided consultancy services to reputable organizations such as the World Bank, USAID, UNDP, DANIDA, Center for Development Research in Bonn, the Economic Research Forum in Egypt, and Ford Foundation. She has published widely on development cooperation management, elections management, decentralization, organizational change, public administration reform, governance, child labor, downsizing, and accountability. Her most recent work has appeared in *International Journal of Public Administration and Journal of Public Affairs Education*. Dr. El Baradei can be reached at: lbrardei@aucegypt.edu

Liza Ceciel van Jaarsvedlt, PhD, associate professor at the University of South Africa's Department of Public Administration and Management, South Africa. Recipient of the UNISA Excellence in Tuition Award (2020), National Research Foundation Y2 rated researcher (2015–2020), ASSAD-PAM Best Paper Award (2018), the UNISA Woman in Research Award (2011), the UNISA Principles Award for Excellence (2010), and Department of Public Administration and Management Researcher of the Year (2008), Dr. Van Jaarsveldt serves as project director of the International Association of Schools and Institutes of Administration (IASIA) Working Group VIII on Public Human Resources Management. She is also an elected board of management member for IASIA, as well as the IASIA vice president: Program since 2022. Her research focuses on public administration education, scholarship, curriculum development, and the development and the use of technology in education and government. She has authored and co-authored reviewed articles in varied accredited journals, as well as conference proceedings and book chapters. She is currently the technical editor of *Administratio Publica*, a quarterly journal that promotes the scholarship of public administration. Dr. van Jaarsveldt can be reached at: vjaarlc@unisa.ac.za

Abraham David Benavides, PhD is currently a professor of public administration at the University of Texas at Dallas. He previously taught for 20 years at the University of North Texas. He is the current president of the North Texas Chapter of the American Society for Public Administration and previously served on the National Council. He is a former commissioner for the Commission on Peer Review and Accreditation and has served on many ad-hoc committees for the Network of Schools of Public Policy, Affairs, and Administration. His research focuses on local government principally but has also published in areas of human resources, public leadership, age friendly policies, those experiencing homelessness, and international issues in Latin America. His research can be found in a number of leading academic journals in public administration. Dr. Benavides can be reached at: Abraham.Benavides@UTDallas.edu

Cristina M. Stanica, PhD, assistant teaching professor public policy & administration, School of Public Policy and Urban Affairs, Northeastern University. Board member of ASPA-SICA and Editor of the *Occasional Paper Series*, Dr. Stanica is a David Gould Scholar Awardee (2019) with a research focus on comparative public policy and administration. Her research focuses on administrative and rules burdens in street-level bureaucracy, new public governance and coproduction, digital transformation and trust in government. Her work appeared in the *International Review of Administrative Sciences*, the *International Journal of Public Leadership*, the *Transylvanian Review of Administrative Sciences*, and the *Journal of Comparative Policy Analysis: Research and Practice*. Dr. Stanica can be reached at: c.stanica@northeastern.edu

ABOUT THE CONTRIBUTORS

Mareve I. M. Biljohn, PhD, is a senior lecturer of public administration and management at the University of the Free State, Bloemfontein, South Africa 9301. She is a recipient of an Erasmus Mundus PhD mobility scholarship to Ghent University in Belgium (2015–2016), and at her university, she received the Economic and Management Sciences Faculty Emerging Researcher's Award (2018). With her research trajectory in the field of social innovation she obtained her National Research Foundation (NRF) rating. Using the lens of social innovation and transformative social innovation, her research aims to contribute solutions to local government service delivery challenges, and approaches to citizen participation in the local governance of service delivery. Maréve has also authored and co-authored book chapters regarding public sector leadership and leading self in South Africa's local government VUCA environment. Her research has been presented at national and international conferences, and she has been an invited speaker and panelist at international conferences and public sector events. Her work has appeared in the *International Journal of Transforming Government: People, Process and Policy* and the *International Journal of Public Administration.* Dr. Biljohn can be reached at: BiljohnMIM@ufs.ac.za

Manlio Felipe Castillo, PhD, associate professor in the Public Administration Department at CIDE, and member of the National System of Researchers, in Mexico. Member of the American Society for Public Administration and the International Network for Social Policy Teaching and Research. His primary research is focused on urban policies, urban and metropolitan governance, and disaster management. He has collaborated on studies evaluating urban programs, the re-densification processes in Mexico City, and local government management practices. His latest books are *La capitalización privada de los bienes públicos. Modelos de precios hedónicos para la vivienda en la Ciudad de México* [The Private Capitalization of Public Goods: Hedonic Pricing Models for Housing in Mexico City] (2020) and *La resbaladilla de la corrupción. Estudios sobre los procesos sociales y organizacionales de la corrupción colusiva en el sector público* [The Slippery Slope of Corruption: An Investigation into the Social and Organizational Processes of Collusive Corruption in the Public Sector] (co-edited in 2019). His research has appeared in *Urban Affairs Review, Public Organization Review, State and Local Government Review,* and *Urban Research & Practice.* Dr. Castillo can be reached at: manlio .castillo@cide.edu

Luis Antonio Castillo-Suarez is a member of the national system of researcher's level 1. He has a degree in biochemical engineering, with a master's degree in science in biochemical engineering from the National Institute of Technology of Mexico. Dr. in water sciences from the Inter-American Institute of Technology and Water Sciences of the Autonomous University of the State of Mexico. He has a postdoctoral stay at the same institute in the framework of the standard project for the generation of a state technical standard for the establishment of physicochemical characteristics for the regulation of wastewater discharge from the textile industry, as well as the establishment of maximum permissible limits. He develops projects generating technology with a sustainable profile for the removal of persistent pollutants present in industrial wastewater and landfill leachates by applying solar radiation to improve the reaction speed. He has participated in research projects with the application of oxidation processes coupled to biological treatments to increase the biodegradability of persistent compounds. Email: lacastillos_s@uaemex.mx

Redemption Chatanga, MPA, is a PhD candidate at the University of Free State, South Africa. Redemption is a former recipient of the Econet Joshua Nkomo Scholarship in Zimbabwe (2006–2012) for both undergraduate and postgraduate studies. She is an emerging researcher who published an article from her PhD studies titled "New Public Governance Theory: A Framework for Lesotho Policymakers to Enhance Community Participation During Climate Change Policy Formulation and Implementation." Her research focuses on public policy, public management, climate governance, and community participation. She has presented her research at national and international conferences. Some of her work has appeared in *Administratio Publica.* Redemption Chatanga can be reached at: redemutoozi@ gmail.com

Alvarez-Bastida Carolina, PhD, is a profesor of chemistry at Universidad Autónoma del Estado de México, Toluca, Estado de México. Dr. Alvarez-Bastida has a PhD in environmental sciences, her research has been related to hydrology, physical chemistry of water, among other topics, she has publications in indexed journals, book chapters, she has participated in national and international conferences as a speaker, she has collaborated in research projects which involve the educational, private and municipal sectors. Dr. Alvarez-Bastida can be reached at: calvarezb@uaemex.mx

Monserrat Castañeda-Juárez, PhD, is a postdoctoral student at Instituto Interamericano de Tecnología y Ciencias del Agua since 2021. Winner of the Ignacio Manuel Altamirano Basilio award in 2020. Since 2021 she belongs to the National System of Researchers in México. Her research focuses on wastewater treatment specially pharmaceuticals, dyes and pesticides wastewater by tertiary process and advanced oxidation process. Her work appeared in indexed journals such as *Environmental Science and Pollution Research, Journal Environmental Technology, Journal of the Photochemistry & Photobiology A: Chemistry*, book chapters, she has participated in national and international conferences as a speaker. Dr. Castañeda-Juárez can be reached at: mcastanedaj_s@uaemex.mx

Miriam Aide Garcia Colindres is a student and researcher attached to the postgraduate program in environmental sciences of the faculty of chemistry of the Autonomous University of the State of Mexico (2021–2023). General director of the Municipal Decentralized Public Organization for the Provision of Drinking Water, Drainage and Residual Water Treatment Services of the Municipality of Almoloya de Juárez (2016–2018) and San Antonio la Isla Estado de México (2013–2015); Mexico. Master in water sciences, Autonomous University of the State of Mexico. Inter-American Center for Water Resources, developing degree thesis "Evaluation of Nitrogenous Species in Water Wells Near Pantheons" (2009–2010). Chemical engineer, specialty in environmental by the Technological Institute of Toluca, developing professional thesis "Implementation of the High Efficiency Liquid Chromatography Technique in the water quality laboratory for Determination of Anions and Cations in water" (1997–2003). Email: mgarciac046@alumno.uaemex.mx

Augusto Victor Ferreros is a contributing faculty member with the School of Public Policy and Administration of the College of Health Sciences and Public Policy at Walden University and was an adjunct faculty member in the Askew School of Public Administration and Policy at Florida State University. He has also taught at Nova Southeastern University and at Howard university. He most recently served on the peer review board for the Fulbright Specialist Program of the Institute for International Education and was recently a Fulbright specialist in math education with the University of Business and Technology in Pristina, Kosovo in 2022. He also served as director of the Northwest Regional Data Center affiliated with Florida State University. He has been a practicing public administrator for most of his career having served in federal, state and local government with consulting engagements in the private sector and higher education. He was a lead consultant in the Citra Metro Manila Toll Project for the South Luzon Expressway in the Philippines. He has also been a consultant to Claremont Graduate University. His service includes director for the U.S. Senate Computer Center in Washington DC., OMB and planning directors for the states of Alaska and Kansas, and prior to that he served as staff economist and senior policy analyst for the states of Kansas and Alaska respectively. He has also served as Chief information officer for the county governments of Palm Beach and Martin counties in Florida. He has also been a project manager in the Office of Research in the Council of State Governments in Lexington, Kentucky. His current research interests include quantitative social science, analytics, applied econometrics and political economy. Dr. Ferreros may be reached at: victor.ferreros@mail.waldenu.edu

Ivonne Linares Hernández, PhD is professor at University of State of Mexico, Mexico, post graduated in environmental sciences. Her research is focused on the electrochemical treatment and advanced oxidation processes of municipal and industrial wastewater, with a recognized prestige for the Galvano Fenton and Solar Galvano Fenton Systems for the elimination of refractory compounds present in wastewater (for the treatment of drugs, herbicides, leachate from landfills and industrial wastewater), processes that do not require electricity and work through solar energy, minimizing the impact of climate change and contribute to the reduction of the water footprint. Her academic productivity is based on the publication of 13 book chapters in renowned publishers; 70 research articles published in international and national scientific journals, with high impact indexes such as JCR and Scopus; articles that have obtained 1,260 citations in Scopus. Her research has been distinguished with the Laudatory Note by the UAEM and recognition for the number of publications in connection with the National Institute of Nuclear Research (ININ). Email: ilinaresh@uaemex.mx

Vanessa González-Hinojosa, PhD in environmental sciences (2015–19). At present is COMECYT (Mexican Council of Science and Technology)—IITCA (Inter-American Institute of Technology and Water Sciences) professor of Autonomous University of the State of Mexico (2022–23), Toluca, State of Mexico 50200. Her research focuses on a critical perspective of the study of bottled water and human consumption since an inter and transdisciplinary field. With areas of interest related to environmental sciences such as treatments to improve the quality of drinking water, political ecology, biopower, environmental history, environmental epistemology, among others. She did a research stay at the Meritorious Autonomous University of Puebla—Postgraduate in sociology from the Institute of Social Sciences and Humanities—(2018). Also, she has promoted the divulgation of science in different academic, governmental, and civic spaces. She collaborates in the research network "Thematic Network of Water Critical Studies" (2015–23). And contributes as activist in the Civil Association "H_2O Lerma con Encanto" (2019–23), which promotes the sanitation of the Lerma River. In addition, she has been invited as a reviewer and commentator on various research papers, coordinator of books, as well as a rapporteur of the state forum "Towards the construction of a new General Water Law" (2019). Dr. Vanessa can be reached at: vangohi13@gmail.com

Anagela Marie Mai, PhD, is a social scientist and research consultant specializing in global positive social change. Dr. Mai has guided and facilitated numerous international public policy studies on topics ranging from environmental sustainability and improving human well-being to local, regional, and global policy development and reformation. Dr. Mai was nominated for Walden University' Dissertation of the Year (2018) and received Western Governors University' National Capstone Mastery Award (2012). Dr. Mai received Honors recognition from the Golden Key International Honours Society; National Honor Society; National Society of Leadership & Success, Sigma Alpha Pi; and School of Public Policy Administrators Honor Society, Pi Alpha Alpha. Dr. Mai also earned Alltel's Pinnacle Achievement award and Phi Beta Lambda's Business & Leadership; Public Presentation; and Extemporaneous Speaking awards. Dr. Mai was recognized by the American Sociological Association, American Society of Public Administrators, Autism Society of America, Global & Regional Asperger Syndrome Partnership, Healing Farms, Society for Human Resource Management, American Evaluation Association, and various other humanitarian and social justice advocacy groups for contributions to social change. Dr. Mai's research passions focus on equitable, equal, diverse, inclusive, and socially just public policy, governance, leadership, administration, education, and public service. Dr. Mai's work has appeared in the *International Journal of Behavioral & Health Science*; *Journal of Social Sciences & Humanities*; *Journal of Social, Behavioral, & Health* Sciences; and *SAGE Open*, among others. Dr. Mai can be reached at: angela@angelamai.com

Verónica Martínez-Miranda, PhD is professor and researcher attached to the Instituto Interamericano de Tecnología y Ciencias del Agua, of the Universidad Autónoma del Estado de México, since 1995, area of water physicochemical development and chemical processes for wastewater treatment and potabilization with PROMEP desirable profile and researcher member of the National System of Researchers, level I. Member of the academic staff of the master's degree in water sciences and postgraduate degree in environmental sciences belonging to the National Postgraduate System. Published articles (76), with 992 citations, h-index of 18, according to the SCOPUS platform. Chapters of books (17), conference proceedings with national and international recognition (42), has given talks, conferences, and workshops (53), television programs (4), dissemination works (3), interviews in printed media, radio, television, and internet (40). Academic career, he has directed and/or advised bachelor's (17), master's (27) and doctorate (12) theses. Abel Ayerza Award for technological and scientific contribution by the Iberoamerican Society of Physics and Chemistry in Spain. Clause 90 of the Collective Labor Contract of FAAPA UAEMEX, as LEADER of the work entitled "CONCANAL 1.0, Web System for quality control in the Water Quality Laboratory of CIRA." Laudatory Note, for outstanding academic activities. Ignacio Ramírez Calzada Award, 2013, for excellence in teaching, research, dissemination and university extension, Higher Level. Recognition UAEMEX-FAAPA, fulfillment of academic activities. Dr. Martínez-Miranda can be reached at: mmirandav@uaemex.mx.

Noluthando Shirley Matsiliza, PhD, is professor of public administration at the University of Fort Hare, Alice, South Africa. She is a member of the South African Association of Public Administration and Association of Schools and Department in Public Administration, a co-founder of the Eastern Cape Chapter for SAAPAM, a recipient of the USAID scholarship in 1997–1999, a recipient of South Africa and Netherlands PhD preparation (SANPAD) in 2004. She is a recipient of the VC award of the 'Best Women Researcher' in 2018 at the Durban University of Technology. Her research focuses on public policy, governance, leadership, development management and monitoring and evaluation. She is responsible for teaching, research, postgraduate supervision, and community engagement. Her work has been published and cited in local and internationally accredited journals. Also, Professor Matsiliza is the assistant editor of the *Journal of Local Government Research and Innovation*. She is involved in several local and international article reviews. She is the editor of the book *Higher Education for Public Good: Perspectives for the New Academic Landscape in Southern Africa*. She can be reached at: nmatsiliza@ufh.ac.za

Henry Kimani Mburu, PhD CPA, is senior lecturer of accounting at the Catholic University of Eastern Africa, Nairobi, Kenya. Founding chair of the Department of Accounting and Finance, (2004–2007), staff exchange fellow of the African Association of Universities at the Ghana Institute of Public Administration, (2010–11). He is also a former dean of the School of Business at the Catholic University of Eastern Africa (2016), a visiting professor at Tangaza University College (2002–2007), and Strathmore University (1997–2004). He is currently the coordinator of curriculum design and review in the School of Business and Economics. He is a member of the American Accounting Association. He has been widely involved in curriculum design and review and was the chair of the Accounting Expert Panel for the Namibia Commission of Higher Education (2022). He has designed curriculum and trained for self-reliance in the Association of Member Episcopal Conferences of Eastern Africa (AMECEA) countries. He has published textbooks with Pauline Publications, Africa, Focus Publishers, and Libros and Aliados Publishers. His research focuses on African business, earnings management, corporate governance, and climate change. His research work has appeared in *Advances in Accounting, Thunderbird International Business Review, Journal of International Business Education, Journal of Corporate Accounting & Finance, Journal of Accounting and Finance*, and *Corporate Ownership & Control*, among others. He is a peer reviewer for the *Journal of International Business Education*. Dr. Mburu can be reached at: henry.mburu@cuea.edu

Joseph Macheru Ngunjiri, PhD, CPA-K is a doctor of philosophy in finance at The Catholic University of Eastern Africa, a Certified Public Accountant of Kenya (CPA-K, and a member in good standing of the Institute of Certified Public Accountants of Kenya (ICPAK). Prior to his current lecturing roles, Dr. Macheru was the country tax and capital planning manager at Barclays Bank of Kenya, now Absa Bank. In his corporate career, he won the Barclays Eagle Award for excellent tax planning and audit at the bank. He has ten years of banking experience, having held several key roles including financial controls analyst, accounts payable assistant, securities analyst and project assistant. His research focusses on international finance, with a focus on capital flights and climate financing. He is a published author of multiple articles with the *International Journal of Finance and Accounting, International Journal of Economics and Finance, International Journal of Economic Policy, Journal of Developing Country Studies, International Journal of Poverty, Investments and Development* and the *International Journal of Finance*. He has also co-authored the *Economic Policy and Financial Performance* book published by the American Journal of Economic Policy. Dr. Macheru can be reached at: jmacheru@gmail.com

Genève Phillip-Durham, PhD, holds a BS in public sector management, postgraduate diploma in international relations, MS in global studies, and a PhD in international relations. She is currently the interim provost and vice president of academic affairs at the University College of the Cayman Islands. Genève has served in academic and administrative management roles, regionally and internationally, working across the private sector, public sector, civil society and multilateral organizations. She currently serves as an associate editor for the *Island Studies Journal*—an international peer-reviewed academic journal, dedicated to the interdisciplinary study of islands and has served as the book reviews editor for the journal from 2021–2023. Dr. Phillip-Durham is a member of the executive council of the Caribbean Studies Association (2023–2025), an academic board member of the Accreditation Agency of Curaçao since 2017, secretary of the board of directors for Solar Head of State—an international NGO which focuses on sustainability and renewable energy (2021–present) and an affiliate of the Resilient and Sustainable Islands Initiative (2023)—a global advisory network which works with small island developing states to find solutions to growing sustainability challenges. She has published peer-reviewed articles and book chapters on topics such as: capacity and institution building, policy practice, good governance, diversification, and resilience in the Caribbean. Dr. Phillip-Durham can be reached at: geneve.phillip@yahoo.com or gdurham@ucci.edu.ky

Hugo Renderos, PhD, is professor of public administration and political science at the University of El Salvador in San Salvador, El Salvador, Central America. Dr. Renderos is an active member of the American Society of Public Administration (ASPA) having served as chair of the section on democracy and social justice and twice served as the international chapter president. He was awarded the Best Professor of the Year Award (2014) at Keiser University. Dr. Renderos is the recipient of the Mary Hamilton Award (2019) for his contributions and leadership as president of the international chapter. Dr. Renderos has taught in Nicaragua, El Salvador, and the United States of America. His research focuses on transnational gangs, immigration, police reforms in Central America, comparative administration, and international relations. Dr. Renderos' research has appeared in *Public Administration Review, Public Integrity, Journal of Cultural Marketing Strategy, International Journal of Innovative Science and Research Technology, International Journal of Education and Social Science, American International Journal of Social Science,* and *Journal of Peace and Conflict Studies.* Additionally, Dr. Renderos has published chapters focused on human resources reforms in Latin America and health disparities on vulnerable populations, published by Routledge and Birkdale, respectively. Dr. Renderos can be contacted at: hr8@1870.uakron.edu

Shaimaa Mahmoud Sabbah, enrolled PhD graduate student at the institute of Global Health & Human Ecology, the American University in Cairo, Cairo, Egypt. MPA, Youssef Jameel Fellow (2018–2022). Her research focuses on governance structures, public health, public policy & administration. Dr. Shaimaa can be reached at: shaimaasabbah@aucegypt.edu

Kalim U. Shah, PhD, is associate professor of energy and environmental policy at the Joseph R. Biden Jr. School of Public Policy and Administration, University of Delaware. His work focuses on public policy, regulation and governance in small states and islands. As an institutional theorist, his research addresses the science-policy interface of energy transitions and climate change; clean technology; and regulatory designs. His research has recently appeared in *International Review of Administrative Sciences, Journal of Cleaner Production, Climatic Change, WIREs Energy and Environment, Current Opinion in Environmental Sustainability* among others. He is the founder of the Island Policy Lab at the University of Delaware, a center that works closely with governments and academia globally, to forward sustainable development. He is coordinator of the United Nations Universities Consortium of Small Island States and has served as a coordinating/lead author of the UN flagship Global Environmental Outlook Reports 6 and 7. Often called to advise multinational institutions, he has designed national climate and energy policies for countries including Guyana, Suriname and the Marshall Islands. Kalim is a Fulbright scholar, originally from Trinidad & Tobago. He can be reached at: kalshah@udel.edu

Brighton Mandizvidza Shoniwa, DPhil, is a full-time lecturer, in the Department of Social Sciences, Faculty of Social and Gender Transformative Sciences, Women's University in Africa, in Harare, Zimbabwe. He is also a part-time lecturer at the University of Zimbabwe, a position held since August 2018. Has a strong research interests in the discourse of sustainable development, in general, and in agriculture and climate governance, in particular. As a public policy and public management expert, he subscribes to the narrative that the contemporary challenges, which are complex, require a multi-disciplinary and multi-stakeholder approach. Accordingly, he seeks to continue contributing to the debate advocating for public-private-community partnerships as developmental options. Some of his work appeared in the *Journal of Humanities, Zambezia,* published by the University of Zimbabwe, and in *Public Administration and Policy* (Emerald Publishing Ltd.). Dr. Shoniwa can be reached at: bmshoniwa@gmail.com or bshoniwa @wua.ac.zw

Heidi Jane M. Smith is research professor in the Department of Economics at the Universidad Iberoamericana, Mexico City. She was named a non-residential fellow, Center for U.S.–Mexican studies, University of California, San Diego from 2022–23, and a Fulbright-García Robles Scholar in Mexico working at the Center for Economic Research and Teaching (CIDE) from 2010–11. In 2016 she was selected to be a member of the peer review Mexican National Research System (SNI II). Dr. Smith has taught courses in political economy, public policy, urban policy, and qualitative and quantitative methods at the Universidad Iberoamericana, Instituto Tecnológico Autónomo de México (ITAM), Florida International University and George Mason University. Her research focuses on subnational debt, capital markets, corruption, public financial management, the bureaucracy, and social equity. She uses comparative cases of the United States, Mexico, Argentina, Korea and China to better understand the political economy of fiscal and debt policy. She has extensively studied the creation of the municipal bond market in Mexico and its regulatory environment. This research has been published in English, French, Chinese, and Spanish as 15 academic peer-reviewed articles, 2 books (in Spanish and English), 12 academic chapters, and 8 policy reports published by international financial institutions. Dr. Smith can be reached by email at heidi.smith@ibero.mx. Her publication can be located here: https://scholar.google.com/citations?user=CdA6wA8 AAAAJ&hl=en

Michael S. Yoder is a human geographer with 34 years of research experience in agricultural, urban and suburban land use change, and the geography of economic development. His regional specialties include Mexico, the U.S. sunbelt, Central America, and the U.S.–Mexico border region. He has held tenured faculty positions at Texas A&M International University and the University of Central Arkansas, where he directed the Master of Science in Community and Economic Development (2010–2017). His most recent publications address freight transportation corridors linking the U.S. and Mexico, urban planning and development in small cities of Texas and Arkansas, community assessments, qualitative research methods, and urban geography of northern Mexico. He is the author of the book *Geographical Scale and Economic Development: Lessons Learned from Texas and Mexico* (2023, Springer). As a research fellow at the University of Texas at Austin, his current research examines transportation and agriculture as features of economic development and political geography of South Texas. He currently teaches geography courses at University of Central Arkansas and Texas A&M International University. Dr. Yoder can be reached by phone at: michael.yoder@tamiu.edu

Printed in the United States
by Baker & Taylor Publisher Services